T0327528

A HANDBOOK FOR DNA-ENCODED CHEMISTRY

A HANDBOOK FOR DNA-ENCODED CHEMISTRY

Theory and Applications for Exploring Chemical Space and Drug Discovery

EDITED BY

Robert A. Goodnow, Jr.

AstraZeneca
Waltham, MA, USA

GoodChem Consulting, LLC
Gillette, NJ, USA

WILEY

Published by John Wiley & Sons, Inc., Hoboken, New Jersey.
Published simultaneously in Canada.

For general information on our other products and services or for technical support, please contact our
Customer Care Department within the United States at (800) 762-2974, outside the United States at
(317) 572-3993 or fax (317) 572-4002.

Wiley also publishes its books in a variety of electronic formats. Some content that appears in print may
not be available in electronic formats. For more information about Wiley products, visit our web site at
www.wiley.com.

Library of Congress Cataloging-in-Publication Data:

A handbook for DNA-encoded chemistry : theory and applications for exploring chemical space
and drug discovery / edited by Robert A. Goodnow, Jr.
    p. ; cm.
  Includes bibliographical references and index.
  ISBN-13: 978-1-118-48768-6 (cloth)
I. Goodnow, Robert A., Jr., editor of compilation.
  [DNLM: 1. Combinatorial Chemistry Techniques–methods. 2. DNA–chemical synthesis.
3. Drug Discovery–methods. 4. Gene Library. 5. Small Molecule Libraries–chemical synthesis.
QV 744] RS420
  615.1′9–dc23
                                                             2013042727

10 9 8 7 6 5 4 3 2 1

CONTENTS

PREFACE

The concept for this book came about after the rejection of an invitation to write a book about combinatorial chemistry. Although a highly interesting field of chemistry, the initial invitation was declined upon the assumption that excellent books already exist in sufficient numbers on various subjects of combinatorial chemistry. However, upon further reflection, the editor realized that a new chapter in the story of combinatorial chemistry had begun with the emergence and development of DNA-encoded chemistry methods. Despite the existence of publications about the concept and practice of DNA-encoded chemistry since 1992 by Brenner and Lerner and DNA-directed chemistry roughly a decade later, the editor found no single, authoritative summary of the theories, practice, and results of DNA-encoded chemistry. Therefore, it seemed a worthy endeavor to recruit experts in the field and create a handbook summarizing theories, methods, and results for this exciting, new field. It is hoped that this handbook will provide a good understanding of the practice of DNA-encoded and DNA-directed chemistry and that such chemistry methods will be more widely embraced and developed by a large community of scientists.

Readers may notice some overlap and/or repetition among various chapters. The editor has tended to allow such commonality as a means not only to highlight the multiple points of view and interpretation on this new technology as it has been applied to organic chemistry and drug discovery, but also as a means to indicate those results which have been received with particular interest by those skilled in the art.

ACKNOWLEDGMENTS

Whenever one is approached by an editor of a project of this sort, a contributing author likely feels an initial sense of recognition, quickly followed by the sobering reality of the work that lies ahead to complete a high-quality chapter. Indeed, there is also an element of trust that the project has been well conceived and appropriately considered. Thus, those authors whose work is reflected here have given not only their insight and expertise on various aspects of DNA-encoded chemistry methods and technology but also their trust and persistence to deliver a finished product. Contributing authors must also wait to see the fruits of their efforts in print. For those reasons, I am deeply grateful to the contributing authors who have dedicated their time, expertise, and patience to this project.

In addition, I am deeply grateful to the proofreaders, whose efforts have immeasurably improved the quality and accuracy of the information and language contained herein. They are Dr. Paul Gillespie of Roche, Dr. Anthony Keefe of X-Chem, Inc., Dr. Brian Moloney of eMolecules Inc., and Dr. Andrew Ferguson of AstraZeneca. Each has contributed in a different way, dedicating his own valuable time to this project. Dr. Ferguson reviewed the entire manuscript with a careful, strict, and critical eye in a short timeframe. Dr. Moloney's review of the small molecule costing analysis of Chapter 18 is much appreciated. Dr. Keefe provided expert scientific criticism and challenge for many chapters. Finally, and above all others, Dr. Gillespie tirelessly provided a startling level of detailed perceptivity on each chapter's logic, compositional style, and representation of scientific literature. I am fortunate to have worked with such diversely skilled reviewers; readers are fortunate to encounter this handbook after their diligent efforts.

Robert A. Goodnow, Jr.
March 2014
GoodChem Consulting, LLC
rgoodnow@hotmail.com

INTRODUCTORY COMMENT 1

The identification of potent and selective lead molecules is the essential first step in any drug discovery research project. Historically, successful drug discovery has focused on a small number of so-called tractable target classes, including G-protein coupled receptors, ion channels, nuclear receptors, kinases, and other enzymes. Until the 1980s, lead molecules were identified through traditional medicinal chemistry approaches, typically through chemical modification of a known bioactive compound. The molecular biology revolution resulted in a huge increase in the number of putative target proteins for drug discovery. This was accompanied by the development of combinatorial chemistry methods to generate very large chemical libraries, which in turn was accompanied by the development of technologies for High-Throughput Screening (HTS) to enable the rapid and cost-effective screening of these large, often several million molecules in size, compound libraries against large numbers of drug targets. HTS, sometimes termed diversity screening, rapidly became embedded as a primary method for lead discovery within the pharmaceutical industry, and more recently there has been the transfer of this technology platform into the academic sector through the huge growth in academic drug discovery centers. Alongside the growth in HTS, advances in biophysics technologies and structural biology have led to the development of methods for the screening of small-molecule fragments and the complementary use of structural biology techniques to guide medicinal chemists in the optimization of such molecules.

The establishment of these technology platforms required huge investment in compound stores and distribution systems, screening automation and detection systems, assay technologies, and systems to generate large quantities of biological reagents to support fragment-based drug discovery and diversity screening. This investment led to the generation of novel, potent, and selective lead molecules, with appropriate physico-chemical and safety properties, for many drug targets. However, there remain a significant number of drug targets for which the identification of novel molecules for use as target validation probes or as the starting points for the development of a drug candidate remains a major challenge. Existing compound collections have been built around the chemistry history of the field, and while successful at identifying lead molecules for the major target classes, in many cases these libraries have not successfully led to the generation of hit molecules for novel target classes or for so-called intractable target families. Advances in fragment screening have provided a mechanism for the design of novel molecules against protein targets, but while there have been recent advances in the development of such methods for screening membrane proteins, the implementation of this methodology remains in its infancy. As a consequence, there continues to be significant interest in the development of novel chemistries and compounds to enhance the quality of existing compound libraries, with a particular focus on physicochemical properties and lead-likeness, and in novel screening paradigms to enhance the overall success of lead discovery.

DNA-encoded library technology involves the creation of huge libraries of molecules covalently attached to DNA tags, using water-based combinatorial chemistry, and the subsequent screening of those libraries against soluble proteins using affinity selection. While DNA-encoded library technology was first described in the early 1990s, it is only in recent years that this technology platform has been considered as an attractive approach for lead discovery. This hugely valuable handbook provides a comprehensive review of the history and capabilities of DNA-encoded library technology. I will not attempt to review these here but would like to highlight the technology developments that have enabled this capability and the potential applications of DNA-encoded library technology as part of a broad portfolio of lead discovery paradigms.

As part of a broad portfolio of lead discovery paradigms, DNA-encoded library technology offers a number of attractions compared to other methods:

- DNA-encoded library selections require a few micrograms of protein; hence they do not require the investments in reagent generation and scale-up associated with other screening paradigms.

- A DNA-encoded library of 100 million or more molecules can be stored in an Eppendorf tube in a standard laboratory freezer; hence it does not require the investment in compound management and distribution infrastructure associated with existing small-molecule compound libraries.

- A DNA-encoded library selection can be performed on the laboratory bench, again avoiding the infrastructure investments required to support high-throughput screening or fragment discovery.

- As a consequence of the simplicity of a DNA-encoded library screen, it becomes possible to run multiple screens in parallel to identify molecules with enriched pharmacology. For example, selectivity can be engineered into hit molecules through the performance of parallel screens against the drug target and a selectivity target and the subsequent identification of molecules for progression with the required pharmacological profile.

- Through affinity-based selection, it is possible to identify molecules that bind to both orthosteric and allosteric sites within the same screen, thus identifying compounds with a novel mechanism of action.

- As a consequence of the use of combinatorial chemistry in library design, it is typical to gain deep insights into the structure–activity relationships of hit molecules generated in a DNA-encoded library selection.

- The combinatorial nature of DNA-encoded library chemistry enables the rapid exploration of new chemistries, leading to the tantalizing prospect that the use of such libraries may increase the success of lead identification for novel, and perhaps so-called intractable, target families.

Considering these attractions of DNA-encoded library technology, one can ask the question as to why the method has not become embedded within the field. The success of DNA-encoded technology relies upon the quality and diversity of the chemical libraries, the availability of next-generation DNA sequencing methods, and the development of informatics tools to identify high-affinity binding molecules from the library. Initially, the size and quality of DNA-encoded libraries were relatively poor, the molecules tended to be large and lipophilic and the libraries relatively small. To a large extent, this has been addressed through the ongoing development of new water-based synthetic chemistry methods, through improvements in library design, and through the availability of larger numbers of chemical building blocks. The ability to identify hit molecules in a DNA-encoded library screen relies upon the power of DNA sequencing to identify hit molecules. The revolution in DNA sequencing methodologies has dramatically reduced the costs and timelines for the analysis of the output of DNA-encoded library screens, enabling the sequencing of many hundred thousand hits for a few hundred dollars. Together with improvements in informatics, this has created a data analysis capability to rapidly understand screening data to identify molecules of interest. These developments are described in detail throughout this handbook. A final limitation to the application of this technology relies upon the defining nature of the selection paradigm. DNA-encoded library screens identify hit molecules through affinity selection. This requires that selections are performed on purified protein. While there have been some reports of the use of DNA-encoded library technology for screening of targets within a membrane or whole cell environment, the primary use of the technology has been for the screening of soluble protein targets, thus limiting the broad application of the platform for all target types.

Looking toward the future, one can anticipate an increasing acceptance of the value of DNA-encoded library technology as part of a portfolio of technologies, alongside high-throughput screening, structure-based drug discovery/fragment screening, virtual screening, and other methods for the generation of lead molecules for drug discovery. This handbook will provide an invaluable guide to scientists interested in learning, developing, and applying this technology.

Stephen Rees
2014
Vice President Screening and Sample Management
AstraZeneca, LLC
steve.rees@astrazeneca.com

INTRODUCTORY COMMENT 2

Medicinal chemistry plays a critical role in the early research essential for the discovery of both lead compounds and the chemical tool compounds that allow us to modulate important protein targets and gain a deeper understanding of disease biology. Many different methods are available for lead identification, and the methods used vary according to the different target classes, gene families, mechanisms of actions, and currently available knowledge. The variety of techniques to identify starting points for drug discovery projects can include some or all of the following: high-throughput, virtual and phenotypic screening, fragment-based design, *de novo* design, and directed screening of compound sets created with specific pharmacophores. Medicinal chemists have become skilled in data analysis, hit evaluation, and prioritization of active compound series based on the physicochemical properties needed for specific biological targets. Although these lead identification techniques are state of the art and often successful, they have not been able to reliably deliver multiple chemical series for every important biological target.

A very exciting technology that has revolutionized combinatorial chemistry, DNA-encoded library technology, is described in this book compiled brilliantly by Robert Goodnow. Although DNA-encoded library technology has been around for over 20 years, only recently has it gotten the attention it deserves within the realm of drug discovery. This technology entails creating libraries with tens to hundreds of millions of small molecules that can be pooled together and screened against protein targets under multiple conditions to obtain active compounds based on target affinity. The DNA encoding allows for the identification of hits that are present in very small amounts. To decode the assay hits, the DNA tags are amplified using PCR technology and then sequenced using one of the quickly evolving techniques for DNA sequencing. The power of using an affinity-based screening technology is that it allows the unbiased discovery of different families of compounds with a variety of mechanisms of modulating the protein. Because the technology requires chemists to expand the synthetic techniques available for generating the libraries in solvents compatible with DNA (e.g., water) and the informatics tools required to interrogate massive, complex data sets can, at first, appear daunting, the uptake of the technology as a universal technique has not yet occurred. *A Handbook for DNA-Encoded Chemistry* aims to provide a tutorial from start to finish on the important aspects of using DNA as a decoding method in the screening of billions of compounds. The experts have done an excellent job of reducing the available information into one reference that will serve to lower the barrier to utilizing this important technology in pursuit of medicines to cure unmet medical needs.

Karen Lackey
2014
Founder & Chief Scientific Officer at JanAush, LLC
karen.lackey@janaush.com

INTRODUCTORY COMMENT 3

A NEW SOLUTION FOR AN OLD PROBLEM: FINDING A NEEDLE IN THE HAYSTACK

In the era of molecular medicine, with an aging population and a briskly increasing demand for more, better, and safer drugs, the identification of suitable bioactive molecules appears to be an insurmountable needle-in-a-haystack problem. Thanks to the striking sensitivity and specificity of small-molecule DNA-encoding/decoding, today DNA-Encoded Chemical Library (DECL) technology holds a concrete promise to elegantly tackle this formidable task. The appeal of DECL technology mainly relies on the unrivalled opportunity to rapidly synthesize and probe by means of simple affinity-capture selection procedures chemical libraries of unprecedented size in a single test tube [1–6].

To date, display techniques employing such principles as phage display [7, 8], ribosome display [9, 10], yeast display [11], covalent display [12], mRNA display [13, 14], and other conceptually analogous methodologies [15, 16] profoundly impact the way novel drugs are discovered, yielding new and more efficacious classes of therapeutic agents (e.g., antibody drug conjugates) [17, 18]. DNA-encoded library technology aims to extend the realm and the potential of these display approaches to the *en masse* interrogation of small *synthetic* organic molecules.

In sharp contrast to traditional *screening* drug discovery methodologies (e.g., high-throughput screening), in which compound libraries are individually probed (i.e., one molecule at a time) in specific functional or binding assays, in display *selection* mode the compound library is interrogated as a whole, imposing the same selection pressure on all library members at the same time. Selection strategies offer massive, practical benefits over conventional approaches. First, time and costs for selection are to a first approximation independent of the library size, since all library members are interrogated simultaneously. By contrast, screening efforts tend to increase linearly with the number of compounds to screen, due to the discrete nature of the assays. Besides this radical minimization of the costs (and robotic needs) for each drug discovery campaign, the affinity-based capture of encoded molecules on a preimmobilized target protein does not require the development of expensive and cumbersome functional assays. Therefore, affinity-based selection procedures are independent of the nature of the target or its biological functions, thus allowing the targeting of components involved in protein–protein interactions or other targets that may be particularly challenging to tackle with conventional drug discovery techniques [19, 20].

Selection experiments (often termed "panning", in analogy to the gold-mining method of washing gravel in a pan to separate gold from contaminants) can be routinely performed in parallel, applying different experimental conditions (e.g., using various immobilization strategies, coating density, washing steps, repanning), or adopting alternative protocols (e.g., presence of competitors, related proteins, cofactors, and other substrates) and hits can be rapidly validated by semiautomated "on-DNA" resynthesis and testing (e.g., biosensor-based hit validation) [21], without facing complex solubility issues of the compound or buffer incompatibility problems.

DECL technology development undoubtedly profited from the recent and stunning high-throughput sequencing advances. Today, cutting-edge, deep-sequencing platforms are capable of collecting millions to billions of DNA-sequence reads per run in just a week [22–24], thus allowing the simultaneous deconvolution of multiple panning experiments using libraries containing millions of compounds in a single shot. Employing DNA-encoded strategies, researchers are quickly provided with instant (structure–activity relationship) databases after each selection experiment: an invaluable set of information for medicinal chemistry optimization of the selected structures and/or the design of successive DNA-encoded affinity maturation libraries [1, 2, 25].

As it is essential to achieve a sufficient degree of sequencing coverage after decoding with respect to library size, sequencing throughput itself poses the natural limit for the largest library that can be conveniently probed [26]. However, the steady increase in deep-sequencing throughput and the jaw-dropping drop in per-base sequencing prices will soon allow the routine interrogation of libraries comprising up to hundreds of millions of compounds.

On the other hand, the performance of a DNA-encoded library ultimately depends on the design and purity of its member compounds. Therefore, while library size exclusively depends on the number of combinatorial split-and-pool steps and building blocks employed, the gain in size and chemical diversity often correspond to an unwanted increase of the average molecular weight, beyond the generally accepted drug-like criteria (e.g., according to Lipinski's rule of 5) [27, 28], and decrease of library quality, due to incomplete reactions [3].

In this light, DECLs synthesized by the combinatorial assembly of two or three different sets of building blocks (typically including up to a few million compounds) usually display structures that are better in line with the current drug-like and medicinal chemistry requirements. In summary, as shown by this book, DNA-encoded chemical library technologies are rapidly moving beyond the proof-of-concept phase. Outstanding developments have been accomplished over the last decade [4].

While we are waiting for the next drug candidate in the clinic stemming from a DNA-encoded chemical library, scientists are already dreaming about a future where ready-to-screen libraries comprising millions of chemical compounds can be routinely designed on-demand, synthesized, and interrogated *en masse*, using fully integrated platforms on which synthetic DNA-encoded molecules such as chemical genes evolve through the panning steps of the process [29–32] as components of an artificial immune system that quickly yields small-molecule hits against exceptionally diverse biomacromolecular targets.

If modern drug discovery is a real needle-in-the-haystack search, so far there have been only two apparent ways out: reduce the size of the haystack or improve our procedure for evaluating more candidates. However, DNA-encoded chemical libraries provide an innovative alternative solution to this old problem: washing the haystack away, the needle(s) always remains at the bottom of the test tube. Only time will tell if DECL technology will fulfill this promise and play a central role in third millennium drug discovery campaigns as well as in pharmaceutical sciences [33, 34]. The availability of this handbook extends the awareness and power of this technique to a wider audience.

Luca Mannocci
2014
Independent Technology Expert & Consultant
luca.mannocci@DECLTechnology.com; web: www.decltechnology.com

REFERENCES

1. Mannocci, L., Zhang, Y., Scheuermann, J., Leimbacher, M., De Bellis, G., Rizzi, E., Dumelin, C., Melkko, S., Neri, D. (2008). High-throughput sequencing allows the identification of binding molecules isolated from DNA-encoded chemical libraries. *Proc. Natl. Acad. Sci. USA*, 105, 17670–17675.

2. Buller, F., Steiner, M., Frey, K., Mircsof, D., Scheuermann, J., Kalisch, M., Buhlmann, P., Supuran, C. T., Neri, D. (2011). Selection of carbonic anhydrase IX inhibitors from one million DNA-encoded compounds. *ACS Chem. Biol.*, 6, 336–344.

3. Clark, M. A., Acharya, R. A., Arico-Muendel, C. C., Belyanskaya, S. L., Benjamin, D. R., Carlson, N. R., Centrella, P. A., Chiu, C. H., Creaser, S. P., Cuozzo, J. W., Davie, C. P., Ding, Y., Franklin, G. J., Franzen, K. D., Gefter, M. L., Hale, S. P., Hansen, N. J., Israel, D. I., Jiang, J., Kavarana, M. J., Kelley, M. S., Kollmann, C. S., Li, F., Lind, K., Mataruse, S., Medeiros, P. F., Messer, J. A., Myers, P., O'Keefe, H., Oliff, M. C., Rise, C. E., Satz, A. L., Skinner, S. R., Svendsen, J. L., Tang, L., van Vloten, K., Wagner, R. W., Yao, G., Zhao, B., Morgan, B. A. (2009). Design, synthesis and selection of DNA-encoded small-molecule libraries. *Nat. Chem. Biol.*, 5, 647–654.

4. Mannocci, L., Leimbacher, M., Wichert, M., Scheuermann, J., Neri, D. (2011). 20 years of DNA-encoded chemical libraries. *Chem. Commun.*, 47, 12747–12753.

5. Podolin, P. L., Bolognese, B. J., Foley, J. F., Long, E., 3rd., Peck, B., Umbrecht, S., Zhang, X., Zhu, P., Schwartz, B., Xie, W., Quinn, C., Qi, H., Sweitzer, S., Chen, S., Galop, M., Ding, Y., Belyanskaya, S. L., Israel, D. I., Morgan, B. A., Behm, D. J., Marino, J. P., Jr., Kurali, E., Barnette, M. S., Mayer, R. J., Booth-Genthe, C. L., Callahan, J. F. (2013). In vitro and in vivo characterization of a novel soluble epoxide hydrolase inhibitor. *Prostaglandins Other Lipid Mediat.* 104–105, 25–31.

6. Leimbacher, M., Zhang, Y., Mannocci, L., Stravs, M., Geppert, T., Scheuermann, J., Schneider, G., Neri, D. (2012). Discovery of small-molecule interleukin-2 inhibitors from a DNA-encoded chemical library. *Chem. Eur. J.*, 18, 7729–7737.

7. Clackson, T., Hoogenboom, H. R., Griffiths, A. D., Winter, G. (1991). Making antibody fragments using phage display libraries. *Nature*, 352, 624–628.

8. Winter, G., Griffiths, A. D., Hawkins, R. E., Hoogenboom, H. R. (1994). Making antibodies by phage display technology. *Annu. Rev. Immunol.*, 12, 433–455.

9. Hanes, J., Plückthun, A. (1997). In vitro selection and evolution of functional proteins by using ribosome display. *Proc. Natl. Acad. Sci. USA*, 94, 4937–4942.

10. Kim, J. M., Shin, H. J., Kim, K., Lee, M. S. (2007). A pseudoknot improves selection efficiency in ribosome display. *Mol. Biotechnol.*, 36, 32–37.

11. Boder, E. T., Wittrup, K. D. (1997). Yeast surface display for screening combinatorial polypeptide libraries. *Nat. Biotechnol.*, 15, 553–557.

12. Bertschinger, J., Grabulovski, D., Neri, D. (2007). Selection of single domain binding proteins by covalent DNA display. *Protein Eng. Des. Sel.* 20, 57–68.

13. Keefe, A. D., Szostak, J. W. (2001). Functional proteins from a random-sequence library. *Nature* 410, 715–718.

14. Wilson, D.S., Keefe, A.D., Szostak, J.W. (2001). The use of mRNA display to select high-affinity protein-binding peptides. *Proc. Natl. Acad. Sci. USA*, 98, 3750–3755.

15. Cull, M. G., Miller, J. F., Schatz, P. J. (1992). Screening for receptor ligands using large libraries of peptides linked to the C terminus of the lac repressor. *Proc. Natl. Acad. Sci. USA*, 89, 1865–1869.

16. Heinis, C., Rutherford, T., Freund, S., Winter, G. (2009). Phage-encoded combinatorial chemical libraries based on bicyclic peptides. *Nat. Chem. Biol.*, 5, 502–507.

17. Sievers, E. L., Senter, P. D. (2013) Antibody-drug conjugates in cancer therapy. *Annu. Rev. Med.*, 64, 15–29.

18. Zolot, R. S., Basu, S., Million, R. P. (2013). Antibody-drug conjugates. *Nat. Rev. Drug Discov.*, 12, 259–260.

19. Buller, F., Zhang, Y., Scheuermann, J., Schäfer, J., Buhlmann, P., Neri, D. (2009). Discovery of TNF inhibitors from a DNA-encoded chemical library based on Diels-Alder cycloaddition. *Chem. Biol.*, 16, 1075–1086.

20. Melkko, S., Mannocci, L., Dumelin, C. E., Villa, A., Sommavilla, R., Zhang, Y., Gruetter, M. G., Keller, N., Jermutus, L., Jackson, R. H., Scheuermann, J., Neri, D. (2010). Isolation of a small-molecule inhibitor of the antiapoptotic protein Bcl-xL from a DNA-encoded chemical library. *ChemMedChem*, 5, 584–590.

21. Zhang, Y., Mannocci, L. et al. (2013). Verfahren und Anordnung zur Erfassung von Bindungsereignissen von Molekülen DE 10 2013 011 304.0.

22. Schuster, S. C. (2008) Next-generation sequencing transforms today's biology. *Nat. Methods*, 5, 16–18.

23. Kircher, M. Kelso, J. (2010). High-throughput DNA sequencing--concepts and limitations. *Bioessays*, 32, 524–536.

24. Mardis, E. R. (2013). Next-generation sequencing platforms. *Annu. Rev. Anal. Chem.*, 6, 287–303.

25. Mannocci, L., Melkko, S., Buller, F., Molnar, I., Gapian Bianke, J. -P., Dumelin, C. E., Scheuermann, J., Neri, D. (2010). Isolation of potent and specific trypsin inhibitors from a DNA-encoded chemical library. *Bioconj. Chem.*, 21, 1836–1841.

26. Buller, F., Steiner, M., Scheuermann, J., Mannocci, L., Nissen, I., Kohler, M., Beisel, C., Neri, D. (2010). High-throughput sequencing for the identification of binding molecules from DNA-encoded chemical libraries. *Bioorg. Med. Chem. Lett.*, 20, 14, 4188–4192.

27. Lipinski, C. A. (2000). Drug-like properties and the causes of poor solubility and poor permeability. *J. Pharm. Toxicol.*, 44, 235–249.

28. Lipinski, C. A., Lombardo, F., Dominy, B. W., Feeney, P. J. (2001). Experimental and computational approaches to estimate solubility and permeability in drug discovery and development settings. *Adv. Drug Deliv. Rev.*, 46, 3–26.

29. Halpin, D. R., Harbury (2004). DNA display II. Genetic manipulation of combinatorial chemistry libraries for small-molecule evolution. *PLoS Biol.*, 2, 1022–1030.

30. Li, X., Liu, D. R. (2004). DNA-templated organic synthesis: nature's strategy for controlling chemical reactivity applied to synthetic molecules. *Angew. Chem. Intl. Ed.*, 43, 4848–4870.

31. Nielsen, J., Brenner, S., Janda, K. D. (1993). Synthetic methods for the implementation of encoded combinatorial chemistry. *J. Am. Chem. Soc.*, 115, 9812–9813.

32. Hansen, M. H., Blakskjaer, P., Petersen, L. K., Hansen, T. H., Hojfeldt, J. W., Gothelf, K. V., Hansen, N. J. (2009). A yoctoliter-scale DNA reactor for small-molecule evolution. *J. Am. Chem. Soc.*, 131, 1322–1327.

33. Haupt, V. J. Schroeder, M. (2011). Old friends in new guise: repositioning of known drugs with structural bioinformatics. *Brief Bioinform.*, 12, 312–326.

34. Paul, S. M., Mytelka, D. S., Dunwiddie, C. T., Persinger, C. C., Munos, B. H., Lindborg, S. R., Schacht, A. L. (2010). How to improve R&D productivity: the pharmaceutical industry's grand challenge. *Nat. Rev. Drug Discov.*, 9, 203–214.

CONTRIBUTORS

Raksha A. Acharya
EnVivo Pharmaceuticals, Inc.
Watertown, MA, USA

Steffen P. Creaser
Genzyme Corp.
Cambridge, MA, USA

John A. Feinberg
Formerly of Roche
Nutley, NJ, USA

Currently of Accelrys, Inc.
Bedminster, NJ, USA

Andrew W. Fraley
Moderna Therapeutics
Cambridge, MA, USA

Robert A. Goodnow, Jr.
Chemistry Innovation Centre
AstraZeneca Pharmaceuticals LP
Waltham, MA, USA

GoodChem Consulting, LLC
Gillette, NJ, USA

Stephen P. Hale
Ensemble Therapeutics
Cambridge, MA, USA

Anthony D. Keefe
X-Chem Pharmaceuticals
Waltham, MA, USA

Agnieszka Kowalczyk
Formerly of Roche
Nutley, NJ, USA

Karen Lackey
JanAush, LLC
Charleston, SC, USA

G. John Langley
Department of Chemistry
University of Southampton
Southampton, UK

David R. Liu
Department of Chemistry and Chemical Biology
and Howard Hughes Medical Institute
Harvard University
Cambridge, MA, USA

Kin-Chun Luk
Discovery Chemistry
Hoffmann-La Roche Inc.
Nutley, NJ, USA

Luca Mannocci
Independent Technology Expert & Consultant
Zurich, Switzerland

Lynn M. McGregor
Department of Chemistry and Chemical Biology
and Howard Hughes Medical Institute
Harvard University
Cambridge, MA, USA

Samu Melkko
Centre for Proteomic Chemistry, Novartis Institutes
for Biomedical Research
Novartis Pharma AG
Basel, Switzerland

Dario Neri
Department of Chemistry and Applied Biosciences
ETH Zurich
Zürich, Switzerland

Johannes Ottl
Centre for Proteomic Chemistry, Novartis Institutes
for Biomedical Research
Novartis Pharma AG
Basel, Switzerland

Zhengwei Peng
Formerly of Roche
Nutley, NJ, USA

Currently of Merck & Co.
Rahway, NJ, USA

George L. Perkins
PerkinElmer, Inc.
Branford, CT, USA

John Proudfoot
Boehringer Ingelheim Pharmaceuticals Inc.
Ridgefield, CT, USA

Stephen Rees
AstraZeneca, LLC
Alderley Park, UK

Alexander Lee Satz
Molecular Design and Chemical Biology
F. Hoffmann-La Roche Ltd.
Basel, Switzerland

Jörg Scheuermann
Department of Chemistry and Applied Biosciences
ETH Zurich
Zürich, Switzerland

Sung-Sau So
Formerly of Hoffmann-La Roche Inc.
Nutley, NJ, USA

Currently of Merck & Co.
Kenilworth, NJ, USA

Charles Wartchow
Formerly of Hoffmann-La Roche Inc.
Roche Discovery Technologies
Nutley, NJ, USA

Currently of Novartis Institute for Biomedical Research
Emeryville, CA, USA

Yixin Zhang
B CUBE, Center for Molecule Bioengineering
Technische Universität Dresden
Dresden, Germany

1

JUST ENOUGH KNOWLEDGE...

Agnieszka Kowalczyk

Formerly of Roche, Nutley, NJ, USA

1.1 INTRODUCTION

At the heart of DNA-Encoded Library (DEL) technology lies Deoxyribonucleic Acid (DNA), a molecule that encodes genetic information in all living organisms. It provides a set of instructions, a blueprint for the development and functioning of an organism. In scientific literature, DNA is too often treated as a schematic, two-ribbon spiral. This model does not convey sufficient information for chemists involved in the production of a DEL. The intent of this chapter is to provide "just enough knowledge" about DNA structure, composition, characteristics, and chemical as well as enzymatic operations so that practitioners may fully embrace DEL technology. Whereas a highly detailed and referenced discussion of these topics is beyond the scope of this chapter, highlighting a few key, basic concepts should assist newcomers to this field. Those wishing for in-depth discussion are advised to search the plethora of textbooks, handbooks, and online materials. For those readers for whom this chapter may seem simplistic and too basic, they are urged to proceed to Chapter 2.

A Handbook for DNA-Encoded Chemistry: Theory and Applications for Exploring Chemical Space and Drug Discovery, First Edition. Edited by Robert A. Goodnow, Jr.
© 2014 John Wiley & Sons, Inc. Published 2014 by John Wiley & Sons, Inc.

1.2 DNA STRUCTURE

DNA was first isolated in 1869 by Friedrich Miescher [1], but its tertiary structure eluded scientists for almost a century. In 1953, James Watson and Francis Crick [2] proposed that DNA exists as a double helix. This discovery led to a period of rapid advances in biology, greatly increasing our knowledge of life processes at the molecular level. The key to understanding how hereditary information is encoded and why DNA is uniquely suited for storage of genetic instructions lies in its composition and structure.

DNA is a linear polymer built from monomers called nucleotides. Synthetic DNA molecules, usually containing fewer than 200 nucleotides, are known as oligonucleotides. Often, oligonucleotides are described in terms of "mers," referring to the number of nucleotides within the oligonucleotide. For example, 12-mer will contain 12 nucleotides in its structure. Nucleotides found in DNA are made of three components: a 2-deoxy-D-ribose unit, a nitrogenous base that is connected to the sugar molecule via a glycosidic bond, and a phosphate group (Fig. 1.1). Whereas a nucleotide is phosphorylated at the 5′-hydroxyl group, a nucleoside has a free 5′-hydroxyl.

All natural DNA nucleosides have β-configuration at the anomeric carbon. The four commonly occurring nitrogenous bases in DNA are adenine, guanine, cytosine, and thymine (Fig. 1.2). One-letter abbreviations A, G, C, and T are commonly used to denote these moieties. To avoid confusion with the numbering of atoms within a nucleotide or a nucleoside, the following convention has been adopted: carbon atoms of a

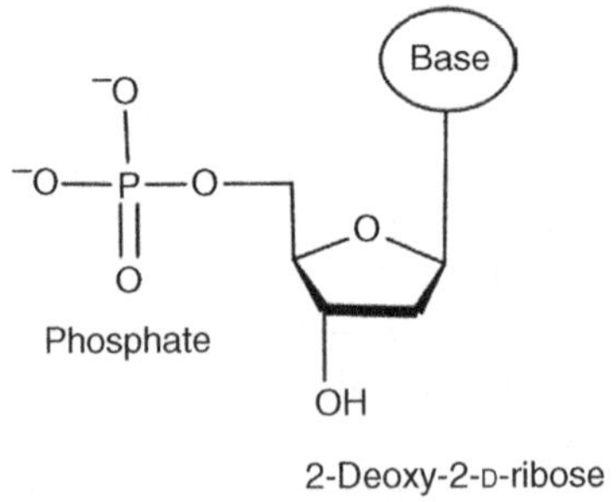

Figure 1.1. Generic structure of a 2-deoxy-2-D-ribose nucleotide.

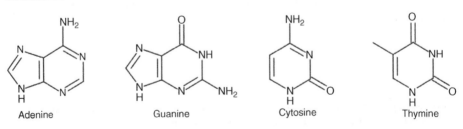

| | | | |
| Adenine | Guanine | Cytosine | Thymine |

Figure 1.2. DNA bases.

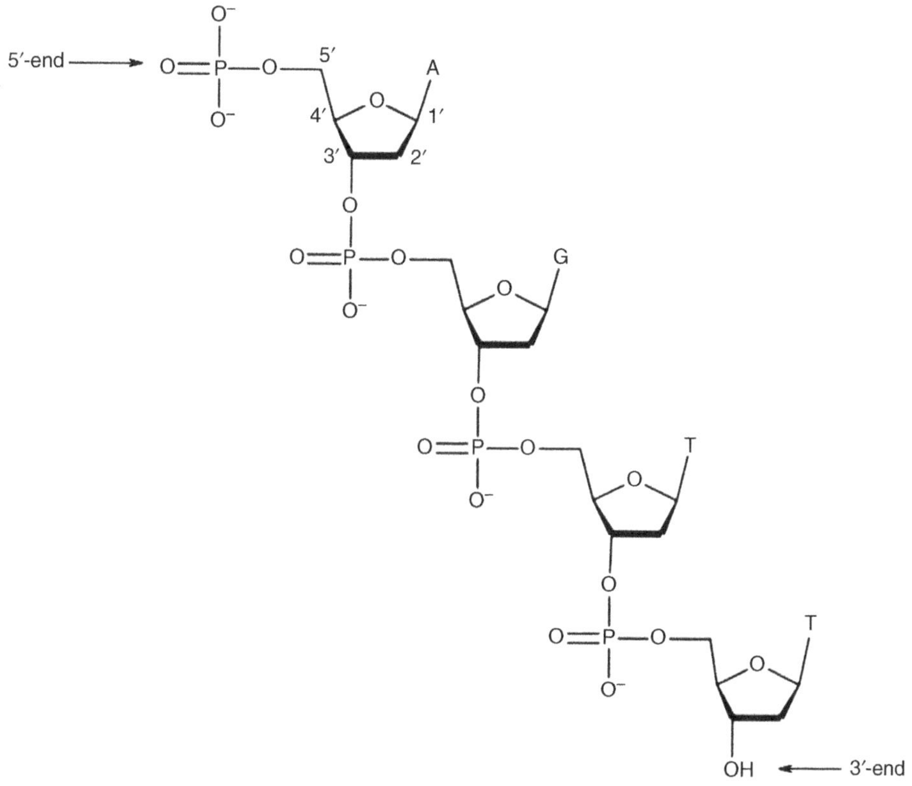

Figure 1.3. Structure of AGTT.

2-deoxy-D-ribose molecule are numbered with prime, 1′, 2′, 3′, etc., while numbers without prime notation are used for atoms in bases. Phosphodiester bonds link 3′- and 5′-hydroxyl groups of neighboring 2-deoxy-D-ribose molecules, forming a DNA backbone. Due to this architecture, a DNA molecule has so-called 5′- and 3′-ends and possesses a defined polarity or direction. The sequence of bases in DNA is conventionally written from left to right starting at the base at the 5′-end and continuing to 3′-end. For simplicity, the prefixes 5′- and 3′- are sometimes dropped. This concept is shown in Figure 1.3. The specific sequence of the bases in DNA, or the order in which they are connected, encodes genetic information.

   DNA forms a double helix, meaning that two polynucleotide chains are twisted together around an axis, forming a double-helical structure. These two chains, also called strands, run antiparallel to each other: one strand running from 5′- to 3′-end and the other one from 3′- to 5′-end. The bases, due to their hydrophobicity, are stacked inside a helix and perpendicular to its axis, while the backbone that contains alternating 2-deoxy-D-ribose and phosphate moieties is located on the outside of the helix. This spatial arrangement makes the bases hard to access and in this way protects them from undesired interactions that could potentially change the genetic instructions they

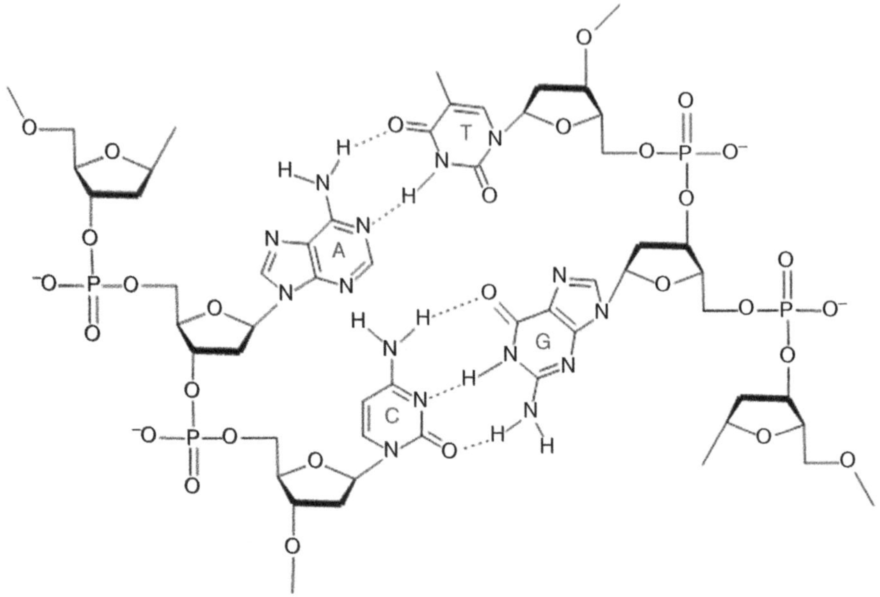

Figure 1.4. Complementary base pairing in DNA.

encode. The formation of a double helix is driven by hydrogen bonding between the bases of opposite strands and van der Waals interactions arising from stacking of the bases. Hydrogen bonding between bases occurs in a specific manner. Adenine forms a base pair with thymine through two H-bonds, while guanine forms a base pair with cytosine through three H-bonds (Fig. 1.4). This means that the opposite strands in a DNA helix are complementary; the sequence of one strand can be used to determine the sequence of the other strand. A DNA molecule that has a double-helical structure is often called a DNA duplex.

Depending on several factors, such as the level of hydration, the salt concentration, and the sequence of bases, the double helix can adopt different conformations while retaining the general spiral-like structure with two antiparallel strands. There are three major conformations of DNA helices: A-, B-, and Z-forms, each form having distinct geometrical features [3]. Both A- and B-forms are right-handed helices, with the anti-conformation around glycosidic bonds. These forms differ in the type of sugar pucker with the A-form adopting a C3′-endo and B-form adopting a C2′-endo conformation (Fig. 1.5). Consequently, the A-form helix is wider and shorter compared to the B-form. The B-form is the most common conformation found under physiological conditions, at low salt concentration and high water content, while the A-form is favored under dehydrated conditions. Z-DNA is strikingly different; it only occurs in DNA with alternating pyrimidine and purine sequences and exists as a left-handed helix. The subtle structural nuances of DNA conformations play an important role in DNA recognition by proteins and other molecules. In DEL chemistry, the most likely form of DNA a chemist will encounter is B-conformation.

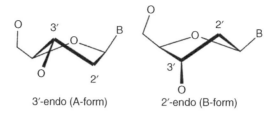

3'-endo (A-form) 2'-endo (B-form)

Figure 1.5. Types of sugar pucker found in DNA.

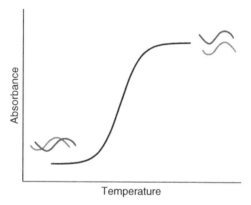

Figure 1.6. DNA melting curve.

1.3 DNA DENATURATION

Due to different physical factors such as temperature, salt concentration, the presence of organic solvents, the presence of chaotropic agents, and pH, the base interactions holding the DNA helix together can be destabilized, resulting in a separation of the duplex into single strands in a process called denaturation. Thermal denaturation, or melting, can be easily followed by measuring absorbance at 260 nm while slowly increasing the temperature of a DNA solution (Fig. 1.6). The absorbance of denatured DNA is higher than that of the corresponding duplex. This phenomenon is known as a hyperchromic shift. The temperature at which half of the DNA molecules exist as a duplex and half in a single-stranded form is referred to as the melting temperature (Tm). The Tm is a measure of duplex stability and depends on both the composition and sequence of the DNA. For example, the Tm of a 10-mer DNA duplex can range from 20°C to 40°C. High GC content stabilizes the duplex, thus increasing the Tm. The sequence also plays a role because the base stacking interactions depend on the neighboring base pairs and some combinations are more energetically favorable than others. Base mismatches, such as those caused by errors during the DNA replication process, destabilize the duplex and lead to local melting, providing a recognition mechanism for DNA repair enzymes. Under appropriate conditions, the single strands of denatured DNA will hybridize with their complementary strands to recreate the double helix in a process called hybridization.

1.4 DNA REPLICATION

For genetic material to be passed on to the next generation, DNA must be copied during cell division. This process is known as DNA replication and involves the complex interplay of several enzymes. Some of these enzymes are utilized by DEL technology; we will look at them and the processes they catalyze in more detail. During replication, the parent DNA molecule is unwound into two single strands that serve as templates and guide the synthesis of two brand-new strands, resulting in the formation of two daughter molecules identical to the parent molecule. Each daughter molecule contains one original parent strand and one newly synthesized strand. This type of replication is known as semiconservative.

The replication is initiated by helicase that unzips the two strands of the double helix creating the replication fork. Proteins known as single-stranded DNA-binding proteins bind to freshly unwound single strands, preventing them from annealing and protecting them from digestion by nucleases. DNA polymerases polymerize nucleotide triphosphates complementary to the template strands. When the polymerase "sees" guanine on the parent strand, it will add cytosine nucleotide to the new strand in the complementary position, and in the case of adenine being on the parent strand, it will add a thymine nucleotide. Different template-dependent DNA polymerases exhibit a range of accuracies with the highest fidelity demonstrating error rates as low as 1 in 10 million [4]. Template-dependent DNA polymerases can only generate DNA chains by adding complementary nucleotides to the free 3′-hydroxyl end of an annealed complementary strand—the primer. Consequently, a new strand is synthesized in the 5′ to 3′ direction. This poses a problem with replication of a duplex because two unwound template strands from the duplex are running antiparallel to each other but both of them can only be synthesized in the 5′ to 3′ direction. One strand, called a leading strand, can be synthesized continuously because its polarity is consistent with the direction of duplex unwinding and polymerase action. It only needs one primer, a short piece of RNA to form an initial duplex that will be continuously extended. On the other hand, the second strand, known as a lagging strand, is synthesized in multiple fragments, called Okazaki fragments, which are later conjoined by a ligase. The lagging strand requires many primers, one for each Okazaki fragment. This process is shown schematically in Figure 1.7.

Two of the classes of enzymes involved in DNA replication, DNA polymerase [5, 6] and DNA ligase [7, 8], play important roles in DEL technology. A DNA polymerase is utilized in the amplification of selected sequences, and a ligase is often used to join encoding tags.

There are several types of DNA polymerases involved in the complex biological processes of DNA replication and repair. The structure of the catalytic unit is highly conserved between different polymerases, indicating that the process they carry out is extremely ancient. All known polymerases catalyze the same reaction—elongation of the DNA chain by addition of a nucleotide to the free 3′-hydroxyl end of the existing chain. Template-dependent DNA polymerases cannot synthesize a new chain de novo; they require both a template strand and a primer for their function.

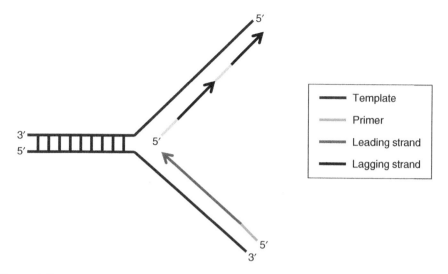

Figure 1.7. DNA replication—synthesis of a leading and a lagging strand.

Some polymerases have a proofreading ability, meaning they are able to detect the incorrectly added nucleotide and replace it with a correct one. A mismatched base pair destabilizes the duplex, causing local melting, and consequently provides a mechanism for its detection. When such a mismatched base pair is found, the polymerase reverses its slide along a template strand and excises the incorrect nucleotide—then, it adds the proper nucleotide as directed by the template strand and resumes its action of chain elongation. This function of a polymerase is also known as 3' to 5' exonuclease activity, and it explains the high fidelity that may be achieved by DNA replication.

For the purpose of *in vitro* DNA amplification, a variety of thermostable template-dependent polymerases are utilized. The most well-known example is Taq polymerase [9], isolated from the bacterium *Thermus aquaticus* found in thermal springs. Taq polymerase has optimal activity between 75°C and 80°C [10]. Due to their heat resistance, Taq polymerase and other thermostable DNA polymerases are widely used in the Polymerase Chain Reaction, abbreviated PCR, a molecular biology technique employed in DNA amplification. PCR methodology allows DNA to be rapidly copied *in vitro* from a single or few DNA fragments many million times. PCR utilizes the enzymatic replication of DNA and therefore requires a polymerase to assemble the new strands, primers to initiate the replication, and free nucleotides to serve as building blocks. PCR is performed in a thermocycler, a programmable apparatus capable of incubating at defined temperatures between 4°C and 100°C and of rapidly transitioning between these temperatures at defined rates. At high temperature, the DNA is denatured. Subsequent cooling allows primers to anneal to the freshly separated DNA strands. Primers are used in excess compared to the sequence being amplified so that the DNA strands will anneal with primers rather

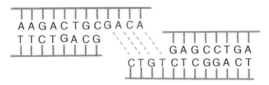

Figure 1.8. Cohesive ends.

than with themselves. The next step is the synthesis of new strands by the thermo-stable polymerase. This sequence of events is repeated a defined number of times. Since newly copied DNA fragments formed in one cycle serve as templates in the next cycle, the amplification process is exponential. It is understandable now why the use of thermostable polymerases, such as Taq polymerase, is very advantageous in the PCR process. Thermostable enzymes are able to survive repeated exposure to elevated temperatures (typically 94°C) and consequently do not require new addition of polymerase with each cycle.

Ligases are a class of enzymes that covalently join, or ligate, two DNA strands together. For example, T4 DNA ligase catalyzes the formation of a phosphodiester bond between a 5′-phosphate and an adjacent 3′-hydroxyl group of a nicked strand in a duplex. Ligases are involved in both DNA replication and DNA repair processes. To be competent for a ligation, the oligonucleotides must be monophosphorylated at their respective 5′-ends and have free 3′-hydroxyl ends. Double-stranded oligonucle-otides with either cohesive or blunt ends can be ligated; however, the latter usually requires much higher ligase concentration. Cohesive ends are overhangs on each oligonucleotide made of unpaired nucleotides that are complementary to each other; thus, they can anneal and hold the two DNA fragments to be ligated together (Fig. 1.8). The term blunt ends means that there are no overhangs and the duplex ends in a complementary base pair. The use of cohesive, or "sticky," ends is preferred because it is more efficient and ensures that the ligation proceeds only in one orien-tation determined by complementarity of the overhangs. There are two major ligase families, NAD+ and ATP dependent, indicating the cofactor needed for their action. NAD+-dependent ligases are found only in bacteria, while eukaryotes and bacterio-phages require ATP.

T4 DNA ligase, isolated from T4 bacteriophage, has been extensively used in many molecular biology applications such as cloning and DEL technology. This ligase operates best in the pH range from 7.5 to 8. Cohesive-end ligations are typically per-formed at room temperature or below to stabilize the transiently annealed oligonucle-otide junctions, although the optimum temperature for the ligase itself is higher. T4 DNA ligase utilizes ATP as a cofactor. The first step in the ligation process involves adenylation of the amino group of a lysine residue at the active site of the ligase with concomitant release of pyrophosphate. Next, AMP is transferred from the ligase to DNA, specifically to the 5′-phosphate group of one of the strands to be ligated. The resulting pyrophosphate is attacked by the 3′-OH group of the other strand, creating a phosphodiester bond and linking two strands covalently together. These steps are shown in Figure 1.9.

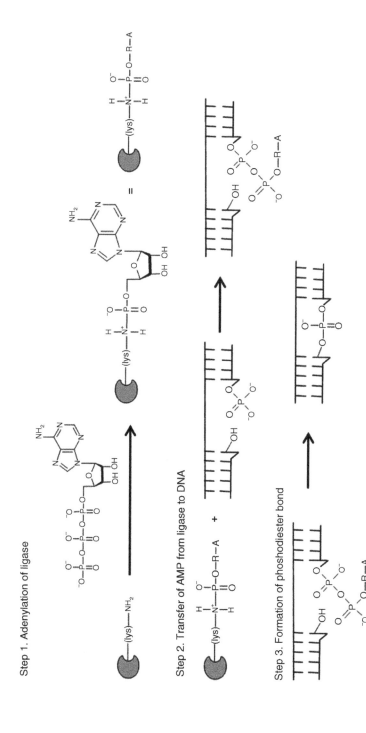

Step 1. Adenylation of ligase

Step 2. Transfer of AMP from ligase to DNA

Step 3. Formation of phoshodiester bond

Figure 1.9. Mechanism of action of the ATP-dependent ligase.

1.5 CHEMICAL SYNTHESIS OF DNA

DNA oligonucleotides can be routinely synthesized chemically using solid-phase methodology [11]. Usually, the encoding tags used in DEL technology are prepared in this way. DNA synthesis proceeds by sequential addition of nucleoside building blocks to a growing oligonucleotide covalently attached to solid support. This process is fully automated in commercially available apparatuses called DNA synthesizers that have now been available for more than 30 years.

The coupling of nucleoside building blocks is based on phosphoramidite chemistry. Briefly, nucleoside phosphoramidites, which are used as nucleotide equivalents, have the following features: the 5′-hydroxyl moiety is protected with a dimethoxytrityl (DMT) group and the 3′-OH is derivatized as a 2-cyanoethyl N, N-diisopropyl phosphoramidite (Fig. 1.10). In addition, the protection of the exocyclic amino functionalities of guanine, adenine, and cytosine is required to avoid side reactions and to improve the solubility of the nucleoside building blocks. For this purpose, an acylation reaction is commonly employed; the N-6 amino group of adenine and the N-4 amino group of cytosine are usually blocked with benzoyl groups, while the N-2 position of guanine is protected with an isobutyryl group. The choice of protecting groups is dictated by the chemistry employed in the synthetic process. Different functionalities must be temporarily blocked and the blocking groups later easily removed at various stages of the oligonucleotide synthesis. This orthogonal protection must be compatible with all reagents and conditions used in the process.

The entire synthesis is implemented as a series of repeated cycles; in each cycle, one nucleoside is added to the oligonucleotide growing on a solid support. The cycle consists of four distinct steps: (i) detritylation, (ii) coupling, (iii) capping, and (iv) oxidation

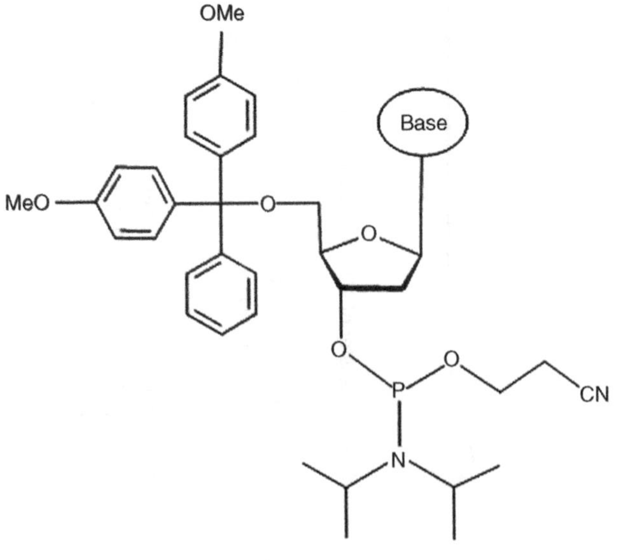

Figure 1.10. Building block for DNA synthesis.

(Fig. 1.11). The very first nucleoside is already attached to the solid support via its 3′-OH group, and after detritylation, its 5′-hydroxyl group will be able to react with the next nucleoside. This design implies that the oligonucleotide is synthesized from the 3′- to the 5′-end, which is opposite to the direction of DNA assembly by a polymerase. The synthesis cycle starts with the removal of the acid-labile DMT group to expose a 5′-hydroxyl group for coupling with the next building block. The DMT cation has a bright orange color in solution. The measurement of its absorbance serves as an indicator of coupling efficiency because the DMT group must be removed from the last nucleoside incorporated into the oligonucleotide prior to coupling with the next nucleoside. The incoming phosphoramidite nucleoside is activated by tetrazole, enabling the formation of an unstable phosphite bond between two nucleosides. Even though the coupling efficiency is usually very high (over 98%), a small amount of uncoupled material may persist. This uncoupled material could potentially react further in subsequent steps forming a sequence that lacks one base and is difficult to separate from the full-length product. To avoid this undesired process, the capping step is implemented immediately after coupling. During capping, all remaining free 5′-hydroxyl groups are acetylated, thus preventing them from

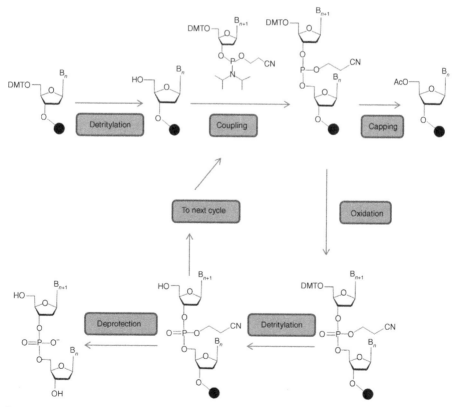

Figure 1.11. Solid-phase oligonucleotide synthesis by phosphoramidite method.

any further reaction. This results in failure, or truncated, sequences that are sufficiently different from the full-length sequence to be easily removed after the synthesis. The phosphite triester that links two newly coupled nucleosides is unstable and following the capping step is oxidized to a more stable phosphate triester. A commonly used oxidizing system is a mixture of iodine and pyridine in water and THF. After all nucleosides have been added, the product needs to be cleaved from the solid support and further processed to remove both the protecting groups from the bases and 2-cyanoethyl groups from the phosphate triesters to yield a functional oligonucleotide. This is achieved by heating a solid-support-bound product in concentrated ammonium hydroxide at 55°C for 1 h. The crude product is then desalted and, if needed, purified further using reversed phase or ion exchange HPLC.

1.6 OLIGONUCLEOTIDE CHARACTERIZATION

Making a DEL involves steps that require assessing the identity and/or purity of oligo-nucleotides. Quality control must be performed after ligation and addition of a chemical building block. Two analytical tools routinely used for these purposes are electrophoresis and Mass Spectrometry (MS).

Electrophoresis [12, 13] allows for the separation of DNA fragments based on their size. DNA molecules have a net negative charge due to the negatively charged phosphate groups of the backbone. This charge enables them to migrate through an inert gel matrix when an electric field is applied across the gel. The rate of migration of charged molecules in electrophoresis is known as electrophoretic mobility. Molecules differing in size move through the pores of the gel with different rates. Smaller DNA fragments migrate faster than larger ones forming a distinct band pattern on the gel. These bands correspond to different lengths of DNA fragments. The gel can be "calibrated" by running a mixture of molecular weight size markers (DNA fragments of known lengths) along with a sample of unknown DNA to estimate its size. Two matrix materials commonly employed in gel electrophoresis are polyacrylamide and agarose. Polyacrylamide is a synthetic polymer prepared from acrylamide monomer and the cross-linking agent N,N'-methylenebisacrylamide. The relative amounts of these two reagents determine the porosity of the gel and can be optimized to obtain the best conditions for a specific separation. When high concentrations of a chaotropic agent such as urea are present in a polyacrylamide gel, hydrogen bonds are destabilized and single-stranded oligonucleotides may be characterized at high resolution; oligonu-cleotides differing in size by only one base pair can be resolved. Such gels are referred to as denaturing. This high resolving power is offset by a low range of polyacrylamide separations, up to a couple of 1000 bp long. Gels made from agarose, a natural polysac-charide isolated from seaweed, have a large pore size and thus are well suited for the separation of much larger DNA fragments than polyacrylamide gels, but their resolu-tion is limited. Agarose gels cannot be made denaturing and are generally only used with double-stranded DNA. Agarose gels can be prepared and poured in the lab prior to use. Alternatively, precast gels, both denaturing and native polyacrylamide and aga-rose, can be purchased from commercial vendors.

DNA on a gel is not visible to the human eye, and thus, it needs to be stained or otherwise rendered visible in order to be detected. Ethidium bromide is widely used for the visualization of double-stranded DNA. It works by intercalating into the DNA duplex and fluorescing under UV light, revealing the localization of the DNA bands on a gel. Similar fluorescent dyes are also available for single-stranded nucleic acids. In general, dyes can be incorporated into either gel matrix prior to the separation or alternatively a gel can be stained with a dye after the separation. The latter method is employed in case of polyacrylamide gels since dyes can interfere with the polymerization reaction.

In the DEL process with double-stranded DNA, agarose gel electrophoresis is used to monitor the efficiency of ligation of the tags. Tag ligation increases the molecular weight of the DNA fragment, and therefore, observed retardation of the ligated product on the gel indicates that the ligation was successful.

The second indispensable analytical tool for a DEL chemist is an MS system suited for oligonucleotide analysis. In most cases, the mass spectrum of an oligonucleotide provides sufficient information about its identity. Two MS methods have been widely used in oligonucleotide analysis: Electrospray Ionization (ESI) [14–16] and Matrix-Assisted Laser Desorption/Ionization–Time Of Flight (MALDI-TOF) [17–19]. MALDI-TOF is slightly more sensitive than ESI for shorter oligos (<50 bp). It produces singly charged species, simplifying the interpretation of the spectrum. Ionization efficiency drops significantly for oligonucleotides longer than 50 bases in MALDI, limiting its use to the shorter sequences. On the other hand, the ESI technique delivers good accuracy and sensitivity over a wide range of oligonucleotide length/mass. In the process of ESI, multiply charged species of a parent molecule are formed, necessitating the use of special algorithms to deconvolute the spectra. Deconvolution allows for the reconstruction of the molecular weight of a parent molecule from the m/z values of its charged ions. It is especially useful for analyzing mixtures of oligonucleotides. Automated deconvolution software packages for processing ESI data are commercially available. The ESI system can be easily coupled with Liquid Chromatography (LC) to provide both sample separation and sample identification.

An LC/MS system optimized for oligonucleotide analysis is an absolutely necessary tool for a DEL chemist. It allows for monitoring chemical reactions on DNA during library development and synthesis. A more detailed discussion of DNA analytical methods is found in Chapter 8.

1.7 DNA SEQUENCING

DNA sequencing makes it possible to determine the exact order of nucleotides within a fragment of DNA. The development of DNA sequencing methodologies was spurred by the need to decipher the functions of single genes and, ultimately, of the entire genome. Eventually, DNA sequencing enabled new applications in diagnostics and forensics as well as various novel biotechnologies such as DEL.

One of the first DNA sequencing methods was developed by Maxam and Gilbert in 1977 [20]. It relies on a series of four carefully selected and performed chemical processes that cleave DNA at specific bases. One of the 5′-ends of the DNA is labeled, usually with radioactive phosphorus (^{32}P). Reaction conditions are optimized in such a way that on average one cleavage occurs per one DNA molecule. It ensures that all possible fragments that include a 5′-labeled end will be represented. Each reaction is run in a separate vessel and generates DNA fragments of varying length. These four sets of fragments are then separated side by side by polyacrylamide gel electrophoresis with a resolution of a single nucleotide, and the sequence is deduced from the pattern on the gel. Maxam–Gilbert methodology is labor-intensive and complex. It was soon replaced by a conceptually very different method established by Sanger [21]. Sanger sequencing, also called the chain-termination method, utilizes an *in vitro* DNA replication process with a small but profound modification. Along with natural 2′-deoxyribonucleotides, their 2′,3′-dideoxy analogs are used. These dideoxy analogs are devoid of a 3′-hydroxyl group, and thus, they can be added to the growing DNA chain via their 5′-ends but cannot form a phosphodiester bond with an incoming nucleotide. Consequently, when a dideoxy analog is incorporated, chain elongation stops generating a terminated DNA fragment. This process is carried out separately for each of the chain-terminating nucleotides. The new chain starts growing from the labeled primer, and therefore, all terminated fragments will be also labeled, enabling their detection. The DNA sequence can be then easily read from a gel electropherogram of the terminated DNA fragments.

Sanger methodology quickly became the method of choice for DNA sequencing. It was simple and amenable to automation. The use of dye terminators, dideoxy analogs labeled with different fluorescent dyes, further simplified the process. With this improvement, the sequencing process could be carried out in one vessel instead of four, since the last nucleotide of the terminated fragment is easily identified by its unique wavelength after excitation. Gel electrophoresis was replaced with capillary gel electrophoresis. Eventually, the DNA sequencing process became fully automated with first-generation sequencers based on the Sanger chain-termination method.

More recently, these sequencing processes have been adapted such that *in vitro* DNA cloning and amplification can be performed relatively simply on a massive scale. Such technologies are known as "next-generation" or "deep" sequencing [22–24]. This allows for parallel sequencing of DNA libraries at high speed and low cost. There are several sequencing platforms commercially available. Despite employing diverse processes and techniques, most next-generation methodologies share a common strategy. DNA fragments are amplified into clusters with each cluster arising from a single DNA molecule. These clusters, made of identical DNA fragments, are sequenced by synthesis. Sequencing by synthesis relies on detecting either labeled nucleotides being incorporated into a new DNA strand or by-products of the synthesis such as pyrophosphate. After each nucleotide incorporation cycle, a synthesis is temporarily halted, and a snapshot image that captures all clusters simultaneously is acquired, revealing the sequence as it is being assembled in each cluster.

DEL technology has been greatly facilitated by the adoption of deep sequencing technologies because it permits the detection of relatively weak signals such as are necessarily found in selection experiments with large libraries. The two platforms commonly used in DEL technology are the Illumina/Solexa and 454/Roche systems.

Illumina is a popular next-generation DNA sequencing platform in the field [25]. It relies on PCR bridge amplification [26, 27] and reversible dye terminator [28, 29] methodologies. Before actual sequencing, all DNA fragments must be equipped with adaptors, short oligonucleotides that are added by ligation or PCR to both ends of a DNA fragment. These adaptors allow for the hybridization of the DNA fragments with primers that are covalently attached to a flow cell surface. There are only two different types of primers present in multiple copies on the flow cell. One type of primer has a cleavable site engineered in its sequence. A denatured DNA library is loaded on a flow cell, allowing single strands of DNA fragments to hybridize with the primers. This step ensures the spatial separation of different DNA fragments on the surface. Next, the complementary strands are synthesized enzymatically starting from the primers. This process leaves newly assembled strands immobilized on the surface since they originated from the tethered primers. The complementary strands, arising from the original DNA fragments, are not tethered and thus are washed off after denaturation. Then, the strands are amplified in a process called bridge amplification. It occurs when an immobilized DNA strand hybridizes with a proximal primer on the slide surface forming a bridge. DNA polymerase then extends the chain from the primer, resulting in two DNA strands covalently attached to the flow cell after denaturation. This cycle of chain extension and denaturation is repeated many times, leading to a formation of a DNA cluster. Each cluster contains up to 1000 copies of a single original DNA fragment in close proximity immobilized on a surface. Several millions of spatially separated clusters can be formed on a single flow cell. In order to enable sequencing, a DNA cluster must contain only one type of single-stranded DNA; however, after amplification, both the original strand and the reverse strand are tethered to the surface in multiple copies. The reverse strand is attached via a primer with a cleavable site and therefore can be easily removed. Next, all 3′-ends of immobilized strands are blocked with dideoxynucleotide analogs to prevent potential extension of DNA strands on each other. The actual sequencing starts with annealing of the sequencing primer and extension of the chain using four reversible dye terminator nucleotides. These reversible terminators are labeled with different fluorescent dyes and transiently blocked at their 3′-hydroxyl groups. The reversible terminators are incorporated one at a time. After each incorporation, the image of the entire flow cell is captured, enabling the identification of the last added nucleotide in all clusters simultaneously. This cycle of chain extension and imaging resumes after the removal of a fluorescent label and unblocking of 3′-hydroxyl functionality of the last added nucleotide and continues until the sequence of the defined region is obtained in each DNA cluster. Transient blocking of the 3′-hydroxyl group of reversible terminators ensures that the synthesis is temporarily halted after single nucleotide incorporation, allowing for image acquisition.

The 454 sequencing platform [30] is based on a similar strategy: separation of DNA fragments and their amplification, followed by parallel sequencing by synthesis. The DNA fragments are amplified by emulsion PCR [31] and then sequenced by pyrosequencing method [32]. At low template concentration, emulsion PCR is a clonal amplification method carried out in a water-in-oil emulsion. Prior to amplification, DNA fragments are ligated with two different adaptors. These adaptor-flanked DNA fragments are combined with two primers complementary to the adaptors, microbeads covered with one of the primers and PCR reagents and then emulsified in oil. In this process, DNA fragments and beads get trapped inside tiny aqueous droplets suspended in oil. The low concentration of DNA fragments ensures that on average only one DNA molecule is encapsulated per droplet, creating separate microreactors for the amplification of each DNA fragment. The DNA molecule inside a droplet is amplified by PCR, coating the beads with multiple copies of itself. Once the emulsion is broken, the complementary strands that are not attached to the beads are denatured and washed off and the sequencing primer is annealed. The beads with amplified DNA fragments are then deposited on a plate with picoliter-sized wells for pyrosequencing. Each well can accommodate only a single bead. Much smaller beads with two immobilized enzymes, namely, ATP sulfurylase and luciferase, are added to all wells, surrounding the bigger beads with clonally amplified DNA. In pyrosequencing, only one of the four nucleotides is added at a time. If the added nucleotide is complementary to the one on the template, it gets incorporated into the growing chain with a concomitant release of a pyrophosphate. The pyrophosphate triggers a series of enzymatic reactions involving ATP sulfurylase and luciferase that result in the emission of a bioluminescence signal. The presence or absence of this signal makes it possible to elucidate the sequence. When using pyrosequencing for sequencing of homopolymer segments, some attention is due. The synthesis is only halted after one nucleotide incorporation if the next nucleotide to be incorporated is of different identity. However, in case of stretches containing the same consecutive nucleotides, for example, AAAAA, the synthesis continues for the entire length of homopolymer segment before it stops. For short nucleotide repeats, up to six nucleotides, there is a direct proportionality between the number of nucleotides incorporated and the intensity of the bioluminescence signal. However, longer nucleotide repeats may result in insertion and deletion errors as in such cases the number of nucleotides cannot be easily deduced from the intensity of light.

DNA sequencing technologies are still rapidly expanding and improving, resulting in increased speed and capacity with reduction of cost. They greatly accelerate progress in biomedical research, forensics, agriculture, and diagnostics and enable novel technologies such as DEL. One expects the impact of high-throughput sequencing to continue to have an impact on the practice and applications of DNA-encoded chemistry methods.

REFERENCES

1. Dahm, R. (2008). Discovering DNA: Friedrich Miescher and the early years of nucleic acid research. *Hum. Genet.*, 122, 565–581.

2. Watson, J. D., Crick, F. H. C. (1953). Molecular structure of nucleic acids: a structure for deoxyribose nucleic acid. *Nature*, 171, 737–738.

3. Dickerson, R. E., Drew, H. R., Conner, B. N., Wing, R. M., Fratini, A. V., Kopka, M. L. (1982). The anatomy of A-, B-, and Z-DNA. *Science*, 216, 475–485.

4. McCulloch, S. D., Kunkel, T. A. (2008). The fidelity of DNA synthesis by eukaryotic replicative and translesion synthesis polymerases. *Cell Res.*, 18, 148–161.

5. Steitz, T. A. (1999). DNA polymerases: structural diversity and common mechanisms. *J. Biol. Chem.*, 274, 17395–17398.

6. Hubscher, U., Spadari, S., Villani, G., Maga, G. (2010). *DNA Polymerases: Discovery, Characterization and Functions in Cellular DNA Transactions.* World Scientific Publishing Company, Hackensack.

7. Weiss, B., Richardson, C. C. (1967). Enzymatic breakage and joining of deoxyribonucleic acid, I. Repair of single-strand breaks in DNA by an enzyme system from Escherichia coli infected with T4 bacteriophage. *Proc. Natl. Acad. Sci. U.S.A.*, 57, 1021–1028.

8. Lehman, I. R. (1974). DNA ligase: structure, mechanism, and function. *Science*, 186, 790–797.

9. Chien, A., Edgar, D. B., Trela, J. M. (1976). Deoxyribonucleic acid polymerase from the extreme thermophile Thermus aquaticus. *J. Bacteriol.*, 127, 1550–1557.

10. Lawyer, F. C., Stoffel, S., Saiki, R. K., Chang, S. Y., Landre, P. A., Abramson, R. D., Gelfand, D. H. (1993). High-level expression, purification, and enzymatic characterization of full-length Thermus aquaticus DNA polymerase and a truncated form deficient in 5′ to 3′ exonuclease activity. *Genome Res.*, 2, 275–287.

11. Pon, R. T. (2003). Chemical synthesis of oligonucleotides: from dream to automation. In Khudyakov, Y. E., Fields, H. A., Eds. *Artificial DNA: Methods and Applications.* CRC Press LLC, Boca Raton, pp. 1–70.

12. Martin, R. (1996). *Gel Electrophoresis: Nucleic Acids (Introduction to Biotechniques).* BIOS Scientific Publishers Ltd., Oxford.

13. Sambrook, J. F., Russell, D. W. (2001). *Molecular Cloning: A Laboratory Manual*, 3rd edition. Cold Spring Harbor Laboratory Press, Cold Spring Harbor.

14. Fenn, J. B., Mann, M., Meng, C. K., Wong, S. F., Whitehouse, C. M. (1989). Electrospray ionization for mass spectrometry of large biomolecules. *Science*, 246, 64–71.

15. Apffel, A., Chakel, J. A., Fischer, S., Lichtenwalter, K., Hancock, W. S. (1997). Analysis of oligonucleotides by HPLC-electrospray ionization mass spectrometry. *Anal. Chem.*, 69, 1320–1325.

16. Potier, N., Van Dorsselaer, A., Cordier, Y., Roch, O., Bischoff, R. (1994). Negative electrospray ionization mass spectrometry of synthetic and chemically modified oligonucleotides. *Nucleic Acids Res.*, 22, 3895–3903.

17. Karas, M., Bahr, U., Giesmann, U. (1991). Matrix-assisted laser desorption ionization mass spectrometry. *Mass Spectrom. Rev.*, 10, 335–357.

18. Wu, K. J., Steding, A., Becker, C. H. (1993). Matrix-assisted laser desorption time-of-flight mass spectrometry of oligonucleotides using 3-hydroxypicolinic acid as an ultraviolet-sensitive matrix. *Rapid Commun. Mass Spectrom.*, 7, 142–146.

19. Stults, J. T. (1995). Matrix-assisted laser desorption/ionization mass spectrometry (MALDI-MS). *Curr. Opin. Struct. Biol.*, 5, 691–698.

20. Maxam, A. M., Gilbert, W. (1977). A new method for sequencing DNA. *Proc. Natl. Acad. Sci. U.S.A.*, 74, 560–564.

21. Sanger, F., Nicklen, S., Coulson, A. R. (1977). DNA sequencing with chain-terminating inhibitors. *Proc. Natl. Acad. Sci. U.S.A.*, 74, 5463–5467.

22. Shendure, J., Ji, H. (2008). Next-generation DNA sequencing. *Nat. Biotechnol.*, 26, 1135–1145.

23. Fuller, C. W., Middendorf, L. R., Benner, S. A., Church, G. M., Harris, T., Huang, X., Jovanovich, S. B., Nelson, J. R., Schloss, J. A., Schwartz, D. C., Vezenov, D. V. (2009). The challenges of sequencing by synthesis. *Nat. Biotechnol.*, 27, 1013–1023.

24. Metzker, M. L. (2010). Sequencing technologies—the next generation. *Nat. Rev. Genet.*, 11, 31–46.

25. Mohamed, S., Syed, B. A. (2013). Commercial prospects for genomic sequencing technologies. *Nat. Rev. Drug Discov.*, 12, 341–342.

26. Adessi, C., Matton, G., Ayala, G., Turcatti, G., Mermod, J. -J., Mayer, P., Kawashima, E. (2000). Solid phase DNA amplification: characterization of primer attachment and amplification mechanism. *Nucleic Acids Res.*, 28, e87.

27. Fedurco, M., Romieu, A., Williams, S., Lawrence, I., Turcatti, G. (2006). BTA, a novel reagent for DNA attachment on glass and efficient generation of solid-phase amplified DNA colonies. *Nucleic Acids Res.*, 34, e22.

28. Ju, J., Kim, D. H., Bi, L., Meng, Q., Bai, X., Li, Z., Li, X., Marma, M. S., Shi, S., Wu, J., Edwards, J. R., Romu, A., Turro, N. J. (2006). Four-color DNA sequencing by synthesis using cleavable fluorescent nucleotide reversible terminators. *Proc. Natl. Acad. Sci. U. S. A.*, 103, 19635–19640.

29. Turcatti, G., Romieu, A., Fedurco, M., Tairi, A. P. (2008). A new class of cleavable fluorescent nucleotides: synthesis and optimization as reversible terminators for DNA sequencing by synthesis. *Nucleic Acids Res.*, 36, e25.

30. Margulies, M., Egholm, M., Altman W. E., Attiya, S., Bader, J. S., Bemben, L. A., Berka, J., Braverman, M. S., Chen, Y. -J., Chen, Z., Dewell, S. B., Du, L., Fierro, J. M., Gomes, X. V., Godwin, B. C., He, W., Helgesen, S., Ho, C. H., Irzyk, G. P., Jando, S. C., Alenquer, M. L. I., Jarvie, T. P., Jirage, K. B., Kim, J.-B., Knight, J. R., Lanza, J. R., Leamon, J. H., Lefkowitz, S. M., Lei, M., Li, J., Lohman, K. L., Lu, H., Makhijani, V. B., McDade, K. E., McKenna, M. P., Myers, E. W., Nickerson, E., Nobile, J. R., Plant, R., Puc, B. P., Ronan, M. T., Roth, G. T., Sarkis, G. J., Simons, J. F., Simpson, J. W., Srinivasan, M., Tartaro, K. R., Tomasz, A., Vogt, K. A., Volkmer, G. A., Wang, S. H., Wang, Y., Weiner, M. P., Yu, P., Begley, R. F., Rothberg, J. M. (2005). Genome sequencing in microfabricated high-density picolitre reactors. *Nature*, 437, 376–380.

31. Dressman, D., Yan, H., Traverso, G., Kinzler, K. W., Vogelstein, B. (2003). Transforming single DNA molecules into fluorescent magnetic particles for detection and enumeration of genetic variations. *Proc. Natl. Acad. Sci. U.S.A.*, 100, 8817–8822.

32. Ronaghi, M., Uhlen, M., Nyren, P. (1998). A sequencing method based on real-time pyrophosphate. *Science*, 281, 363–365.

2

A BRIEF HISTORY OF THE DEVELOPMENT OF COMBINATORIAL CHEMISTRY AND THE EMERGING NEED FOR DNA-ENCODED CHEMISTRY

Robert A. Goodnow, Jr.

Chemistry Innovation Centre,
AstraZeneca Pharmaceuticals LP, Waltham, MA, USA
GoodChem Consulting, LLC
Gillette, NJ, USA

2.1 INTRODUCTION

Small-molecule combinatorial chemistry has been the focus of much research and innovation during the last 20 years. The roots of this chemistry go back much further. With great effort by thousands of chemists focused on high-throughput methodology, innovations have been brought about in solid- and solution-phase chemistry, in design, and in purification. The effort began with peptide synthesis but evolved quickly to include the synthesis of small molecules, followed by catalysts and substances of interest in the field of materials science. Many highly creative efforts were made to develop synthetic schemes amenable to a combinatorial format. However, the development and practice of combinatorial chemistry has not been without problems. With respect to the application of combinatorial chemistry to small-molecule drug discovery, an initial great excitement gave way to more conservative applications as well as outright skepticism. Often, drug discovery scientists were disappointed with the poor drug-like properties of molecules that resulted in many combinatorial libraries. An inability to effectively describe and fill a drug-like diversity space has led to the

A Handbook for DNA-Encoded Chemistry: Theory and Applications for Exploring Chemical Space and Drug Discovery, First Edition. Edited by Robert A. Goodnow, Jr.
© 2014 John Wiley & Sons, Inc. Published 2014 by John Wiley & Sons, Inc.

idea that simply making more compounds will "fill the gap." This practice has often proved unsatisfactory.

When using combinatorial chemistry for lead identification and lead optimization, many pharma organizations have struggled to find the right strategy regarding library size and design, diversity coverage, screening capacity, and evaluation of the impact of this approach. The application of combinatorial methods for the development of new catalysis and the exploration and development of process methods and in materials research has continued steadily. It is from this background of vigorous innovation and problem solving that DNA-encoded chemistry methods have emerged. With the emergence of such methods, a reenergization of combinatorial chemistry is coming about.

Progress in drug discovery, which requires the design, synthesis, and testing of novel substances, has usually been slow, costly, and prone to high rates of failure. Therefore, many chemists stay attuned to opportunities for enhancing efficiency. Many have sought advantage through the exploitation of the dynamic between different approaches spanning from the purely rational up to and including entirely serendipitous discovery. A great example of serendipitous discovery is the story of the discovery of penicillin. Although it is now often noted that Fleming was neither the first to observe nor the first to apply this discovery to therapeutic effect [1], his isolation of this important natural product continues to be a compelling account because it highlights the great impact that may begin from an observant scientist taking advantage of serendipity. In contrast, accounts of the discovery of H2 antagonists [2] celebrate the power of a sequence of interpretation of data and subsequent rational decisions resulting in highly successful results.

As often happens when two opposing extremes exist, many scientists aim for a balance between the rational and the serendipitous. The ideal lies in testing many rational approaches and rationally following up on serendipitous findings. The evolution of combinatorial chemistry has come about as explorations along the serendipitous path that lies between these two extremes. At the same time, one must not lose sight that combinatorial chemistry came about with the application of new technologies (e.g., automated synthesis and purification) and materials (e.g., solid-phase supports and scavenger resins) to more than a century of progress in synthetic chemistry. Each significant advance has been associated with the application of a recently accessible technology. DNA-encoded libraries of combinatorially generated small molecules are not different in this respect.

Recently, the emergence of readily accessible "deep" or "next-generation" DNA cloning and sequencing technologies has greatly facilitated the evaluation of DNA-encoded libraries enriched for binding to various protein targets. To put into context the fundamental details, the advantages and challenges of DNA-encoded library synthesis as well as its applications and potential, one must have some understanding of the historical developments of combinatorial chemistry.

2.2 DEFINITIONS

In the course of the evolution of combinatorial chemistry methods, a diverse terminology has arisen. Some authors interchange or conflate these terms to the extent that confusion becomes inevitable. In order to avoid any ambiguity, concise definitions for several commonly employed terms are given below.

- **Combinatorial chemistry** is the performing of a chemistry process whereby some or all combinations of diverse reagents are reacted according to a common synthetic scheme. In the simplest example, a central template is diversified with m reagents at one position and n reagents at a second position, giving a total of m × n synthetic products.
- **High-Throughput Chemistry** (HTC) is a general term referring to various aspects of a chemistry process (i.e., analog design, synthesis, purification, compound handling) that are performed such that multiple operations are conducted in batch mode. HTC, as a general term, includes the concepts of parallel synthesis and combinatorial chemistry, among others. High-throughput purification is an aspect of this process.
- **Parallel synthesis** is the performance of a common synthetic step or steps while simultaneously varying a single diversity reagent. Often, a common core synthon is used in this approach. Parallel chemistry is one of the simplest methods for obtaining small numbers of compounds. It requires little more preparation than would be necessary for the standard analog synthesis that most organic chemists perform.
- **Solid-phase synthesis** refers to synthetic procedures that are conducted by attaching at least one reaction component to a substance that is insoluble in the reaction solvent. The approach is appealing for several reasons: (i) applicability to split-and-pool synthesis, (ii) ease of purification by filtration and rinsing of the derivatized solid support, and (iii) application of excess reagents to drive reactions to completion. On the other hand, the approach has several technical challenges, including the quantity of material that can be loaded to the polymer and its properties while suspended in solution. In addition, the adaptation of many chemical reactions to a solid-phase format is rarely routine.
- **Solution-phase chemistry** represents the running of chemical reactions in homogeneous solutions. Although this definition may not differentiate the method from standard organic chemistry operations, the term is usually employed in reference to parallel or combinatorial synthesis processes.
- **Polymer-assisted chemistry or solid-phase reagent chemistry** indicates a chemical process that employs a reagent bound to solid phase. The reagent is sometimes common to the synthetic procedure, not necessary related to the incorporation of a diversity reagent into the final product (e.g., base, scavenger, activation reagent) [3].
- **Diversity-Oriented Synthesis (DOS)** represents a synthetic approach such that a single starting material reacts with a range of reagents to give products with different molecular skeletons, thus providing rapid access to a library of compounds with significant skeletal diversity [4].

2.3 BRIEF HISTORY OF THE EVOLUTION OF COMBINATORIAL CHEMISTRY

A full history of the developments and evolution of combinatorial chemistry is beyond the scope of this chapter. Further, excellent coverage already exists [5]. In an attempt to summarize critical highlights, a tabulation of highlights of developments of combinatorial

TABLE 2.1. A tabulation of highlights in developments of combinatorial chemistry

#	Year	Notable discovery or advance	Innovator	Significance	Ref.
1.	1963	Solid-phase peptide chemistry	R. B. Merrifield	Early reports of the use of polymer support for synthetic peptides	[6]
2.	1965	Solid-phase oligonucleotide synthesis	R. L. Letsinger	Early reports of the use of polymer support for synthetic oligonucleotides	[7]
3.	1970	Solid-phase carbohydrate modification and oligosaccharide synthesis	C. Schuerch and J. M. Fréchet	Early reports of the use of polymer support for synthetic oligosaccharides	[8]
4.	1984	Peptide library synthesis on solid-phase pins	H. M. Geysen	Enhancement of process efficiency towards a synthesis systemization	[9]
5.	1985	Use of polystyrene resin in "tea bags"	R. A. Houghten	Additional enhancement of process efficiency for peptide synthesis	[10]
6.	1992	Solid-phase synthesis of benzodiazepines	J. A. Ellman	Clear demonstration of the potential for pharmaceutical small-molecule research	[11]
7.	1988–1991	Split-and-pool	A. Furka	Theoretical basis to create large numbers of compounds	[12]
8.	1992	Encoding with DNA	S. Brenner and R. A. Lerner	Concept experiment for using DNA to encode small-molecule synthesis	[13]
9.	1993	Encoding split-and-pool libraries with DNA	M. A. Gallop	Synthesis and encoding of approximately one million peptides with DNA-based encoding	[14]
10.	1993	Haloaromatic binary coding	W. C. Still	Split-and-pool, solid-phase synthesis with haloaromatic binary encoding	[15]
11.	1995	Radio frequency	K. C. Nicolaou and E. J. Moran	Commercially available encoding method for solid-phase synthesis	[16, 17]
12.	1995	Materials science applications	P. Schultz et al.	Spatially addressable combinatorial libraries of copper oxide films for superconducting materials research	[18]

TABLE 2.1. (cont'd).

#	Year	Notable discovery or advance	Innovator	Significance	Ref.
13.	1996	Catalyst development with high-throughput methods	M. L. Snapper, A. H. Hoveyda, et al.	Early example of solid-phase synthesis to prepare discovery of new catalysts for enantioselective synthesis	[19]
14.	1998	High-throughput purification	D. Kassel	High-throughput means of purifying compounds libraries	[20]
15.	2000	DOS concept	S. Schreiber et al.	A strategy for expanding structural diversity of small-molecule libraries	[4]
16.	2008	Development of cheap high-throughput sequencing technology		Transition from Sanger-based to second-generation sequencing technology	[21]
17.	2009	Report of first 800 million DNA-encoded compound library	M. Clark et al.	Exemplification of DNA-encoded combinatorial library synthesis and assay with very large library	[22]

chemistry is shown in Table 2.1. The following commentary of significant events attempts to put into context important discoveries that have paved the way for the emergence of DNA-encoded combinatorial chemistry. Given that innovation evolved along different themes, the years of such discovery and subsequent applications were hardly sequential, but rather often overlapping.

Many consider the beginnings of combinatorial chemistry to have occurred in the laboratories of Bruce Merrifield at Rockefeller University in 1963 with the report of the synthesis of a tetrapeptide (H-Leu-Ala-Gly-Val-OH (**1**), Fig. 2.1) on 200-mesh polystyrene using Cbz-protected amino acids [6]. Despite modest yields and laborious starting material preparations, the author foreshadowed the application of automation to this process. The author went on to publish many other reports of solid-phase synthesis. In 1965, Letsinger and colleagues reported the solid-phase synthesis of a dimer of deoxycytidine (**2**) [7]. Six years later, Fréchet and Schuerch reported several reactions performed on polystyrene including the formation of a 1,6-linked glucose–glucose disaccharide (**3**) [8]. In each of these early solid-phase synthesis reports, the authors focused on single-target synthesis and characterization; as such, they are not the first reports of combinatorial chemistry, but it should be clear that these methods established the thinking and some early methodology for the transformation of a chemistry technology into high-throughput and eventually combinatorial endeavors. These three examples highlighted the potential for the synthesis of peptides, oligonucleotides, and

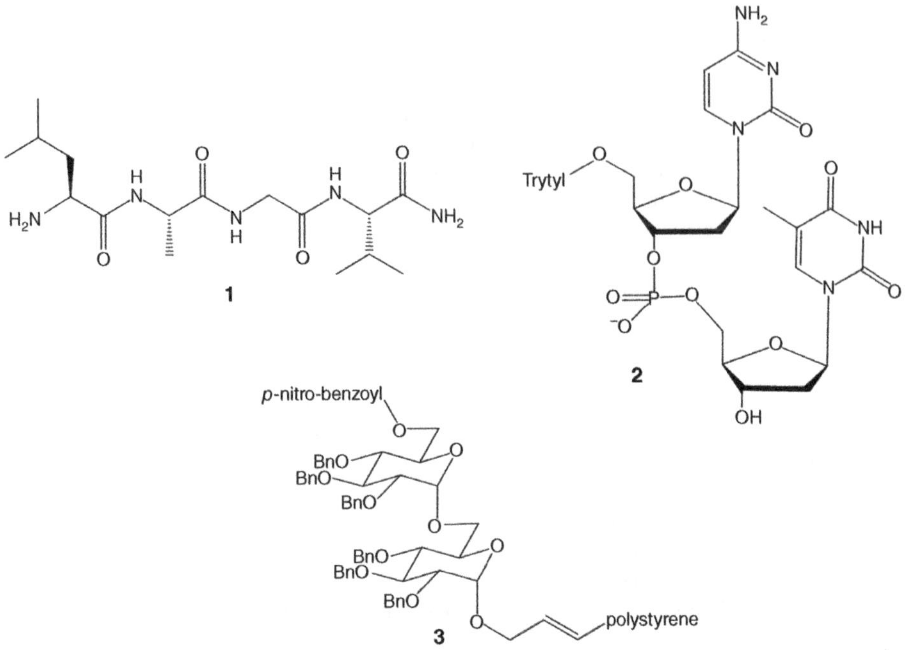

Figure 2.1. Structures of early solid-phase synthesis targets: Merrifield's tetrapeptide, Letsinger's oligonucleotide, and Fréchet's disaccharide.

oligosaccharides, each having linear structures. These structures result, of course, from the repetition of the same type of chemical operations and, thus, had the potential to be readily adapted to an automated, solid-phase method.

The extension of solid-phase synthesis to drug-like small molecules came in 1992, with the publication of a solid-phase synthesis of substituted benzodiazepines by Ellman and Bunin [11] (Fig. 2.2). This report drew much attention to the idea that combinatorial chemistry could produce large numbers of not just peptides and oligo-nucleotides, but of compounds that were among prototypical drug-like structures. An important aspect of the Ellman work was not only the type of target molecules that could be made but also the type of reactions that could be applied to the solid phase; in addition to amidations, the reaction sequence included carbonyl-amine conden-sations and N-alkylations. Since that time, tremendous efforts have gone into devel-oping elegant and intricate synthetic procedures on various solid-phase supports. The diversity of possible structures has been widely reported and reviewed [23].

2.3.1 Industrialization of Combinatorial Chemistry

The commentary on the productivity of drug discovery as measured by NDA filings has often been negative. The declining productivity of the pharmaceutical industry in providing new drugs relative to a high point in 1996 is a familiar introduction to

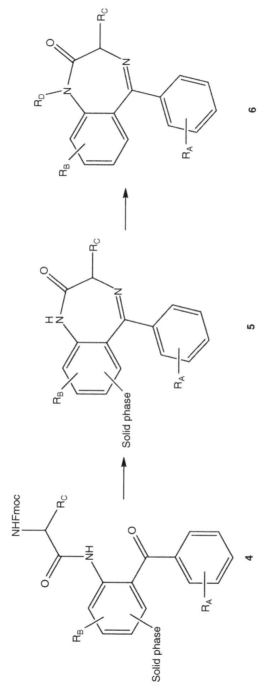

Figure 2.2. Ellman and Bunin's solid-phase synthesis of 1,4-benzodiazepine derivatives.

25

many articles focused on strategies for improving the speed and quality of drug discovery. In such papers, one often reads of the need for new technology and approaches to enhance productivity and efficiency in drug discovery [24]. In the 1990s, the association of HTC methods and High-Throughput Screening (HTS) approaches to hit identification in the pharma industry seemed exactly the strategy to increase the productivity of the drug discovery process. HTS technology evolved to make the assay of one million compounds a fairly routine (although expensive, ~US$ 200,000) event. Many pharma research organizations during the later 1990s and early 2000s embarked on library-building campaigns [25] to install large numbers of highly drug-like compounds with diverse structures. The source of much of this growth initially came from HTC methods. However, aspects of HTC were not readily compatible with the rigorous requirements for many organizations' compound collections. Traditionally, synthesized, single compounds, accumulated over long years of medicinal chemistry research, have often been included in an organization's compound collection. Such compounds are often recrystallized solids, highly pure, of unquestioned identity, well characterized, and available in multiple milligram quantities. Given the one-at-a-time synthetic approach, synthetic chemists applied much effort to make molecules with diverse substitution patterns for which reagents may not have been readily available and, therefore, required additional synthesis. Such requirements set a high hurdle for the combinatorial chemists hoping to use the power of an organization's HTS engines to test library designs. While the requirements of chemical diversity, identity, purity, and quantity proved daunting challenges for HTC methods, such requirements also prompted great innovation and invention by chemists.

The research, design, development, and application of automation systems for chemistry evolved vigorously during the 1990s, largely in response to the enthusiasm around combinatorial chemistry. Many chemists were excited by the concept of making greater numbers of compounds and considered the application of automated chemical synthesis systems to be the answer to the challenges. Peptide and oligonucleotide synthesis systems served as useful starting points for the creation of synthesis platforms that could accommodate a wider variety of chemical reactions. An excellent review describes early systems such as the ACT496 by Advanced ChemTech, the Nautilus by Argonaut Technologies, and the Zymate by Zymark [26]. These systems were configured around a number of reaction cells (24 to ~800). Clever engineering was applied to create systems that could accommodate different chemistries using solution- and solid-phase formats and conditions. However, a general trend emerged that automation of specific repetitive tasks (e.g., vial weighing and labeling, liquid arraying) in a workstation approach was ultimately more easily implemented and applied than a fully automated, complex system. Fully automated, complex systems require frequent care and maintenance and are often fixed around a set workflow; by contrast, focused workstations are often robust and easily reconfigured to accommodate different workflows. For example, the use of microwave technology integrated with a vial-changing robot has found widespread application in the creation of small libraries (i.e., hundreds of compounds), but an entirely automated system built around a microwave reactor has not been commercialized.

2.3.2 Expectations of Purity, Identity, and Diversity: Trying to Do More with a Parallel Action

As research efforts worked to integrate the promise of combinatorial chemistry into the workflows of a modern drug discovery organization, several disconnects became evident. The identity and purity of the final compound for testing was a focus of significant concern. Many scientists were attempting to integrate the compounds resulting from high-throughput synthesis methods into the established workflows that had been developed over many years; such workflows were based on compounds of known identity and high purity and available in tens of milligram quantities in many cases. The challenge to develop HTC methods that produced substances of such quality and quantity created not only the need for better chemistry but also for means of purifying large numbers of compounds in a high-throughput manner. Kassel and coworkers at Combichem reported modifications of liquid chromatography conditions (e.g., high flow rates) that allowed for the purification of compounds, particularly by Mass Spectrometry (MS)-triggered collections after short HPLC separation times [21]. The intent of this work was to develop a chromatographic method that could be commonly applied to a series of compounds in a collection, thereby rapidly purifying a set of crude combinatorial chemistry products into products of acceptable purity. The development of these analytical and preparative methods has been well reviewed [27]. It is interesting to note that the great need for high-throughput purification of combinatorial libraries resulted in a plethora of high-throughput analytical and preparative systems that are now more likely employed for the purification of single compounds or small sets resulting from non-high-throughput methods.

2.4 SPLIT-AND-POOL SYNTHESIS AND THE ENCODING SOLUTIONS

2.4.1 The Potential of Split-and-Pool Synthesis

Given the well-established, widely practiced solid-phase peptide synthesis approach for which a diversity of building blocks were readily available, it should not come as a surprise that the synthesis of peptides was the first chemistry to expand in a combinatorial manner. Indeed, Furka's initial disclosure of the split-and-pool concept was entitled "More peptides by less labour" [13]. As many now know well, a combinatorial distribution of products is created by iterations of splitting portions of solid-phase intermediates for separate reactions with different diversity reagents followed by "pooling" or mixing to create the solid-phase intermediate ready for the next step (Fig. 2.3). The generation of products is easily predicted as the product of the numbers of each diversity reagent set. It is the multiplicative increase in the number of products that makes this approach to synthesis so powerful. With hindsight, like many concepts that have far-reaching consequences, the logical simplicity of the split-and-pool method now seems unsurprising and, frankly, obvious, yet the implications continue to be profound and far-reaching.

When applying this method to drug discovery, several options for assay were used: binding experiments of products still bound to the polymer bead, assay of single-bead cleavage liquors, and assay of cleavage liquors from pooled beads. Despite the conceptual simplicity of the split-and-pool synthesis method, it creates

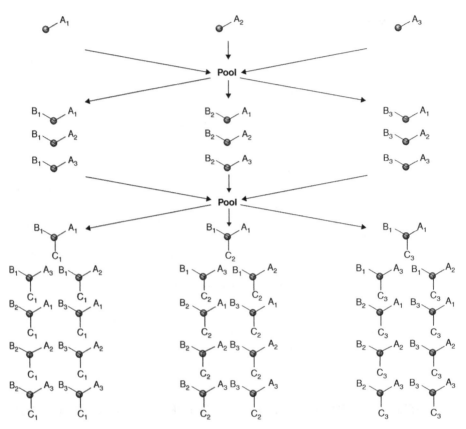

<u>Figure 2.3.</u> Scheme of a 3 × 3 split-and-pool synthesis around a central template or solid-phase support.

a problem of substantial complexity: differentiation of the active from inactive compounds in a complex product mixture.

2.4.2 Addressing the Problem of Compound Identity and Encoding: Early Attempts at Encoding Split-and-Pool

The potential of creating large numbers of analogs using a split-and-pool method is obvious. The multiplicative advantage escapes no researcher attempting to explore the vastness of chemistry space with a mixture, despite the problem of identifying the active compound once a hit has been detected. Evolving directly from the split-and-pool method was the concept of "one bead, one compound." In this concept, a free-flowing mixture of solid-phase synthesis beads allows for the creation of many copies of the same compound on a single bead. Lam and Lebl calculated that up to 10^{13} copies of a single compound could exist on a single 100 μm bead [28]. Colorimetric assay formats exist whereby it is possible to detect a bead that contains an active compound based on a difference in color of that particular bead relative to other beads that do not display active compounds. The subsequent problem becomes the identification of the active substance. It is feasible that peptide sequences from active beads could be sequenced using Edman degradation. However, such sequencing is possible only for peptides; with small molecules, some other method of hit identification becomes necessary. Among several methods for encoding [29], some have proposed identification by MS, but reports of this method have not been widespread. Other methods are needed to address this problem of hit identification from a combinatorial library.

Combinatorial libraries have been generated in two distinct approaches on solid phase: spatially addressable synthesis and split-and-pool synthesis. In the former, it is the maintenance or indexing of distinct batches of solid-phase resin that permits the eventually identification of the final product with that batch. Such a method is most commonly associated with Chiron Mimotopes [23]. A specific resin pin location could be indexed with a particular synthesis history, and thus, a potentially active compound coming from that pin was easily identified. Although simple in concept, the 96-pin format in an array compatible with 96-well plates may become challenging when contemplating the synthesis of tens of thousands of compounds. Further, although the original loading capacity of the resin tip of 10–100 nmol was later increased, the quantity expected to come from a tip was still low. For example, in the case of 100 nmol of a compound of 500 Da/mol, theoretically, 0.5 mg is possible for a 100% efficient synthetic sequence.

In 1985, Houghten used polypropylene mesh "tea bags" to synthesize a library of 248 13-mer peptides and assay them for binding to an antibody raised against a segment of the hemagglutinin protein [9]. In this approach, a permeable yet insoluble polypropylene mesh was used to contain the solid-phase polystyrene support. Each bag containing some 20 mg of resin was uniquely labeled. Each bag could then be put through a split-and-pool method of peptide synthesis for coupling, deprotection, and washing steps. Using this method, the author was able to identify three amino acids that are important for binding to the hemagglutinin protein–antibody interaction. This method is simple, readily implemented, and flexible as to the quantity of resin used. However, the manual sorting of bags can become laborious once the number of bags exceeds several hundred. The volumes associated with each bag depend on the size of the bead and quantity of resin.

The numeric advantages of split-and-pool have been highlighted previously, but the problem of identification of the active substance remained largely intractable for any large number (>10,000) of compounds until the advent of binary encoding with electrophoric molecular tags as reported in 1993 by Still and coworkers [15]. In this solid-phase, split-and-pool method called the ECLIPS technology, the attachment of haloaromatic tags to the solid support was coordinated with the reaction with diversity reagents in a combinatorial sequence. The tags (Fig. 2.4) were attached by an amide bond formation or by a carbene insertion [30].

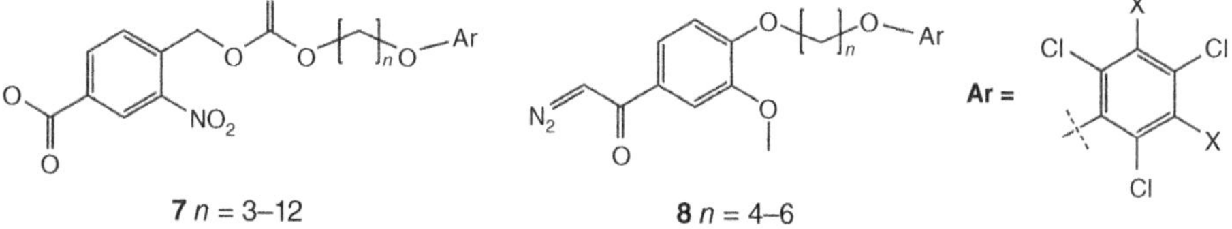

7 $n = 3–12$ **8** $n = 4–6$

Figure 2.4. Reagents for haloaromatic encoding of solid phases with amide formation (**7**) and by carbene insertion (**8**).

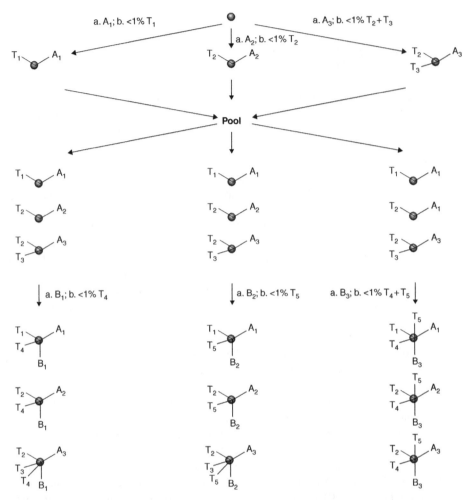

Figure 2.5. Haloaromatic binary encoding of a two-cycle split-and-pool combinatorial library.

Tags attached by amide bond formation (**7**, Fig. 2.4) can be removed photolytically, whereas those installed by carbene insertion (**8**) can be removed by treatment with ceric ammonium nitrate. In both cases, the liberated primary alcohol was silylated to generate the trimethylsilyl ether (**11**) (Fig. 2.5 and Fig. 2.6). Such compounds were then detected using a gas chromatograph having electron-capture detection (ECGC). ECGC detection operates at a very high degree of sensitivity (e.g., 5 femtograms per second (fg/s)). Due to such a high sensitivity, it was necessary to attach only 1% loading equivalents of each bead. Further, the high sensitivity of ECGC permits the association of the absence of a tag with an encoding event. Thus, in a binary format, with n tags, a total of $2^n - 1$ encoding events were possible. For 20 tags, a library could be encoded with 1,048,575 reagents. The interleaving of the tagging and diverse combinatorial chemistry is represented as shown in Figure 2.5.

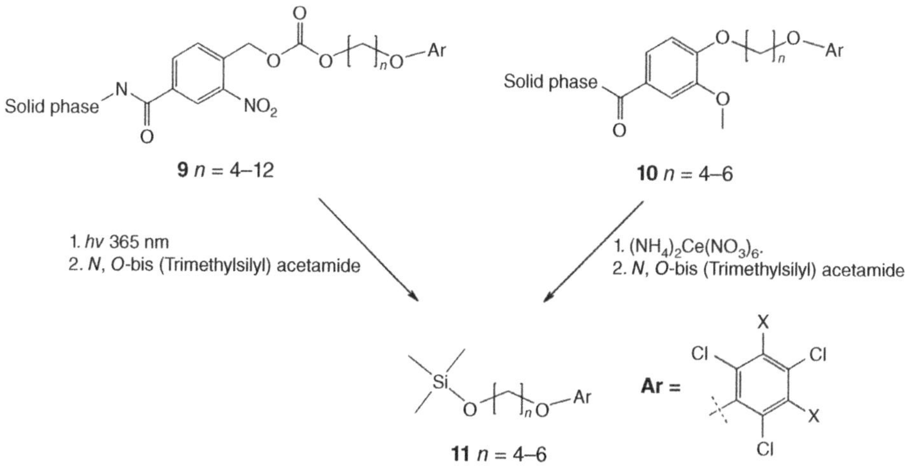

Figure 2.6. Release of haloaromatic encoding tags under photolytic or oxidative conditions followed by silylation.

This method was very successful for creating large libraries, up to approximately 100,000 members. Several publications describe the successful identification of hits using this technology (e.g., carbonic anhydrase [31] and p38 kinase [32]). However, an important aspect of this technology was the redundancy in the compounds that were theoretically synthesized. The fact that there are multiple beads having the same compound is important from a sampling perspective; if there were only one bead per compound, the chances of detecting a single active compound is potentially lower. In addition, libraries containing single-copy beads could be used only once for an assay; using part of a library for an assay would risk missing some of the chemical diversity present in the entire collection. For example, to assume the presence of 200 copies of each compound for a 100,000-member library, it was necessary to begin with 20,000,000 beads. This equates to some 180 g of resin when working with 230 μm ArgoGel® for which there are 110,000 beads/g. Assuming the loading of each bead was 1 nmol of material, the starting number of molar equivalents was then 0.02. Assuming that one uses an excess of reagent to drive the reaction to completion, a fivefold excess on a single 0.1 mol of the first reagent is required. Where diversity reagents are plentiful and cheap, such a quantity is not an issue. Where diversity reagents are not available and must be synthesized, this quantity becomes a significant detail for advance planning. Should a library size grow from 100,000 members, the need for reagents in quantity is all the greater. For example, in order to make 1,000,000 compounds, one would need to start with nearly 2 kg of resin and 1 mol of a single, first reagent. Solvent consumption for reaction and washing steps also becomes an aspect for operational planning, cost, and mechanization.

Another important consideration relates to the amount of material that can be derived from a bead. In the ECLIPS process, several beads were arrayed per microtiter plate well, followed by a partial release of compound. For each well in which a biologically active

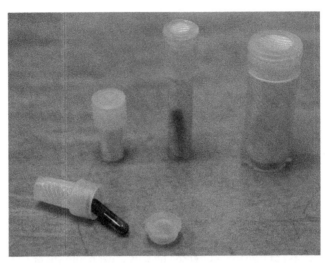

Figure 2.7. From left to right: a MicroKan® with a microfrequency tag and a cap, a closed MicroKan®, a MiniKan® containing a microfrequency tag, and a MacroKans®. For color detail, please see color plate section.

compound was detected, the beads were then separated and more compound was released, followed by decoding now of the single, active bead. When using 130 μm poly(ethylene glycol)-grafted polystyrene beads, it is estimated that approximately 300 pmol is the maximum amount of releasable compound [33]. When released into 100 μL, a volume not uncommon for a 96-well, this results in 3 μM solution; such a concentration is not excessive for many bioassay formats. When released into 10 μL, the concentration becomes 30 μM, a number more manageable with respect to expected assay-related dilutions. Such details regarding the quantity of release material inform on the general applicability and adaptability of this method to the requirements of the HTS process. This method was practiced mainly by Pharmacopeia until the mid-2000s. The technology continues to be applied in some laboratories [34].

Another practical solution for encoding split-and-mix combinatorial synthesis was reported in 1995 by scientists at IRORI [16] and at Ontogen [17]. The basic technology eventually commercialized by IRORI was the use of a porous polymer mesh vessel, which contained solid-phase synthesis resin and a radio-frequency tag. The radio-frequency tag was a small piece of circuitry contained within an impermeable glass shell. When subjecting the microfrequency tag to an electromagnetic field of appropriate strength, the radio-frequency tag would transmit a distinct code. A software system was integrated with the tag-reading circuitry so that a synthetic history could be associated with each tag-associated resin batch. Such tags were sold with MicroKans®, MiniKans®, and MacroKans® (Fig. 2.7). Although the Kans were usually not reusable, the tags were durable to most reaction conditions and could be reused many times. Table 2.2 later indicates the approximate amount of resin and compound loading possible with these systems. Prior to the performance of each reaction, tags were read and assigned to an appropriate reaction vessel according to the synthetic plan. After the reaction, the tags

TABLE 2.2. The quantity of resin and potential quantity of a target compound derived from solid phase

	Resin capacity (mg)	Weight of 500 Da compound possible assuming 1 mmol/g resin and 100% chemical and purification yields (mg)	Weight of 500 Da compound possible assuming 1 mmol/g resin and 50% chemical yield and 50% purification yield (mg)
NanoKans®	~5	2.5	0.63
MicroKans®	~30	15	3.8
MiniKans®	~150	75	18.8
MacroKans®	~500	250	62.5

were rinsed with solvent, either in separate vessels or collectively, lightly centrifuged and dried. At the conclusion of the reaction sequence, the tags were read again, and each Kan was assigned to an individual cleavage well for treatment with a resin cleavage cocktail (e.g., TFA/CH_2Cl_2). The cleavage liquors were collected into microtiter plates, which had been mapped to a particular tag ID, thereby identifying the presumed compound. The usual challenges of solid-phase synthesis with respect to reaction development, efficiency, and yield could be expected for such systems. For libraries composed of some hundreds of Kans, manual manipulation was possible. For libraries composed of thousands of Kans, an automated sorter was developed and sold, the IRORI AutoSort®. Where reaction chemistry was not completely efficient, purification of the products was often performed before assay. The advantage of this technology was the relative ease to set up and run encoded, solid-phase combinatorial library synthesis. In order to accommodate the synthesis of libraries composed of tens of thousands of compounds, IRORI developed the NanoKans®, a small porous vessel with approximately 5 mg of resin associated with a 2D bar code, which was read by an optical system. Although the tag identification technology is different, the concept and practice was similar to that for the MicroKan® system. This technology was used for the synthesis of a 10,000-member benzopyran library [35]. The NanoKans® systems represented substantial investment in capital equipment and infrastructure.

2.5 ENCODING WITH OLIGONUCLEOTIDES

Brenner and Lerner first proposed the use of DNA to encode the synthesis of a peptide in 1992 [13]. In this theoretical paper, the authors laid out the principles of encoding the attachment of each amino acid on an unspecified linker to a solid phase that would also be used for the addition of six encoding oligonucleotides. The authors describe a design for a "chemical gene" composed of PCR primer sequences flanking the reagent-encoding sequences. They also designed in cleavage sites for Sty I and Apa I restriction enzymes for manipulation of these constructs. The report foreshadows many of the basic concepts of DNA-encoded combinatorial synthesis. This encoding strategy and others that have developed subsequently are covered in more detail in later chapters.

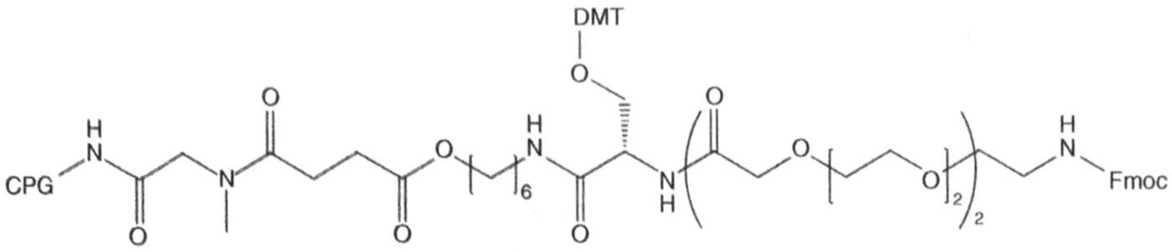

Figure 2.8. Basic schema for DNA-encoded peptide synthesis reported by Nielsen et al.

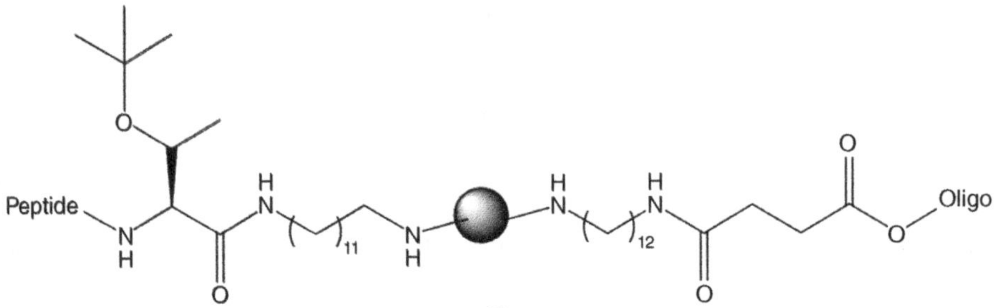

Figure 2.9. Structure of oligonucleotide-encoded peptide support by Gallop et al.

The first implementation of DNA-encoded peptide synthesis was reported by Nielsen et al. in 1993 [36]. In this method, Controlled Pore Glass (CPG) support was used to construct a template system that was compatible with the orthogonal chemistry for peptide couplings and oligonucleotide synthesis (Fig. 2.8). Each amino acid was encoded with trinucleotide tags after the installation of necessary primer sequences, similar to what had been described in the Brenner paper 1 year earlier. The authors addressed important design considerations including the detachment of the peptide/oligo construct, steric effects of the linker components, as well as the potential for steric interferences during peptide-binding assays. A series of short (~5-mer) beta-endorphin epitope-containing peptides were synthesized with this encoding system. The amino acid sequence was confirmed by Edman degradation; the oligonucleotides were competent as PCR templates.

Also in 2003, Gallop and coworkers at Affymax synthesized some 800,000 heptapeptides, encoding each amino acid monomer with a dinucleotide using monodisperse polystyrene beads (Fig. 2.9) [14]. In this work, the authors considered the important concept of starting with a sufficient number of beads such that there were some 200 copies of each sequence in the library. The peptides were present at 20-fold the quantity of the oligonucleotide on each bead. In this method, encoding sequences were framed with PCR primer sequences. The library was selected against fluorescently labeled anti-dynorphin B antibody. Beads containing a high-affinity peptide bound the fluorescently labeled antibody and were separated from nonbinding beads by FACS cell sorting. PCR was then used to amplify the encoding DNA on the beads; this was followed by sequencing to identify the sequences of the high-affinity heptamers.

2.5.1 Limitations of the Combinatorial Chemistry Approach That DNA-Encoded Libraries Begin to Address

Chemical diversity is huge. Many papers cite the number of compounds that could be made within a space defined by the Lipinski rules [37] as 10^{60} molecules [38]. This number appears as a footnote in a paper by Bohacek et al. and is based on rough estimates of the number of compounds by resulting from a fixed number of rings and branch points. More recent and rigorous estimates focusing on molecules of smaller-molecule weight still indicate many, many billions of possible structures [39]. A persistent question for any practitioner of combinatorial chemistry is "How many compounds are needed to sufficiently cover a particular chemistry space?" An interesting response was proposed by Jacoby et al. [40]. In an analysis of combinatorial libraries that had provided hits in biological screens at the Novartis Institutes for Biomedical Research, computational chemists were able to plot the "statistical variance of the information content described by Daylight fingerprints." In this analysis, such statistical variance leveled off after 5000 molecules. The authors argued that the change in variance was larger for an initial set of 1000 well-selected compounds and, thus, such a number was a good balance between potential cost and benefit. Between 1000 and 5000 compounds, the statistical variance in Daylight fingerprints decreased. This analysis presumes that statistical variance in Daylight fingerprints informs on meaningful diversity of molecules available for interaction with biological targets. The authors also noted at the time that Dolle had observed that three-quarters of the libraries reported in the literature were composed of fewer than 100 compounds [41]. The differences by orders of magnitudes of these numbers indicate great gaps in the actual practice of combinatorial chemistry compared to theoretically defined limits. Perhaps these gaps are a result of current practical limitations of HTC methods. DNA-encoded combinatorial libraries offer a means to narrow such a gap.

Perhaps the question of how many compounds are needed to adequately fill a chemistry space is unanswerable. Some 10–100 million molecules have been disclosed in the public domain [42]. Recent publications have focused rather on how many molecules can be made with existing chemistry and building blocks. For example, Nikitin and coworkers reported in 2005 their enumeration of some 10^{13} compound resulting from approximately 400 published chemical reactions and diversified with commercially available reagents [43]. At this point, several statements are unavoidable. Chemical space is still vast and beyond the reach of current technology. Combinatorial chemistry to date has made a modest exploration into this vast space. As will be discussed, the DNA-encoded combinatorial chemistry approach is a means for increasing the numbers of synthesized compounds by several orders of magnitude.

Some of the larger libraries reported at the time numbered in the tens of thousands [33]. Even large combinatorial chemistry libraries are a modest sampling of the possible space. It was often proposed that parallel chemistry libraries offer greater diversity coverage due to the ready implementation of different templates, but such theories were rarely tested and never convincingly verified. The term "sparse matrix" describes the synthesis of a subset of all the compounds that can be made with a given set of building blocks, on the premise that those synthesized sufficiently provided "good diversity coverage." This was despite the fact that "good diversity coverage" could not be well defined. Some may well conclude that such terminology belies a regrettable state of affairs: because one cannot define diversity, because chemical diversity space is huge, because one could never make such a number of compounds, and because parallel and combinatorial chemistry as they have been developed could only make from several hundreds to several tens of thousands of compounds in the extreme, one must compromise with the fact that a "sparse matrix may offer good diversity coverage."

Not too many years after the initial excitement surrounding the advantages of combinatorial chemistry, there came some disappointment [44]. Concerns about the diversity and drug-likeness of the compounds that could be produced in any significant number were widely expressed. Naturally, it was noted that chemistries that were amenable to the generation of a large number of compounds because of multiple reagent inputs (e.g., Ugi condensations) often generated compounds that were excessively heavy and lipophilic. The lack of useful structural diversity and drug-likeness was highlighted as a particular failing of combinatorial chemistry. The problems with the libraries that had been produced are possibly a manifestation of the amount of effort necessary to develop chemistry that is compatible with a high-throughput methodology as well as the limited chemical diversity that exists for most reagent classes. As chemists develop ever more creative synthetic schemes and greater numbers of uniquely functionalized building blocks become readily available, greater and greater numbers of high-value compounds become possible. As noted previously, the publication of a solid-phase synthesis of substituted benzodiazepines by Ellman and Bunin serves as good example of the sorts of molecules that are possible with organic synthesis adapted to solid-phase supports. Tremendous efforts have gone into developing elegant and intricate synthetic procedures on various solid-phase supports. The diversity of possible structures has been widely reported and reviewed [45]. Although often elegant, there are several aspects that may temper the enthusiasm for developing solid-phase synthetic methods: (i) time and effort to develop solid-phase-compatible chemistry; (ii) quantity of compound, which may ultimately be obtained from a solid phase; and (iii) availability of diverse reagents in useful quantities.

Adaptation of a solution-phase synthesis to the solid phase must include considerations of solvents, which will be compatible with the solid phase. For example, some solvents create a swelling effect on the solid phase (THF and polystyrene), and this is useful for the permeation of reagents reacting with functional groups attached to the solid phase. Other solvents result in the resin contracting (e.g., methanol or water and polystyrene), and this may hinder the progress of a particular reaction. Since solid-phase syntheses often involve some sort of labile linker (e.g., cleavage by treatment with strong acid), care must be taken to avoid such conditions until the end of the synthesis. Reaction rates on solid phase can be quite different as compared to those for solution-phase reactions.

As shown in Table 2.2, the amount of reactive functionality of commonly available polystyrene solid phase (i.e., the loading) is often limited to 1 mmol/g of resin. There are specialty polymers, which may be as high as 5 mmol/g [46]. One must factor in the yields of each step in a synthesis, including the initial loading and final cleavage to determine a potential yield. With the use of excess reagent, it may be possible to push the efficiency of a transformation to a nearly quantitative transformation. Certainly, high-yielding synthetic conditions (>99%) have been developed for peptide and oligo-nucleotide synthesis, but there are only a few reactions that have been studied and refined to accommodate a broad chemical reaction diversity (e.g., amide formation, reductive amination). For each new chemical transformation, a chemist must determine the right balance of effort needed to optimize chemistry on a solid-phase format versus accommodating a low-yielding step into the overall synthesis scheme. Ultimately, for most synthesis on solid support, modest quantities of product are synthesized relative to the quantities of resin and reagents that have been employed (i.e., milligrams vs. grams). The solution-phase format of DNA-encoded libraries exchanges the problem of devel-oping reactions on a solid phase with the challenge of developing chemistry in the presence of DNA. This topic will be treated in greater detail in Chapter 4.

In the production of a library of compounds according to a common procedure, the diversity of products correlates with the relative diversity of the reagents. While there are many thousand amines and acids, there are relatively few isothiocyanates, for example. Given the putative advantage in solid-phase chemistry of driving reactions to completion by addition of a large excess of a particular reagent, it is important to have a large diver-sity of reagents in good supply. Limitations in reagent number and cost are usually seen in the number of diverse compounds resulting from a particular reagent class.

One approach to address the first and second concerns has been the development of solution-phase synthetic methods [47]. With solution phase, there is less adaptation of the reaction conditions used in traditional, one-at-a-time reactions. There are no extra steps to load and release compounds from the solid phase. Reagent quantities are com-mensurate with the compound present. It is conceivable with a solution-phase reaction to obtain several hundred milligrams from a few milliliters of reaction solution. For example, it is routine to consider the reaction of 100 mg of a 400 Da compound in 5 mL of solvent; such a reaction concentration is thus 50 mM. By contrast, in a solid-phase format, one may have difficulty to suspend efficiently 100 mg of 1 mmol/g polystyrene resin in 5 mL of solvent. Where this is possible, it would only result in a 20 mM reaction. Loading the same 400 Da compound to such a resin would not exceed 40 mg of the said compound. While such quantities are vastly greater than that produced by DNA-encoded methods, it sheds light on some of the limitations of solid-phase methods.

As a result of the inherent limitations of solid-phase and solution-phase synthesis, many combinatorial chemists have turned to the idea of exploring chemical diversity with "focused libraries." In this concept, a relatively small number of compounds (scores to hundreds) are synthesized around a common template by the installation of one or two diversity components. Such libraries are usually synthesized in parallel as discrete com-pounds, followed by purification and characterization. Such designs were often closely associated with a particular design motif (e.g., kinase inhibitor) or central template. Often, such designs were intended for solution-phase synthesis. In summary, these designs rep-resented a practical compromise between the desire to explore chemical space and the need to use a readily feasible and low-technology method. An excellent example of focused libraries is the BioFocus SoftFocus® designs [48]. Such libraries of some one to two thousand compounds were specifically designed and synthesized after extensive computational design for a particular protein target subfamily (e.g., tyrosine kinase, ion channels, GPCRs). Such compounds have been acquired by many drug discovery organizations to augment HTS collections. Many contend that focusing such designs to a target class will increase the chances of finding an active compound when screening against a related protein family target. The performance of these libraries over time in corporate HTS campaigns will eventually provide a measure of success for this approach.

As will be discussed in subsequent chapters, DNA-encoded combinatorial chemistry libraries address several of these problems. The practice of solution-phase synthesis obviates the need to develop solid-phase chemistry. Greater compound numbers are possible through the split-and-pool methodology. Detection limits are much lower because PCR amplification can be used to amplify encoding oligonucleotides, and thus, working with smaller quantities is less of a problem. The copy number of each structure present is also an advantage with DNA-encoded combinatorial libraries.

It is interesting to note that the extension of combinatorial chemistry to other fields such as materials [49] and catalysis [50] has enjoyed good success. In 1995, Schultz and coworkers published on the synthesis of superconducting materials in an array format. Since that time, the use of combinatorial and high-throughput methodology for the discovery of catalysts and new substances in materials science has seen vigorous development. DNA-encoded methods may also bring great benefits in the future to the areas of catalysis and materials science. Affinity-based selection methods will have to be developed when attempting to screen DNA-encoded libraries as better catalysts.

2.5.2 Lessons Learned from the Experience of Combinatorial Chemistry Applied to Drug Discovery

There are several general lessons that one can take from a brief history of combinatorial chemistry and these lessons can be applied to the practice of DNA-encoded combinatorial synthesis. First, no matter what approach to combinatorial chemistry one takes, the need to develop useful chemistry has been constant. Large numbers do not remediate poor designs. Perusal of the 100 most prescribed drugs helps one to understand the structural diversity among many drugs; many are derived from natural products, which are unrivaled in structural diversity. To date, many such structures cannot be reduced to a simple, commonly applied chemistry with available building blocks. Much of the disappointment that followed the heralding of combinatorial chemistry was the result of the production of too many molecules having poor drug-like properties. Those hoping to practice DNA-encoded approaches successfully must be mindful of this lesson.

Second, in addition to unique and innovative chemistry, the availability of necessary building blocks must be established to create diverse libraries. Strategies that put in place not only collections of commercially available reagents but also custom designs, which may also include functional groups masked by protective groups for subsequent revelation, are more likely to be successful.

Third, the lack of definition of true drug-likeness or diversity and the relation of these abstract measures to pharmacological utility is a continuing hindrance for designing new molecules with drug-like properties. Despite huge investment in computational analysis and simulation, medicinal chemists are challenged to define convincingly and generally a drug-like chemistry space. If, indeed, such a space cannot be defined unambiguously, then the best approach may be to cast a broader net in the most efficient manner possible. The possibility to create dramatically larger numbers of compounds with DNA-encoded combinatorial libraries has progressed to a broader exploration of chemical space in a cost-effective manner (see Chapter 18). Concepts of drug-likeness must also take into account different target types, modes of administration other than oral, and examples of large molecules that have become drugs.

Fourth, the extent to which it has been necessary to develop costly equipment and labor-intensive processes to meet the challenge of creating multiple small-molecule libraries has limited the practice of high-throughput synthesis by most chemists. Complex, fully automated synthesis systems are rarely employed today, despite the great costs that went into their development and installation. Although such systems highlight clever engineering, because they required too much time to set up and maintain, the fully automated approach was generally abandoned in favor of simply making the few compounds that seemed to matter most. The creation of simpler processes using commercially available reagents and materials leads to more efficient practices and broader participation. Simpler, readily available expertise avoids the need for specialized practitioners. Although specialized "super users" may create some of the most sophisticated applications for any method, the innovation needed for the discovery of new drugs occurs from broad participation by many scientists.

Fifth, the amount of time to run the full cycle for hit and lead discovery (synthesis, purification, registration, assay, and follow-up on actives) for complex combinatorial systems is not compatible with all phases of the discovery process. Using combinatorial chemistry to prepare and archive compounds well in advance of testing is a useful, strategic approach. The use of combinatorial and/or parallel chemistry to follow-up on HTS hits has also been reported to have been successful. At this stage in lead discovery, the testing of many structural ideas is an important step before narrowing the focus on a few structural classes for detailed study. However, as projects move into and through the lead optimization phase, combinatorial and/or parallel chemistry becomes less useful due to the need to develop new chemistry and perhaps prepare unavailable building blocks in order to synthesize highly specific target compounds. Testing ideas about structural design and diversity is most productive when it is part of a rapid feedback loop. Thus, the chemistry approach must meet the needs of the project in a timely manner; the reverse relationship has rarely been successful. As will be discussed in subsequent chapters, the ready integration of library synthesis, selection, and decoding is a significant advantage for DNA-encoded combinatorial chemistry.

Sixth, with the synthesis and subsequent manipulation (e.g., purification, registration, distribution) of more molecules, data systems capable of tracking-related information become increasingly important. It must be assumed that Information Management (IM) is an integral component of any process whereby many molecules will be made and from which a few actives will be found. Too often, organizations attempting to implement a combinatorial chemistry process found that custom-built information modules were required. Custom-built systems are usually expensive and have limited adaptability to the changing needs of an evolving workflow. The application of existing IM tools to the new problem of working with DNA-encoded libraries will be an advantage for starting quickly and for a sustained workflow evolution.

Finally, it is helpful to keep in mind that a scientific innovation initially captures attention in part due to its novelty. However, sustained enthusiasm for a new approach is based on the utility it delivers; in other words, it is the Return On Investment (ROI) that sustains commitment and investment. Combinatorial chemistry heralded an increase in efficiency to make new and useful molecules. In many cases, combinatorial chemistry required too much effort for the development of the necessary chemistry, at the expense of delivering useful compounds in a timely rate. Although many scientists continue to maintain that the ROI for combinatorial chemistry has been disappointing, it has become a standard approach to address many problems in drug discovery and other chemical sciences. The development of many innovations followed this course: early enthusiasm is often followed by disappointment, which eventually gives way to a realistic, useful application. Currently, DNA-encoded chemistry, still in its early days, has many aspects of appeal based on its innovation, yet it too, in time, will be judged on the discoveries of new molecules that it makes possible. Perhaps the proper understanding of its potential and limitations will facilitate the realistic and most useful application of this approach.

REFERENCES

1. Meger, S. (2011) Discovery of penicillin. *Am. Biol. Teach. 73*, 441.
2. Ganellin, C.R. (2006) Discovery of the antiulcer drug Tagamet. Edited by Chorghade, M.S. *Drug Discovery and Development.* John Wiley & Sons, Inc., Hoboken, NJ, *1*, 295–311.
3. Ley, S.V., Baxendale, I.R., Bream, R.N., Jackson, P.S., Leach, A.G., Longbottom, D.A., Nesi, M., Scott, J.S., Storer, R.I., Taylor, S.J. (2000) Multi-step organic synthesis using solid-supported reagents and scavengers: a new paradigm in chemical library generation. *J. Chem. Soc. Perkin Trans. 1*, 3815–4195.
4. Schreiber, S.L. (2000) Target-oriented and diversity-oriented organic synthesis in drug discovery. *Science. 287*, 1964–1969.
5. Nicolaou, K.C., Hanko, R., Hartwig, W. Eds. (2002) *Handbook of Combinatorial Chemistry: Drugs, Catalysts, Materials.* Wiley-VCH Verlag GMBH, Weinheim, Germany.
6. Merrifield, R.B. (1963) Solid phase peptide synthesis. I. The synthesis of a tetrapeptide. *J. Am. Chem. Soc. 85*, 2149–2154.
7. Letsinger, R.L., Mahadevan, V. (1965) Oligonucleotide synthesis on a polymer support. *J. Am. Chem. Soc. 87*, 3526–3527.
8. Fréchet, J.M., Schuerch, C. (1971) Solid-phase synthesis of oligosaccharides. I. Preparation of the solid support. Poly[p-(1-propen-3-ol-1-yl)styrene]. *J. Am. Chem. Soc. 93*, 492–496.
9. Geysen, H.M., Meloen, R.H., Barteling, S.J. (1984) Use of peptide synthesis to probe viral antigens for epitopes to a resolution of a single amino acid. *Proc. Natl. Acad. Sci. U.S.A. 81*, 3998–4002.
10. Houghten, R.A. (1985) General method for the rapid solid-phase synthesis of large numbers of peptides: specificity of antigen-antibody interaction at the level of individual amino acids. *Proc. Natl. Acad. Sci. U.S.A. 82*, 5131–5135.
11. Bunin, B.A., Ellman, J.A. (1992) A general and expedient method for the solid-phase synthesis of 1,4-benzodiazepine derivatives. *J. Am. Chem. Soc. 114*, 10997–10998.

12. Furka, A., Sebestyen, F., Asgedom, M., Dibo, G. (1988) More peptides by less labour. *Proceedings of the 10th International Symposium of Medicinal Chemistry*, Budapest, 288; Furka, A. (2002) Combinatorial chemistry: 20 years on... *Drug Discov. Today. 7*, 1–4; Furka, A. (2005) Combinatorial chemistry: from split-mix to discrete. In Edited by Fassina, G., Miertus, S. *Combinatorial Chemistry and Technologies Methods and Applications*. CRC Press/Taylor & Francis, Boca Raton, FL, 7–32; Furka, Á. (2001) Visszapillantás a kombinatorikus kémia korai időszakára. *Magyar Kémikusok Lapja. 56*, 250–254.

13. Brenner, S., Lerner, R.A. (1992) Encoded combinatorial chemistry. *Proc. Natl. Acad. Sci. U.S.A. 89*, 5381–5383.

14. Needels, M.C., Jones, D.G., Tate, E.H., Heinkel, G.L., Kochersperger, L.M., Dower, W.J., Barrett, R.W., Gallop, M.A. (1993) Generation and screening of an oligonucleotide-encoded synthetic peptide library. *Proc. Natl. Acad. Sci. U.S.A. 90*, 10700–10704.

15. Ohlmeyer, M.H.J., Swanson, R.N., Dillard, L.W., Reader, J.C., Asouline, G., Kokayashi, R., Wigler, M., Still, W.C. (1993) Complex synthetic chemical libraries indexed with molecular tags. *Proc. Natl. Acad. Sci. U.S.A. 90*, 10922–10926.

16. Nicolaou, K.C., Xiao, X.-Y., Parandoosh, Z., Senyei, A., Nova, M.P. (1995) Radiofrequency encoded combinatorial chemistry. *Angew. Chem. Int. Ed. 34*, 2289–2291.

17. Moran, E.J., Sarshar, S., Cargill, J.F., Shahbaz, M.M., Lio, A., Mjalli, A.M.M., Armstrong, R.W. (1995) Radio frequency tag encoded combinatorial library method for the discovery of tripeptide-substituted cinnamic acid inhibitors of the protein tyrosine phosphatase PTP1B. *J. Am. Chem. Soc. 117*, 10787–10788.

18. Xiang, X.-D., Sun, X., Briceño, G., Lou, Y., Wang, K.-A., Chang, H., Wallace-Freedman, W.G., Chen, S.-W., Schultz, P.G. (1995) A combinatorial approach to materials discovery. *Science. 268*, 1738–1740.

19. Cole, B.M., Shimizu, K.D., Krueger, C.A., Harrity, J.P.A., Snapper, M.L., Hoveyda, A.H. (1996) Discovery of chiral catalysts through ligand diversity: Ti-catalyzed enantioselective addition of TMSCN to *meso* epoxides. *Angew. Chem. Int. Ed. 35*, 1668–1671.

20. Zeng, L., Wang, X., Wang, T., Kassel, D.B. (1998) New developments in automated PrepLCMS extends the robustness and utility of the method for compound library analysis and purification. *Comb. Chem. High Throughput Screen. 1*, 101–111.

21. Wetterstrand, K. (2012) DNA sequencing costs, data from the NHGRI large-scale genome sequencing program. http://www.genome.gov/sequencingcosts/ (accessed on November 21, 2013).

22. Clark, M.A., Acharya, R.A., Arico-Muendel, C.C., Belyanskaya, S.L., Benjamin, D.R., Carlson, N.R., Centrella, P.A., Chiu, C.H., Creaser, S.P., Cuozzo, J.W., Davie, C.P., Ding, Y., Franklin, G.J., Franzen, K.D., Gefter, M.L., Hale, S.P., Hansen, N.J.V., Israel, D.I., Jiang, J., Kavarana, M.J., Kelley, M.S., Kollmann, C.S., Li, F., Lind, K., Mataruse, S., Medeiros, P.F., Messer, J.A., Myers, P., O'Keefe, H.,Oliff, M.C., Rise, C.E., Satz, A.L., Skinner, S.R., Svendsen, J.L., Tang, L., van Vloten, K., Wagner, R.W., Yao, G., Zhao, B., Morgan, B.A. (2009) Design, synthesis and selection of DNA-encoded small-molecule libraries. *Nat. Chem. Biol. 5*, 647–654.

23. Hlaváč, J., Soural, M., Krchňák, V. (2012) Practical aspects of combinatorial solid-phase synthesis. Edited by Toy, P.H., Lam, Y. *Solid-Phase Organic Synthesis*. John Wiley & Sons, Inc., Hoboken, NJ, 95–130.

24. Booth, B., Zemmel, R. (2004) Prospects for productivity. *Nat. Rev. Drug Discov. 3*, 451–456.

25. Schuffenhauer, A., Popov, M., Schopfer, U., Acklin, P., Stanek, J., Jacoby, E. (2004) Molecular diversity management strategies for building and enhancement of diverse and focused lead discovery compound screening collections. *Comb. Chem. High Throughput Screen.* 7, 771–781.

26. Hird, N.W. (1999) Automated synthesis: new tools for the organic chemist. *Drug Discov. Today.* 4, 265–274.

27. Kassel, D.B. (2007) High-throughput purification. *Compr. Med. Chem. II.* 3, 861–874.

28. Lam, K.S., Lebl, M., Krchňák, V. (1997) "The one-bead-one-compound" combinatorial library method. *Chem. Rev.* 97, 411–448.

29. Czarnik, A.W. (1997) Encoding methods for combinatorial chemistry. *Curr. Opin. Chem. Biol.* 1, 60–66.

30. Nestler, H.P., Bartlett, P.A., Still, W.C. (1994) A general method for molecular tagging of encoded combinatorial chemistry libraries. *J. Org. Chem.* 59, 4723–4724.

31. Burbaum, J.J., Ohlmeyer, M.H.J., Reader, J.C., Henderson, I., Dillard, L.W., Li, G., Randle, T.L., Sigal, N.H., Chelsky, D., Baldwin, J.J. (1995) A paradigm for drug discovery employing encoded combinatorial libraries. *Proc. Natl. Acad. Sci. U.S.A.* 92, 6027–6031.

32. Leftheris, K., Ahmed, G., Chan, R., Dyckman, A.J., Hussain, Z., Ho, K., Hynes, J., Jr., Letourneau, J., Li, W., Lin, S., Metzger, A., Moriarty, K.J., Riviello, C., Shimshock, Y., Wen, J., Wityak, J., Wrobleski, S.T., Wu, H., Wu, J., Desai, M., Gillooly, K.M., Lin, T.H., Loo, D., McIntyre, K.W., Pitt, S., Shen, D.R., Shuster, D.J., Zhang, R., Diller, D., Doweyko, A., Sack, J., Baldwin, J., Barrish, J., Dodd, J., Henderson, I., Kanner, S., Schieven, G.L., Webb, M. (2004) The discovery of orally active triaminotriazine aniline amides as inhibitors of p38 MAP kinase. *J. Med. Chem.* 47, 6283–6291.

33. Baldwin, J.J., Henderson, I. (1996) Recent advances in the generation of small-molecule combinatorial libraries: encoded split synthesis and solid-phase synthetic methodology. *Med. Res. Rev.* 16, 391–405.

34. Krattiger, P., McCarthy, C., Pfaltz, A., Wennemers, H. (2003) Catalyst-substrate coimmobilization: a strategy for catalysts discovery in split-and-mix libraries. *Angew. Chem. Int. Ed. Engl.* 42, 1722–1724.

35. Nicolaou, K.C., Pfefferkorn, J.A., Mitchell, H.J., Roecker, A.J., Barluenga, S., Cao, G.-Q., Affleck, R.L., Lillig, J.E. (2000) Natural product-like combinatorial libraries based on privileged structures. 2. Construction of a 10,000-membered benzopyran library by directed split-and-pool chemistry using NanoKans and optical encoding. *J. Am. Chem. Soc.* 122, 9954–9967.

36. Nielsen, J., Brenner, S., Janda, K.D. (1993) Synthetic methods for the implementation of encoded combinatorial chemistry. *J. Am. Chem. Soc.* 115, 9812–9813.

37. Lipinski, C.A., Lombardo, F., Dominy, B.W., Feeney, P.J. (1997) Experimental and computational approaches to estimate solubility and permeability in drug discovery and development settings. *Adv. Drug Deliv. Rev.* 23, 3–25.

38. Bohacek, R.S., McMartin, C., Guida, W.C. (1996) The art and practice of structure-based drug design: a molecular modeling perspective. *Med. Res. Rev.* 16, 3–50.

39. Reymond, J.-L., van Deursen, R., Blum, L.C., Ruddigkeit, L. (2010) Chemical space as a source for new drugs. *Med. Chem. Commun.* 1, 30–38.

40. Jacoby, E., Schuffenhauer, A., Popov, M., Azzaoui, K., Havill, B., Schopfer, U., Engeloch, C., Stanek, J., Acklin, O., Rigollier, P., Stoll, F., Koch, G., Meier, P., Orain, D., Giger, R., Hinrichs, J., Malagu, K., Zimmermann, J., Roth, H.-J. (2005) Key aspects of the Novartis compound

collection enhancement project for the compilation of a comprehensive chemogenomics drug discovery screening collection. *Curr. Top. Med. Chem. 5*, 397–411.

41. Dolle, R.E. (2002) Comprehensive survey of combinatorial library synthesis: 2001. *J. Comb. Chem. 4*, 369–418.

42. (a) Chemical Abstract Service: http://www.cas.org/ for substances count. (b) PubChem: http://www.ncbi.nlm.nih.gov/sites/entrez?cmd=search&db=pccompound&term=allfilt. (c) http://zinc.docking.org/ for available compounds for virtual screening (accessed on November 25, 2013).

43. Sergey, N., Natalia, Z., Olga, D., Vera, S., Evgeny, M., Sergey, M., Maxim, S., Anatoly, R., Peter, V., Dmitry, L., Denis, K., Valery, F., Cary, Q., Viktor, Z. (2005) A very large diversity space of synthetically accessible compounds for use with drug design programs. *J. Comput. Aided Mol. Des. 19*, 47–63.

44. Black, J. (1999) Future perspectives in pharmaceutical research. *Pharm. Policy Law. 1*, 85–92.

45. Kapeller, D.C., Bräse, S. (2012) Toward tomorrow's drugs: the synthesis of compound libraries by solid-phase chemistry. Edited by Pignataro, B. *New Strategies in Chemical Synthesis and Catalysis*. Wiley-VCH Verlag GmbH & Co. KGaA, Weinheim, Germany, 343–375.

46. Haag, R., Hebel, A., Stumbé, J.-F. (2002) Solid phase and soluble polymers for combinatorial synthesis. Edited by Nicolaou, K.C., Hanko, R., Hartwig, W. *Handbook of Combinatorial Chemistry, Drugs, Catalysis, Materials*. Wiley-VCH Verlag GMBH, Weinheim, Germany, 24–58.

47. Orru, R.V.A., de Greef, M. (2003) Recent advances in solution-phase multicomponent methodology for the synthesis of heterocyclic compounds. *Synthesis. 10*, 1471–1499.

48. Birault, V., Harris, C.J., Le, J., Lipkin, M., Nerella, R., Stevens, A. (2006) Bringing kinases into focus: efficient drug design through the use of chemogenomic toolkits. *Curr. Med. Chem. 13*, 1735–1748.

49. Potyrailo, R., Rajan, K., Stoewe, K., Takeuchi, I., Chisholm, B., Lam, H. (2011) Combinatorial and high-throughput screening of materials libraries: review of the state of the art. *ACS Comb. Sci. 13*, 579–633; Narasimhan, B., Mallapragada, S., Porter, M.D. Eds. (2007) *Combinatorial Materials Science*, John Wiley & Sons Inc., New York.

50. Maier, W. F. Ed. (2003) *Applied Catalysis A*, Volume 254, Issue 1, Pages 1–170 (November 10, 2003).

3

A BRIEF HISTORY OF DNA-ENCODED CHEMISTRY

Anthony D. Keefe

X-Chem Pharmaceuticals, Waltham, MA, USA

3.1 BEFORE 1992: THE INSPIRATION FOR DNA-ENCODED CHEMISTRY

Looking back to the conception of DNA-encoded chemistry in 1992, it is clear that this event was inspired by the then recent development of phage display [1, 2] and of SELEX/*in vitro* selection [3–5]. Both of these technologies utilize numerically large stochastically generated libraries of variants and use enrichment by selection and amplification to discover individual members with desired characteristics. Phage display is currently the most widely used *in vitro* selection technique and is depicted in Figure 3.1. In phage display, the identity of each individual library member is encoded by the sequence of the associated genetic material. Library generation followed by iterative cycles of selection for a property such as binding to a desired target and amplification and cloning and sequencing of the associated genetic material permits the identification of library members that were selectively enriched and may bind to the target. Phage display and SELEX/*in vitro* selection were in turn enabled by a range of other technologies that had been developed just prior to 1992 including the automated chemical synthesis of DNA oligonucleotides [6], the *in vitro* amplification of DNA (PCR) [7, 8], chain-termination methods for the sequencing of

A Handbook for DNA-Encoded Chemistry: Theory and Applications for Exploring Chemical Space and Drug Discovery, First Edition. Edited by Robert A. Goodnow, Jr.
© 2014 John Wiley & Sons, Inc. Published 2014 by John Wiley & Sons, Inc.

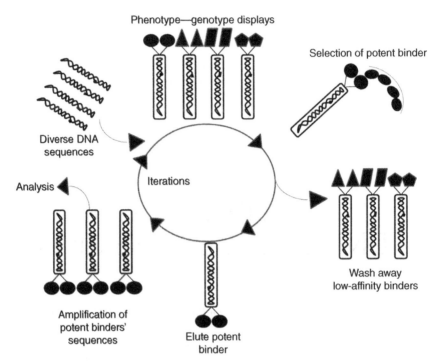

Figure 3.1. Phage are viral particles that display coat proteins that encapsulate the phage genome. Standard molecular biology techniques can be used to introduce variable regions into phage genomes such that variable protein regions are displayed upon the coat. Affinity-mediated selection techniques coupled with phage amplification in their host bacteria and followed by cloning and sequencing can be used to identify displayed proteins that bind to arbitrary targets.

DNA [9], and the enzymatic ligation of DNA [10]. DNA-recorded DNA-encoded chemistry is a variant of "split-and-pool" chemical synthesis, first reported in 1991 [11]. More recently, progress with the utilization of DNA-encoded chemistry has been greatly facilitated by the introduction of massively parallel cell-free cloning and "deep" or "next-generation" sequencing technologies developed by Roche/454 [12] and Illumina/Solexa [13].

Phage display and SELEX/*in vitro* selection are limited to the discovery of very specific kinds of molecule—proteins or peptides and oligonucleotides, respectively. These limitations exist because of the specificity of the enzymes used to generate or sequence the library members. DNA-encoded chemically generated libraries enable the input of a much broader range of categories of molecules into *in vitro* selection experiments, including those that more closely resemble traditional orally available cell-permeable therapeutics. Once methods had been developed for the generation of such libraries, they could then be utilized by similar affinity-mediated *in vitro* selection processes for the discovery of small-molecule therapeutic leads.

3.2 1992–1995: THE CONCEPTION OF DNA-ENCODED CHEMISTRY

The first suggestion of how chemically synthesized libraries could be generated in a DNA-encoded form came from Brenner and Lerner at the Scripps Research Institute in 1992 [14]. They described the concept of the coupling of solid-phase peptide and oligo-nucleotide synthesis using a "split-and-pool" methodology. The oligonucleotide–peptide conjugates thus synthesized would subsequently be liberated from the controlled pore glass synthesis matrix, thereby generating a library of peptides, each of which is covalently attached to an encoding oligonucleotide. The cloning and sequencing of such a library would reveal the identity of the encoded peptides. The liberation of the peptides from the solid-state surface (bead) distinguishes this approach from other bead-based split-and-pool peptide discovery technologies, including those using oligo-nucleotides for encoding [15]. This library-generation process is shown in Figure 3.2.

Brenner and Lerner published a second paper in 1993 [16] in which experimental work was performed to demonstrate the orthogonal nature of the conditions chosen for peptide and oligonucleotide synthesis such that cosynthesis could be performed, that is, that the proposed approach could work. For cosynthesized oligonucleotide–peptide constructs, they demonstrated that known peptide ligands for antibodies retained the ability to bind and that the covalently linked oligonucleotides could be amplified by PCR.

The method Brenner and Lerner presented is only suitable for chemical synthesis that is compatible with oligonucleotide synthesis, its diversity is limited to the number of beads that can be utilized for split-and-pool solid-phase synthesis, and it does not appear to have ever been used to discover a novel ligand. However, this method was the first to describe the generation of DNA-encoded chemical entities, and other workers were inspired to extend and improve upon it.

The next publication in the field appeared in 1995 from Kinoshita and Nishigaki at Saitama University in Japan [17]. The authors demonstrated an enzymatic oligo-nucleotide tagging method and suggested that it could be used to record (encode) chemical entities generated in a combinatorial fashion, also utilizing a split-and-pool strategy. This strategy is depicted in Figure 3.3. Using this approach, the diversity of a chemically encoded library is not limited by the number of beads that they are attached to (there are no beads), and there is no requirement that the chemistry used to construct the library be compatible with the oligonucleotide synthesis process. The chemistry utilized only needs to be compatible with the presence of the oligonu-cleotide tags and the maintenance of their ability to be ligated and to act as templates for a template-dependent polymerase, for example, in PCR. Since this paper appeared in 1995, no further related work appears to have been reported by this group.

3.3 2001–2004: THE BIRTH OF SEQUENCE-DIRECTED DNA-ENCODED CHEMISTRY

In 2001, the Liu group at Harvard University in Cambridge, Massachusetts, demonstrated a new approach within the field—that of oligonucleotide template-directed chemistry. They showed that oligonucleotides could be used to program the solution-phase chemistry of

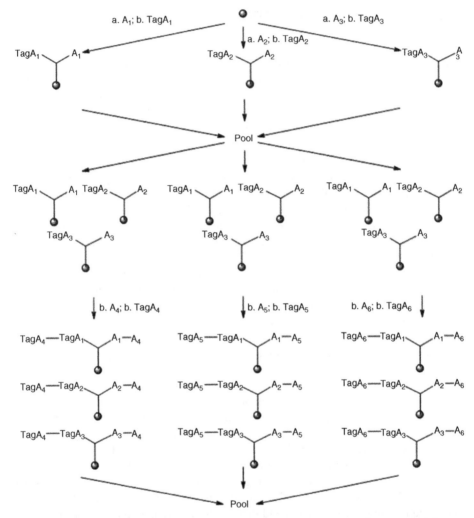

Figure 3.2. Controlled pore glass beads with a bifunctional linker are split into multiple synthesis columns, and within each column, chemical coupling occurs to a different amino acid and a different corresponding sequence of nucleotides followed by pooling of the beads. Multiple iterations of this split-and-pool process yield a diversity of beads each of which displays a stochastically generated peptide along with a corresponding encoding oligonucleotide sequence. Finally, a constant oligonucleotide sequence is synthesized such that all members have the same primer-binding sequence and may be coamplified by PCR. Cleavage from the beads yields a library of DNA-encoded peptides.

oligonucleotide-conjugated small molecules using the complementary relationship between the oligonucleotide sequences they are conjugated to [18]. The authors showed that reaction rates can be accelerated several hundredfold when the reactants are attached to oligonucleotides that are complementary to each other or are both complementary to a third template oligonucleotide, so long as they anneal within a minimum distance of each other

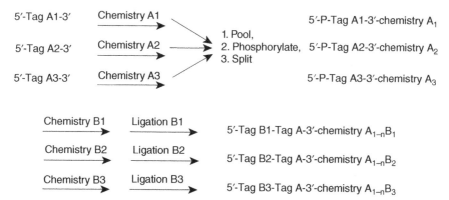

5'-Tag A1-3'   Chemistry A1 →

5'-Tag A2-3'   Chemistry A2 →

5'-Tag A3-3'   Chemistry A3 →

1. Pool,
2. Phosphorylate,
3. Split

5'-P-Tag A1-3'-chemistry A_1

5'-P-Tag A2-3'-chemistry A_2

5'-P-Tag A3-3'-chemistry A_3

Chemistry B1 →   Ligation B1 →   5'-Tag B1-Tag A-3'-chemistry $A_{1-n}B_1$

Chemistry B2 →   Ligation B2 →   5'-Tag B2-Tag A-3'-chemistry $A_{1-n}B_2$

Chemistry B3 →   Ligation B3 →   5'-Tag B3-Tag A-3'-chemistry $A_{1-n}B_3$

Figure 3.3. The figure shows the scheme proposed by Kinoshita and Nishigaki [17] for the generation of a DNA-encoded chemical library using the enzymatic ligation of encoding oligonucleotide tags. Each encoding oligonucleotide A is transformed chemically in a tag-specific manner. The resultant conjugates are pooled, phosphorylated, and split, and then, each split is subjected to chemical condition B along with ligation to a B-encoding oligonucleotide. Further cycles of chemistry and ligation may be performed. The end result is a library of stochastically generated chemical entities covalently joined to a sequence of concatenated tags from which chemical history may be inferred.

HS-C-C-A-T-G-C-T-T-A-A5'
| | | | | | | | | | |
T-G-G-T-A-C-G-A-A-T-T....3'

E-template

O || NH SH
T-G-C-G-A-G-C-G-C-T   C-C-A-T-G-C-T-T-A-A 5'
| | | | | | | | | | | | | | | | | | | |
A-C-G-C-T-C-G-C-G-A-T-G-G-T-A-C-G-A-A-T-T....3'

H-template

Figure 3.4. Two examples of different oligonucleotide architectures that support hybridization-dependent chemistry.

of approximately 30 nucleotides. Two different oligonucleotide architectures that both support hybridization-dependent chemistry are shown in Figure 3.4. The authors also showed that the presence of a single mismatch could reduce the rate of reaction by 200-fold and that a biotin-conjugated oligonucleotide could be enriched by 500-fold from a library of 1025 members using *in vitro* selection for immobilized streptavidin binding. The Liu group demonstrated several DNA-compatible chemistries including amide bond formation, reductive amination, and carbon–carbon bond-forming reactions including the Heck reaction, Wittig olefination, dipolar cycloaddition, nitro-aldol (Henry), and nitro-Michael addition [19]. The use of such a library to discover a novel ligand was first demonstrated by the same group in 2004 [20]. A 65-member peptide macrocycle library was generated using three acylation steps with 4 or 5 diversity elements in each, followed by a Wittig olefination macrocyclization. A member of this library that bound to carbonic anhydrase was then identified using two cycles of *in vitro* selection for binding to the immobilized target followed by the amplification, cloning, and sequencing of the output DNA.

In 2004, the Harbury group at Stanford University in California published three papers introducing an alternative variant of oligonucleotide template-directed chemistry. In this approach, libraries of concatenated DNA tags are routed via hybridization to immobilized complementary oligonucleotides, and different chemical conditions are applied to differently routed library members. In the first of these papers [21], the authors introduce the routing process in which "concatenated codons" are presynthesized prior to any chemistry occurring; these are sequentially passed over multiple "anticodon" capture matrices. Chemical transformations are performed upon the 5'-primary amino terminus of each partitioned sublibrary after elution from the anticodon support and capture upon DEAE-Sepharose. Solubility concerns during the chemical steps are largely avoided owing to the immobilized state of the oligonucleotide conjugates. Multiple cycles of split, chemistry, and pool were used to generate the DNA-encoded chemical library. The library-generation process is shown in Figure 3.5, reproduced with permission from [22]. In the second of the three papers [22], the authors demonstrated that the DNA-encoded chemical libraries synthesized using their routing methodology may be used to support the various molecular biological steps (selection, amplification, and translation of chemical structures) that are required for the discovery of ligands within them. The authors generated a library of one million distinct six-codon oligonucleotides using recombination and amplification. This library was then routed over 10 anticodon columns and through successive specific acylations was translated into one million different hexapeptides. Affinity-based selection was then shown to enrich for peptide sequences that bind to a known antibody with an enrichment of greater than 1000-fold in a single step. In the third of the three simultaneously published papers [23], the authors discuss in greater depth their development of on-DNA Fmoc peptide and peptoid synthesis.

Also in 2004, the Neri group at the ETH in Zurich, Switzerland, introduced the concept of Encoded Self-Assembling Chemical (ESAC) libraries along with experimental proof that this approach could be used for the affinity maturation of known small-molecule ligands for protein targets [24]. The ESAC libraries are created by conjugating individual small molecules to either the 3'- or the 5'-termini of encoding oligonucleotides that contain sufficient constant complementary sequence for them to hybridize together in pairs. These proximate pairs of small molecules can be subjected to affinity selection against immobilized targets for their ability to bind the target when held together. By using one of the encoding oligonucleotides to function as a template for the enzymatic extension of the other, all of the encoding information can be captured in a single oligonucleotide such that it can be amplified, cloned, and sequenced and used to infer the identities of the proximal pairs of small molecules present in a selection output. Various ESAC architectures are shown in Figure 3.6. The authors demonstrated an approximately 100-fold enrichment of biotin from a 138-member library using a single cycle of *in vitro* selection for binding to immobilized streptavidin. Additional work paired known binders to serum albumin (dansyl amide) and to

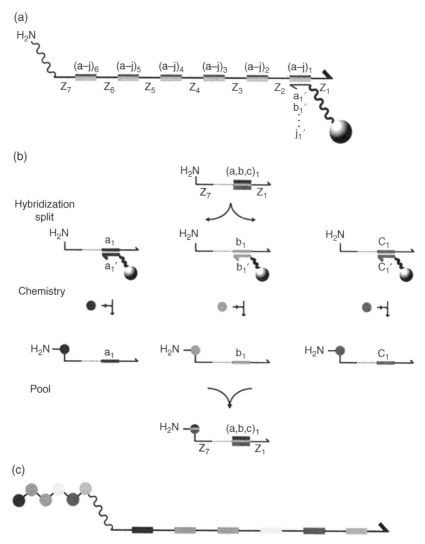

Figure 3.5. The figure shows the method devised by Harbury that uses the capture of concatenated encoding oligonucleotides upon immobilized complementary sequences to generate DNA-encoded combinatorially generated chemical entities. During serial capture, chemical transformation, and elution events, the identity of the encoding oligonucleotide tags determines the routes the conjugate take, and thereby, the tag sequence is translated into chemical history that is ultimately inferable from the oligonucleotide tag sequence. (a) The oligonucleotide template containing six encoding "codon" sequences. (b) The routing methodology in which variant codons at one encoding position are each captured by their complementary immobilized anticodon sequences and subject to route-dependent chemistry such that the chemical conditions become encoded by the codon sequence. (c) One fully elaborated library member in which each of the six codons encode each of the six cycles of chemistry, as indicated by the color scheme. Figure reproduced from PLoS publication Halpin and Harbury [22].

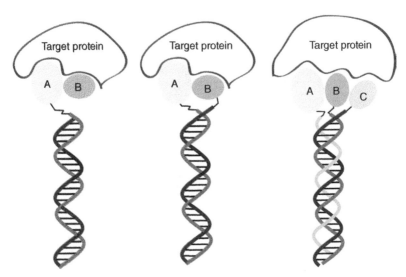

Figure 3.6. Alternative architectures of ESAC libraries are shown. The figure was kindly provided by Dr. Jörg Scheuermann. For color detail, please see color plate section.

carbonic anhydrase (4-carboxybenzenesulfonamide) with 137 other small molecules for affinity maturation. A single cycle of selection was performed in which the libraries were incubated with 100 nM of the biotinylated targets, the output was visualized using hybridization to an oligonucleotide array, and selected combinations were then synthesized off-DNA in covalently joined forms using linkers of various lengths and tested for increased potency. For serum albumin, a conjugate with a coumarin was observed to increase affinity from K_D values of 146 and 41 µM for the isolated entities to 4 µM for the linked combination. For carbonic anhydrase, conjugates were identified with IC_{50} values for esterase activity as low as 25 nM, compared with 1 µM for the 4-carboxybenzenesulfonamide alone.

Despite the novelty of these methods, the sequence-directed technologies of Liu, Harbury, and Neri face challenges when confronted with large split sizes (number of different chemistries encoded within a single cycle of chemistry). The Neri ESAC method requires that each encoded chemical entity be separately linked to a unique encoding oligonucleotide, and the Liu method requires the same for each chemical diversity element (building block.) This limits the sizes of libraries that can be generated. Additionally, the Neri ESAC method requires significant postselection work be performed to evaluate chemical linking strategies. The Liu and Harbury methods both require that mistranslation by inappropriate hybridization be kept to a minimum, a phenomenon that will increase linearly with split size and results in miscoding. Typical reported split sizes for libraries generated by these two methods are in the range of 4–24. Because the number of entities within these libraries is the product of the split sizes and the average molecular weight of the encoded entity increases linearly with the number of synthesis cycles, this greatly limits the numeric size of libraries that can be generated if it is desired to keep the encoded entities' molecular weights within a moderate range.

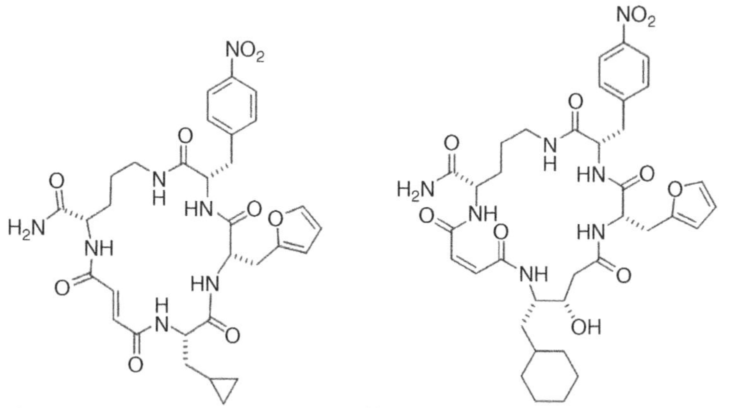

Src kinase inhibitor, IC50 680 nM   VEGFR2 activator, 70% activation at 10 μM

Figure 3.7. The structures of a Src kinase inhibitor and a VEGFR2 activator discovered using the affinity-mediated selection of library of small molecules constructed using sequence-directed DNA-encoded chemistry [26].

3.4 2005–2012: THE FURTHER DEVELOPMENT OF SEQUENCE-DIRECTED DNA-ENCODED CHEMISTRY

Since the original publications, the Liu, Neri, and Harbury groups have continued to make further progress with their variants of sequence-directed chemistry and encoded fragment proximity, and an additional method has been presented by Hansen and coworkers of Vipergen ApS and Aarhus University, Denmark.

The Liu group has reported a larger library of 13,284 peptide macrocycles [25]. This library was constructed using three codons, each encoding 12 possibilities that were in turn installed upon 8 scaffolds. The single-cycle affinity selection of this library against 36 different therapeutic protein targets was then reported in 2010 [26]. Inhibitors were reported for four of these targets (Src, Akt3, MAPKAPK2, p38α), and an activator was reported for one (VEGFR2). The most potent macrocycle inhibited Src with an IC_{50} of 680 nM, and the VEGFR2 activator activated the kinase activity of VEGFR2 by 70% at 10 μM. The structures of these two compounds are shown in Figure 3.7. Eight of nine of the resynthesized compounds were active, and significant affinity differences were observed between the cis- and trans-isomers of the macrocycles with both the linear versions and point mutants being inactive. This technology has been licensed to Ensemble Therapeutics where it is being used to discover macrocycles that inhibit protein–protein interactions. Ensemble Therapeutics has reported in a press release made on January 5, 2012, that it has discovered an orally available inhibitor of IL-17a and that it expects that this will advance to drug candidate nomination in a timely manner.

In 2007, the Harbury group reported the generation of an eight-cycle peptoid library with 10 diversity elements at each cycle, yielding 100 million eight-mer peptoids [27]. Three cycles of affinity-based selection were performed for binding to an SH3 domain target for which peptide ligands were known. The population converged to five distinct groups of related sequences. Resynthesis of library members identified after the sixth round of selection produced multiple binding peptoids with K_D values in the 10–100 μM range.

In 2012, the Harbury group reported a demonstration that larger split sizes than the 9–20 described in their earlier papers can be utilized by their routing methodology in order that larger libraries of lower MW compounds can be created [28]. The authors used split sizes of 384 and showed that using a single biotinylated building block led to 13,000-fold enrichment for the biotin-encoding sequence after a single selection step for immobilized streptavidin binding. Here they showed that a single individual split of biotin became greater than 95% of the selected library after a single selection step for streptavidin binding. The use of such a library to discover novel functional molecules does not appear to have yet been reported.

The Neri group has generated multiple ESAC libraries, both for the affinity maturation of known binders as well as for the discovery of novel ligands. In one affinity maturation study [29], benzamidine, a trypsin inhibitor with an IC_{50} in the 100 μM range, was used as a starting point. This compound was conjugated to an oligonucleotide, paired with a library of 620 other compounds, and subjected to a single cycle of selection for binding to trypsin with a hybridization-based approach used for output evaluation. Among the selected proximal pairs, a phenylthiourea motif was common, and one of the partners, when appropriately conjugated to benzamidine, yielded a compound with an IC_{50} value of 98 nM.

In addition to the ESAC approach in which pairs of individually encoded entities are coassociated, the Neri group has also pioneered the "single-pharmacophore" approach in which presynthesized molecules are installed upon encoding oligonucleotides. In one example of the use of such a strategy [30], a library of 477 compounds with molecular weights ranging between 100 and 500 was displayed upon 477 different encoding oligonucleotides using various conjugation chemistries, pooled, and incubated with streptavidin immobilized upon Sepharose. Output evaluation using PCR amplification followed by hybridization with a microarray of complementary oligonucleotides identified two streptavidin binders with K_D values close to 1 mM along with the recovery of a positive control compound.

In a hybrid approach of these two strategies, the Neri group has also reported a study in which a novel MMP-3 binder was identified using the single-pharmacophore approach, which was then affinity matured using ESAC [31]. A library of 550 compounds was incubated with MMP3 covalently immobilized upon Sepharose, and binders were identified after PCR and hybridization-based output evaluation. One member of the library, a phenol ether, was identified as being enriched by a factor of 7 over the unselected library. An ESAC follow-up library was then prepared in which the selected phenol ether was displayed alongside each of the 550 members of the original library. A suitable pairing with a stilbene disulfonate was discovered. Subsequent preparation of linked versions of the two pharmacophores yielded compounds with IC_{50} values as low as 10 μM for MMP-3 and with little inhibitory activity observed for related targets.

In 2009, Hansen and colleagues at Aarhus University and Vipergen ApS in Denmark reported the "YoctoReactor," an additional variant within the oligonucleotide template-directed synthesis theme [32]. In this variant, oligonucleotide–small-molecule conjugates self-assemble into three-way junctions in a one-pot reaction in which the chemical entities are brought together in the center of the junction with colocalization driven by hybridization to constant sequences. After the chemical reactions are complete, all but one of the oligonucleotide linkages are severed, and the encoded chemical entity is displayed upon encoding linearized double-stranded DNA and is accessible for affinity-mediated selection. The encoding DNA may be amplified by PCR, and the chemical entity regenerated after the amplification product is restricted and converted into single-stranded DNA. This method is both sequence-directed and sequence-recorded. It is included in this section because the chemistry is hybridization-mediated. The authors describe the generation of a 100-member peptide library with three cycles of encoded chemistry and the successful enrichment of a known binding pentapeptide that was diluted into this library by a factor of 10 million. After two cycles of selection and amplification, the known binder constituted 1.7% of the library, an enrichment of 170,000-fold.

3.5 2004–2012: SEQUENCE-RECORDED DNA-ENCODED CHEMISTRY

The sequence-directed DNA-encoded chemical library-generation methods developed by Hansen, Harbury, and Liu all support the amplification of library members, including of the encoded chemical entity. However, for reported screens in which novel binders were sought and found, this property is rarely utilized. Usually, the libraries are small enough or the depth of sequencing is high enough such that a single cycle of selection for binding is sufficient to enrich binders to the point of identification, for example [26]. Accordingly, alternative DNA-encoded chemistry approaches have been developed and successfully applied in which the encoding DNA is amplifiable, but the encoded chemical entity is not. These libraries are generated by "split-and-pool" methodologies in which the successive chemistries occurring in each split are encoded by the successive concatenation of split-specific DNA tags using enzymatic ligation. Such libraries may be subjected to multiple sequential cycles of enrichment by binding and elution. Subsequently, the encoding oligonucleotide sequences may be amplified, but the encoded chemical entities may not. Because of the absence of hybridization-mediated control during chemical synthesis, the limits to split sizes are removed, and very high split sizes and large libraries have been reported. The first reports of the generation of sequence-recorded libraries occur in patent applications. Patent applications are cited here solely for the purpose of discussion of the experimental achievements reported in the examples sections therein.

In December of 2004, Praecis Pharmaceuticals of Waltham, Massachusetts, filed US Patent Application 11/015,458 [33]. This application describes a method for the recording of chemical information by the enzymatic ligation of encoding duplexed oligonucleotides to a bifunctional display duplex oligonucleotide that additionally contains a site for the construction of a small molecule. In the examples presented in the application, several libraries are described including five-cycle 249,000-member and four-cycle 85-million-member libraries in which the bond-forming steps are all acylations and additionally macrocyclization to yield a macrocyclic version of the four-cycle library. A known Abl-kinase-binding compound was conjugated to DNA, diluted 1000-fold into a library, and was observed to comprise half of the output, as assessed by sequencing, after a single cycle of affinity selection, an enrichment of 500-fold.

In December of 2005, Nuevolution A/S of Copenhagen, Denmark, filed US Patent Application 60/741,490 [34]. This application also describes a method for the recording of chemical information by the enzymatic ligation of encoding duplexed oligonucleotides to a bifunctional display duplex oligonucleotide that additionally contains a site for the construction of a small molecule. In the examples presented in the application, they describe the construction of a four-cycle pilot library of 65,000 members using a split-and-pool methodology with a range of bond-forming reactions including acylation, reductive amination, isocyanate addition, sulfonylation, and nucleophilic substitution. A known thrombin binder was enriched from this library using affinity-based selection for binding to biotinylated thrombin immobilized upon streptavidin–agarose followed by elution with a known thrombin-binding small molecule. After four cycles of selection, the known binder comprised half of the selection output as assessed by sequencing, an enrichment of 30,000-fold. Subsequently, a larger four-cycle library of 85 million compounds was prepared using a similar synthetic approach, from which the same thrombin-binding compound was identified by selection. This patent application also describes the preparation of a 3.5-million-member triazine library in which each triazine carbon bears a diversity group. The triazine library was selected against p38 kinase and generated several hits with IC_{50} values in the nanomolar range. Structures were not disclosed.

In 2008, the Neri group reported the formation of two libraries using sequence-recorded chemistry. A 4000-member two-cycle DNA-encoded chemical library assembled using Diels–Alder cycloaddition [35] and a second 4000-member two-cycle DNA-encoded chemical library assembled using acylation [36]. The Diels–Alder library was used for affinity-based selection using streptavidin, HSA, trypsin, Bcl-xL, MMP-3, and TNF [37]. Sequencing of the selected and naïve libraries was performed using the Roche/454 platform with tens of thousands of sequence reads being collected for each experiment and up to a few hundred reads being observed for the most enriched individual library members. Novel ligands discovered in this report include those for streptavidin with K_D values as low as 185 nM, for HSA with K_D values as low as 440 nM, for Bcl-xL with K_D values as low as 10 µM, for trypsin with K_i values as low as 11 µM, and for TNF with K_D values as low as 20 µM. Follow-up compounds for TNF had K_D values as low as 10 µM and completely inhibited TNF-induced apoptosis at concentrations greater than 300 µM. For the amide library, single-cycle selection was performed against immobilized streptavidin, MMP-3, and IgG. Novel and known streptavidin binders were found with K_D values as low as 350 nM. Resynthesized immobilized IgG binders were shown to be able to completely and specifically capture IgG from solution under affinity-purification conditions. MMP-3 binders were discovered with K_D values as low as 11 µM. DNA recording was also used to encode the identity of the individual experiments in order that multiple output samples could be combined for sequencing. Read numbers of up to 40,000 per sample were observed, and as many as 100 reads were observed for the most enriched individual compounds. In further selection experiments with the amide library against Bcl-xL [38], a number of compounds were observed to be enriched for binding and upon resynthesis as fluorescein conjugates K_D values as low as 930 nM were observed. The most potent compound incorporated indomethacin and naphthalene building blocks and bound to Bcl-xL in a manner that is competitive with the Bcl-xL-binding peptide, Bak.

The Neri group has also reported a three-cycle one-million-member library in which both amide bond and Diels–Alder cycloaddition were used for library synthesis. Five thousand diene/amine-reactive combinations were reacted with 200 third-cycle N-modified maleimides [39]. Single-cycle affinity-based selections were performed for binding to immobilized carbonic anhydrase and streptavidin. After PCR and sequencing using the

Roche/454 platform, a total of 1.8 million reads were generated with 0.64 million unique reads including a few hundred reads for the most enriched individual library members. Enriched on-DNA compounds exhibited IC_{50} values as low as $0.24\,\mu M$ with carbonic anhydrase for a family of related bis-sulfonamides, and fluorescein conjugates demonstrated tumor accumulation in a murine subcutaneous xenograft.

In a more recent report from the Neri group [40], a new 30,000-member two-cycle amide-linked library was generated with Lipinski-compliance in mind [41]. One hundred N-protected amino acids were conjugated to 300 carboxylic acids with about half of the resulting library members being Lipinski-compliant. This library was used for selection for binding to IL-2 and carbonic anhydrase. Illumina/Solexa sequencing generated 120 million sequence reads for 170 distinct mixed DNA-encoded experiments with individual experiments represented by between approximately one hundred thousand and one million reads with the most strongly enriched individual library members being represented by a few hundred reads. Two phenyl sulfonamides were strongly enriched for carbonic anhydrase; upon resynthesis, these were observed to have K_i values of 74 and 76 nM. Two of the most strongly enriched compounds for IL-2 were resynthesized and observed to have K_D values as low as $2.5\,\mu M$ in a fluorescence polarization assay; the same compounds were active in a T-cell proliferation assay.

The Neri group has also demonstrated the use of this approach for the affinity maturation of a known binder [42]. Benzamidine was selected for binding to trypsin in an earlier DNA-encoded chemistry experiment, and subsequently, an 8000-member library was generated in which 40 benzamidine-containing amino acids were reacted with 200 amine-reactive compounds. Selection for trypsin binding at reduced target concentration yielded substituted benzamidines with enrichment values approaching 100-fold. Resynthesized off-DNA compounds had IC_{50} values as low as 3 nM, a four-log improvement over the parental compound.

An 800-million-member four-cycle library was generated by Clark and coworkers at Praecis Pharmaceuticals prior to its early 2007 acquisition by GSK [43]. This library was generated using the ligation of short synthetic DNA duplexes to a primary-amine-functionalized duplexed oligonucleotide headpiece in order to record the chemical steps used to elaborate the amine. The library was constructed by decorating a triazine at each of its three carbons. Initially, 192 Fmoc amino acids were installed upon the amine, followed by triazine installation, followed by the installation of 32 bifunctional acids upon the triazine followed by a third and a fourth cycle in which 340 and 384 amines were installed upon the bifunctional acids and the remaining triazine carbon, respectively. A related seven-million-member three-cycle library was also created in which the bifunctional acids were omitted. Potential building blocks were each first evaluated for yield in a model reaction, with only those with synthesis yields above an arbitrary threshold being utilized for library generation—fewer than half of the evaluated building blocks were ultimately included. After the successful observation of a 100,000-fold enrichment of known binders to Aurora A kinase that had been spiked into the library, these libraries were used to discover novel inhibitors of Aurora A and p38 MAP kinases. Selections were performed for binding to His_6 tagged recombinant proteins with library member–target complexes being captured on Ni–NTA followed by elution using heat denaturation of the protein targets, usually with three successive cycles of binding and elution being performed. The ultimate selection outputs were evaluated using Roche/454 sequencing followed by visualization using 3D plots in which each of the three plot axes represent building block identities at different cycles of library synthesis. For Aurora A kinase, multiple compounds were found within the seven-million-member three-cycle library that showed preference at only two of the cycles of chemistry (allowing the other to be absent). Compounds synthesized off-DNA exhibited IC_{50} values as low as 270 nM. Selection in the presence of ATP indicated that these compounds bound in an ATP-competitive manner, a mechanism that was subsequently supported by both biochemical assay results and crystallography. Only approximately 1% of the 65,000 sequence reads corresponded to Aurora A binders, which lead the authors to emphasize the importance of deep sequencing for the evaluation of libraries of this size. For p38 MAP kinase, several features were detected within the 3D plots of the 800-million-member four-cycle library. Focusing upon a three-cycle feature, off-DNA compounds were synthesized with IC_{50} values as low as 7 nM, and as for the Aurora A kinase compounds, selection with ATP indicated ATP competitiveness, subsequently also supported by crystallography. In both instances, the crystal structures indicated a clear path to solvent for the DNA attachment point of the discovered inhibitors.

In a second publication from the Praecis/GSK group [44], the authors describe the largest library yet reported—a 4.05-billion-member four-cycle triazine library. This library was incubated with chemically biotinylated metalloprotease ADAMTS-5 immobilized upon streptavidin–agarose with elution at 80°C for three successive selection cycles. Evaluation of the selection output was performed by sequencing with the 454/Roche platform. A family of active compounds was identified with IC_{50} values as low as 30 nM. Several variants of these compounds were explored as part of a medicinal chemistry study. This study indicated a range of positions where substitution was tolerated, but most substitutions decreased activity as might be expected for compounds identified from a library of this size. Preservation of activity required an (R)-pyrolidin-2-ylmethanimine linked to a parasubstituted phenyl sulfonamide at one position on the triazine and a nitrogen-containing aryl at another; substitution of the latter with an N-methyl imidazole was the only reported substitution that increased potency. Variants are described with rat oral bioavailability (F) of 8%, a rat circulatory half-life of 8 h, and dose-dependent activity in an *ex vivo* human osteoarthritis cartilage explant disease model.

Additional achievements of the Praecis/GSK group were described in posters presented at the 2010 ACS symposium in Boston. These show that further compounds have been isolated from their triazine libraries including a two-cycle compound for EGLN1 with an IC_{50} of 10 nM [45] and a three-cycle compound for ADAMTS-4 with an IC_{50} of 5 nM [46]. Another poster presented at the same meeting [47] introduced a 16-million-member library constructed by capping 22 diamino acids with 800 carboxylic acids, sulfonyl chlorides, isocyanates, and aldehydes at each of the amines. A compound discovered within this library inhibits Mtb InhA with an IC_{50} of 14 nM while having a molecular weight of 433 and a $cLogP$ of 1.6. An article published in Chemistry and Engineering News in December 2010 [48] indicates that at GSK DNA-encoded chemistry has been used upon 140 targets in 40 programs and 25 disease areas with targets that include metalloenzymes, kinases, cellular and viral surface receptors, and protein–protein interactions and that the first compound discovered using DNA-encoded chemistry to enter clinical trials was expected to do so in January of 2011.

Sequence-recorded DNA-encoded chemical libraries are able to support the creation of larger libraries than have been reported for sequence-directed libraries. However, constraints that continue to apply include the instability of DNA under some chemical synthesis conditions, for example, low pH and strongly oxidizing conditions and in the presence of nucleases such as are present in most biological milieu. One possible strategy to circumvent these limitations is to use Peptide Nucleic Acid (PNA) as an alternative to DNA for encoding chemical history. A wide range of chemistries have recently been reported on PNA [49], and while PNA does not function as a template for template-dependent polymerases, strategies for the indirect evaluation of encoded chemical information recorded in PNA have been reported [50, 51]. Another recent innovation uses chemical or photochemical methods to ligate successive encoding oligonucleotides during library generation [52]. This avoids enzymatic ligation methods and the thousands of individual buffer exchanges that these require during the split of split-and-mix.

Various lead discovery achievements of DNA-encoded chemistry were reviewed in 2010 [53] and will also be discussed in succeeding chapters.

3.6 2012 AND BEYOND: THE FUTURE OF DNA-ENCODED CHEMISTRY

DNA-encoded chemistry offers a complementary approach for the discovery of therapeutic hits and leads when compared to high-throughput screening. Very significant achievements are being made using DNA-encoded chemistry within the pharmaceutical and biotech industries, including Ensemble, GSK, Nuevolution, Philochem, Vipergen, and X-Chem. It appears that most of this progress has not been publicly disclosed. It is very much hoped that as this technology becomes more widely utilized, more of these successes are publicly disclosed. Six recently published papers may be marking the beginning of this era. In January of 2012, Gentile and coworkers at GSK in Verona, Italy, and Stevenage, UK, reported a series of novel potent inhibitors of the kinase GSK3β (Fig. 3.8) discovered using sequence-recorded DNA-encoded chemistry [54]. Compound (**1**) (Fig. 3.8) is Lipinski compliant with an IC_{50} of 0.3 μM in a biochemical assay and was also shown to be active in a cell-based assay. The crystal structure of a member of this series in complex with GSK3β shows the oxazole nitrogen hydrogen bonding to the hinge residue and the 4-methoxy phenyl group lying within the ATP site. This structural information was then used to aid optimization that yielded Compound (**29**) (Fig. 3.8) with an IC_{50} of 0.03 μM, a reduced polar surface area, and sufficient oral availability and brain penetrance to permit evaluation in an animal model. In February of 2013, Podolin and coworkers at GSK in Collegeville and King of Prussia, Pennsylvania, and Waltham, Massachusetts, reported the discovery and evaluation of GSK2256294A (Fig. 3.8), a novel Soluble Epoxide Hydrolase (sEH) inhibitor [55] that was discovered using sequence-recorded DNA-encoded chemistry with the library reported in [43]. This Lipinski-compliant molecule has very high oral availability and appears to be a clinical candidate. GSK2256294A inhibits sEH with an IC_{50} of 27 pM in a biochemical assay and with an IC_{50} of 0.66 nM in a cell-based assay, is active in human whole blood, and reduces pulmonary inflammation in a mouse model. GSK2256294A does not inhibit the off-targets EPHX1 or EPHX2. In April of 2013, Disch and coworkers at Sirtris/GSK in Upper Providence, Pennsylvania, and Cambridge and Waltham, Massachusetts, reported the discovery and characterization of a series of pan inhibitors of the deacetylases Sirtuin-1, Sirtuin-2, and Sirtuin-3 (Fig. 3.8)

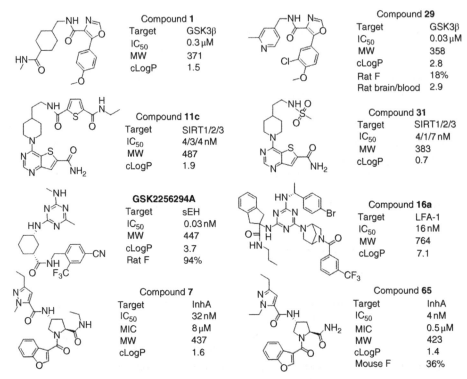

Compound 1

Target	GSK3β
IC$_{50}$	0.3 μM
MW	371
cLogP	1.5

Compound 29

Target	GSK3β
IC$_{50}$	0.03 μM
MW	358
cLogP	2.8
Rat F	18%
Rat brain/blood	2.9

Compound 11c

Target	SIRT1/2/3
IC$_{50}$	4/3/4 nM
MW	487
cLogP	1.9

Compound 31

Target	SIRT1/2/3
IC$_{50}$	4/1/7 nM
MW	383
cLogP	0.7

GSK2256294A

Target	sEH
IC$_{50}$	0.03 nM
MW	447
cLogP	3.7
Rat F	94%

Compound 16a

Target	LFA-1
IC$_{50}$	16 nM
MW	764
cLogP	7.1

Compound 7

Target	InhA
IC$_{50}$	32 nM
MIC	8 μM
MW	437
cLogP	1.6

Compound 65

Target	InhA
IC$_{50}$	4 nM
MIC	0.5 μM
MW	423
cLogP	1.4
Mouse F	36%

Figure 3.8. The structures and some physicochemical properties of recently published therapeutic candidate and tool compounds discovered using the affinity-mediated selection of libraries of small molecules constructed using DNA-encoded sequence-recorded chemistry. Compound (**1**) is a GSK3β inhibitor that was optimized to yield Compound (**29**) with improved affinity [54]. Compound (**11c**) is a pan inhibitor of Sirtuin-1, Sirtuin-2, and Sirtuin-3 that was optimized to yield Compound (**31**) with improved physicochemical properties [55]. GSK2256294A is a sEH inhibitor [56]. Compound (**16a**) is an inhibitor of the ICAM-1/LFA-1 interaction [57]. Compounds (**7**) and (**65**) are inhibitors of the TB target InhA [58].

[56] using sequence-recorded DNA-encoded chemistry followed by lead optimization. The library was constructed with three cycles of chemistry in which 16 bis-acids were conjugated to encoding DNA and then further elaborated with 134 diamines that were in turn capped with 570 heteroaryls to generate a library of 1.22 million compounds. Compound (**11c**) (Fig. 3.8) binds each of Sirtuin-1, Sirtuin-2, and Sirtuin-3 with a K_D in the range of 2.7–4.0 nM, is Lipinski compliant, and was optimized to yield Compound (**31**) (Fig. 3.8), which appears likely to be useful as a target validation tool. The authors crystallized Compound (**11c**) with Sirtuin-3, and the crystal structure of the complex shows the carboxamide binding in the nicotinamide C-pocket and the aliphatic part of the molecule extending through the substrate channel [56]. In early 2014, Kollmann and coworkers at GSK in Waltham Massachusetts and Harvard Medical School in Boston, Massachusetts, reported Compound **16a** (Fig. 3.8), an LFA-1 binder that inhibits the LFA-1–ICAM-1 interaction [57]. Compound **16a** appears to have been discovered from

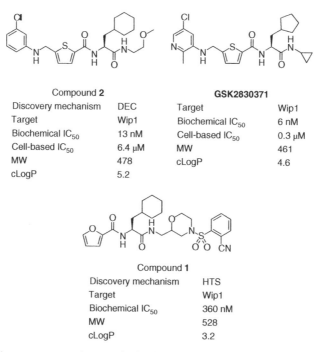

Compound **2**

Discovery mechanism	DEC
Target	Wip1
Biochemical IC$_{50}$	13 nM
Cell-based IC$_{50}$	6.4 μM
MW	478
cLogP	5.2

GSK2830371

Target	Wip1
Biochemical IC$_{50}$	6 nM
Cell-based IC$_{50}$	0.3 μM
MW	461
cLogP	4.6

Compound **1**

Discovery mechanism	HTS
Target	Wip1
Biochemical IC$_{50}$	360 nM
MW	528
cLogP	3.2

Figure 3.9. The structures and some physicochemical properties of compounds discovered in a pair of parallel screens conducted against Wip1 phosphatase [59]. Compound (**1**) was discovered in an affinity-based screen of a DNA-encoded library and Compound (**2**) was discovered in an activity-based high-throughput screen of a nonencoded library. Optimization yielded (**GSK2830371**), which is orally available and active *in vivo*.

the 4.1-billion-member triazine library reported in Ref. [44]; it has an IC$_{50}$ value of 16 nM and shows activity in a cell-adhesion assay. In common with most disclosed protein–protein interaction inhibitors discovered by any method, Compound 16a is not Lipinski-compliant. Also published in early 2014, Encinas and coworkers at GSK in Waltham, Massachusetts, Stevenage, UK, and Madrid, Spain, reported a series of compounds that inhibit the TB target InhA discovered using sequence-recorded DNA-encoded chemistry [58]. Selections were conducted against a panel of DNA-encoded libraries using biotinylated InhA. The reported compounds were derived from a previously undisclosed library in which each amine on 22 diamino acids were successively functionalized with a set of amine-capping reagents comprising 857 carboxylic acids, aldehydes, sulfonyl chlorides, and isocyanates for a total of 16.1 million three-cycle compounds. Compound (**7**) (Fig. 3.8) has an IC$_{50}$ value of 32 nM in a biochemical assay of InhA function, an MIC of 8 μM against *Mycobacterium tuberculosis*, and binds to the active site while hydrogen-bonding to NADH as determined by crystallography. Lead optimization yielded Compound (**65**) (Fig. 3.8) with an IC$_{50}$ value of 4 nM and an MIC of 0.5 μM. Unfortunately, Compound (**65**) was not active in a murine TB acute infection model.

Also early in 2014 a very interesting report appeared in which the authors ran the affinity-mediated selection of a library of sequence-recorded DNA-encoded small molecules and a biochemical high-throughput screen side-by-side against the same target. Gilmartin and coauthors from GSK in Collegeville, Pennsylvania, found and studied inhibitors of Wip1 phosphatase [59]. Both HTS and DEC were successful and yielded hits with related chemical structures. The DNA-encoded library they used is previously undisclosed and appears to be an N-terminally capped dipeptide linked to DNA via its C-terminus. This library yielded Compound (**2**) (Fig. 3.9) with a biochemical IC_{50} of 13 nM, and the HTS library yielded Compound (**1**) (Fig. 3.9) with a biochemical IC_{50} of 360 nM. These inhibitors bind to an allosteric site on the "flap" subdomain of Wip1, and because this site is not structurally conserved in other related phosphatases these inhibitors are highly selective for Wip1. No inhibition was detected across a panel of 21 other phosphatases indicating a specificity of at least 5000-fold. Lead optimization yielded (**GSK2830371**) (Fig. 3.9) with a biochemical IC_{50} of 6 nM and of 0.3 μM in the cell-based assay. (**GSK2830371**) is orally available and showed activity in a murine-xenografted tumor model of B-cell lymphoma when dosed via this route. Wip1 dephosphorylates p53 and p38MAPK, and p53 responds to stresses such as DNA damage. Accordingly a Wip1 inhibitor such as (**GSK2830371**) may be a viable strategy for treating tumors that retain p53 activity.

ACKNOWLEDGMENTS

With grateful thanks to Matthew Clark, John Cuozzo, Christoph Dumelin, Robert Goodnow, Daniel Resnicow, Richard Wagner, and Ying Zhang for their helpful comments.

REFERENCES

1. Smith GP. (1985). Filamentous fusion phage: novel expression vectors that display cloned antigens on the virion surface. *Science, 228*, 1315–1317.

2. Scott JK, Smith GP. (1990). Searching for peptide ligands with an epitope library. *Science, 249*, 386–390.

3. Ellington AD, Szostak JW. (1990). In vitro selection of RNA molecules that bind specific ligands. *Nature, 346*, 818–822.

4. Robertson DL, Joyce GF. (1990). Selection in vitro of an RNA enzyme that specifically cleaves single-stranded DNA. *Nature, 344*, 467–468.

5. Tuerk C, Gold L. (1990). Systematic evolution of ligands by exponential enrichment: RNA ligands to bacteriophage T4 DNA polymerase. *Science, 249*, 505–510.

6. Caruthers MH. (1985). Gene synthesis machines: DNA chemistry and its uses. *Science, 230*, 281–285.

7. Saiki RK, Scharf S, Faloona F, Mullis KB, Horn GT, Erlich HA, Arnheim N. (1985). Enzymatic amplification of beta-globin genomic sequences and restriction site analysis for diagnosis of sickle cell anemia. *Science, 230*, 1350–1354.

8. Mullis K, Faloona F, Scharf S, Saiki R, Horn G, Erlich H. (1986). Specific enzymatic amplification of DNA in vitro: the polymerase chain reaction. *Cold Spring Harb Symp Quant Biol, 51*, 263–273.

9. Sanger F, Nicklen S, Coulson AR. (1977). DNA sequencing with chain-terminating inhibitors. *Proc Natl Acad Sci USA, 74*, 5463–5467.

10. Olivera BM, Lehman IR. (1967). Linkage of polynucleotides through phosphodiester bonds by an enzyme from Escherichia coli. *Proc Natl Acad Sci USA*, *57*, 1426–1433.

11. Furka A, Sebestyén F, Asgedom M, Dibó G. (1991). General method for rapid synthesis of multicomponent peptide mixtures. *Int J Pept Protein Res*, *37*, 487–493.

12. Margulies M, Egholm M, Altman WE, Attiya S, Bader JS, Bemben LA, Berka J, Braverman MS, Chen YJ, Chen Z, Dewell SB, Du L, Fierro JM, Gomes XV, Godwin BC, He W, Helgesen S, Ho CH, Irzyk GP, Jando SC, Alenquer MLI, Jarvie TP, Jirage KB, Kim JB, Knight JR, Lanza JR, Leamon JH, Lefkowitz SM, Lei M, Li J, Lohman KL, Lu H, Makhijani VB, McDade KE, McKenna MP, Myers EW, Nickerson E, Nobile JR, Plant R, Puc BP, Ronan MT, Roth GT, Sarkis GJ, Simons JF, Simpson JW, Srinivasan M, Tartaro KR, Tomasz A, Vogt KA, Volkmer GA, Wang SH, Wang Y, Weiner MP, Yu P, Begley RF, Rothberg JM. (2005). Genome sequencing in open microfabricated high density picoliter reactors. *Nature*, *437*, 376–380.

13. Bentley DR, Balasubramanian S, Swerdlow HP, Smith GP, Milton J, Brown CG, Hall KP, Evers DJ, Barnes CL, Bignell HR, Boutell JM, Bryant J, Carter RJ, Cheetham RK, Cox AJ, Ellis DJ, Flatbush MR, Gormley NA, Humphray SJ, Irving LJ, Karbelashvili MS, Kirk SM, Li H, Liu X, Maisinger KS, Murray LJ, Obradovic B, Ost T, Parkinson ML, Pratt MR, Rasolonjatovo IMJ, Reed MT, Rigatti R, Rodighiero C, Ross MT, Sabot A, Sankar SV, Scally A, Schroth GP, Smith ME, Smith VP, Spiridou A, Torrance PE, Tzonev SS, Vermaas EH, Walter K, Wu X, Zhang L, Alam MD, Anastasi C, Aniebo IC, Bailey DMD, Bancarz IR, Banerjee S, Barbour SG, Baybayan PA, Benoit VA, Benson KF, Bevis C, Black PJ, Boodhun A, Brennan JS, Bridgham JA, Brown RC, Brown AA, Buermann DH, Bundu AA, Burrows JC, Carter NP, Castillo N, Catenazzi MCE, Chang S, Cooley RN, Crake NR, Dada OO, Diakoumakos KD, Dominguez-Fernandez B, Earnshaw DJ, Egbujor UC, Elmore DW, Etchin SS, Ewan MR, Fedurco M, Fraser LJ, Fuentes Fajardo KV, Furey WS, George D, Gietzen KJ, Goddard CP, Golda GS, Granieri PA, Green DE, Gustafson DL, Hansen NF, Harnish K, Haudenschild CD, Heyer NI, Hims MM, Ho JT, Horgan AM, Hoschler K, Hurwitz S, Ivanov DV, Johnson MQ, James T, Jones TAH, Kang GD, Kerelska TH, Kersey AD, Khrebtukova I, Kindwall AP, Kingsbury Z, Kokko-Gonzales PI, Kumar A, Laurent MA, Lawley CT, Lee SE, Lee X, Liao AK, Loch JA, Lok M, Luo S, Mammen RM, Martin JW, McCauley PG, McNitt P, Mehta P, Moon KW, Mullens JW, Newington T, Ning Z, Ng BL, Novo SM, O'Neill MJ, Osborne MA, Osnowski A, Ostadan O, Paraschos LL, Pickering L, Pike AC, Pike AC, Pinkard DC, Pliskin DP, Podhasky J, Quijano VJ, Raczy C, Rae VH, Rawlings SR, Chiva Rodriguez A, Roe PM, Rogers J, Rogert Bacigalupo MC, Romanov N, Romieu A, Roth RK, Rourke NJ, Ruediger ST, Rusman E, Sanches-Kuiper RM, Schenker MR, Seoane JM, Shaw RJ, Shiver MK, Short SW, Sizto NL, Sluis JP, Smith MA, Sohna Sohna JE, Spence EJ, Stevens K, Sutton N, Szajkowski L, Tregidgo CL, Turcatti G, Vandevondele S, Verhovsky Y, Virk SM, Wakelin S, Walcott GC, Wang J, Worsley GJ, Yan J, Yau L, Zuerlein M, Rogers J, Mullikin JC, Hurles ME, McCooke NJ, West JS, Oaks FL, Lundberg PL, Klenerman D, Durbin R, Smith AJ. (2008). Accurate whole human genome sequencing using reversible terminator chemistry. *Nature*, *456*, 53–59.

14. Brenner S, Lerner RA. (1992). Encoded combinatorial chemistry. *Proc Natl Acad Sci USA*, *89*, 5381–5383.

15. Needels MC, Jones DG, Tate EH, Heinkel GL, Kochersperger LM, Dower WJ, Barrett RW, Gallop MA. (1993). Generation and screening of an oligonucleotide-encoded synthetic peptide library. *Proc Natl Acad Sci USA*, *90*, 10700–10704.

16. Nielsen J, Brenner S, Janda KD. (1993). Synthetic methods for the implementation of encoded combinatorial chemistry. *J Am Chem Soc*, *115*, 9812–9813.

17. Kinoshita Y, Nishigaki K. (1995). Enzymatic synthesis of code regions for encoded combinatorial chemistry (ECC). *Nucleic Acids Symp Ser, 34,* 201–202.

18. Gartner ZJ, Liu DR. (2001). The generality of DNA-templated synthesis as a basis for evolving non-natural small molecules. *J Am Chem Soc, 123,* 6961–6963.

19. Gartner ZJ, Kanan MW, Liu DR. (2002). Expanding the reaction scope of DNA-templated synthesis. *Angew Chem Int Ed Engl, 41,* 1796–1800.

20. Gartner ZJ, Tse BN, Grubina R, Doyon JB, Snyder TM, Liu DR. (2004). DNA-templated organic synthesis and selection of a library of macrocycles. *Science, 305,* 1601–1605.

21. Halpin DR, Harbury PB. (2004). DNA display I. Sequence-encoded routing of DNA populations. *PLoS Biol, 2,* E173.

22. Halpin DR, Harbury PB. (2004). DNA display II. Genetic manipulation of combinatorial chemistry libraries for small-molecule evolution. *PLoS Biol, 2,* E174.

23. Halpin DR, Lee JA, Wrenn SJ, Harbury PB. (2004). DNA display III. Solid-phase organic synthesis on unprotected DNA. *PLoS Biol, 2,* E175.

24. Melkko S, Scheuermann J, Dumelin CE, Neri D. (2004). Encoded self-assembling chemical libraries. *Nat Biotechnol, 22,* 568–574.

25. Tse BN, Snyder TM, Shen Y, Liu DR. (2008). Translation of DNA into a library of 13,000 synthetic small-molecule macrocycles suitable for in vitro selection. *J Am Chem Soc, 130,* 15611–15626.

26. Kleiner RE, Dumelin CE, Tiu GC, Sakurai K, Liu DR. (2010). In vitro selection of a DNA-templated small-molecule library reveals a class of macrocyclic kinase inhibitors. *J Am Chem Soc, 132,* 11779–11791.

27. Wrenn SJ, Weisinger RM, Halpin DR, Harbury PB. (2007). Synthetic ligands discovered by in vitro selection. *J Am Chem Soc, 129,* 13137–13143.

28. Weisinger RM, Wrenn SJ, Harbury PB. (2012). Highly parallel translation of DNA sequences into small molecules. *PLoS One, 7,* e28056.

29. Melkko S, Zhang Y, Dumelin CE, Scheuermann J, Neri D. (2007). Isolation of high-affinity trypsin inhibitors from a DNA-encoded chemical library. *Angew Chem Int Ed Engl, 46,* 4671–4674.

30. Dumelin CE, Scheuermann J, Melkko S, Neri D. (2006). Selection of streptavidin binders from a DNA-encoded chemical library. *Bioconjug Chem, 17,* 366–370.

31. Scheuermann J, Dumelin CE, Melkko S, Zhang Y, Mannocci L, Jaggi M, Sobek J, Neri D. (2008). DNA-encoded chemical libraries for the discovery of MMP-3 inhibitors. *Bioconjug Chem, 19,* 778–785.

32. Hansen MH, Blakskjaer P, Petersen LK, Hansen TH, Højfeldt JW, Gothelf KV, Hansen NJV. (2009). A yoctoliter-scale DNA reactor for small-molecule evolution. *J Am Chem Soc, 131,* 1322–1327.

33. Morgan B, Hale S, Arico-Muendel CC, Clark M, Wagner R, Israel DI, Gefter ML, Benjamin D, Hansen NJV, Kavarana MJ, Creaser SP, Franklin GJ, Centrella PA, Acharya RA. (2005). Methods for synthesis of encoded libraries. US Patent Application US 0,158,765.

34. Franch T, Lundorf MD, Jacobsen SN, Olsen EK, Andersen AL, Holtmann A, Hansen AH, Sorensen AM, Goldbech A, de Leon D, Kaldor DK, Slok FA, Husemoen GN, Dolberg J, Jensen KB, Pedersen L, Norregaard-Madsen M, Godskesen MA, Glad SS, Neve S, Thisted T, Kronborg TTA, Sams CK, Felding J, Freskgard P-O, Gouliaev AH, Pedersen H. (2009). Enzymatic encoding methods for efficient synthesis of large libraries. US Patent Application US 0,264,300.

35. Buller F, Mannocci L, Zhang Y, Dumelin CE, Scheuermann J, Neri D. (2008). Design and synthesis of a novel DNA-encoded chemical library using Diels-Alder cycloadditions. *Bioorg Med Chem Lett, 18,* 5926–5931.

36. Mannocci L, Zhang Y, Scheuermann J, Leimbacher M, De Bellis G, Rizzi E, Dumelin C, Melkko S, Neri D. (2008). High-throughput sequencing allows the identification of binding molecules isolated from DNA-encoded chemical libraries. *Proc Natl Acad Sci USA*, *105*, 17670–17675.

37. Buller F, Zhang Y, Scheuermann J, Schäfer J, Bühlmann P, Neri D. (2009). Discovery of TNF inhibitors from a DNA-encoded chemical library based on Diels-Alder cycloaddition. *Chem Biol*, *16*, 1075–1086.

38. Melkko S, Mannocci L, Dumelin CE, Villa A, Sommavilla R, Zhang Y, Grütter MG, Keller N, Jermutus L, Jackson RH, Scheuermann J, Neri D. (2010). Isolation of a small-molecule inhibitor of the antiapoptotic protein Bcl-xL from a DNA-encoded chemical library. *ChemMedChem*, *5*, 584–590.

39. Buller F, Steiner M, Frey K, Mircsof D, Scheuermann J, Kalisch M, Buhlmann P, Supuran CT, Neri D. (2011). Selection of carbonic anhydrase IX inhibitors from one million DNA-encoded compounds. *ACS Chem Biol*, *6*, 336–344.

40. Leimbacher M, Zhang Y, Mannocci L, Stravs M, Geppert T, Scheuermann J, Schneider G, Neri D. (2012). Discovery of small-molecule interleukin-2 inhibitors from a DNA-encoded chemical library. *Chem Eur J*, *18*, 7729–7737.

41. Lipinski CA, Lombardo F, Dominy BW, Feeney PJ. (2001). Experimental and computational approaches to estimate solubility and permeability in drug discovery and development settings. *Adv Drug Deliv Rev*, *46*, 3–26.

42. Mannocci L, Melkko S, Buller F, Molnàr I, Bianké JPG, Dumelin CE, Scheuermann J, Neri D. (2010). Isolation of potent and specific trypsin inhibitors from a DNA-encoded chemical library. *Bioconjug Chem*, *21*, 1836–1841.

43. Clark MA, Acharya RA, Arico-Muendel CC, Belyanskaya SL, Benjamin DR, Carlson NR, Centrella PA, Chiu CH, Creaser SP, Cuozzo JW, Davie CP, Ding Y, Franklin GJ, Franzen KD, Gefter ML, Hale SP, Hansen NJV, Israel DI, Jiang J, Kavarana MJ, Kelley MS, Kollmann CS, Li F, Lind K, Mataruse S, Medeiros PF, Messer JA, Myers P, O'Keefe H, Oliff MC, Rise CE, Satz AL, Skinner SR, Svendsen JL, Tang L, van Vloten K, Wagner RW, Yao G, Zhao B, Morgan BA. (2009). Design, synthesis and selection of DNA-encoded small-molecule libraries. *Nat Chem Biol*, *5*, 647–654.

44. Deng H, O'Keefe H, Davie CP, Lind KE, Acharya RA, Franklin GJ, Larkin J, Matico R, Neeb M, Thompson MM, Lohr T, Gross JW, Centrella PA, O'Donovan GK, Bedard KL, van Vloten K, Mataruse S, Skinner SR, Belyanskaya SL, Carpenter TY, Shearer TW, Clark MA, Cuozzo JW, Arico-Muendel CC, Morgan BA. (2012). Discovery of highly potent and selective small molecule ADAMTS-5 inhibitors that inhibit human cartilage degradation via encoded library technology (ELT). *J Med Chem*, *55*, 7061–7079.

45. Fosbenner DT, King BW, Brown KK, Keller P, McKenzie K, Siegfried B, Mitchell L, Smallwood A, Concha N, Miller L, Xue Y, Franklin GJ, Centrella PA, Svendsen J, Mataruse S, Clark MA. (2010). Utilization of DNA-encoded libraries for the identification of novel inhibitors of prolyl hydroxylases. MEDI-239. Poster presented at 2010 ACS symposium, August 22–26, Boston, MA.

46. Ding Y, O'Keefe H, Svendsen JL, Franklin GJ, Centrella PA, Clark MA, Acharya RA, Li F, Messer JA, Chiu CH, Matico RE, Murray-Thompson MF, Skinner SR, Belyanskaya SL, Israel DI, Cuozzo JW, Arico-Muendel C, Morgan BA. (2010). Discovery of inhibitors for ADAMTS-4 using DNA-encoded library technology (DEL). MEDI-460. Poster presented at 2010 ACS symposium, August 22–26, Boston, MA.

47. Davie CP, Centrella PA, O'Donovan GK, Evindar G, O'Keefe H, Patel AM, Clark MA. (2010). Design and synthesis of a 16 million member DNA-encoded library (DEL) and discovery of

potent Mycobacterium tuberculosis (*Mtb*) *InhA* inhibitors. MEDI-150. Poster presented at 2010 ACS symposium, August 22–26, Boston, MA.

48. Jarvis LM. (2010). Tech Test Drive: early success with a screening technique highlights GSK's external partnering strategy. *Chem Eng News*, *88*, 22.

49. Chouikhi D, Ciobanu M, Zambaldo C, Duplan V, Barluenga S, Winssinger N. (2012). Expanding the scope of PNA-encoded synthesis (PES): Mtt-protected PNA fully orthogonal to Fmoc chemistry and a broad array of diversity-generating reactions. *Chem Eur J*, *18*, 12698–12704.

50. Brudno Y, Birnbaum ME, Kleiner RE, Liu DR. (2009). An in vitro translation, selection and amplification system for peptide nucleic acids. *Nat Chem Biol*, *6*, 148–155.

51. Svensen N, Díaz-Mochón JJ, Bradley M. (2011). Decoding a PNA encoded peptide library by PCR: the discovery of new cell surface receptor ligands. *Chem Biol*, *18*, 1284–1289.

52. Litovchick A, Clark MA, Keefe AD. (2014). Universal strategies for the DNA-encoding of libraries of small molecules using the chemical ligation of oligonucleotide tags. *Artificial DNA: PNA & XNA*, *5*, e27896-1–e27896-11.

53. Clark MA. (2010). Selecting chemicals: the emerging utility of DNA-encoded libraries. *Curr Opin Chem Biol*, *14*, 396–403.

54. Gentile G, Merlo G, Pozzan A, Bernasconi G, Bax B, Bamborough P, Bridges A, Carter P, Neu M, Yao G, Brough C, Cutler G, Coffin A, Belyanskaya S. (2012). 5-Aryl-4-carboxamide-1,3-oxazoles: potent and selective GSK-3 inhibitors. *Bioorg Med Chem Lett*, *22*, 1989–1994.

55. Podolin PL, Bolognese BJ, Foley JF, Long E 3rd, Peck B, Umbrecht S, Zhang X, Zhu P, Schwartz B, Xie W, Quinn C, Qi H, Sweitzer S, Chen S, Galop M, Morgan BA, Behm DJ, Marino JP Jr., Kurali E, Barnette MS, Mayer RJ, Booth-Genthe CL, Callahan JF. (2013). In vitro and in vivo characterization of a novel soluble epoxide hydrolase inhibitor. *Prostaglandins Other Lipid Mediat*, *104–105*, 25–31.

56. Disch JS, Evindar G, Chiu CH, Blum CA, Dai H, Jin L, Schuman E, Lind KE, Belyanskaya SL, Deng J, Coppo F, Aquilani L, Graybill TL, Cuozzo JW, Lavu S, Mao C, Vlasuk GP, Perni RB. (2013). Discovery of thieno[3,2-d]pyrimidine-6-carboxamides as potent inhibitors of SIRT1, SIRT2, and SIRT3. *J Med Chem*, *56*, 3666–3679.

57. Kollmann CS, Bai X, Tsai CH, Yang H, Lind KE, Skinner SR, Zhu Z, Israel DI, Cuozzo JW, Morgan BA, Yuki K, Xie C, Springer TA, Shimaoka M, Evindar G. (2014). Application of encoded library technology (ELT) to a protein–protein interaction target: Discovery of a potent class of integrin lymphocyte function-associated antigen 1 (LFA-1) antagonists. *Bioorg Med Chem*, in press. http://www.sciencedirect.com/science/article/pii/S096808961400087X. Accessed on March 1, 2014.

58. Encinas L, O'Keefe H, Neu M, Remuiñán MJ, Patel AM, Guardia A, Davie CP, Pérez-Macías N, Yang H, Convery MA, Messer JA, Pérez-Herrán E, Centrella PA, Alvarez-Gómez D, Clark MA, Huss S, O'Donovan GK, Ortega-Muro F, McDowell W, Castañeda P, Arico-Muendel CC, Pajk S, Rullás J, Angulo-Barturen I, Alvarez-Ruíz E, Mendoza-Losana A, Pages LB, Castro-Pichel J, Evindar, G. (2014). Encoded library technology as a source of hits for the discovery and lead optimization of a potent and selective class of bactericidal direct inhibitors of Mycobacterium tuberculosis InhA. *J Med Chem*, in press. http://pubs.acs.org/doi/abs/10.1021/jm401326j. Accessed on March 1, 2014.

59. Gilmartin AG, Faitg TH, Richter M, Groy A, Seefeld MA, Darcy MG, Peng X, Federowicz K, Yang J, Zhang SY, Minthorn E, Jaworski JP, Schaber M, Martens S, McNulty DE, Sinnamon RH, Zhang H, Kirkpatrick RB, Nevins N, Cui G, Pietrak B, Diaz E, Jones A, Brandt M, Schwartz B, Heerding DA, Kumar R. (2014). Allosteric Wip1 phosphatase inhibition through flap-subdomain interaction. *Nat Chem Biol*, *10*, 181–187.

4

DNA-COMPATIBLE CHEMISTRY

Kin-Chun Luk[1] and Alexander Lee Satz[2]

[1]Discovery Chemistry, Hoffmann-La Roche Inc, Nutley, NJ, USA
[2]Molecular Design and Chemical Biology,
F. Hoffmann-La Roche Ltd, Basel, Switzerland

Combinatorial libraries are derived from building blocks and the chemistry used to connect the building blocks to each other. Chapter 5 describes what a building block collection should look like. This chapter will discuss chemistries used to piece the building blocks together. The first part of this chapter will focus on chemistries proven to work in the presence of DNA oligomers, while the second part will focus on chemistries thought to have potential value. Taken together, Chapter 5 and this chapter provide the foundation needed to design DNA-Encoded Libraries (DELs) capable of providing valuable ligands in a drug-discovery setting.

DEL-compatible chemistry requires methodologies that are robust, work in the presence of protic solvents, do not require strong acidic conditions, and will not degrade DNA. These requirements exclude many common reagents and synthetic transformations including the use of very strong bases (e.g., t-butyl lithium) or strong reducing reagents (e.g., lithium aluminum hydride), highly reactive reagents or intermediates that degrade instantaneously in water (acyl chlorides, Grignard reagents, Mitsunobu intermediates, etc.), reagents that react with DNA bases (some alkyl halides), oxidants that degrade DNA (DNA is generally not stable in the presence of oxidizing agents), and acid-catalyzed

A Handbook for DNA-Encoded Chemistry: Theory and Applications for Exploring Chemical Space and Drug Discovery, First Edition. Edited by Robert A. Goodnow, Jr.
© 2014 John Wiley & Sons, Inc. Published 2014 by John Wiley & Sons, Inc.

cyclizations or rearrangements (e.g., Pomeranz–Fritsch reaction, Fischer indole synthesis, or Beckmann rearrangement). The reactions that tend to be compatible with DEL chemistry possess reagents that are reasonably stable in aqueous conditions and require basic, not acidic, conditions. DNA oligomers are extremely stable in basic aqueous conditions even with prolonged heating (for instance, 1 M NaOH at 80°C for 18 h causes no significant degradation). DNA is usually soluble in aqueous solutions, especially under basic conditions, but is poorly soluble or insoluble in organic solvents. Many organic building blocks needed for DEL preparation are poorly soluble in aqueous solutions. For ease of running reactions and recovery of DNA-containing products in the presence of organic building blocks, a mixture of aqueous buffer together with a water-miscible organic solvent should be considered. Some useful solvent mixtures that could be used in such organic reactions in the presence of DNA have been studied [1–6].

Using the aforementioned criteria, it is possible to predict if a known synthetic organic transformation has a realistic chance of being compatible with DEL production. However, prediction is no replacement for experimentation. For this reason, we have divided this chapter into two parts. The first part provides methodologies for reactions shown to work in the presence of DNA. This section also provides an example that utilized these known reactions to produce a structurally diverse DEL. The second part surveys the literature and provides examples of reactions expected to have a reasonable likelihood of success.

4.1 LITERATURE EXAMPLES OF DNA-COMPATIBLE CHEMISTRY AND THEIR USE IN LIBRARY DESIGN

4.1.1 Practical Issues for DNA-Conjugated Organic Transformations

Thirty different reactions are listed in Table 4.1. Entries 1–19 provide DEL-compatible methodologies for nontemplated transformations (i.e., split-and-pool). Entries 20–31 provide DEL-compatible methodologies solely reported for DNA-templated transformations. A brief summary of nontemplated and templated DEL synthesis is provided in Figure 4.1 and Figure 4.2, respectively. In Sections 4.1.1.1–4.1.3.5 we discuss details of select reactions from Table 4.1 to illustrate some practical issues when conducting reactions with DNA conjugates. In order to build DELs, there needs to be a way of linking the library elements to the encoding DNA. One such method is to employ a bifunctional DNA conjugate starting material and compatible chemistry to synthesize the library.

4.1.1.1 Bifunctional DNA Conjugate Starting Material. Starting materials for all DEL chemistry works consist of three parts: (i) a DNA-encoding region to which further DNA sequences may be appended as discussed in Chapter 6; (ii) a synthetically useful chemical handle, usually an aliphatic primary amine; and (iii) a linker or spacer between the primary amine handle and the DNA-encoding region. The size of the DNA conjugate starting material may be relatively small (<4000 Da) depending upon the encoding strategy; however, it must be large enough in mass to undergo the precipitation procedure provided in Section 4.1.2.1. Figure 4.3 provides an example of one bifunctional starting material for the synthesis of a DEL [6].

TABLE 4.1. Reported synthetic transformations employing DNA conjugates[a]

	Reaction	Scheme[b]	Conditions[c]	Comments
1.	Acylation [2–6]		pH 9.4 buffer (150 mM borate), 40 equiv. carboxylic acid (2×), 40 equiv. DMT-MM (2×), 18 h, RT [6]	Cosolvent such as DMA or CH$_3$CN may be used
2.	Acylation [6]		pH 5.5 buffer (150 mM phosphate) 50 equiv. amine (200 mM 1:1 CH$_3$CN–water stock) (2×), 100 equiv. HCl (200 mM stock), 100 equiv. DMT-MM (200 mM stock) (2×), 72 h, RT	
3a.	S$_N$Ar (triazine) [6]		pH 9.4 buffer (150 mM borate) 1. 10 equiv. cyanuric chloride (200 mM CH$_3$CN stock), 4°C, 1 h 2. 50 equiv. R^1-NH$_2$ (200 mM DMA stock), RT, 18 h 3. 46 equiv. R^2-NH$_2$ (200 mM DMA stock), 80°C, 6 h	
3b.	S$_N$Ar [7]		[d]2500 equiv. amine (from a 100 mM DMSO stock), ~50 mM final conc. pH 9 buffer (100 mM borate), ~20 μM DNA, 90°C, 16 h	[e]No analytical data or yields provided
4.	Fmoc removal [6]		Water, 1 mM DNA, 20% piperidine, RT, 4 h	
5.	Boc removal [7]		38 mM NaOAc, 200 mM MgCl$_2$, 70°C, 18 h	[e]No analytical data or yields provided

(Continued)

TABLE 4.1. (cont'd)

	Reaction	Scheme[b]	Conditions[d,c]	Comments
6.	Suzuki reaction [8, 9]		1–10 equiv. boronic acid, 0.6 equiv. Pd(OAc)₂, 15 equiv. TPPTS, 2 equiv. Na₂CO₃, 2:1 H₂O/CH₃CN. 0.1 μM DNA. 70°C, 24 h, under argon, all stock solutions 2:1 H₂O/CH₃CN [8]	Scope limited to 5 boronic acids and 8-Br-G substrate. [e]Templated version also reported [9]
7.	Diol oxidation to glyoxal-amide [10]		50 mM NaIO₄, pH 3.7 buffer (500 mM NaOAc), 1 min, RT	
8.	Hydrolysis, *t*Bu, Me, Et esters [7]		100 mM LiOH, ~20 μM DNA, 80°C, 30 min	[e]No analytical data or yields provided
9.	Azide reduction [11]		[d]2000 equiv. Ru(bpy)₃Cl₂ (1 mM), 50 mM ascorbate, pH 7.4 (200 mM Tris), 0.5 μM DNA, *hv*, 10 min, 25°C	Also done with 2 × 10⁵ equiv. TCEP (100 mM)
10.	Triazole formation (click) [11, 12]		10 equiv. azide, 5 equiv. CuSO₄, ascorbate, and (tris-((1-benzyltriazol-4-yl)methyl)amine), 1 mM DNA, RT. 1–2 h [12]	[e]No analytical data or product yields provided Alternative method uses 50,000 equiv. azide [11]
11a.	Carbamoylation with isocyanates [7]		[d]1500 equiv. isocyanate (from a 300 mM CH₃CN stock), 30 mM final conc., pH 8 buffer (100 mM borate, 100 mM phosphate), ~20 μM DNA, 50°C, 16 h	[e]No analytical data or yields provided

11b.	Carbamoylation with isocyanates [13]		26 nmol of DNA-NH$_2$ in 26 µL pH 9.4 buffer (150 mM Na borate) add 40 equiv. RNCO (2.1 µL 500 mM in CH$_3$CN) incubated at RT overnight	
12a.	Sulfonylation with sulfonyl chlorides [7]		[d]1000 equiv. sulfonyl chloride (from a 200 mM THF stock), 20 mM final conc., pH 9.2 buffer (100 mM borate), ~20 µM DNA, 50°C, 16 h	[e]No analytical data or yields provided
12b.	Sulfonylation [13]		26 nmol of DNA-NH$_2$ in 26 µL pH 9.4 buffer (150 mM Na borate) add 40 equiv. RSO$_2$Cl (2.1 µL 500 mM in CH$_3$CN) incubated at RT overnight	
13.	Msec deprotection [7]		pH 10 buffer (100 mM borate), [d]~20 µM DNA, 40°C, 3 h	[e]No analytical data or yields provided
14.	Ns deprotection [7]		DNA loaded onto DEAE resin; incubated with 0.5 M mercaptoethanol and 0.25 M DIPEA in DMF for 24 h at 25°C with shaking	[e]No analytical data or yields provided
15.	Nvoc deprotection [13]		pH 4.5 AcOH buffer (100 mM AcOH in water), 4°C, 365 nm UV irradiation for 16 h	

(Continued)

TABLE 4.1. (cont'd)

Reaction	Scheme[b]	Conditions[d]	Comments
16a. Reductive alkylation [7]	DNA—NH₂ + R-CHO → DNA—N(CH₂R)₂	[d]2500 equiv. aldehyde (from a 200 mM DMSO stock), ~50 mM final conc. [d]1400 equiv. NaCNBH₃, ~28 mM pH 5 buffer (200 mM NaOAc), ~20 µM DNA. 30°C, 16 h	[e]May yield dialkylated product with primary amines. No analytical data or yields provided
16b. Reductive alkylation of secondary amines [13]	DNA—NH—C(=O)—(piperidine NH) + R-CHO → DNA—NH—C(=O)—(piperidine N—R)	To 26 nmol of DNA-NH₂ in 26 µL pH 5.5 buffer (250 mM Na phosphate) is added 100 equiv. RCHO (13 µL 200 mM in DMF) followed by 100 equiv. NaCNBH₃ solution (13 µL 200 mM in CH₃CN) and incubated at RT overnight	
17. Pentenoyl deprotection [7]	DNA—NH—C(=O)CH₂CH₂CH=CH₂ → DNA—NH₂	pH 5 buffer (10 mM NaOAc), ~80 µM DNA, I₂, 0.25 volumes of 25 mM solution (1:1 THF–water), 37°C, 2 h Quenched by addition of Na₂S₂O₃	[e]No analytical data or yields provided
18. Indole–styrene coupling [14]	DNA—NH—C(=O)—indole + R-styrene → DNA—NH—C(=O)—indole-CH(CH₃)(C₆H₄-R)	1:8 water–CH₃CN, 10 mM AuCl₃, 2.2 µM DNA, 1 h, 25°C	Yield not provided, no scope provided

No.	Reaction	Scheme	Conditions	Notes
19.	Diels–Alder reaction employing *in situ*-generated maleimides [15, 16]		d~1300 equiv. crude maleimide (26 mM), ~<0.2 mM DNA, 16h, 30°C [15]	54 maleimide derivatives were prepared in parallel and used without purification
20.	Templated Wittig reaction [2, 5, 17]		pH 5 (100 mM HEPES, 1 M NaCl), 60 nM DNA, 1.5h, 55°C [5]	
21.	Templated S$_N$2 and Michael addition (amine and thiol) [2, 4]		pH 7.5 (50 mM MOPS, 250 mM NaCl), 60 nM DNA, 37°C, 16h [2] [See entry 22 for Michael addition conditions]	Scope of electrophiles includes maleimide, vinyl sulfones, and acyl bromide/chloride [4]
22.	Michael reaction with acrylamide [18]		Equimolar quantities of reagents, pH 8.5 (50 mM MOPS, 250 mM NaCl), 25°C, 10–75 min [18]	Also with vinyl sulfones
23.	Templated reductive amination [2]		3 mM NaCNBH$_3$, pH 6 buffer (100 mM MES, 500 mM NaCl), 60 nM DNA, 1.5h, 25°C	

(Continued)

TABLE 4.1. (cont'd)

	Reaction	Scheme[b]	Conditions[d,c]	Comments
24.	Templated Heck reaction [2]		0.3 mM Na$_2$PdCl$_4$/P(p-SO$_3$C$_6$H$_4$)$_3$ 1:2, 50 mM pH 7.5 buffer (MOPS), 2.8 M NaCl, 120 nM DNA, 22 h, 25°C	Scope includes vinyl sulfone, maleimide, and acrylamide. Yields 26–54%
25.	Templated nitro-aldol reaction (Henry reaction) [2]		100 mM pH 8.5 buffer (TAPS), 300 mM NaCl, 60 mM DNA, 12 h, 25°C	Scope includes glyoxal-amide and maleimide as the electrophile Yields 20–50%
26.	Templated nitrone 1,3-dipolar cycloadditions with activated alkenes [2]		50 mM pH 7.5 buffer (MOPS) 2.8 M NaCl, 60 mM DNA, 22 h, 25°C	Scope includes maleimide and vinyl sulfone Yields 40–50%

27.	Templated oxazolidine formation [19]		100 mM pH 7.5 buffer (MOPS) 60–120 nM DNA 6 h, 25°C	Resulting oxazolidine was then acylated via template-directed DMT-MM acylation
28.	Trifluoroacetamide deprotection [7]		Water, ~20 µM, 45°C, 18 h	[e]No analytical data or yields provided
29.	Templated alkene–alkyne oxidative coupling [1, 20]		pH 7 buffer (100 mM MOPS, 1 M NaCl), 0.5 µM DNA [d]50,000 equiv. Na$_2$PdCl$_4$, final conc. 25 mM, 37°C, 4 h [1]	No scope provided, yield = 35% [20]
30.	Templated aldol reaction [1]		95:5 CH$_3$CN–water, 50 mM pyrrolidine, 33 nM DNA 25°C, 16 h	High concentration of organic solvent is required

(Continued)

TABLE 4.1. (cont'd)

31. Templated Sonogashira reaction [20]	DNA 	pH 7 buffer (50 mM MOPS, 0.5 M NaCl) 12 nM DNA ^{d}42,000 equiv. Na$_2$PdCl$_4$, final conc. 500 µM, 37°C, 1 h	No scope provided, 36% yield Other Pd-catalyzed carbon–carbon bond-forming reactions provided [20]

Abbreviations: RT, room temperature; TCEP, tris(2-carboxyethyl)phosphine, a mild reducing agent; DEAE, diethylaminoethyl cellulose, a positively charged ion exchange resin; MOPS, 3-(N-morpholino)propanesulfonic acid; DMSO, dimethyl sulfoxide; TPPTS, 3,3′,3″-phosphinidynetris(benzenesulfonic acid) trisodium salt, a water-soluble phosphine; DMA, dimethylacetamide; Tris, tris(hydroxymethyl)aminomethane; DMT-MM, 4-(4,6-dimethoxy-1,3,5-triazin-2-yl)-4-methylmorpholinium chloride; HEPES, (4-(2-hydroxyethyl)-1-piperazineethanesulfonic acid); MES, 2-(N-morpholino)ethanesulfonic acid; TAPS, 3-[[1,3-dihydroxy-2-(hydroxymethyl)propan-2-yl]amino]propane-1-sulfonic acid.

^{a}Examples of reactions done on solid phase using organic solvents are not listed [9]. Some carbon–carbon bond-forming reactions related to examples 29–31 [20] are not listed in Table 4.1.

^{b}"DNA-NH2" represents the DNA piece encoding the library members; DNA, as presented in the reaction schemes, may represent double- or single-stranded DNA including that of a template-directed reaction complex.

^{c}When multiple literature reports exist, selected experimental procedures are shown. See original publication for details.

^{d}Equivalents of reagents and concentration of DNA were estimated from reported methods.

^{e}Method provided in a patent. No analytical data or yields of the reactions were provided.

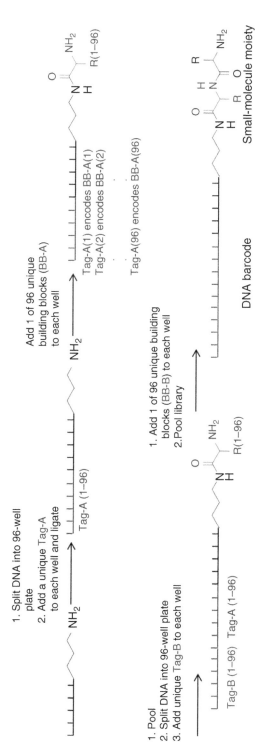

Figure 4.1. Scheme for split-and-pool library synthesis. Cycle A: DNA bioconjugate starting material consists of a DNA-encoding region conjugated to an aliphatic amine. The DNA is equally divided among 96 wells and to each well is added a unique DNA tag (Tag-A), which is ligated to the DNA bioconjugate starting material. To each well is then added 1 of 96 building blocks (BB-A), in this case Fmoc amino acids. After acylation, the library is pooled and the Fmoc-protecting group removed. As shown in the figure, each different building block is encoded by a unique DNA sequence. Cycle B: the product of cycle A, a crude mixture of 96 different structures, is split again into 96 wells. To each well is added a unique DNA tag (Tag-B), which is ligated to the terminus of Tag-A, thus extending the DEL "barcode." After ligation, 1 of 96 different building blocks (BB-B) is added to each well. A DNA barcode encodes each synthetic step for each library member. The split-and-pool steps may be repeated as needed, building a larger and more complex molecule with each successive cycle of chemistry.

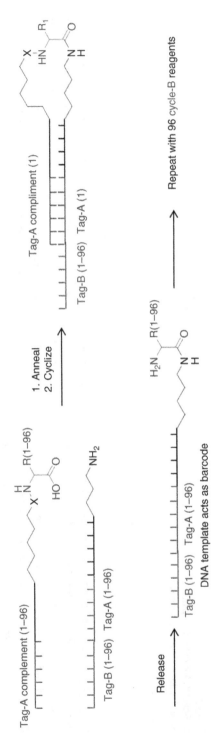

Figure 4.2. Scheme for template-directed DNA synthesis. 96 DNA-conjugated reagents are prepared, each with a unique DNA sequence. 9216 library templates are prepared, the product of 96 different cycle A and 96 cycle B tags. The 96 tagged reagents are combined and added to a single solution containing all 9216 library templates. Each reagent anneals to its appropriate library template. The formation of the double-stranded complex presumably results in fidelity between the template sequence and the composition of the resulting small-molecule moiety. Reagent DNA tags are removed by employing a cleavable linker "X." The single-stranded library template sequence encodes the structure of the small-molecule moiety. A multistep synthesis can be accomplished via consecutive cycles of annealing, reacting, and removal of reagent tags. Reaction methods optimized for template-directed synthesis are not necessarily applicable to nontemplated (split-and-pool) library synthesis.

78

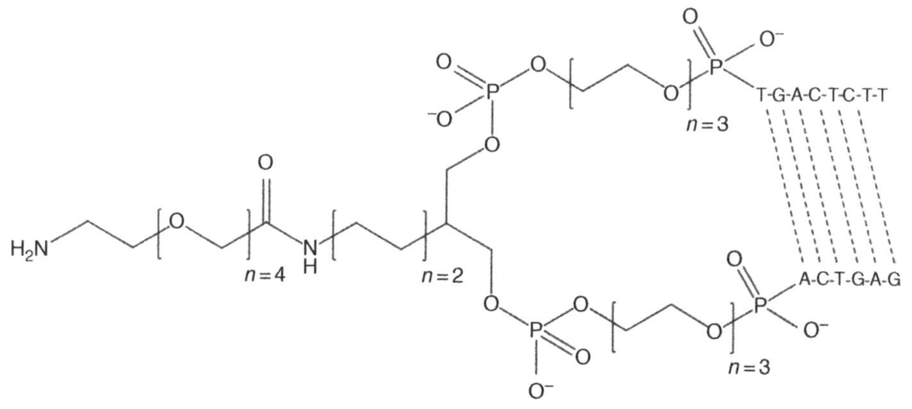

Figure 4.3. Example of a bifunctional starting material for synthesis of a DEL (MW = 5184) [6].

4.1.2 General Methods for Reactions on DNA

4.1.2.1 Preparation of DNA Conjugates for Reaction. The preparation starts with precipitation of the DNA or DNA conjugate from solution. To a solution of 1 mM DNA or functionalized conjugate in buffer are added 5 M NaCl (10% by volume) and 2.5 volumes of ethanol. Chill sample in freezer, or flash freeze with liquid nitrogen, and centrifuge at 3,000–14,000 rpm (depending upon the sample container type and instrument) for 30 min. After the pellet is formed, the ethanol may be decanted (in some cases, the pellet may be rinsed with clean ethanol) and residual solvent left to evaporate to provide the pellet as a mixture of DNA conjugate and precipitated salt. This process is also performed after each reaction cycle.

Troubleshooting

Pellet not forming: Pellets should be visible to the naked eye (they primarily consist of precipitated salts). If pellets are not readily observed, then an additional volume of ethanol and 5 M NaCl (10% by volume) can be added and the centrifugation process repeated.

Further purification needed: When the precipitation step is performed after chemistry, issues may arise because some reagents or by-products fail to dissolve in ethanol. These leftover reagents can interfere with later chemistry steps. There are simple solutions to these problems that provide an alternative to time-consuming HPLC. Approaches to purification include [4–7, 12, 15]:

1. Reconstitution of the pellet in a small amount of water and repetition of the precipitation process a second or third time.
2. Centrifugation of the aqueous DNA sample prior to ethanol precipitation. (As a general rule, DNA will dissolve in a solution that is at least 50% water. All solids in these aqueous solutions can be centrifuged to a pellet and discarded. The solids are usually excess buffer salts and/or water-insoluble reagents.)

3. The use of desalting gels (e.g., Bio-Rad P6 DG) has been reported as a method for removing excess reagents [21]. (DNA passes through the column material and lower-molecular-weight reagents and salts are trapped.)

4. The use of centrifugal filter units (e.g., Amicon Ultra with 3–50 kDa membranes) that can quickly desalt and purify large batches of DNA (>100 μmol).

4.1.2.2 Monitoring Progress of DNA-Conjugated Reactions by LC/MS.
Analytical techniques are detailed in Chapter 8. An example of a deconvoluted mass spectrum of a DNA-conjugated reaction is provided in Figure 6.7. DNA conjugates are readily distinguished from small-molecule reagents by mass spectrometry. Thus, crude reaction mixtures containing a large excess of reagent (even hundreds of equivalents) can still be analyzed for product formation without any workup. The analytical monitoring of DNA-conjugated reactions is generally limited to LC/MS.

4.1.3 Selected Examples of Methods for Reactions on DNA

A few examples of reaction procedures from Table 4.1 are given in the following to illustrate general techniques for conducting chemistry on DNA.

4.1.3.1 Amide Formation (Table 4.1, Entry 1). The DNA pellet is reconstituted after centrifugation in 250 mM borate pH 9.5 buffer to give a 1 mM DNA solution. To this solution is added 80 equiv. of a carboxylic acid (from a 200 mM *N,N*-Dimethylacetamide (DMA) stock solution), followed by 80 equiv. of 4-(4,6-Dimethoxy-1,3,5-Triazin-2-yl)-4-Methylmorpholinium Chloride (DMT-MM) (from a 250 mM aqueous solution). The acylation reactions may proceed for 18 h.

Troubleshooting

Reaction does not proceed to completion: If the pH of the reaction mixture drops below 8, then the reaction will not proceed. It is best to test the pH of the reaction following the addition of the DMT-MM coupling reagent by dispensing a very small amount (~0.5 μL) of the reaction mixture onto pH paper.

The carboxylic acid reagent appears to be insoluble in the aqueous reaction mixture: Many building blocks, particularly lipophilic Fmoc amino acids, will crash out of solution upon reaction with DMT-MM in water. The precipitate is presumed to be the activated ester that is formed. Many of these reactions will still proceed to completion even if the reaction mixture becomes a gel-like. Some lipophilic carboxylic acids do not give products with DMT-MM as coupling reagent. Alternative methods for acylation are provided in references [3–5].

4.1.3.2 Amide Formation (Table 4.1, Entry 2). The DNA pellet is reconstituted after centrifugation in 250 mM sodium phosphate pH 5.5 buffer to give a 1 mM DNA solution. To this solution are added 100 equiv. of amine building block (200 mM in DMA) and 100 equiv. of DMT-MM. If the pH of the reaction mixture has increased to 7, then it will need to be readjusted using 100 equiv. of HCl (200 mM in water). This reaction usually takes 18 h.

4.1.3.3 Nucleophilic Aromatic Substitution (Table 4.1, Entries 3a and 3b).
The DNA pellet (DNA conjugated to a reactive aliphatic amine) is reconstituted after centrifugation in 250 mM borate pH 9.5 buffer to give a 1 mM DNA solution. To this solution is added 40–80 equiv. of heteroaryl chloride. The reaction time and temperature depend upon the reactivity of the heteroaryl chloride. For example, cyanuric chloride is highly reactive and needs to be added at 4°C to limit the quenching of the first chloride by water, while the second and third amine substitutions in this heteroaryl system are accomplished at Room Temperature (RT) and 80°C, respectively. The success of this method is limited by the reactivity of the heteroaryl chloride. Relatively unreactive reagents such as 2-chloropyridine will yield little product even after 18 h heating at 80°C.

4.1.3.4 Fmoc Deprotection (Table 4.1, Entry 4).
The removal of the Fmoc group is facile and occurs in minutes in a 5–10% aqueous piperidine solution. However, the fluorenyl by-products of Fmoc removal will be generally insoluble in both water and ethanol. These by-products may be removed by centrifuging the aqueous solution and discarding any solids prior to ethanol precipitation of the DNA. When using centrifugal filtration to purify the reaction mixture after this chemistry step, the insoluble fluorenyl by-products have a tendency to clog the centrifugal filter units. It may be necessary to precipitate and reconstitute the DNA sample several times prior to purification by this method.

4.1.3.5 Boc Deprotection (Table 4.1, Entry 5).
The usual acidic conditions employed in traditional Boc deprotection are not compatible with the presence of DNA. Thermal removal of the Boc group in traditional organic synthesis is known [22] but less often reported. The stability of DNA in aqueous solution during prolonged heating allows for the removal of the Boc group to proceed smoothly. One such procedure is to heat a mixture of 1 nmol of Boc-protected DNA conjugate in 20 μL of buffer (37.5 mM sodium acetate buffer and 5 μL of 1 M magnesium chloride) at 70°C overnight. This can conveniently be done in a PCR tube in a PCR machine [7].

4.1.4 Employing Known DNA-Compatible Chemistries to Construct DNA-Encoded Libraries (DELs)

In this section, we describe how the lists of available building block classes and DNA-compatible chemistries have been combined to design and synthesize a diverse DEL. The numbers of commercially available monofunctional building blocks are provided in Chapter 5, Table 5.1. There is an abundance of available amine (6306), boronate (3246), carboxylic acid (2501), and aldehyde (1308) containing building blocks. Available difunctional building blocks are provided in Chapter 5, Table 5.2. Amino acids (896), diamines (644), diacids (264), and Boc amino acids (261) are the most prevalent. Some examples of library design schemes are provided in Chapter 5. Figure 5.1 in Chapter 5 illustrates how Fmoc amino acids may be joined to form dipeptides that can then be capped with carboxylic acid building blocks. Figure 5.1 in Chapter 5 also shows the use of cyanuric chloride with amine-containing building blocks to create a two-cycle branched DEL. Examples of trifunctional cores are shown in Chapter 5, Figure 5.2.

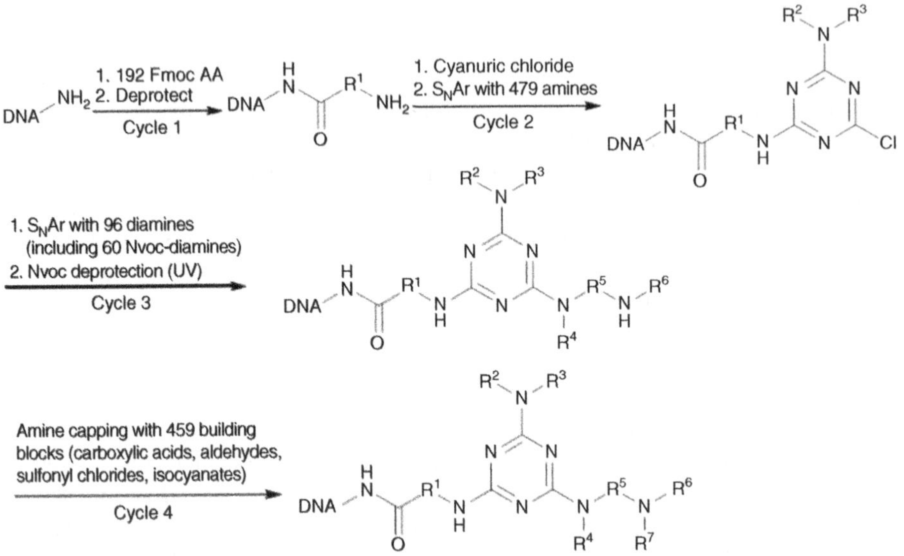

Figure 4.4. Example of a triazine-based DEL using some of the chemistries listed in Table 4.1 and commercially available building blocks. Here, for simplicity and to focus on the chemical reactions employed, the DNA-encoding steps are omitted. Please see original publication for details [13].

Figure 4.4 shows one DEL derived from the commercially available trifunctional scaffold, cyanuric chloride, and a subset of commercially available building blocks [13]. The library in Figure 4.4 is a four-cycle library employing a set of bifunctional building blocks (Fmoc amino acids) at cycle 1. Cycle 1 acylation is followed by a S_NAr reaction with a core that has three reactive groups (cyanuric chloride): the first chloride to react with the amino group from the amino acid, the second with another anime (cycle 2 chemistry), and the third with a diamine (cycle 3 chemistry). The deprotected diamines are then reacted with various reagents to form amides, ureas, sulfonamides, and alkyl amines via known chemistry (cycle 4 chemistry).

We abbreviate the DNA bioconjugate starting material ending in an amino group as DNA-NH$_2$ (Fig. 4.4). For simplicity and to focus on the chemical reactions employed, the DNA-encoding steps are omitted in this discussion and are implied. In addition, after each step, the DNA-containing material is recovered by cold ethanol-induced precipitation and centrifugation, discarding the supernatant material. After dissolving the pellet in the appropriate buffer solution, this material is divided into the appropriate number of wells for the next step. For each cycle, the DNA conjugate is divided into a number of wells greater than the number of building blocks planned for that cycle, usually one extra. The extra wells are treated together with the rest of the wells under identical conditions with the exception that no building blocks are added to these wells (blank wells). These blank wells may be used to later identify products where incomplete reaction occurred.

Deng et al. [13] acylated the DNA bioconjugate with 192 Fmoc-protected amino acids in individual wells (40 equiv. of amino acids and 40 equiv. of DMT-MM); the additions of both reagents were repeated after 2 h to push the acylation reaction to completion. The reaction was performed at RT overnight. After pooling and collecting the DNA-containing product, this material was purified by reverse-phased HPLC. The Fmoc-protecting groups were then removed by treatment with 10% aqueous piperidine solution to complete the first cycle of chemistry. After splitting into separate wells, intermediates were treated with 10 equiv. of cyanuric chloride for the first S_NAr reaction. After 1 h, the reaction mixtures were cooled to 4°C and each well treated with 50 equiv. of 1 of 479 amines for 40 h to complete the second cycle of chemistry.

Following pooling and splitting at the end of the second cycle, for the third cycle of chemistry, the individual wells were treated with 50 equiv. of 1 of 60 Nvoc-protected diamines, 20 symmetrical diamines (such as piperazine), and 16 unprotected aromatic–aliphatic diamines (such as 4-aminobenzylamine) at 80°C for 8 h. After standard workup to collect the DNA-containing material, the product mixture was dissolved in 100 mM aqueous acetic acid and irradiated with UV light for 16 h to remove the Nvoc-protecting group followed by HPLC purification. The purified material was then divided into 459 wells, which were used in four different capping reactions for the fourth cycle of chemistry. The first set of wells of these third-cycle products was treated with 40 equiv. of 1 of 173 carboxylic acids and DMT-MM (a second portion of both reagents was added after 2 h). The second set of wells was treated with 40 equiv. of 1 of 85 isocyanates or 1 of 107 sulfonyl chlorides at RT overnight to form ureas or sulfonamides. The third set of wells was treated with 100 equiv. of 1 of 94 aldehydes followed by 100 equiv. of NaCNBH$_3$ to give the N-alkylated product. All the fourth-cycle DNA-containing products were precipitated, combined, and purified by HPLC followed by the standard library-closing activities (see Chapter 6) to complete the DEL synthesis. Using these sets of building blocks and this reaction sequence in the four cycles, Deng *et al.* produced a DEL with four billion components.

As discussed in 4.1.4, the reactions provided in Table 4.1, coupled with commercially available building blocks, easily allow for the production of tens of millions to billions of diverse small molecules with reasonable molecular weights for drug discovery.

4.2 REACTIONS THAT HAVE POTENTIAL FOR USE ON DNA

One of the keys to the synthesis of DELs is the development of chemical reactions compatible with the presence of DNA oligomers and water. A number of reactions that have been used successfully in the synthesis of DELs have already been discussed in the previous section. Unfortunately, many common synthetic reactions lead to degradation of the DNA-encoding region, are incompatible with aqueous conditions, or require harsh reaction conditions such as pH < 3. Although not many reactions have been demonstrated or claimed in patents to be compatible with DNA, a significant number of reactions that can be performed with water as solvent or cosolvent under relatively mild conditions are known in the literature. These reactions have a chance of being optimized to work in the presence of DNA and thus could be useful in DEL synthesis. A good starting point is to look at reactions that can work in the presence of water. A number of recent reviews of organic reactions in water have been published [23–27]. The following is by no means a comprehensive review of such reactions. However, a few reactions have been chosen as examples to illustrate the potential of water-compatible procedures. Obviously, other reactions could be considered. Synthetic chemists should be able to design DELs by mixing and matching reaction sequences with appropriate building blocks to generate libraries of interesting molecules. Three examples of such theoretical library designs are discussed. The design of actual libraries will be left to the imagination of the reader. Whether the reactions described in this chapter can be used successfully in a DEL synthesis is left to the practitioner to perform the tasks of validating and optimizing such reaction conditions.

4.2.1 Examples of Potential Libraries

As in the previous section, in the following reaction schemes, "DNA-NH$_2$," represents the DNA piece encoding the library members. It is understood that the necessary linkers and sequential ligation of DNA to encode each subsequent cycle of reaction in the library building are included and not shown explicitly. LC/MS should routinely be used to demonstrate that the appropriate reactions have occurred. The use of LC/MS for this purpose will be discussed in Chapter 8 of this book. Also in putting together a sequence of reaction steps, there is no requirement that all the reactions are novel. Once a reaction has been demonstrated to work in the presence of DNA, it can easily be combined with well-known reactions that have already been used in earlier DEL synthesis in the design of new DELs. The three examples discussed in the following illustrate this point.

4.2.1.1 Potential Library 1 Yan et al. have demonstrated that gold(III) chloride catalyzes a three-component (amine, aldehyde, and terminal alkyne) reaction with water as solvent [28]. Furthermore, Rozenman et al. have shown that gold(III)-catalyzed reactions can be done in the presence of DNA [14]. These two results together suggest that this reaction could be a good candidate for incorporation into a library design (Fig. 4.5).

In this specific example, one can envision a four-cycle library by first coupling Fmoc-protected amino acids (Fmoc AA) to the DNA piece, followed by the removal of the Fmoc-protecting group. One of the reaction wells may contain no protected amino acid, as the amino group of DNA-NH$_2$ can be used to react with aldehydes, and the resulting DEL would then include a sublibrary where the first cycle reaction is omitted by design. These two reactions are well documented in the literature and mentioned in the previous section. The condensation of aryl aldehydes with amines to form imines in the presence of DNA is also well known and discussed in the aforementioned section. Thus, by the use of appropriately substituted 2-formyl-pyridines (X = CH) or 2-formyl-pyrimidines (X = N) to form the necessary imines, the stage is set for the gold(III)-catalyzed aminoindolizine formation. If this reaction is successful, and the appropriately substituted terminal alkynes are used, additional cycles can be incorporated. For example, if the alkyne contained an ester function (Y = ester), it could be hydrolyzed to the free acid, which could then be further diversified by the formation of amides. On the other hand, if the alkyne contained a Boc-protected

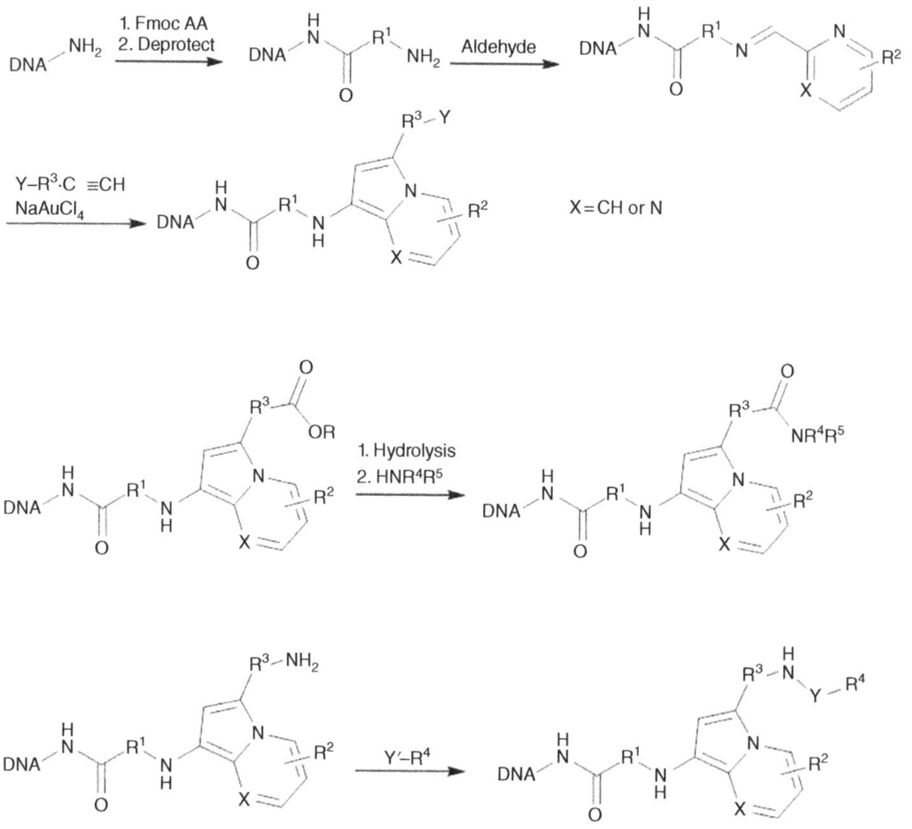

Figure 4.5. Example of potential library 1 (aminoindolizine library).

amine (Y=N-Boc), it could be deprotected to give the free amine. These amines could then be divided into four or more parts and each reacted with a different set of building blocks to form a part of the overall library [13]. The functionalization of amines attached to DNA is well known and discussed earlier. For example, the conversion of such DNA-linked amines to amides, ureas, sulfonamides, and alkyl amines (via reductive amination) is possible [13]. The number of commercially available, properly functionalized terminal alkynes may not be high, but employing custom synthesis of a number of well-designed building blocks should be well worth the effort.

Potentially, this theoretical library could possess a significant number of members as shown in Table 4.2. For example, one can assemble a set of 200 Fmoc-protected amino acids for the first cycle, 100 aldehydes for the second cycle, and 100 alkynes (50 with ester and 50 with Boc-protected amine) for the third cycle. After the first three cycles of chemistry and ligation of the appropriate DNA sequences, the samples reacting with alkynes bearing ester groups could be pooled for use as part A of cycle four, while the samples reacting with alkynes bearing Boc amines could be pooled separately

TABLE 4.2. Summary of potential library 1

Number of building blocks				Number of library members
First cycle	Second cycle	Third cycle	Fourth cycle	
200 amino acids	100 aldehydes	50 alkyne–ester	10,000 amines (part A)	10 billion
200 amino acids	100 aldehydes	50 alkyne–amine	10,000 acids (part B)	10 billion
200 amino acids	100 aldehydes	50 alkyne–amine	5000 aldehydes (part C)	5 billion
200 amino acids	100 aldehydes	50 alkyne–amine	300 sulfonyl chlorides (part D)	0.3 billion
200 amino acids	100 aldehydes	50 alkyne–amine	100 isocyanates (part E)	0.1 billion
Library grand total				25.4 billion

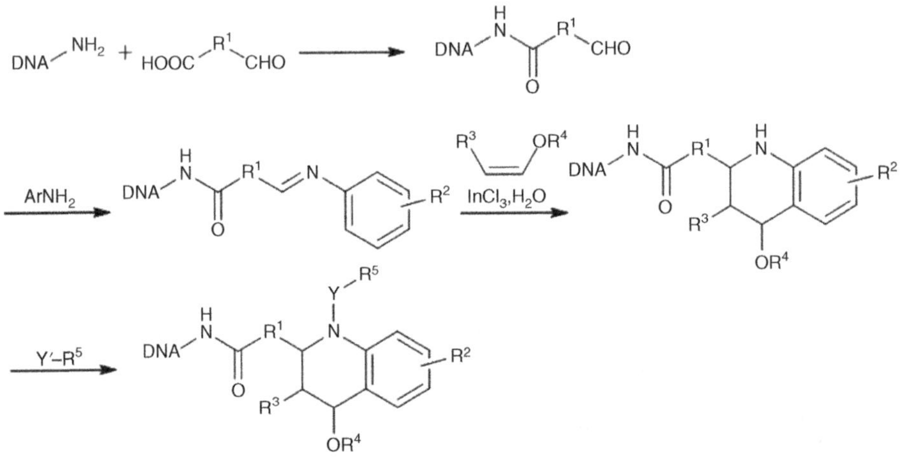

Figure 4.6. Example of potential library 2 (aza-Diels–Alder tetrahydroquinoline library).

and divided into four parts (B–E). Part A (ester) could be hydrolyzed to the free acid and then reacted with 10,000 amines. The other parts from the deprotected Boc amine could be used for reactions with 10,000 acids (part B), 5000 aldehydes (part C), 300 sulfonyl chlorides (part D), and 100 isocyanates (part E) to complete the fourth cycle of the library. Employing these sets of building blocks and four cycles of chemistry, the total number of compounds in this library would be over 25 billion.

4.2.1.2 Potential Library 2 In a second example of a potential library (Fig. 4.6), the selected reaction is the indium(III) chloride-catalyzed aza-Diels–Alder reaction in water [29, 30]. Here, the stereochemistry around the newly formed ring is expected to be *cis* for R^3 and OR^4, while R^1 will be a mixture of both *syn* and *anti* to R^3 and OR^4. The key here is to demonstrate that indium(III) chloride-catalyzed reactions are compatible with DNA. If the indium(III) chloride reaction does cause damage to DNA, there are probably other Lewis acids that are compatible with on-DNA reactions that can catalyze this reaction. Therefore, a screen for potential catalysts for this reaction is in order.

TABLE 4.3. Summary of potential library 2

Number of building blocks				Total library members
First cycle	Second cycle	Third cycle	Fourth cycle	
200 FA	200 anilines	50 enol ethers	10,000 acids (part A)	20 billion
200 FA	200 anilines	50 enol ethers	300 sulfonyl chlorides (part B)	0.6 billion
Library grand total				20.6 billion

Here, again, combining reactions that are known to work in the presence of DNA together with this potential reaction, we can formulate an interesting library. The acylation of amines and imine formation on DNA are well known. With these two as the first two cycles of this library, the stage is set to employ the aza-Diels–Alder reaction with the appropriately functionalized enol ether. Like the first example earlier, the library can be capped by derivatizing the resultant amine in four parts to complete the library. A reasonable selection of building blocks for this library could easily be 200 Formyl Acids (FA) for the first cycle, 200 anilines for the second cycle, and 50 enol ethers for the third cycle. Since the product after the third cycle will be a collection of anilines, we can use two reactions from the four-part capping strategy discussed in potential library 1 earlier to complete this library. (Reductive amination and reaction of isocyanates with substituted anilines may not work well, so these two sets are not included.) With this selection of building blocks, the total number of this library would be over 20 billion members (see Table 4.3).

4.2.1.3 Potential Library 3 In a third example (Fig. 4.7), we propose to use an aqueous-based aldol reaction [31] together with a Michael reaction followed by intramolecular condensation to form benzothiazepines [32]. The ability to perform aldol reactions with water as solvent is well known. Michael addition of a thiol to an enone on DNA is also known [9, 33]. Intramolecular condensation to form the thiazepine ring should be possible [34]. The reduction of the resulting imine with borohydride to give the saturated ring with conditions identical to reductive aminations on DNA is also possible [35] with R^1 and Ar being *trans*. As in the example earlier, this library can be capped by derivatizing the resultant aniline after three cycles and reduction in two parts as in example 2 earlier to complete the library. Table 4.4 shows a possible selection of building blocks that could include 100 Acetyl Acids (AcA) for the first cycle, 500 aldehydes for the second cycle, 50 1,2-aminothiols for the third cycle, and the same set for the fourth cycle as in library 2 earlier to give a library of over 25 billion members.

4.2.2 Additional Potential Reactions

Table 4.5 illustrates a few examples of additional potential reactions that could be developed for on-DNA chemistry. As illustrated in the aforementioned three examples, these reactions can be linked with other known DNA-compatible chemistry to

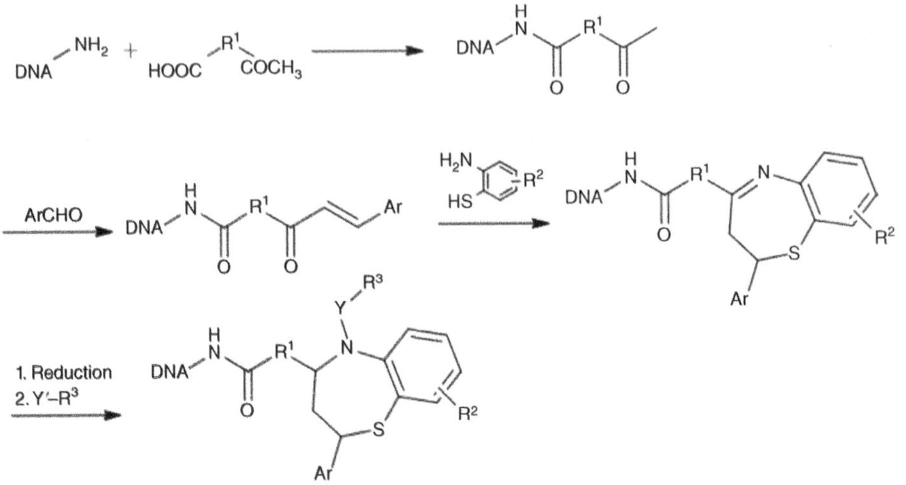

Figure 4.7. Example of potential library 3 (benzothiazepine library).

TABLE 4.4. Summary of potential library 3

Number of building blocks				Total library members
First cycle	Second cycle	Third cycle	Fourth cycle	
100 AcA	500 aldehydes	50 aminothiols	10,000 acids (part A)	25 billion
100 AcA	500 aldehydes	50 aminothiols	300 sulfonyl chlorides (part B)	0.75 billion
Library grand total				25.75 billion

make interesting libraries. Again, the focus is on reactions that can be performed in the presence of water with reagents, catalysts, or conditions similar to those that are known to work in the presence of DNA. Before these reactions can be used successfully in a DEL synthesis, it must be shown that they work in the presence of DNA in a solvent system containing water, with broad reagent scope and followed by optimizing the reaction conditions. While a reaction is being demonstrated to work in this context, one should also pay attention to the availability of the desired building blocks to go with the reaction sequence. In cases where there are few commercially available building blocks, it may be profitable to custom design and synthesize a number of such reagents. With a good collection of building blocks and an ever-increasing number of successful reactions, innovative chemists will be able to synthesize many very interesting DELs.

TABLE 4.5. Reactions that have potential of working in the presence of DNA

	Reaction	Scheme	Conditions	Comments
1.	Robinson annulation [36]		Tb-PEG-Li 50°C, 48 h THF, H$_2$O (6 equiv.)	
2.	Proline Mannich reaction [37]		The reaction is usually done in 1:9 H$_2$O–THF. In at least one example, high ee is obtained in a 1:1 H$_2$O–THF solvent system	The effect of DNA on ee is unknown. For extra diversity in library, one can always use racemic proline as catalyst
3.	Addition to alkylidene Meldrum acid [38]		Cu(OAc)$_2$, Na-Ascorbate H$_2$O, tBuOH (10:1), rt	
4.	Addition to activated alkynes to form enynes [39]		CuBr, PdCl$_2$(PPh$_3$)$_2$ H$_2$O, rt, 48 h	EWG = ketone or ester

(Continued)

TABLE 4.5. (cont'd)

	Reaction	Scheme	Conditions	Comments
5.	1,4-Addition to ynones [40]	R^1–≡–H + (enone with R^2) →($Pd(OAc)_2$, PMe_3, H_2O, 60°C, 40h) 50–90% → product with R^1, R^2		Acetone can also be used as solvent with slightly lower yield
6.	Formation of chalcone [41, 42]	Ar—Hal + (alkyne with OH, Ar') →([Pd^0, Cu^1]) product (Ar, Ar' enone)	$Pd(Ph_3P)_2Cl_2$, CuI, Et_3N, THF, 10h reflux [45]	
7.	Addition of hydrazine to enones [43–45]	(enone Ar, Ar') + $RNHNH_2$ → pyrazoline (R, N–N, Ar, Ar')	NaOAc, EtOH, H_2O, 9h reflux [45]	Depending on reactivity, lower temperature and shorter time are possible
8.	Formation of pyrimidine from enones [46, 47]	(enone Ar^1, Ar^2) + H_2N–C(=$N^+H_2X^-$)–Ar^3 → pyrimidine (Ar^1, Ar^2, Ar^3)	NaOH, EtOH, H_2O, 2h reflux [47]	
9.	Formation of benzohetero-azepine from enones [32, 48]	(enone Ar^1, Ar^2) + (benzene with HX and H_2N) → benzohetero-azepine (X, Ar^1, Ar^2)	Piperidine, EtOH, 10h reflux [48]	X=N, O, S

#	Reaction	Scheme	Notes	Comments
10.	Enone condensation with boronic acids [49]		In a direct comparison of running the reaction in dichloroethane versus in water–DMA (1:1), similar yields were obtained with both solvent systems. With the aqueous system, the reaction was complete in less than 5 h at 60°C, while in dichloroethane, the reaction took overnight at the same temperature to complete (Luk, K. unpublished result)	Kikushima *et al.* "The reaction displays a remarkable tolerance to water and oxygen, and reactions are typically performed under an atmosphere of air in screw-top vials" [49]
11.	Aldehyde–alkyne–amine condensation [50, 51]		10 mol% Cu(I)OTf and 10 mol% pybox ligand in H_2O at 35°C for 2 days. The use of chiral pybox ligand resulted in 78–96% ee	The effect of DNA on ee is unknown
12.	Hantzsch dihydropyridine synthesis [52]		Aldehyde (10 mmol), ethyl acetoacetate (22 mmol), EtOH (10 mL), 28% aq. NH_3 (12 mmol) in sealed tube at 110°C, 17–45 h	When optimized on DNA, such high temperature may not be necessary
13.	*N*-Aminopyrrolidine-2,5-dione synthesis [53]			The imine needed for this reaction could easily be generated as the first cycle of the library

(*Continued*)

TABLE 4.5. (cont'd)

Reaction	Scheme	Conditions	Comments
14. Aza-Diels–Alder reaction [54]		Imine (1 mmol), diene (1–3 mmol), Cu(OTf)$_2$ (1 mol%), sodium dodecyl sulfate (SDS) (1 mol%), in H$_2$O (2 mL) at 30°C for 10 min	The imine needed for this reaction could easily be generated as the first cycle of the library
15. Nitrile oxide cycloaddition [55]		Hydroximoyl chloride (2.5 equiv.) added to mixture of alkene and NaHCO$_3$ in aqueous EtOAc at 0°C and then stirred at RT for 3 h	The two building blocks for this reaction are not that common, but diverse sets of both can be generated by simple processes
16. S$_N$Ar [56]		40% methylamine solution, 0°C, 1.5 h (99%)	For amines that are less soluble in H$_2$O, a cosolvent can be used
17. Aqueous olefin metathesis [57–59]		HGII = Hoveyda–Grubbs II catalyst	The possibility of using metathesis in library should be considered for olefin-containing building blocks

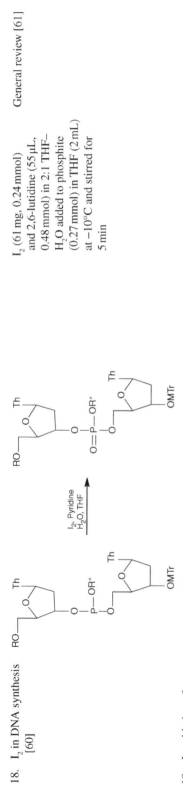

18.	I_2 in DNA synthesis [60]	I_2 (61 mg, 0.24 mmol) and 2,6-lutidine (55 μL, 0.48 mmol) in 2:1 THF–H_2O added to phosphite (0.27 mmol) in THF (2 mL) at −10°C and stirred for 5 min	General review [61]
19.	I_2 oxidation of dihydropyridine to pyridine [62]	The solutions of I_2 (2 mmol) and KOH (each 2 mL) in MeOH (4 mmol) were successively added to a solution of dihydropyridine (2 mmol) in MeOH (10 mL) at 0°C and stirred for 15–35 min	
20.	I_2 oxidation of benzimidazoline to benzimidazole [63]	Aldehyde (1 mmol) and diamine (1 mmol) added to solution of I_2 (4 mL, THF–H_2O, 1:1) and stirred at RT for 1.5 h	
21.	Nitro reduction to amine [64, 65]		Catalytic hydrogenation

(Continued)

TABLE 4.5. (cont'd)

	Reaction	Scheme	Conditions	Comments
22.	Alkene reduction [66–68]			Catalytic hydrogenation
23.	Alkyne reduction [69]			Catalytic hydrogenation
24.	Transfer hydrogenation: azide reduction [70]			

REFERENCES

1. Rozenman, M. M., Liu, D. R. (2006). DNA-templated synthesis in organic solvents. *ChemBioChem.*, *7*, 253–256.

2. Li, X., Liu, D. R. (2004). DNA-templated organic synthesis: nature's strategy for controlling chemical reactivity applied to synthetic molecules. *Angew. Chem. Int. Ed.*, *43*, 4848–4870.

3. He, Y., Liu, D. R. (2011). A sequential strand-displacement strategy enables efficient six-step DNA-templated synthesis. *J. Am. Chem. Soc.*, *133*, 9972–9975.

4. Wrenn, S. J., Weisinger, R. M., Halpin, D. R., Harbury, P. B. (2007). Synthetic ligands discovered by in vitro selection. *J. Am. Chem. Soc.*, *129*, 13137–13143.

5. Gartner, Z. J., Kanan, M. W., Liu, D. R. (2002). Expanding the reaction scope of DNA-templated synthesis. *Angew. Chem. Int. Ed.*, *41*, 1796–1800.

6. Clark, M. A., Acharya, R. A., Arico-Muendel, C. C., Belyanskaya, S. L., Benjamin, D. R., Carlson, N. R., Centrella, P. A., Chiu, C. H., Creaser, S. P., Cuozzo, J. W., Davie, C. P., Ding, Y., Franklin, G. J., Franzen, K. D., Gefter, M. L., Hale, S. P., Hansen, N. J. V., Israel, D. I., Jiang, J., Kavarana, M. J., Kelley, M. S., Kollmann, C. S., Li, F., Lind, K., Mataruse, S., Medeiros, P. F., Messer, J. A., Myers, P., O'Keefe, H., Oliff, M. C., Rise, C. E., Satz, A. L., Skinner, S. R., Svendsen, J. L., Tang, L., van Vloten, K., Wagner, R. W., Yao, G., Zhao, B., Morgan, B. A. (2009). Design, synthesis and selection of DNA-encoded small-molecule libraries. *Nat. Chem. Biol.*, *5*, 647–654.

7. Franch, T., Lundorf, M. D., Jacobsen, S. N., Olsen, E. K., Andersen, A. L., Holtmann, A., Hansen, A. H., Sorensen, A. M., Goldbech, A., De Leon, D., Kaldor, D. K., Slok, F. A., Husemoen, G. N., Dolberg, J., Jensen, K. B., Petersen, L., Nørregaard-Madsen, M., Godskesen, M. A., Glad, S. S., Neve, S., Thisted, T., Kronborg, T. T. A., Sams, C., Felding, J., Freskgaard, P. -O., Gouliaev, A. H., Pedersen, H. (2007). Enzymatic encoding methods for efficient synthesis of large libraries. Patent Application WO 2007/062664 A2.

8. Gouliaev, A. H., Franch, T., Godskesen, M. A., Jensen, K. B. (2012). Bi-functional complexes and methods for making and using such complexes. Patent Application WO 2011/127933 A1.

9. Franch, T., Lundorf, M. D., Jakobsen, S. N., Olsen, E. K., Andersen, A. L., Holtmann, A., Hansen, A. H., Sørensen, A. M., Goldbech, A., Kaldor, D. K., Sløk, F. A., Husemoen, B. N., Dolberg, J., Jensen, K. B., Petersen, L., Nørregaard-Madsen, M., Godskesen, M. A., Glad, S. S., Neve, S., Thisted, T., Kronberg, T. T., Sams, C., Felding, J., Freskgård, P., Gouliaev, A. H., Pedersen, H. (2011). Enzymatic encoding methods for efficient synthesis of large libraries. European Patent Application EP2336315.

10. Gartner, Z. J., Tse, B. N., Grubina, R., Doyon, J. B., Snyder, T. M., Liu, D. R. (2004). DNA-templated organic synthesis and selection of a library of macrocycles. *Science*, *305*, 1601–1605.

11. Chen, Y., Kamlet, A. S., Steinman, J. B., Liu, D. R. (2011). A biomolecule-compatible visible-light-induced azide reduction from a DNA-encoded reaction-discovery system. *Nat. Chem.*, *3*, 146–153.

12. Rasmussen, P. B. (2012). Template directed split and mix synthesis of small molecule libraries. US Patent US 8,168,381, May 1, 2012.

13. Deng, H., O'Keefe, H., Davie, C. P., Lind, K. E., Acharya, R. A., Franklin, G. J., Larkin, J., Matico, R., Neeb, M., Thompson, M. M., Lohr, T., Gross, J. W., Centrella, P. A., O'Donovan, G. K., Bedard (Sargent), K. L., van Vloten, K., Mataruse, S., Skinner, S. R., Belyanskaya, S. L., Carpenter, T. Y., Shearer, T. W., Clark, M. A., Cuozzo, J. W., Arico-Muendel, C. C.,

Morgan, B. A. (2012). Discovery of highly potent and selective small molecule ADAMTS-5 inhibitors that inhibit human cartilage degradation via encoded library technology (ELT). *J. Med. Chem.*, *55*, 7061–7079.

14. Rozenman, M. M., Kanan, M. W., Liu, D. R. (2007). Development and initial application of a hybridization-independent, DNA-encoded reaction discovery system compatible with organic solvents. *J. Am. Chem. Soc.*, *129*, 14933–14938.

15. Buller, F., Mannocci, L., Zhang, Y., Dumelin, C. E., Scheuermann, J., Neri, D. (2008). Design and synthesis of a novel DNA-encoded chemical library using Diels-Alder cycloadditions. *Bioorg. Med. Chem. Lett.*, *18*, 5926–5931.

16. Buller, F., Steiner, M., Frey, K., Mircsof, D., Scheuermann, J., Kalisch, M., Bühlmann, P., Supuran, C. T., Neri, D. (2011). Selection of carbonic anhydrase IX inhibitors from one million DNA-encoded compounds. *ACS Chem. Biol.*, *6*, 336–344.

17. Chen, X. -H., Roloff, A., Seitz, O. (2012). Consecutive signal amplification for DNA detection based on de novo fluorophore synthesis and host-guest chemistry. *Angew. Chem. Int. Ed.*, *51*, 4479–4483.

18. Gartner, Z. J., Liu, D. R. (2001). The generality of DNA-templated synthesis as a basis for evolving non-natural small molecules. *J. Am. Chem. Soc.*, *123*, 6961–6963.

19. Li, X., Gartner, Z. J., Tse, B. N., Liu, D. R. (2004). Translation of DNA into synthetic N-acyloxazolidines. *J. Am. Chem. Soc.*, *126*, 5090–5092.

20. Kanan, M. W., Rozenman, M. M., Sakurai, K., Snyder, T. M., Liu, D. R., (2004). Reaction discovery enabled by DNA-templated synthesis and in vitro selection. *Nature*, *431*, 545–549.

21. Freskgård, P. -O., Gouliaev, A. H., Thisted, T., Olsen, E. K. (2004). Method for producing second-generation library. Patent Application WO 2004/074429.

22. Wasserman, H. H., Berger, G. D., Cho, K. R. (1982). Transamidation reactions using β-lactams. The synthesis of homaline. *Tetrahedron Lett.*, *23*, 465–468.

23. Li, C. -J. (2005). Organic reactions in aqueous media with a focus on carbon-carbon bond formations: a decade update. *Chem. Rev.*, *105*, 3095–3165.

24. Lipshutz, B. H., Ghorai, S. (2008). Transition-metal-catalyzed cross-couplings going green: in water at room temperature. *Aldrichim. Acta*, *41*, 59–72.

25. Chanda, A., Fokin, V. V. (2009). Organic synthesis "on water". *Chem. Rev.*, *109*, 725–748.

26. Raj, M., Singh, V. K. (2009). Organocatalytic reactions in water. *Chem. Commun.*, *2009*, 6687–6703.

27. Lombardo, M., Trombini, C. (2010). Catalysis in aqueous media for the synthesis of drug-like molecules. *Curr. Opin. Drug Discov. Devel.*, *13*, 717–732.

28. Yan, B., Liu, Y. (2007). Gold-catalyzed multicomponent synthesis of aminoindolizines from aldehydes, amines, and alkynes under solvent-free conditions or in water. *Org. Lett.*, *9*, 4323–4326.

29. Li, Z., Zhang, J., Li, C.-J. (2003). InCl$_3$-catalyzed reaction of aromatic amines with cyclic hemiacetals in water: facile synthesis of 1,2,3,4-tetrahydroquinoline derivatives. *Tetrahedron Lett.*, *44*, 153–156.

30. Glushkov, V. A., Tolstikov, A. G. (2008). Synthesis of substituted 1,2,3,4-tetrahydroquinones by the Povarov reaction. New potentials of the classical reaction. *Russ. Chem. Rev.*, *77*, 137–159.

31. Gilman, H., Blatt, A. H. (1941). *Organic Syntheses, Collective*, Volume *I*, 2nd Edition, John Wiley & Sons, Inc. New York, pp. 77–80.

32. Braun, R. U., Zeitler, K., Müller, T. J. J. (2000). A novel 1,5-benzoheteroazepine synthesis via a one-pot coupling-isomerization-cyclocondensation sequence. *Org. Lett.*, *2*, 4181–4184.

33. Ranu, B. C., Mandal, T. (2007). A simple, efficient, and green procedure for the 1,4-addition of thiols to conjugated alkenes and alkynes catalyzed by sodium acetate in aqueous medium. *Aust. J. Chem.*, *60*, 223–227.

34. Sharma, G., Kumar, R., Chakraborti, A. K. (2008). 'On water' synthesis of 2,4-diaryl-2,3-dihydro-1,5-benzothiazepines catalysed by sodium dodecyl sulfate (SDS). *Tetrahedron Lett.*, *49*, 4269–4271.

35. Xu, J., Jin, S., Xing, Q. (1998). Cycloaddition reaction of benzoheteroazepine: synthesis of 4a,5,6,12-tetrahydro-1H-1,3-oxazino[3,2-d][1,5] benzothiazepin-1-ones and 1H,7H-1,3-oxazino[3,2-d][1,5]benzodiazepin-1-ones. *Phosphorus Sulfur*, *141*, 57–70.

36. Kamaura, M., Daikai, K., Hanamoto, T., Inanaga, J. (1998). Preparation and utility of novel water-tolerant higher-order lanthanoid alkoxide complexes as a base catalyst. *Chem. Lett.*, *27*, 697–698.

37. Córdova, A., Barbas, C. F. III. (2003). Direct organocatalytic asymmetric Mannich-type reactions in aqueous media: one-pot Mannich-allylation reactions. *Tetrahedron Lett.*, *44*, 1923–1926.

38. Knöpfel, T. F., Carreira, E. M. (2003). The first conjugate addition reaction of terminal alkynes catalytic in copper: conjugate addition of alkynes in water. *J. Am. Chem. Soc.*, *125*, 6054–6055.

39. Chen, L., Li, C. -J. (2004). Facile and selective copper–palladium catalyzed addition of terminal alkynes to activated alkynes in water. *Tetrahedron Lett.*, *45*, 2771–2774.

40. Chen, L., Li, C. -J. (2004). The first palladium-catalyzed 1,4-addition of terminal alkynes to conjugated enones. *Chem. Commun.*, 2362–2364.

41. Müller, T. J. J., Ansorge, M., Aktah, D. (2000). An unexpected coupling-isomerization sequence as an entry to novel three-component-pyrazoline syntheses. *Angew. Chem. Int. Ed.*, *39*, 1253–1256.

42. Review: Müller, T. J. J. (2012). Synthesis of carbo- and heterocycles via coupling-isomerization reactions. *Synthesis*, *44*, 159–174.

43. Rivett, D. E., Rosevear, J., Wilshire, J. F. K. (1979). The preparation and spectral properties of some monosubstituted 1,3,5-triphenyl-2-pyrazolines. *Aust. J. Chem.*, *32*, 1601–1612.

44. Haunert, F., Bolli, M. H., Hinzen, B., Ley, S. V. (1998). Clean three-step synthesis of 4,5-dihydro-1H-pyrazoles starting from alcohols using polymer supported reagents. *J. Chem. Soc. Perk. Trans.*, *1*, 2235–2238.

45. von Auwers, K., Kreuder, A. (1925). Influence of constitution on the transformation of phenylhydrazones of unsaturated compounds into pyrazolines. II. *Ber. Dtsch. Chem. Ges.*, *58B*, 1974–1986.

46. Müller, T. J. J., Braun, R., Ansorge, M. (2000). A novel three-component one-pot pyrimidine synthesis based upon a coupling-isomerization sequence. *Org. Lett.*, *2*, 1967–1970.

47. Hicks, R. G., Koivisto, B. D., Lemaire, M. T. (2004). Synthesis of multitopic verdazyl radical ligands. Paramagnetic supramolecular synthons. *Org. Lett.*, *6*, 1887–1890.

48. Baktir, Z., Akkurt, M., Samshuddin, S., Narayana, B., Yathirajan, H. S. (2011). 2,4-Bis (4-fluorophenyl)-2,3-dihydro-1H-1,5-benzodiazepine. *Acta Crystallogr.*, *E67*, o1262–o1263.

49. Kikushima, K., Holder, J. C., Gatti, M., Stoltz, B. M. (2011). Palladium-catalyzed asymmetric conjugate addition of arylboronic acids to five-, six-, and seven-membered β-substituted cyclic enones: enantioselective construction of all-carbon quaternary stereocenters. *J. Am. Chem. Soc.*, *133*, 6902–6905.

50. Wei, C., Li, C.-J. (2002). Enantioselective direct-addition of terminal alkynes to imines catalyzed by copper(I)pybox complex in water and in toluene. *J. Am. Chem. Soc.*, *124*, 5638–5639.

51. Wei, C., Mague, J. T., Li C. -J. (2004). Cu(I)-catalyzed direct addition and asymmetric addition of terminal alkynes to imines. *Proc. Natl. Acad. Sci. U.S.A.*, *101*, 5749–5754.

52. Watanabe, Y., Shiota, K., Hoshiko, T., Ozaki, S. (1983). An efficient procedure for the Hantzsch dihydropyridine synthesis. *Synthesis*, 761.

53. Adib, M., Ansari, S., Bijanzadeh, H. R. (2011). Reaction between N-isocyanimino-triphenylphosphorane, aldimines, Meldrum's aid and water: diastereoselective synthesis of 3,4-disubstituted N-aminopyrrolidine-2,5-diones. *Synlett*, 619–622.

54. Lanari, D., Piermatti, O., Pizzo, F., Vaccaro, L. (2012). Copper(II) triflate-sodium dodecyl sulfate catalyzed preparation of 1,2-diphenyl-2,3-dihydro-4-pyridones in aqueous acidic medium. *Synthesis*, *44*, 2181–2184.

55. Beattie, N. J., Francis, C. L., Liepa, A. J., Savage, G. P. (2010). Spiroheterocycles via regioselective cycloaddition reactions of nitrile oxides with 5-methylene-1H-pyrrol-2(5h)-ones. *Aust. J. Chem.*, *63*, 445–451.

56. Okaniwa, M., Hirose, M., Imada, T., Ohashi, T., Hayashi, Y., Miyazaki, T., Arita, T., Yabuki, M., Kakoi, K., Kato, J., Takagi, T., Kawamoto, T., Yao, S., Sumita, A., Tsutsumi, S., Tottori, T., Oki, H., Sang, B. -C., Yano, J., Aertgeerts, K., Yoshida, S., Ishikawa, T. (2012). Design and synthesis of novel DFG-Out RAF/Vascular Endothelial Growth Factor Receptor 2 (VEGFR2) inhibitors. 1. Exploration of [5,6]-fused bicyclic scaffolds. *J. Med. Chem.*, *55*, 3452–3478.

57. Connon, S. J., Rivard, M., Zaja, M., Blechert, S. (2003). Practical olefin metathesis in protic media under an air atmosphere. *Adv. Synth. Catal.*, *345*, 572–575.

58. Binder, J. B., Blank, J. J., Raines, R. T. (2007). Olefin metathesis in homogeneous aqueous media catalyzed by conventional ruthenium catalysts. *Org. Lett.*, *9*, 4885–4888.

59. Burtscher, D., Grela, K. (2009). Aqueous olefin metathesis. *Angew. Chem. Int. Ed.*, *48*, 442–454.

60. Letsinger, R. L., Lunsford, W. B. (1976). Synthesis of thymidine oligonucleotides by phosphite triester intermediates. *J. Am. Chem. Soc.*, *98*, 3655–3661.

61. General review: Togo, H., Iida, S. (2006). Synthetic use of molecular iodine for organic synthesis. *Synlett*, 2159–2175.

62. Yadav, J. S., Reddy, B. V. S., Sabitha, G., Reddy, G. S. K. K. (2000). Aromatization of Hantzsch 1,4-dihydropyridines with I_2-MeOH. *Synthesis*, 1532–1534.

63. Sun, P., Hu, Z. (2006). The convenient synthesis of benzimidazole derivatives catalyzed by I2 in aqueous media. *J. Heterocycl. Chem.*, *43*, 773–775.

64. Gilman, H., Blatt, A. H. (1941). *Organic Syntheses, Collective*, Volume *I*, 2nd Edition, John Wiley & Sons, Inc. New York, pp. 240–241.

65. Furniss, B. S., Hannaford, A. J., Smith, P. W. G., Tatchell, A. R. (1989). *Vogel's Textbook of Practical Organic Chemistry*, 5th Edition, John Wiley & Sons, Inc. New York, pp. 890–897.

66. Gilman, H., Blatt, A. H. (1941). *Organic Syntheses, Collective*, Volume *I*, 2nd Edition, John Wiley & Sons, Inc. New York, pp. 101–102.

67. Furniss, B. S., Hannaford, A. J., Smith, P. W. G., Tatchell, A. R. (1989). *Vogel's Textbook of Practical Organic Chemistry*, 5th Edition, John Wiley & Sons, Inc. New York, pp. 472–474.

68. Lyga, J. W. (1992). Synthesis of 2,3-dideoxy-D-arabino-heptono-1,4-lactone via a Wittig reaction of unprotected D-arabinose. *Org. Prep. Proced. Int.*, *24*, 73–76.

69. Furniss, B. S., Hannaford, A. J., Smith, P. W. G., Tatchell, A. R. (1989). *Vogel's Textbook of Practical Organic Chemistry*, 5th Edition, John Wiley & Sons, Inc. New York, pp. 493–495.

70. Carofiglio, T., Cordioli, M., Fornasier, R., Jicsinszky, L., Tonellato, U. (2004). Synthesis of 6[I]-amino-6[I]-deoxy-2[I–VII],3[I–VII]-tetradeca-O-methyl-cyclomaltoheptaose. *Carbohydr. Res.*, *339*, 1361–1366.

5

FOUNDATIONS OF A DNA-ENCODED LIBRARY (DEL)

Alexander Lee Satz

*Molecular Design and Chemical Biology,
F. Hoffmann-La Roche Ltd, Basel, Switzerland*

As discussed in previous chapters, DNA-Encoded Libraries (DELs) provide an unprecedented opportunity to interrogate vast swaths of chemical space for activity against targets of interest. However, it is not clear from the literature how one proceeds to build a portfolio of DELs. What libraries does one make? Do some libraries produce more hits than others and can this be predicted? What is required to produce DELs capable of consistently providing hit molecules in a drug discovery setting? Split-and-pool libraries are the product of the Building Blocks (BBs) used at each split and the chemistry that links them. The role that BBs play in the design of a DEL is addressed in this chapter. Chapter 4 will discuss different methodologies for piecing the BBs together.

5.1 INTRODUCTION TO THE REQUIREMENTS FOR BUILDING BLOCKS

Any chemical compound that may be covalently connected to a DNA-encoding unit is defined as a BB. All BBs must contain at least one reactive functional chemical group or so-called handle, such as a reactive amine or a carboxylic acid. Additionally, some

A Handbook for DNA-Encoded Chemistry: Theory and Applications for Exploring Chemical Space and Drug Discovery, First Edition. Edited by Robert A. Goodnow, Jr.
© 2014 John Wiley & Sons, Inc. Published 2014 by John Wiley & Sons, Inc.

TABLE 5.1. Exemplar monofunctional BB classes and number available from commercial vendors

Functional class[a]	Number available[b]
Amines	6306
Aldehydes	1308
Carboxylic acids	2501
Alkynes	88
Aryl iodides	367
Boronates	3246
Activated aryl halides	213
Sulfonyl chlorides	321

[a] Aryl iodo, suitable for Suzuki coupling; alkyne, terminal alkyne suitable for Sonogashira coupling; activated aryl halide, suitable for $S_N Ar$ reaction with a reactive amine, for example, 2-chloropyrimidine.

[b] ACD database search and limited to select vendors; Aldrich, Alfa, Fluka, Acros, Maybr-Int, Matrix, Enamine, Oakwood, TCI-US, Frontier, Boron Molecular, and Combi-Blocks. Price filter set at <= 200 $/g and application of property and structural filters as described herein.

BBs may contain two functional handles, for example, an amino acid. BBs with three functional handles, for instance, cyanuric chloride [1,4], also provide value as scaffolding. In this section, useful BB classes and their availability from commercial sources are described. This section also illustrates how the different BB classes may be used to construct various DELs.

BBs may be purchased from commercial vendors, specially synthesized for use in a DEL, or gathered from a proprietary collection of compounds from a corporate inventory. Table 5.1 provides a breakdown of the most useful monofunctional BBs available through commercial vendors, at reasonable costs and current availability. From commercial sources, the amines, boronates, carboxylic acids, and aldehydes are most abundant with over a thousand available in each class. Of these, the amines are most abundant with over 6000 available. Note that a short list of preferred vendors is used for this analysis and that the number of available compounds could be significantly higher with a more comprehensive list.

Table 5.1, Table 5.2, and Table 5.3 provide a breakdown of potential BBs that can be used in the synthesis of a DEL. The total number of BBs is approximately 13,000. As previously mentioned, Table 5.1 shows a breakdown of the most useful monofunctional BB classes. Amines, aldehydes, carboxylic acids, and boronic acids/esters make up the vast majority of all BBs, with each numbering >1000 compounds. These four large BB sets are followed by a second tier of smaller sets that include activated aryl halides (primarily useful in $S_N Ar$ reactions), alkynes, aryl iodides, and sulfonyl chlorides, each numbering less than 500 compounds.

Constructing libraries from the most abundant BB classes presumably provides increased structural diversity. Momentarily ignoring synthetic chemistry issues, covalently connecting all available amines with all available carboxylic acids in a split-and-pool

TABLE 5.2. Exemplar difunctional BB classes and numbers available from commercial sources[a]

	Acid	Aldehyde	Alkyne	Amine	Aryl halide	Aryl iodide	BOC	Boronates	Ester	Fmoc	Nitro	Sulfonyl Chloride
Acid	264	33	11	869	41	63	261	93	76	138	138	11
Aldehyde			0			21	21		47		82	
Alkyne				14								
Amine				644		100	149	151			293	
Aryl iodide									52		49	2

Acid, carboxylic acid; aryl iodo, suitable for Suzuki coupling; BOC, *tert*-butyloxycarbonyl-protected amine; Fmoc, fluorenylmethyloxycarbonyl-protected amine; alkyne, terminal alkyne suitable for Sonogashira coupling; ester; generally methyl or ethyl ester; nitro, aryl nitro group; aryl halide, suitable for S_NAr reaction with a reactive amine, for example, 2-chloropyrimidine.

[a]See Table 5.1 for description of which commercial BBs were flagged as available. Each value represents the number of available BBs containing two useful functional groups. For instance, the upper left-hand corner of this table states that 264 BBs are available from commercial sources that contain two or more carboxylic acids. The upper right-hand corner of this table states that 11 BBs are available that contain at least one sulfonyl chloride and one carboxylic acid.

TABLE 5.3. Exemplar trifunctional building block classes and number available from commercial sources.[a]

Functional groups*	Number available:	Functional groups	Number available:
Acid_aldehyde_nitro	4	Acid_Fmoc_ester	0[b]
Acid_ester_nitro	2	Acid_Fmoc_nitro	0[b]
Acid_alkyne_BOC	2	Acid_arylhalide_nitro	0[b]
Acid_alkyne_Fmoc	2	Acid_aldehyde_iodo	0[b]
Amine_acid_alkyne	1	Arylhalide_acid_nitro	0[b]

[a]See Table 5.1 for description of which commercial BBs were flagged as available. Each BB type contains at least three useful functional groups.
[b]Potentially useful tri-functional BBs are listed in the table which may not be commercially available. However, these BBs could likely be procured through specialty vendors or contract synthesis.

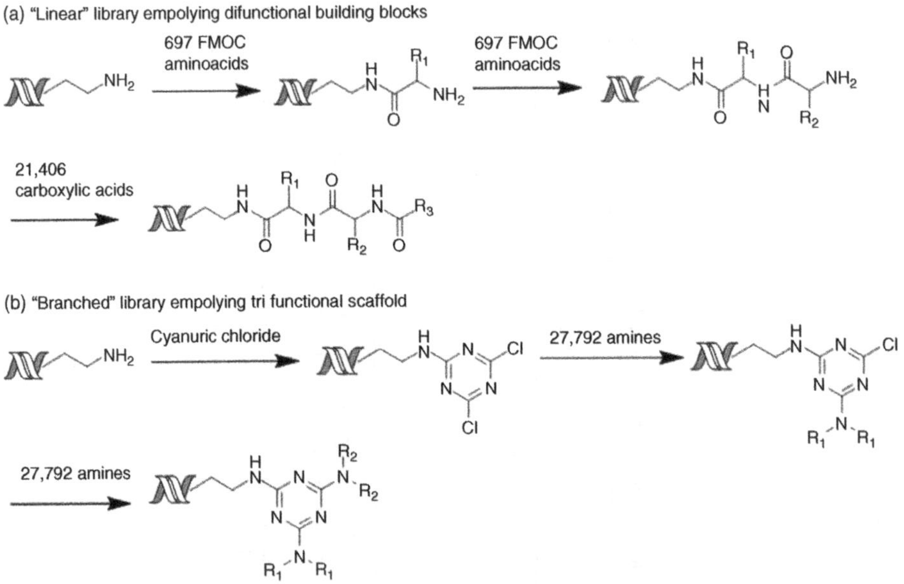

Figure 5.1. Hypothetical schema for (a) linear and (b) branched DELs. Schema does not show DNA-encoding steps. For a description of how a split-and-pool DEL is produced, see Chapter 6 or Clark et al. [1]. The number of BBs used in each cycle is taken from Table 5.1.

format yields $6306 \times 2501 = 1.58 \times 10^7$ structures comprised of 8807 different BBs. Alternatively, linking all available aldehydes with all available alkynes yields $1308 \times 88 = 1.15 \times 10^5$ structures comprised of 1396 different BBs. Employing more abundant BB classes yields not only a larger number of distinct products but perhaps more importantly increases diversity as measured by the number of BBs used.

Table 5.2 provides the number of difunctional BBs available from commercial sources (using price, availability, and property filters as described in Table 5.1). Some difunctional BB classes are abundant and provide a relatively high level of structural diversity. For example, 264 BBs are available that contain two carboxylic acids, 644 that

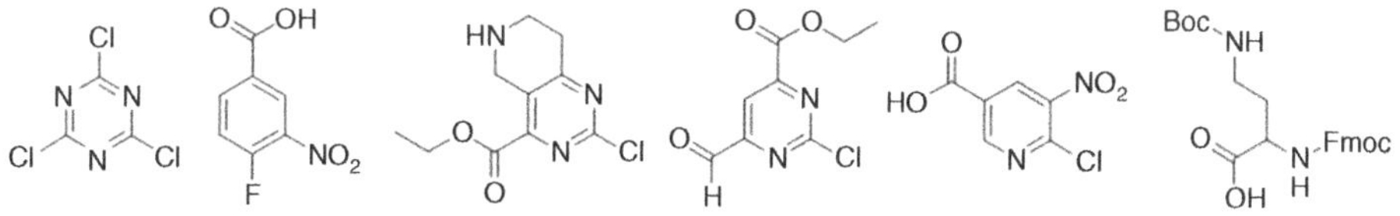

Figure 5.2. Examples of commercially available trifunctional BBs.

contain two amines, 869 that are unprotected amino acids, 293 that contain an amine and an aryl nitro group, and 261 that are BOC-protected amino acids. As for the monofunctional BBs, libraries that employ more abundant BB classes will display greater diversity and, therefore, an increased chance of providing valuable hit molecules. Difunctional BBs can be utilized to build linear molecules, in a manner similar to constructing peptides from amino acids. Figure 5.1a illustrates how difunctional and monofunctional BBs may be combined to form a linear peptide-like DEL. This library design is quite versatile and employs one or two cycles of a difunctional BB followed by capping of the reactive terminus with a diverse monofunctional BB class such as carboxylic acids.

The library design shown in Figure 5.1b makes use of a trifunctional core that yields molecules with a "branched" structure. Table 5.3 provides the availability of trifunctional scaffolds from commercial sources. A single trifunctional core can provide the basis for an entire library, and so, these BBs are particularly valuable [1]. The BBs in Figure 5.2, which are commercially available, provide an easy way to access varied chemical space. If making several libraries, it is advised to make both "linear" and "branched" DELs since this may provide greater structural diversity in any collection of DELs.

5.2 PRACTICAL ASPECTS OF BUILDING BLOCK ACQUISITION

Purchasing compounds from a commercial vendor is an easy way to build a large and diverse collection of BBs. Contract service providers absorb many of the logistical issues involving the acquisition of large numbers of different compounds. In addition to purchasing BBs from various suppliers, many contract services will also transfer the desired quantities to preferred containers such as 2D barcoded tubes. This option is particularly attractive if an internal compound management infrastructure is lacking. The task of receiving and unpacking hundreds or thousands of bottles of chemicals is difficult.

It is also possible to have specialty BBs synthesized for use in a DEL. Contract synthesis is expensive but necessary to gain access to particular trifunctional BBs. As shown in Table 5.4, trifunctional BBs are far less common than trifunctional BBs, which in turn are far less common than monofunctional BBs. Examples of trifunctional BBs are shown in Figure 5.2. For this reason, contract synthesis resources are best used for the synthesis of the less common trifunctional scaffolds.

BBs may be available through existing corporate collections. Pharmaceutical companies often have a wealth of chemical diversity stored in their proprietary high-throughput screening collections and/or collections of intermediates left over from decades of small-molecule discovery programs. These proprietary compound collections offer many benefits including no upfront purchasing costs, inherent

TABLE 5.4. Number of commercially available BBs binned by number of functional groups

BB type	Number available
Monofunctional	13,224
Difunctional	2,981
Trifunctional	11

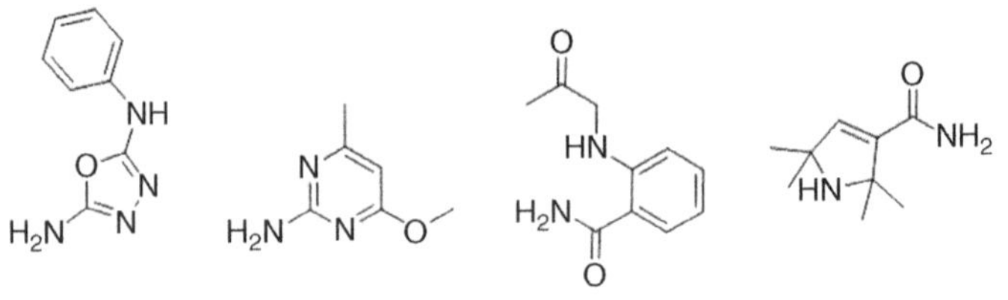

Figure 5.3. Exemplar amine BBs that lack reactivity for use in on-DNA chemistry reactions.

drug-likeness of the structures, and an increased likelihood of structural novelty relative to commercial purchases.

Generally, only 5–10 mg of any particular BB is needed to produce a single DEL. A 250 mg bottle of most compounds may provide enough material for the production of approximately 25–50 different DELs. This is an important point when considering how much of each BB to purchase and/or how quickly proprietary stocks of compounds might be depleted. Purchasing larger amounts of each BB may allow faster hit follow-up (synthesis of a potential "hit" without its DNA tag attached). However, the vast majority of BBs employed in a DEL will never occur in a selected feature. Additionally, it may be preferable to have hit follow-up work done at a remote site (i.e., this task may be best delegated to an external contract agency), in which case possessing extra quantities on-site becomes less valuable.

5.3 SIMPLE FILTERS USED TO PRIORITIZE BUILDING BLOCK ACQUISITION

5.3.1 Reactivity

To avoid wasting resources, it is best to avoid acquisition of BBs that lack appropriate reactivity to yield desired product (see examples in Fig. 5.3). Every BB acquired for use in a DEL should be validated in an on-DNA test reaction (a reaction between a reagent and an appropriately functionalized DNA conjugate) to ensure appropriate reactivity and quality control of the weighed-out compound (as discussed in Section 5.7). Preliminary validation results can provide clear structure–reactivity relationships, eliminating certain structural motifs from future acquisition. For instance, the exocyclic amines of adenine, guanine, and cytosine nucleobases are unreactive under the conditions employed when

TABLE 5.5. Examples of functionalities to exclude from the final products of a DEL designed to provide hits for drug discovery programs [2]

Reactive/unstable	Acid halides, aldehydes, alkyl halides, anhydrides, aziridines, beta-lactams, boronic acid, carbodiimides, imines, iso(thio) cyanates, ketones, Michael acceptors, N-methylol derivatives, peroxides, epoxides, reactive heterohalides, thiols
Interact with protein capture resins	Quaternary amines, nicotinic acids, biotin substructure, N-oxides
DNA intercalators	Polycyclic aromatic systems
Structural alerts due to carcinogenicity	Nitro, N-nitroso, hydrazines (and other unsubstituted heteroatom-bonded heteroatom), hydroxylamines, vinyl halides, azides, organophosphorus, diazonium

synthesizing DELs. Similar deactivated aryl amines such as 2-aminothiazoles, 2-amino-pyridines, and aminopyrimidines are also likely to be unreactive. Additionally, a large number of sterically hindered amines are available from commercial sources that are presumably too hindered to be of use in DEL-compatible chemistry, for example, 2,2,6,6-tetramethylpiperidine. Again, preliminary on-DNA test reactions will help to delineate those types of amines that are too hindered from those that may provide value.

5.3.2 Stability and Structural Alerts

When designing libraries for hit discovery, a list of functional groups as well as other filters are often used to identify BBs for automatic exclusion from acquisition should such functionality persist after the reaction sequence (Table 5.5). BBs to potentially exclude include those with reactive moieties, inherent metabolic liabilities, electrophilic sites, and functional groups that may bind to the resins most often used to immobilize the target protein during the selection process (see Chapter 13 for a detailed discussion of these topics). BBs with electrophilic groups such as acrylamides may form covalent interactions with target proteins. Libraries rich with electrophiles may lead to complications due to polymerization of library members. BBs that possess groups linked to mutagenicity and toxicity may also be excluded, for instance, nitro-containing compounds [2]. However, hits containing these groups may still provide value against certain targets, particularly as probes, and so their acquisition might be deprioritized instead of completely excluded. BBs containing DNA intercalators and binders should also be excluded as these molecules could indiscriminately interact with other library members.

5.4 THE EFFECT OF NESTED LIBRARIES ON LIBRARY DESIGN AND BUILDING BLOCK ACQUISITION

Literature reports have noted the correlation between Molecular Weight (MW) and desired drug-like properties [3]. In a small-molecule drug discovery setting, consistently finding hit molecules with MWs <500 is fundamental to the success of a new technology

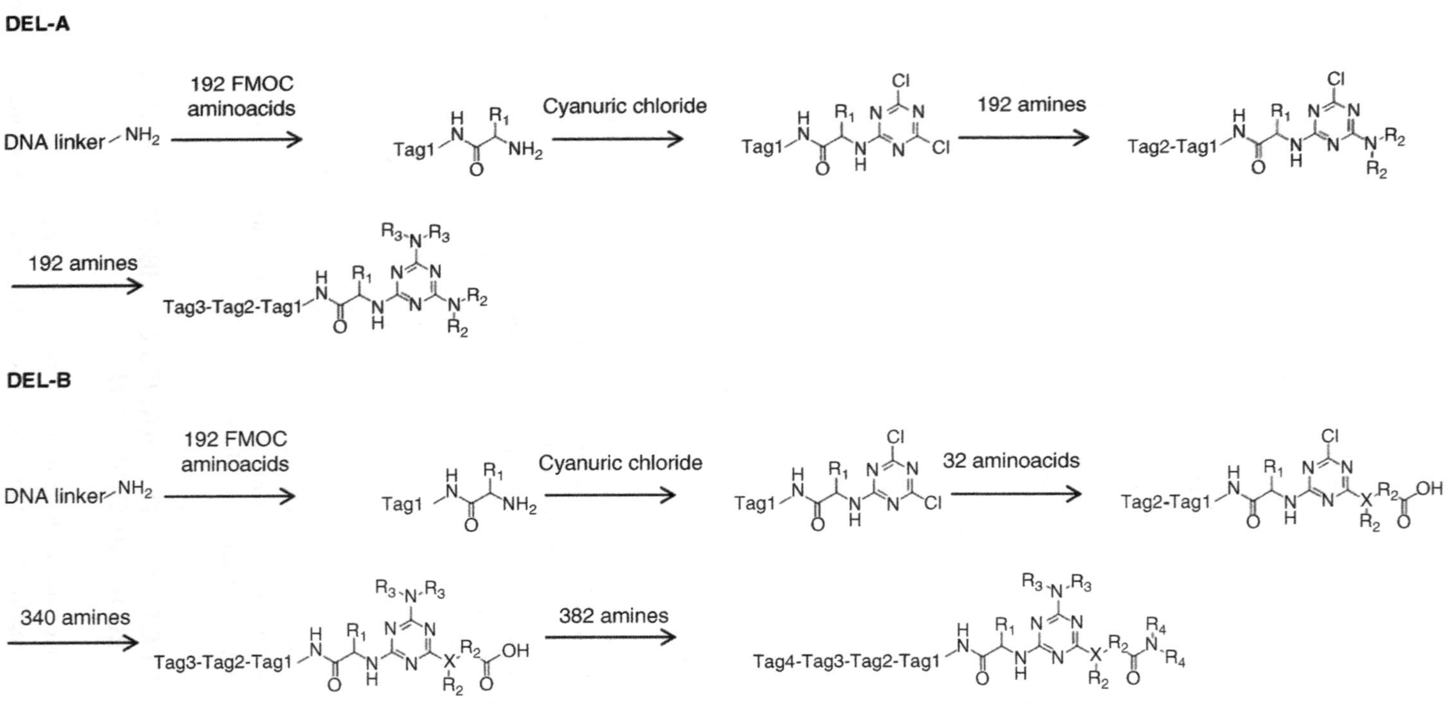

Figure 5.4. Schema for DEL-A and DEL-B [1]. Library synthesis starts with a bifunctional piece of DNA that possesses a site for ligation to encoding tagging regions and an amine handle that can undergo chemical reactions. For a thorough description of library synthesis, see Chapter 6 and/or Clark et al. [1]. For each diversity cycle of chemistry, there is an associated cycle of DNA encoding. In this manner, the growing DNA tagging region describes the BBs used during the chemical synthesis steps.

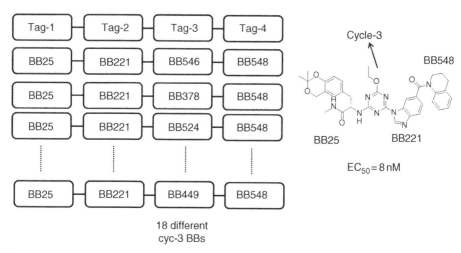

Figure 5.5. Output of selection of DEL-B against p38 MAPK [1]. After removing all unique sequences that only occurred once, 355 reads were left encoding 153 unique sequences. The sequences describing one of the more interesting features are shown earlier. The feature was described by 18 different unique sequences, each sequence occurring between two and four times. DEL-B contains 732 unique BBs, and for the sake of this example, these BBs were indexed according to the order given in the original publication. Each of the 18 unique sequences encodes BB25, BB221, and BB548 at cycles 1, 2, and 4, respectively. Cycle 3 however varies, 18 different BBs being encoded at that position. Many molecules based upon this feature were synthesized and investigated for *in vitro* activity. The most active molecule was a cycle 3 truncate (shown in the figure) that likely results from ethanolysis following cycle 2 chemistry.

such as DELs. However, the relationship between library design, BBs used, and resulting hit properties appears complex with little data published.

For instructive purposes, the selection of a DEL against p38 MAPK reported by Clark et al. [1] is discussed herein. The DEL (herein referred to as DEL-B) contains four diversity cycles of chemistry and an invariant triazine core (Fig. 5.4). The first cycle consists of 192 Fmoc amino acids (Fmoc AA), the second cycle 32 amino acids, the third cycle 340 amines, and the fourth cycle 382 amines. The selection of DEL-B against p38 (the enzyme target) yielded a feature as described by the DNA-encoding regions shown in Figure 5.5.

Note that a "feature" is defined as a library molecule, or series of structurally related library molecules, which enrich upon selection against a particular protein target. Features are usually followed up by synthesizing the appropriate small molecules without the presence of the DNA tagging region. These "off-DNA" compounds are then tested for activity in the appropriate assay. If the off-DNA compounds are active against the target protein, then they may be referred to as a "hit molecule." Further SAR and optimization may then result in discovery of a lead compound.

The feature consists of 18 unique sequences where cycles 1, 2, and 4 are held constant, while cycle 3 varies over 18 different and seemingly unrelated BBs. There are

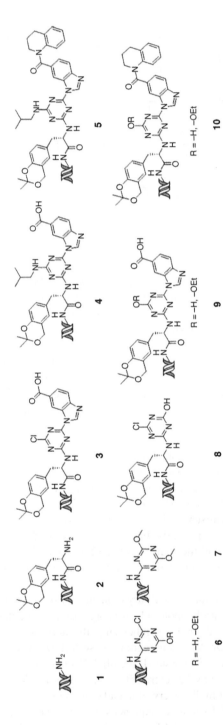

Figure 5.6. Partial list of products, side products, and leftover starting materials that may be encoded by a single DEL sequence read. Due to the nature of split-and-pool combinatorial synthesis and the inability to remove side products following chemistry steps, a unique DNA sequence may be conjugated to a number of different small molecules. For this reason, it is best to view the DNA encoding as a recipe for a series of synthetic steps as opposed to defining a single final product.

two possible reasons why cycle 3 varies among structurally unrelated BBs. The first reason is that no meaningful interaction occurs between the moiety at cycle 3 and the protein binding site. Hence, the 18 different cycle 3 BBs contained in the feature are essentially random. The second possibility arises since synthetic reactions usually yield a mixture of product along with side products such as hydrolysis, ethanolysis, and unreacted starting materials. However, since side products are not removed during library synthesis, they become encoded in the same manner as the desired products. Thus, the observed sequences in Figure 5.5 may be the result of side products, which lack any BB at cycle 3, binding to the target protein. Again, the BBs observed at cycle 3 in Figure 5.5 would be expected to be essentially random.

Figure 5.6 provides an example of how a single unique sequence can encode several different products. Side product 7 is due to the reaction of the coupling reagent DMT-MM with the library molecule. Side products 6, 8, 9, and 10 are due to hydrolysis or ethanolysis of the triazine intermediate. Side products 2, 3, and 4 are unreacted starting materials. Side products 1, 6, and 7 are common to every full-length DNA-encoding region, since they are independent of any particular BB. For this reason, 1, 6, and 7 cannot be enriched during a selection since they are not unique to any particular full-length sequence. All other side products are intermediates along the pathway to synthesis of the final product 5.

The number of potential side products that might be encoded by any particular DNA sequence read may appear overly complex. However, during feature follow-up efforts, these intermediates will generally be synthesized in the process of making the final product (the off-DNA version of **5**; see Fig. 5.6) with only modest additional effort. Literature reports show that it may be the side product that results in the active hit molecule and not the presumed final product (as exemplified in Fig. 5.5) [1, 4]. Thus, the removal of side products following split-and-pool chemistry steps, even if technically feasible, would only result in lessening the chance of finding valuable hit molecules.

The discussion earlier culminates in an important conclusion: a library with four cycles of chemistry displays several truncated nested libraries simultaneously during any selection. The nested libraries for DEL-B are described in Figure 5.7. Thus, in the process of making the 4-cycle library described by the tetrasynthon illustration, three different trisynthon libraries, four different disynthon libraries, and four different monosynthon libraries are simultaneously produced. As will be discussed shortly, the concept of nested libraries complicates strategies regarding BB acquisition and library design.

Many DEL features will arise from a series of structurally related library molecules [1, 4]. The ability to track which nested libraries provide valuable hit molecules is desirable for optimization of future library design and BB selection. Figure 5.8 shows how the p38 MAPK hit molecule from DEL-B is mapped to its nested library as shown in Figure 5.7. One method of mapping a feature to a single nested library involves choosing a single exemplar from the series. Plotting the ligand efficiencies of the various products/truncates as shown in Figure 5.8 provides a consistent method for designating the exemplar. As shown in Figure 5.8, it is the trisynthon with a ligand efficiency of 0.17 that is designated the exemplar molecule of the feature (we assume the mono- and

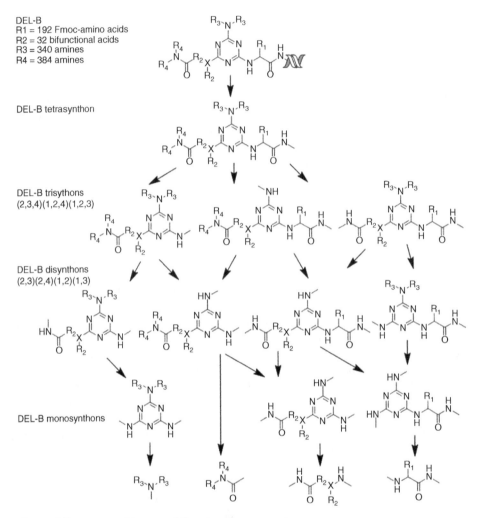

Figure 5.7. DEL-B and its nested libraries of compounds.

disynthon compounds to be inactive). The exemplar is then mapped to its appropriate nested library in Figure 5.7 (the 1,2,4-trisynthon). Note that using ligand efficiency and not potency to choose the exemplar focuses attention on those BBs making key protein interactions.

Momentarily ignoring the concept of nested libraries, the relationship between library cycle number and BB MW is easily calculated. For instance, if a library contains four cycles of chemistry, then each BB should have a MW < 125 g/mol for all final products to have a final MW <500 g/mol. Table 5.6 summarizes MW limits that could be employed to ensure that only drug-like final products are formed. Also provided in Table 5.6 are the numbers of commercially available BBs after binning according to MW limits. Roughly 13,000 BBs have a MW <500 g/mol, while only approximately

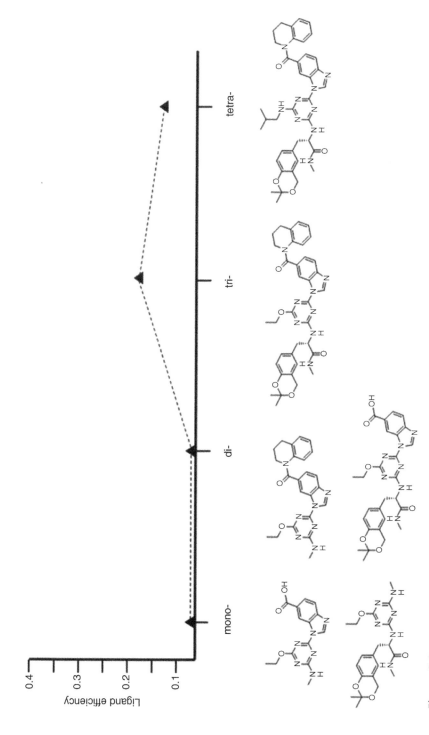

Figure 5.8. Defining an exemplar hit molecule and mapping it to a nested library. Investigation of a DEL feature requires the synthesis of the small-molecule moieties of the enriched library molecules as shown earlier. These small molecules can then be tested in the appropriate assay and their ligand efficiencies calculated (ligand efficiency = pIC50 ÷ heavy atom count). Final products and synthetic intermediates are submitted for assay to determine if a truncation or side product is the cause of activity.

111

TABLE 5.6. Relationship between cycles of chemistry and BB MW and breakdown of commercially available compounds

Number cycles in DEL	MW limit for BBs	Number BBs commercially available
Monosynthon	500	13,224
Disynthon	250	10,023
Trisynthon	167	3,565
Tetrasynthon	125	990

TABLE 5.7. Effect on the composition of DEL-B when including high-MW BBs

	[a]DEL-B			[b]DEL-B(HMW)		
	Avg MW	%<500 g/mol	# molecules MW <500	Avg MW	% <500 g/mol	# molecules MW <500
Tetra	713	~0	~0	812	~0	~0
Tri	567	17	5×10^6	656	4	1×10^9
Di	446	87	6×10^4	495	56	7×10^6
Mono	296	100	948	328	99	13,536

[a]Values calculated for DEL-B using BBs provided by Clark et al. [1]. DEL-B split sizes are 192 Fmoc AAs × 32 amino acids × 340 amines × 384 amines = 8×10^8 tetrasynthon products (synthetic schema provided in Fig. 5.4).
[b]Hypothetical DEL similar to DEL-B except employing all the BBs available from commercial sources including those with MWs up to 500 g/mol. DEL-B(HMW) = 192 Fmoc AAs × 869 amino acids × 6306 amines × 6306 amines = 6.6×10^{12} tetrasynthons. Refer to Figure 5.8 for descriptions of tetra-, tri-, di-, and monosynthon structures.

1,000 meet the threshold of <125 g/mol. The cost of maximizing the percentage of drug-like final products by excluding higher-MW BBs is that much of the chemical diversity contained in the BB collection is forgone.

In Table 5.7, an analysis of DEL-B [1] and its nested libraries is provided. The tetrasynthon products have an average MW of 713 g/mol. None of the tetrasynthons possess a MW<500 g/mol. For the discovery of drug-like hits, this cohort of DEL-B provides little value. The three nested DEL-B trisynthon libraries possess an average MW of 567 g/mol, and 17% of these molecules possess a MW of <500. The number of trisynthons in DEL-B with a MW <500 g/mol is calculated to be approximately 5.3×10^6. The four nested DEL-B disynthon libraries have an average MW of 446, and 87% (6×10^4 molecules in total) of these disynthons meet the <500 g/mol threshold. Lastly, the monosynthons have an average MW of 296 (including the invariant triazine core as shown in Fig. 5.7). All 948 of these monosynthons have a MW <500. As cycle number increases, each group of nested libraries has a lower percentage of members that meet the 500 g/mol threshold. However, larger cycle numbers also yield greater numbers of final products and can offset the lower pass rate. Another conclusion reached from this analysis is that even though the average MW of the BBs used in cycles 3 and 4 of DEL-B was kept relatively low at 154 g/mol, only roughly 1 in 160 products meets

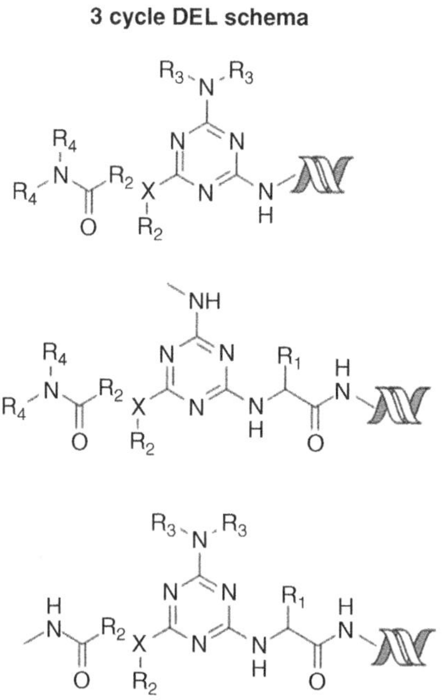

3 cycle DEL schema

Figure 5.9. Schema for increasing ratio of low- to high-MW products by eliminating production of tetrasythons. Cycle 1 can be truncated by directly reacting cyanuric chloride with the bifunctional DNA linker, whereas cycles 3 and 4 are replaced with methylamine.

the <500 g/mol threshold. Hence, less than 1% of the library has a realistic chance of being a valuable hit for a drug discovery program due to high MW.

Also included in Table 5.7 is an analysis of a hypothetical DEL-B (herein referred to as DEL-B (High MW) (DEL-B(HMW))) constructed in the same manner as DEL-B but employing all the chemically appropriate BBs available from commercial sources (see Table 5.1), including those with high MWs (DEL-B(HMW) = 192 Fmoc AAs × 869 amino acids × 6306 amines × 6306 amines = 6.6×10^{12} tetrasynthons). In this case, the average MW of cycles 3 and 4 BBs is 190 g/mol, compared to 154 g/mol for DEL-B (as reported by Clark et al. [1]). DEL-B(HMW) possesses all the BBs contained within DEL-B, along with many thousands of additional BBs. This hypothetical library has a lower percentage of products with MW<500 g/mol. However, its split sizes are significantly greater resulting in approximately 200-fold more products that meet the <500 g/mol threshold. The nested disynthon libraries of DEL-B(HMW) possess 100-fold more products that meet the desired cutoff. However, the ratio of desirable to undesirable products drops precipitously to roughly 1 out of every 10^4 library members. Thus, Table 5.7 shows that including high-MW BBs yields DELs with greater total numbers of drug-like products while simultaneously increasing the percentage of undesirable molecules.

It is possible to employ library design strategies to lessen the number of high-MW products formed while still maintaining BB diversity. One possible Schema is provided in Figure 5.9 where three DELs are independently synthesized. The three resulting DELs are identical to the three nested trisynthon libraries shown in Figure 5.7. In this

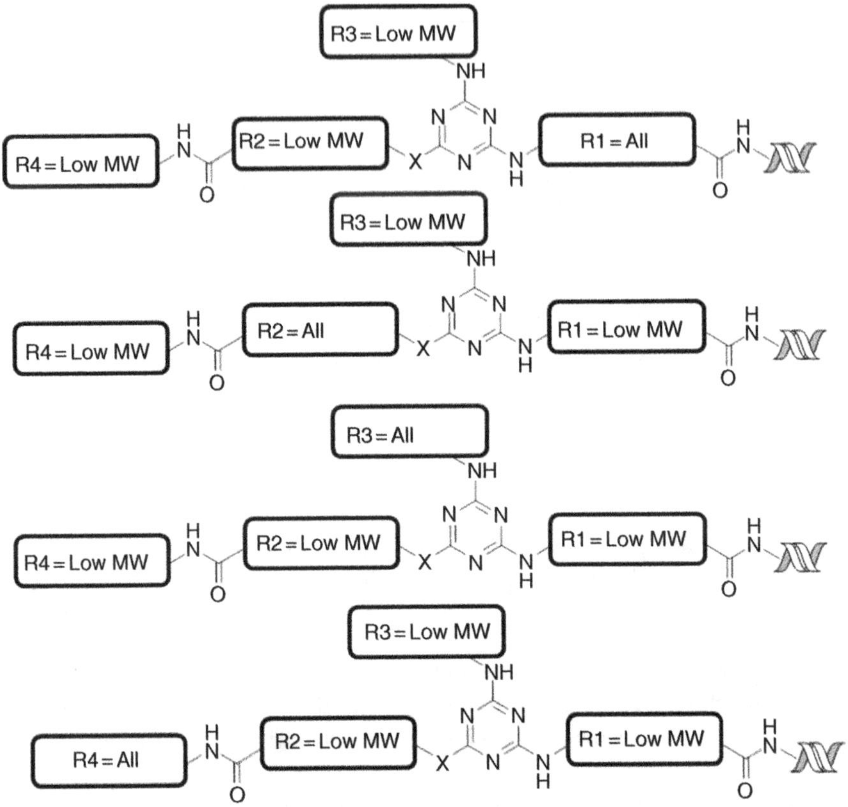

Figure 5.10. Schema for increasing ratio of low- to high-MW products by eliminating the combination of high-MW BBs. Four separate DELs are produced, each one in turn having one cycle of chemistry employing all available BBs while limiting the other three cycles of chemistry to only lower-MW BBs. In this manner, high-MW BBs are never combined together to form a high-MW product.

manner, all the trisynthon, disynthon, and monosynthon nested library products of DEL-B are produced while having the potential advantage of not producing any tetra-synthons. In the case of the hypothetical DEL-B(HMW), employing the schema shown in Figure 5.9 would improve the ratio of low- to high-MW products by approximately 200-fold (~4% of all products would meet the <500 g/mol threshold).

A second alternative toward the goal of increasing the ratio of low- to high-MW prod-ucts is to avoid producing molecules that contain more than one high-MW BB. The nature of split-and-pool chemistry makes this task impossible unless you divide a library such as DEL-B into four separate DELs as outlined in Figure 5.10. In this schema, one cycle in each DEL will employ all available BBs regardless of MW, while the other 3 cycles of chemistry would exclude BBs with high MWs. In this manner, all available BBs are used at least once, yet no product is ever produced that contains more than one high-MW BB. In the case of the hypothetical DEL-B(HMW), employing the schema shown in Figure 5.10 would increase the ratio of low- to high-MW products by approximately five-fold.

The library designs shown in Figure 5.9 and Figure 5.10 offer the possibility of maintaining product diversity while lessening the percentage of high-MW products being formed. However, implementing these plans requires greater chemistry resources compared to simply following the reported 4-cycle DEL-B schema [1]. Before committing to the increased workload, it makes sense to consider the value of increasing the ratio of low- to high-MW products. This question is discussed in the next section. The published data shows little reason to believe that increasing this ratio will result in a corresponding increase in value.

5.5 THE EFFECT OF HIGH-MW PRODUCTS ON DEL SELECTIONS

As outlined in the previous section, there are several strategies that can improve a library's ratio of low-MW to high-MW products. Strategies include excluding all higher-MW BBs, preferentially producing libraries with low cycle numbers (Fig. 5.9), and/or taking steps to prevent the formation of products with multiple high-MW BBs as outlined in Figure 5.10. All these strategies suffer drawbacks including lessening diversity and/or greatly increasing chemistry and informatics workload. In this section, the minimization of the percentage of high-MW products as a worthwhile pursuit is discussed.

The most obvious reason why a large percentage of high-MW products may be detrimental is the presumed tendency of large lipophilic molecules to promiscuously bind to protein targets. If true, then high-MW products may disproportionately consume sequence depth resulting in decreased signal to noise and preventing the discovery of more desirable low-MW ligands. To determine the validity of this concern, library populations before and after selections need to be directly compared and disproportional enrichment of large lipophilic library members assessed. Unfortunately, very little of the required selection data has been reported by practitioners in the field. However, Clark et al. [1] do provide some sequence data for each of their three reported features. The use of this reported data to determine if large lipophilic molecules are disproportionately enriched by a DEL selection is discussed.

The selection of DEL-A (Fig. 5.11) against p38 MAP kinase was previously reported (DEL-A is a 3-cycle DEL that like DEL-B contains a triazine core (see Fig. 5.4 for a description of DEL-A)) [1]. The selection yielded a trisynthon hit where cycles 1 and 3 varied widely, while cycle 2 was fixed with the BB 3-amino-4-methyl-N-methoxybenzamide. The selection thus yielded a feature comprised of 1,970 different small molecules or "warheads" encoded by 61,292 sequence reads (the complete list of smiles strings are provided in the supplemental data of Clark et al. [1]). The MW of each of the 1970 trisynthon products were calculated and plotted against their normalized enrichment (i.e., the relative number of times a particular small molecule was observed following the selection). This MW distribution was overlaid with the MW distribution of the library prior to selection (holding the conserved 3-amino-4-methyl-N-methoxybenzamide BB constant). As shown in Figure 5.11, the MW distribution after the selection does not shift significantly, with both MW distributions averaging ~600 g/mol. The selection does not appear to favor enrichment of higher-MW products.

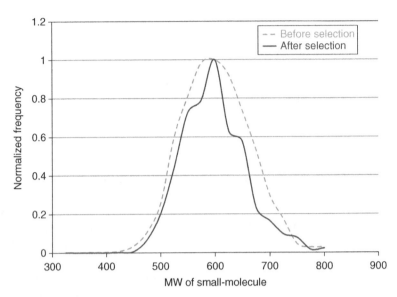

Figure 5.11. MW distribution of DEL-A [1] before and after selection against p38 MAPK (cycle 2 BB is fixed as 3-amino-4-methyl-*N*-methoxybenzamide).

TABLE 5.8. Selections of DEL-A against p38 MAP and Aurora A kinase [1]

	[a]BB properties before selection		[b]BB properties of enriched features after selection	
	Avg MW	Avg *c*Log*P*	Avg MW	Avg *c*Log*P*
Cycle 1 (Fmoc AA)	402	4.0	405 (*n* = 164)	4.0 (*n* = 164)
Cycle 2 (amines)	153	1.6	152 (*n* = 56)	1.6 (*n* = 56)
Cycle 3 (amines)	156	1.3	154 (*n* = 184)	1.2 (*n* = 184)

[a]The schema for synthesis of DEL-A is provided in Figure 5.4. The average MW (g/mol) and *c*Log*P* of all BBs used in DEL-A, binned by cycle, was calculated. See Clark et al. for library details [1]. Masses (g/mol) are of the BBs as used in library production, including the Fmoc group in the case of cycle 1.

[b]Selections of DEL-A against p38 MAP and Aurora A kinase resulted in two reported features comprised of 1970 and 148 distinct library molecules, respectively [1]. These 2118 enriched library molecules are derived from 404 different BBs. The 404 BBs were binned according to cycle # and average MW and *c*Log*P* calculated. The number (*n*) of different BBs at each cycle # is provided.

An alternative method of interrogating the data provided by Clark et al. [1] is shown in Table 5.8. Selection data was reported for three features: DEL-A against p38 MAP and Aurora kinase, and selection of DEL-B against p38 MAP. Table 5.8 and Table 5.9 provide the average MW and calculated Log*P* for the BBs at each cycle of DEL-A and DEL-B (i.e., the average values before selection). The average MW and *c*Log*P* at each chemistry cycle for the three reported "enriched" features are also provided for direct comparison. For instance, the selection of DEL-A against p38 MAPK yielded a feature comprised of 1970 different small molecules, while the selection against Aurora A kinase yielded a feature

TABLE 5.9. Selection of DEL-B against p38 MAP kinase [1]

	BB properties before selection		[a]BB properties of enriched features after selection	
	Avg MW	Avg cLogP	Avg MW	Avg cLogP
Cycle 1 (Fmoc AA)	392	3.9	434 ($n = 3$)	3.0 ($n = 3$)
Cycle 2 (amino acid)	157	0.05	162 ($n = 1$)	1.0 ($n = 1$)
Cycle 3 (amines)	153	1.2	142 ($n = 18$)	1.2 ($n = 18$)
Cycle 4 (amines)	156	1.4	156 ($n = 2$)	1.47 ($n = 2$)

[a]The schema for synthesis of DEL-B is provided in Figure 5.4. Selections of DEL-B against p38 MAP kinase result in a single reported feature comprised of 18 distinct library members and 24 different BBs. The BBs were binned according to cycle number and average MW and cLogP calculated. The number (n) of different BBs at each cycle number is provided.

comprised of 148 distinct small molecules. Together, these two features derived from DEL-A include 2118 different small-molecule "warheads" constructed from 404 different BBs. Broken down by cycle number, the two features include 164 different amino acids at cycle 1, 56 different amines at cycle 2, and 184 different amines at cycle 3. The average MW of DEL-A BBs at cycle 1 is 402 g/mol (including the Fmoc protecting group) prior to selection (note that the standard deviation for each BB set ranges from 40 to 50 g/mol and ~1 log unit with respect to MW and cLogP, respectively). The average MW of the 164 cycle 1 amino acids contained in the two enriched features is 405 g/mol, a difference of only 3 g/mol from the preselection average. As shown in Table 5.8 and Table 5.9, direct comparison of average BB physical properties before selection and in reported features is strikingly similar at every cycle # for both DEL-A and DEL-B. In conclusion, no reported data to date shows a disproportionate enrichment of large or lipophilic molecules or component BBs.

5.6 STORAGE AND HANDLING OF BUILDING BLOCKS

Large numbers of BBs require storage and tracking solutions. 2D barcoded storage tubes provide the basis for handling BBs (and may be used for storage and tracking of libraries too). A specific BB is weighed out into each unique tube and thereafter associated with that tube's unique identifier. Even if the tube is later mixed with other tubes, the tubes can always be put back into a 96-place rack and rescanned to determine the exact location of each particular BB. The fidelity of future libraries is dependent upon correct record keeping. The following is an overview of the equipment needed to track and manage collections of BBs.

5.6.1 Equipment

2D barcoded tubes are available from several vendors (e.g., Thermo Scientific Matrix and Abgene 2D barcoded storage tubes and Micronic 2D storage tubes). The tubes generally have a volume of 1–2 mL and may be capped with screw-on caps or a rubber plug.

The screw-on caps may be preferred since storage of organic solutions can result in the rubber plugs popping off due to vapor pressure. The bottoms of the tubes are marked with a 2D barcode that can be visually noted upon flipping the tube upside down. It is also possible to purchase 2D tubes where an alphanumeric code is present alongside the barcode. This is useful for manual identification of tubes. Barcodes usually encode approximately ten characters. Manufacturers guarantee that every tube produced will receive a unique identifier, eliminating concern about potential overlap. Lastly, depending upon manufacturer, tube caps may be purchased in several different colors, allowing for rapid visual identification. This may be a useful way to segregate different BB classes. For instance, all amino acid BBs may be stored in 2D barcoded tubes with red caps, while all carboxylic acids are stored using yellow caps.

There are several devices that are useful for tracking and storing the 2D barcoded tubes. Open bottom racks that can hold 96 2D barcoded tubes allow for easy organization of BBs. The open bottoms allow the racks to be scanned on a 2D barcode reader. Some racks also possess lids that snap on, securely keeping the 2D tubes within from falling out if the rack is accidentally knocked over. Many racks also possess easily identified markings to help the scientist track which side of the rack is the "top" and avoid aspirating from the rack after rotation of 180 degrees.

A 2D barcode reader (similar to a standard document scanner) is used to record the position of each unique tube in any particular rack. It is usually best to label each rack with a unique plate identifier and save the plate map under that name. Caps may be manually placed upon and removed from the 2D barcodes. However, this action can be tedious for large numbers of plates. Automated capping/decapping equipment may be purchased, which allows for handling of large numbers of tubes. Cappers and decappers are sold that interact with one row (12 tubes) at a time (automatic and manual devices) or the entire 96-tube rack simultaneously.

5.6.2 Preparing Building Block Tubes for Library Production

BBs may be tracked and sorted using only Excel spreadsheets, tweezers, and manual labor. The steps outlined in Table 5.10 are relatively tedious but also relatively mistake proof due to the ease of rescanning plates prior to library synthesis.

5.6.3 Informatic Sorting of Building Blocks

As suggested by Table 5.1, large databases of compounds will need to be interrogated to select BBs for acquisition and bin them according to physical properties and/or chemical functionality. For the purposes of writing this chapter, the scientific informatics platform Pipeline Pilot [5] was used. A detailed description of Pipeline Pilot falls out of the scope of this handbook; however, a very brief overview follows. Pipeline pilot allows for a list of BBs, input as smiles strings or an SD file, to be operated upon by any number of premade components. Operations include calculating physical properties such as MW and *c*Log*P*. Also, filters based on the SMARTS language allow the user to select for any desired substructure. A working knowledge of writing in the SMARTS language is helpful.

TABLE 5.10. Steps required to sort BB 2D barcoded tubes for library production

Needed

- List of desired BB ID#s (or smile strings)
- List of BB ID# to 2D barcoded tube conversions
- 2D barcoded scanner
- Racks of BBs (stored in 2D barcoded tubes)

Steps

1. Assemble racks of 2D barcoded tubes (containing BBs) from storage
2. Scan racks using 2D barcoded scanner and export as CSV files
3. Convert 2D barcodes of scanned tubes into BB IDs using *vlookup* function in Excel
4. Use *vlookup* function to identify physical location of desired BB tubes
5. Flag position of desired tubes in racks and print spreadsheet
6. Use tweezers to select desired tubes and transfer to new racks
7. Scan new racks using 2D barcoded scanner and export as CSV file
8. Use *vlookup* function to compare new plate maps against desired list of BBs
9. Flag any BBs that are missing or mistakenly transferred and print list
10. Remove incorrect tubes, transfer missing tubes, and rescan racks
11. Repeat steps 8–10 until plate maps match desired list
12. Rescan all racks shortly before use in library production to ensure fidelity

TABLE 5.11. HPLC coupling yields and recovery assessed after peptide bond formation reaction between three selected Fmoc AAs and a model 5'-amino-oligonucleotide (DEL_O_1) [6]

Fmoc AA		DEL_O_1 conjugated (5'-amino-C12-GGA GCT TGT GAA TTC TGG ATC TTA GGA CGT GTG TGA ATT GTC-3')
	Yield[a] %	97
	Recovery[b] %	53
	ESI-MS (Da)	13,474 (13,473)
	(in brackets is the expected MS)	
	Yield[a] %	90
	Recovery[b] %	65
	ESI-MS (Da)	13,439 (13,437)
	(in brackets is the expected MS)	
	Yield[a] %	73
	Recovery[b] %	55
	ESI-MS (Da)	13,457 (13,459)
	(in brackets is the expected MS)	

[a]Determined by HPLC after Fmoc deprotection of the oligonucleotide-conjugated compound.
[b]Evaluated measuring the absorption at 260 nm using a NanoDrop instrument (ND-1000 UV–vis spectrophotometer) following HPLC purification.

5.7 VALIDATIONS

Weighing out BBs into 2D barcoded tubes is one of the first steps in the production of a DEL. Validations are used to help ensure that this process is done correctly. Validations also help ensure that BBs have the appropriate reactivity and/or purity for a given reaction.

Validation generally involves a model reaction such as described in Table 5.11 where peptide bond formation between three different Fmoc AAs and a model 5'amino-oligonucleotide is assessed. It is common to validate every BB used in the production of a DEL [1,5]. Generally, BBs that show >50% conversion to desired product are considered to have passed validation [1, 6]. Depending upon the reaction and cutoff used, roughly 50–70% of appropriate BBs will pass any given validation [1].

There are many reasons why BBs may fail a validation reaction including a lack of reactivity, side reactions, or simply that the 2D barcoded tube does not contain the expected compound. Additionally, decomposition of the BB during storage can result in validation failure. Assessing BB purity by NMR, LCMS, or other methods can be done to remove BBs suffering from decomposition. However, the value of doing so is questionable since most DEL synthetic methodologies use large excess of reagents (40 equivalents or greater) and pure starting materials are not necessarily required. Hence, the validation reaction itself is usually the most pragmatic method of measuring a BB's purity (if the BB passes validation, then it is pure enough).

5.8 CONCLUSIONS

Split-and-pool libraries are the product of the BBs used at each split and the chemistry that links them together. This chapter described what a DEL BB collection should look like and how to handle and store it. This chapter also describes the concept of nested libraries and its implication on BB selection and library design. Chapter 4 provides synthetic methodologies for linking BBs together. These two chapters combine to provide a formula for designing DELs capable of providing drug-like small-molecule ligands. The quantity of information presented in this book illustrates the complexity of designing, building, and selecting against DELs. Additionally, it is philosophically impressive that millions and even billions of different chemical structures can be produced via DNA-encoded combinatorial chemistry. However, at its root, the ability of DEL technology to provide value likely resides in the diversity of the BB starting materials and not in the total number of compounds made.

REFERENCES

1. Clark, M. A., Acharya, R. A., Arico-Muendel, C. C., Belyanskaya, S. L., Benjamin, D. R., Carlson, N. R., Centrella, P. A., Chiu, C. H., Creaser, S. P., Cuozzo, J. W., Davie, C. P., Ding, Y., Franklin, G. J., Franzen, K. D., Gefter, M. L., Hale, S. P., Hansen, N. J. V., Israel, D. I., Jiang, J., Kavarana, M. J., Kelley, M. S., Kollmann, C. S., Li, F., Lind, K., Mataruse, S., Medeiros, P. F., Messer, J. A., Myers, P., O'Keefe, H., Oliff, M. C., Rise, C. E., Satz, A. L., Skinner, S. R.,

Svendsen, J. L., Tang, L., van Vloten, K., Wagner, R. W., Yao, G., Zhao, B., Morgan, B. A. (2009). Design, synthesis and selection of DNA-encoded small-molecule libraries. *Nat. Chem. Biol.*, *5*, 647–654.

2. Benigni, R., Bossa, C. (2006). Structural alerts of mutagens and carcinogens. *Curr. Comput.-Aid. Drug Design*, *2*(2), 169–176.

3. Lipinski, C. A., Lombardo, F., Dominy, B. W., Feeney, P. J. (2001). Experimental and computational approaches to estimate solubility and permeability in drug discovery and development settings. *Advanced Drug Delivery Reviews*, *46*, 3–26.

4. Deng, H., O'Keefe, H., Davie, C. P., Lind, K. E., Acharya, R. A., Franklin, G. J., Larkin, J., Matico, R., Neeb, M., Thompson, M. M., Lohr, T., Gross, J. W., Centrella, P. A., O'Donovan, G. K., Bedard, K. L., van Vloten, K., Mataruse, S., Skinner, S. R., Belyanskaya, S. L., Carpenter, T. Y., Shearer, T. W., Clark, M. A., Cuozzo, J. W., Arico-Muendel, C. C., Morgan, B. A. (2012). Discovery of highly potent and selective small molecule ADAMTS-5 inhibitors that inhibit human cartilage degradation via encoded library technology (ELT). *J. Med. Chem.*, *55*, 7061–7079.

5. http://accelrys.com/products/pipeline-pilot/. Accessed on November 25, 2013.

6. Mannocci, L. (2009). DNA-Encoded Chemical Libraries. Doctoral Dissertation. Retrieved from e-collection, ETH Institutional Repository. http://dx.doi.org/10.3929/ethz-a-005783014. Accessed on November 25, 2013.

6

EXERCISES IN THE SYNTHESIS OF DNA-ENCODED LIBRARIES

Steffen P. Creaser[1] and Raksha A. Acharya[2]

[1]*Genzyme Corp., Cambridge, MA, USA*
[2]*EnVivo Pharmaceuticals, Inc., Watertown, MA, USA*

If you have never made a DNA-Encoded Library (DEL) or performed any on-DNA reactions, you may have many preconceptions that this type of chemistry is excessively complex and would require specialized skills outside the realm of most chemists. This could not be further from the truth. The techniques used to create a DEL are easily learned and can be applied by chemists and biologists alike. By making it this far in the book, you will have hopefully come to appreciate the immense potential of this technology and its growing popularity as a drug discovery tool.

There are many ways to build a DEL. Each of the various techniques found in the literature offers unique advantages and outcomes, which should be considered when designing your target library. Learning every DEL method at once would be a formidable task, and it may be beneficial to first understand the principles of a single method before exploring others. It is also important to know that this field is still relatively young and has considerable potential for improvement and innovation.

A Handbook for DNA-Encoded Chemistry: Theory and Applications for Exploring Chemical Space and Drug Discovery, First Edition. Edited by Robert A. Goodnow, Jr.
© 2014 John Wiley & Sons, Inc. Published 2014 by John Wiley & Sons, Inc.

6.1 WHAT KIND OF DEL?

If a graduate student were assigned the task of creating a DEL by their university professor, the student would undoubtedly be left scratching their head. Not only does the concept of performing such a complex and multidisciplinary task seem overwhelming, there are many different ways to do it. The first question the student should ask is, "What kind of DEL?" There are a number of different methodologies being applied in the field using different encoding and synthesis strategies, and there are already many excellent reviews in the literature [1–6]. Rather than try to review each one, the purpose of this chapter is to teach the untrained scientist how to synthesize a DEL from start to finish. To do this effectively, this chapter will focus on one method only. By understanding the process in detail, you will hopefully gain the confidence and knowledge to build your own DELs and contribute to this growing field.

6.2 BACKGROUND

The method reported by scientists at GlaxoSmithKline (GSK) illustrates many principles of the DNA-encoded synthesis process [7]. One of the appealing aspects of this technique is that the synthesis of library molecules can be achieved with commercially available building blocks. Furthermore, despite the many steps involved, the strategy is relatively simple and easy to learn.

In order to understand the principles behind this method, it is useful to consider that its development had arisen with a focus on peptidomimetic drug design, and the chemical synthesis strategy owes much to the principles of Solid-Phase Peptide Synthesis (SPPS) described in Chapter 2. Pioneered by Robert Bruce Merrifield, SPPS created a paradigm shift in the way peptides are synthesized [8, 9]. Small solid beads treated with functional units or "linkers" serve as substrates to build peptide chains that remain covalently attached to the bead until chemically cleaved. The general principle of SPPS involves repeated cycles of coupling–washing–deprotection–washing. The superiority of this technique partially lies in the ability to use large excesses of reagents to drive a reaction to completion and the ease of purification after each step. The growing peptide is purified by simply washing away the waste reagents and solvent.

So how does DNA-encoded chemistry resemble SPPS? DNA readily precipitates from solution when ethanol and NaCl are added, providing insoluble sodium phosphate adducts that can be easily isolated by centrifugation (Fig. 6.1). This simple trick enables the scientist to easily separate DNA from soluble reagents or exchange the reaction buffer by precipitating the DNA into an insoluble pellet and decanting away the solution.

The ability to precipitate or "crash" DNA on demand allows the use of excess reactants, similar to SPPS, to drive the reaction further to completion. Even when using a mixture of water and organic solvent, such as DMF, the DNA crashes readily from solution. However, the one thing that cannot be removed during an ethanol crash is unwanted DNA. As described later in this chapter, the buildup of excess DNA tags can be problematic, and HPLC purification is also necessary to clean up the reaction mixture.

Figure 6.1. Ethanol precipitation of DNA. By adding two volumetric equivalents of ethanol followed by 5 M NaCl to a solution of DNA, the DNA will precipitate from solution and can later be separated as a pellet by centrifugation. For color detail, please see color plate section.

6.3 DIVERSITY

One of the great advantages of DNA-encoded chemistry over traditional techniques is the ability to quickly generate chemical diversity, on a massive scale. A recent publication by GSK reported a DEL containing over four billion individual components [10]. These extraordinary numbers are achieved using split-and-pool chemistry.

Imagine a single 96-well plate, wherein each well contains the same DNA substrate; after completing a synthesis cycle involving DNA ligations followed by the addition of building blocks, 96 unique components are produced across the plate. If the contents of each well are then pooled into a single reservoir (Fig. 6.2) and then split evenly into a new 96-well plate, each well now contains the same 96 components. If the process is repeated, each well will contain 9,216 species, and after a fourth cycle, 84,934,656 individual components will have been synthesized. The number of components synthesized is dependent on the number of split-and-pool cycles and also on the number of wells (Table 6.1). Of course, after several split-and-pool cycles, the concentration of each molecule would also be lessened so quickly that it would normally be impossible to identify the structures of any of the molecules produced. This is where the encoding DNA becomes useful; through PCR amplification, the encoding region can later be read and the identity of the attached molecule deduced by piecing together the encoded building blocks.

Obviously, diversity is not everything. When designing a DEL, careful consideration should be given to the overall size of the library and the complexity of the molecules generated. Diversity created by using fewer plates but more synthesis rounds is likely to produce molecules with higher Molecular Weights (MW) that may have less desirable ADME properties [11].

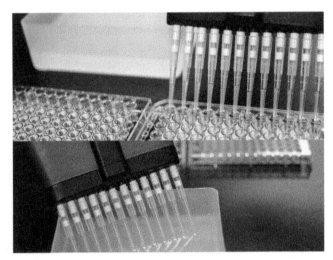

Figure 6.2. Split-and-pool. By splitting the growing library into a 96-well plate at the beginning of each ligation and synthesis cycle, a unique DNA tag is paired with a unique building block and 96 new encoded molecules can be generated. When the products of each cycle are collectively pooled into a single reservoir and split once more, the number of components in each well is multiplied by 96n (n=number of plates used per cycle). For color detail, please see color plate section.

TABLE 6.1. The number of components generated through split-and-pool synthesis based on the number of 96-well plates and the number of synthesis cycles

Split size	Cycle 1	Cycle 2	Cycle 3	Cycle 4
1 × 96	96	9,216	884,736	>84 million
2 × 96	192	36,864	>7 million	>1.3 billion
3 × 96	288	82,944	>23 million	>6.8 billion
4 × 96	384	147,456	>56 million	>21 billion

Three factors should be considered when setting up a synthesis strategy:

1. The structure of the target scaffold
2. The number of available building blocks
3. The number of cycles (consider complexity and MW)

It is also important to remember that with each additional cycle comes a higher probability of error due to lower reaction yields. Deletions and mistagging will ultimately lower the quality of your DEL and your confidence in the selection data.

TABLE 6.2. Examples of DELs showing split sizes, number of cycles, and final number of components generated

Library	Cycle 1	Cycle 2	Cycle 3	Cycle 4	Library size
Test library [7]	1	1	4	4	10^a
Del A [7]	192	192	192	–	7,077,888
Del B [7]	192	32	340	384	802,160,640
ELT [10]	192	479	96	459	4,052,477,952

^{a} Includes six duplicates.

The diversity of four separate DELs is outlined in Table 6.2. The synthesis details of a library as reported in Ref. [7] are used to demonstrate the DEL synthesis process. The size of a library can usually be calculated by simply multiplying the number of building blocks used per cycle. For example, Del A [7] represents $192 \times 192 \times 192 = 7,077,888$ components. An exception to this rule is when duplicates arise. The initial test library [7] would suggest that 16 unique compounds would be synthesized ($1 \times 1 \times 4 \times 4$), but because Cycle 3 and Cycle 4 used the same building blocks on a symmetrical triazine scaffold, six compounds are duplicates and the true number of unique components is only 10.

6.4 DNA LIGATION

When learning how to create a DEL, we must first understand the DNA-encoding strategy. The encoding DNA serves two functions: to record information (i.e., building blocks) and to enable PCR amplification. In many cases, practitioners have chosen to buy their DNA tags from specialty vendors rather than synthesize them in-house. The benefit of using a vendor to supply your DNA is convenience, but the disadvantage may be the cost depending upon the research strategy. Alternatively, with a DNA synthesizer, it is possible to make your own tags using standard phosphoramidite chemistry (see Chapter 1). Table 6.3 shows some commercial sources of DNA tags that may be useful for DEL technology.

In the DEL synthesis method reported in Ref. [7], the starting point for both ligations and chemistry is a unique DNA-linker species named the "Headpiece" (HP). It consists of two single strands of DNA covalently attached at opposing ends via a linker containing a free amine. This creates a DNA duplex with a 2-base-pair overhang that can be elongated with encoding double-stranded DNA tags and also modified chemically at the amine end. When designing the HP, some contend that double-stranded DNA is preferred over single-stranded DNA as it is more stable, giving chemists more flexibility with their reaction conditions. Not only are the nucleobases protected from chemical modification while in a duplex, if the two strands do denature under harsh conditions, each single strand will reanneal with its covalently coupled complementary

TABLE 6.3. Several commercial sources of DNA tags

DNA vendor	Location	Website
Biosearch Technologies	81 Digital Drive Novato, CA, USA	www.biosearchtech.com
DNA Technology A/S	Voldbjergvej 16B DK-8240 Risskov, Denmark	www.dna-technology.dk
IBA GmbH	Rudolf-Wissell-Str. 28 D-37079 Göttingen, Germany	www.iba-lifesciences.com
IDT	1710 Commercial Park, Coralville, IA, USA	www.idtdna.com
TAG Copenhagen A/S	Kong Georgs Vej 12 DK-2000 Frederiksberg, Denmark	www.tagc.com

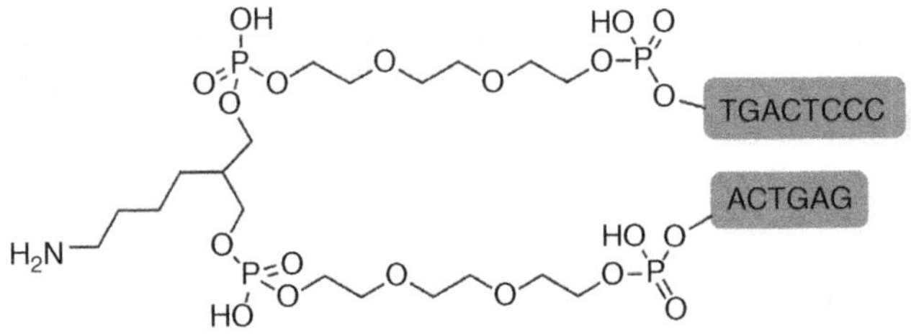

Figure 6.3. The structure of the "HP" [7]. The HP is comprised of two strands of DNA covalently joined through a linker region, which contains a primary amine. This structure links the encoding double-stranded DNA tags to the molecule synthesized during DEL production.

strand when conditions become favorable, preventing tag loss or annealing with foreign strands. The structure of a reported HP is shown in Figure 6.3 [7].

If the aim of creating a DEL is to screen it against an immobilized protein, it is important that the encoding DNA does not interfere with the binding of the synthesized molecule to the protein target. To provide greater distance between the molecule and the encoding DNA, the amine of the HP in Ref. [7] can be further extended by an additional 16 atoms by reacting it with Fmoc-15-amino-4,7,10,13-tetraoxapentadecanoic acid (AOP) followed by treatment with piperidine (Table 6.4). Following ethanol precipitation and isolation, this material is then reconstituted in water to a make a 50 μM solution. This AOP-Headpiece (AOP-HP) solution serves as the starting material for DEL synthesis, and large quantities can be made and stored frozen at −80°C.

6.4.1 Tagging

The tagging strategy reported in Ref. [7] relies on short 7-base-pair double-stranded DNA with a 2-base overhang at each 3′-end. These overhangs serve as "sticky ends" for

TABLE 6.4. Stepwise synthesis conditions used to attach the AOP linker to the HP [7]

Step					
1	**Synthesis of AOP-HP**				
1	**Add**	**Volume**	**Mols**	**Equiv.**	**Total volume**
1.1	1.0 mM (HP) in sodium borate buffer (150 mM, pH 9.4)	100 mL	100 µmol	1	100 mL
1.2	Fmoc-AOP (400 mM in DMF)	10 mL	4 mmol	40	110 mL
1.3	DMT-MM (500 mM in H₂O)	8 mL	4 mmol	40	118 mL
1.4	Agitate for 4 h at 4°C				
1.5	Analyze by LC/MS				
Step	**Remove Fmoc**				
2	**Add**	**Volume**		**Total volume**	
2.1	10% piperidine in H₂O	12 mL		130 mL	
2.2	Agitate for 1 h at room temperature				
2.3	Analyze by LC/MS				

ligation to the next cycle of tags (Fig. 6.4). There are examples of longer tags used in other DELs [2], but short tags have the advantage that they can be more easily separated from the library during HPLC purification.

Generally, the tags arrive from the vendor as lyophilized solids in 96-well plates as single-stranded DNA. These single-stranded tags must first be hybridized to their complementary partner tag to form double-stranded DNA. Each single strand is dissolved in water to a concentration of 2 mM, with careful adjustments made via Optical Density (OD) measurements, and then combined in a 1:1 ratio on a 96-well plate. These plates are used as the source of DEL tags for each ligation round and are dispensed onto reaction plates at the beginning of each synthesis cycle. The first set of plates should be treated with great care as they are expensive and hold the key information for future DEL codes. Tag-containing plates can be stored in the freezer at −80°C until needed again.

An encoding tag will generally be added to the DEL moiety at a ratio of 1.5:1 to ensure complete ligation. However, by using excess tags, there will inevitably be unused DNA remaining at the end of a ligation cycle. These free tags can be problematic as they may participate in the following round of ligations to form a growing DNA chain that

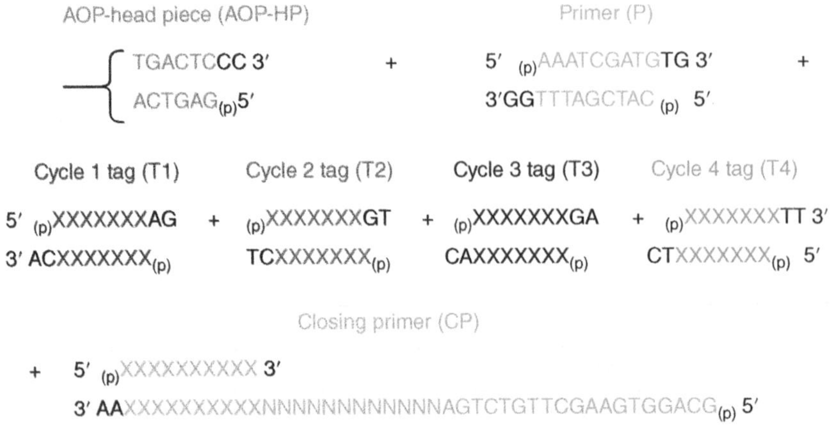

AOP-head piece (AOP-HP)   Primer (P)

┌ TGACTCCC 3'   +   5' (p)AAATCGATGTG 3'   +
│ ACTGAG(p)5'   3'GGTTTAGCTAC (p) 5'
└

Cycle 1 tag (T1)   Cycle 2 tag (T2)   Cycle 3 tag (T3)   Cycle 4 tag (T4)

5' (p)XXXXXXXAG   +   (p)XXXXXXXGT   +   (p)XXXXXXXGA   +   (p)XXXXXXXTT 3'
3' ACXXXXXXX(p)   TCXXXXXXX(p)   CAXXXXXXX(p)   CTXXXXXXX(p) 5'

Closing primer (CP)

+   5' (p)XXXXXXXXXXX 3'
     3' AAXXXXXXXXXXXXNNNNNNNNNNNNNNAGTCTGTTCGAAGTGGACG(p) 5'

Figure 6.4. Example of a DNA tagging strategy [7]. Each segment will enzymatically ligate to the growing DEL structure via 2-base-pair "sticky ends" at the 3'-ends. The 5'-ends contain a free phosphate group (p). The X regions in the encoding tags (T1, T2, T3, and T4) represent undefined oligonucleotides that will encode the building blocks used in each synthesis cycle. The closing primer (CP) contains a long single-stranded segment, which includes a degenerate region (**N**), that will later be filled in by Klenow reaction at the completion of the DEL. For color detail, please see color plate section.

has no molecule associated with it. Also, as these tags grow longer, they become harder to separate from the legitimate DEL components.

The first segment to be ligated to the HP is the Primer (P). This sequence is used to initiate PCR reactions when later amplifying hits from the DEL selection process. Similarly, a Closing Primer (CP) is also attached at the end of the final synthesis cycle. Due to its length, the CP can be used to encode additional information such as the library size and conditions using a variable region, denoted by X. The variable N region in the CP can be a randomized sequence used to identify PCR duplication in the sequence data. Later, during PCR amplification of potential hits from the screening process, duplications in the N region will help resolve true binders (no duplication) from noise (duplication) that may arise from overamplification of nonbinders.

Although a number of ligases are available such as Taq DNA ligase, T7 DNA ligase, and *E. coli* DNA ligase, the preferred ligase in Ref. [7] is T4 DNA ligase. Reference [12] has suggested that Taq DNA ligase may be preferred for sticky-end ligation and T4 DNA ligase for blunt-end ligation.

6.4.2 Ligation Analysis

Ligations are performed in 96-well plates by adding individual tags to each well with T4 DNA ligase and ligation buffer. After standing overnight at room temperature, the ligations should be complete. However, it is important to analyze each ligation cycle to determine completeness and identify any potential errors. The best way to analyze the

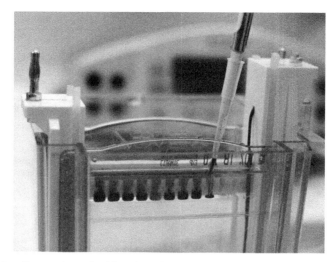

Figure 6.5. Loading a polyacrylamide gel. Performing gel electrophoresis on samples taken from representative wells after each ligation cycle allows you to follow the progression of the lengthening DEL and also look for problematic ligations. For color detail, please see color plate section.

tagging process is by gel electrophoresis using either agarose or polyacrylamide gel plates (Fig. 6.5). Typically, a diagonal cross section of a 96-well plate is sufficient to gauge the extent of ligation of the entire plate. If it becomes apparent that the ligations did not proceed to completion, the best remedy is to add more ligase and ligation buffer and let the plate stand overnight again.

An example of a stained agarose gel is seen in Figure 6.6. Agarose gels are typically easier and faster to run and are not denaturing. Samples from each cycle of a five-cycle DEL including the ligation of the CP are represented alongside MW standards. Lanes 9–12 in this gel represent a hyperladder that helps to quantitate the amount of DNA in each DEL sample. Each cycle should result in an MW increase consistent with the average MW of the tags. The gel also shows that the products of each cycle are essentially homogeneous with regard to MW.

6.4.3 HPLC Purification

Although ethanol precipitation provides an efficient method of removing excess building blocks, reagents, and exchanging solvent, it does not remove excess DNA tags. These tags can continue to ligate with DNA from the next cycle, and it is important to remove them. This is achieved using reverse-phase liquid chromatography on a preparative HPLC system as outlined in Table 6.5.

The frequency of HPLC purification will vary with each DEL synthesis and is best determined by examining both electrophoresis gels and LC/MS chromatograms for tag buildup. HPLC purification after each cycle will ensure a high-quality DEL with high-efficiency ligations, but DNA will inevitably be lost during HPLC fractionation. Table 6.6 shows the frequency at which HPLC purification was done during the preparation of three separate DELs.

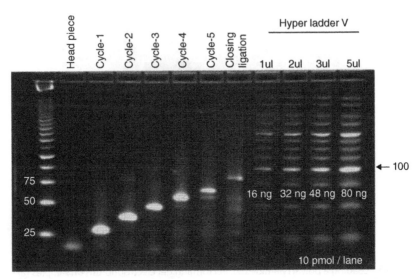

Figure 6.6. A stained agarose gel* [13]. This gel shows samples taken from a DEL involving five tagging cycles. The MW of the DEL increases with every ligation event, while the concentration is seen to decrease slightly. *2% agarose gel stained with ethidium bromide and exposed to UV light.

TABLE 6.5. Quantitative HPLC conditions [7]

Flow rate	Time	Solvent A (80–30%)	Solvent B (20–70%)
0.36 mL/min	22 min linear gradient	50 mM TEAA (triethylamine acetate) in H_2O (pH 7.5)	1% ACN in H_2O

Purifications can be performed on a reverse-phase column (Phenomenex Gemini, 5 μm C18, 100×21.20 mm, 100 Å) using the above conditions.

TABLE 6.6. The frequency of HPLC purification during the preparation of three DELs and the % yield of recovered material

	P	Cycle 1 (%)	Cycle 2	Cycle 3 (%)	Cycle 4	CP	Final DEL yield (%)
DEL A [7]	–	38		52	N/A[a]	–	20%
DEL B [7]	–	54	–	64	58%	–	20%
ELT [10]	–	75	–	47	24%	–	8.5%

In the three-cycle synthesis (DEL A), purification was performed after Cycle 1 and Cycle 3. The four-cycle syntheses (DEL B and ELT) required an additional purification step at the end of Cycle 4. HPLC purification was not deemed necessary after the ligation of the final CP. The yields were calculated by OD measurements.
[a] Cycle 3 only.

6.4.4 Optical Density (OD) Measurements

Throughout the synthesis of your DEL, you will need to measure the concentration of DNA in order to ensure that the correct quantities of DNA tags, building blocks, and other reagents are added. This is most effectively done using an Ultraviolet (UV) spectrophotometer to measure the amount of UV radiation absorbed by the bases. You can calculate the concentration of double- or single-stranded DNA in your sample as follows [14]:

$$\text{dsDNA conc. } (\mu g/mL) = (OD_{260}) \times (DF^*) \times (50\,\mu g \text{ dsDNA/mL})/(1\ OD_{260}\ \text{unit})$$

$$\text{ssDNA conc. } (\mu g/mL) = (OD_{260}) \times (DF^*) \times (33\,\mu g \text{ ssDNA/mL})/(1\ OD_{260}\ \text{unit})$$

* DF, dilution factor.

6.5 CHEMISTRY

One of the desirable aspects of this DEL method is that the synthesis can be achieved using commercially available building blocks. As outlined in Section 5.2, a single DEL can be synthesized using only a few milligrams of any particular building block. Although the price and available quantity of any suitable building blocks can vary dramatically, a diverse collection of building blocks can quickly be amassed by purchasing minimal quantities from suppliers. Before the start of each synthesis cycle, the chosen building blocks are dissolved in an appropriate solvent to create 200–500 mM stock solutions and distributed to labeled vials or Eppendorf tubes. This is typically done the day before a synthesis cycle because it is a time-consuming process. A list of appropriate solvents and concentrations of various building blocks is shown in Table 6.7. The reagents for ligations and synthesis should also be prepared in advance. The one

TABLE 6.7. Solvents and concentrations typically used for building blocks and reagents used in DEL production [7, 10]

Building blocks	Solvent	Conc.	Reagent	Solvent	Conc.
Acids	DMF	200 mM	DMT-MM	H_2O	200 mM
Aldehydes	DMF	200 mM	DNA tags	H_2O	1 mM
Amines	1:1 ACN:H_2O or DMA	200 mM	NaCl	H_2O	5 M
Cyanuric chloride	ACN	200 mM	NaCNBH$_3$	ACN	200 mM
Diamines	1:1 ACN:H_2O	200 mM			
Fmoc-AAs	DMF	200 mM			
Isocyanates	ACN	500 mM			
Sulfonyl chlorides	ACN	500 mM			

exception is the DMT-MM solution (used for amide bond formation in an aqueous environment), which should always be made fresh because it is prone to hydrolyze in water.

When considering your synthesis strategy to build your chemical scaffold, it is essential that your reactions can occur in the presence of water (see Chapter 4). By choosing DNA to encode your chemical synthesis, you have immediately imposed limitations on the type of chemistry at your disposal. By nature, DNA is a water-soluble macromolecule and is prone to decomposition at extreme temperatures and under acidic conditions. That being said, DNA is remarkably robust and can tolerate a range of chemical conditions. Typically, on-DNA reactions are performed in water or in a water/DMF mixture. Chapter 4 highlights many of the chemical reactions that can be performed in the presence of water. A further list of useful reactions and conditions is outlined in Table 6.8 [7, 10].

TABLE 6.8. Examples of reactions and conditions suitable for aqueous DEL synthesis [7, 10]

Reaction	Building block	Equiv.	Reagents	Temp.	Buffer (pH)	Time
Acylation	Fmoc-AA	40	DMT-MM	4°C	Sodium borate (pH 9.4)	18 h
	RCO_2H					
N-Acyl capping	Succinimidyl Acetate	10	–	4°C–RT	Sodium borate (pH 9.4)	10 m
Cycloaddition to an Alkyne	Azide	10	Copper Sulfate/ $(BimH)_3$	RT	Phosphate (pH 8.0)	1–2 h
Reductive alkylation	Aldehyde	40–100	$NaBH_3CN$	RT–80°C	Sodium phosphate (pH 5.5)	2 h
Reductive amination	Amine	40	$NaBH_3CN$	80°C	Sodium phosphate (pH 5.5)	2 h
Sulfonation	Sulfonyl Chloride	40	–	RT	Sodium borate (pH 9.4)	16 h
Triazine-Cl amination	Amine	40	–	80°C	Sodium borate (pH 9.4)	6 h
Triazine-Cl_2 amination	Amine	40	–	4°C	Sodium borate (pH 9.4)	1 h
Triazine arylation	Cyanuric Chloride	10	–	4°C	Sodium borate (pH 9.4)	16 h
Urea formation	Isocyanate	40	–	RT	Sodium borate (pH 9.4)	16 h

6.5.1 Building Block Validation

To ensure that chemical reactions perform well during library production, every aspect of the DEL synthesis should be individually tested and optimized before initiating your synthesis. This is especially true for building blocks. Before deciding to use a particular building block, it should be tested to measure its effectiveness in the reaction in question. As outlined in Section 5.7, validations were performed on the AOP-HP, and the suitability of each building block was determined by LC/MS analysis. Generally, building blocks that give less than 70% yield in their desired role(s) when tested on mock scaffolds should be excluded from the synthesis. The process of building block validation may take some time, but it is important and necessary. Fortunately, a large number of building blocks have already been validated and can be found in the literature [2, 5, 7, 13, 15–17], although bear in mind that these have been validated for specific scaffolds.

6.5.2 LC/MS Analysis

The most practical way of measuring the outcome of chemical transformations during DEL synthesis is LC/MS analysis. Most on-DNA reactions that involve acylation or amine deprotection will cause a noticeable change in peak retention time when viewing the UV chromatogram. When performing building block validations, reaction yields can be determined by comparing the area of the starting material peak against the product peak. The identity of both the starting material and product peaks can be confirmed by Total Ion Current (TIC) analysis of each peak.

Chemists unfamiliar with DNA may at first be surprised to learn that mass analysis can be achieved using a standard LC/MS. Most commercially available quadrupole mass spectrometers have a mass-to-charge (m/z) range less than $3000\,m/z$ units, while the AOP-HP has an MW of 4937. So how can this be detected? When using the negative mode, it is possible to observe multiply charged species of the DNA compound at $[(MW/3) - 1]$, $[(MW/4) - 1]$, $[(MW/5) - 1]$, and $[(MW/6) - 1]$, confirming its identity. When deconvolution becomes more difficult in low signal-to-noise spectra, specialized software is available for producing excellent artifact-free deconvoluted mass spectra [18]. See Chapter 8 for more details regarding analytical methods for oligonucleotides. The conditions for analyzing DNA on an LC/MS are shown in Table 6.9, and useful chromatograms can be obtained by injecting samples of less than 100 pmol on a HPLC/electrospray mass spectrometer in negative ion mode.

TABLE 6.9. Analytical LC/MS conditions used to analyze on-DNA reactions

Flow rate	Time	Sample volume	Solvent A (85–15%)	Solvent B (15–70%)	UV
0.36 mL/min	7 min	1 µL in 40 µL of H$_2$O	1. 0.75% HFIP 2. 0.38% TEAA 3. 5–10 µM EDTA in H$_2$O	1. 0.75% HFIP 2. 0.38% TEAA 3. 5 µM EDTA MeOH:H$_2$O (9:1)	260 nm

The electrospray mass spectrometer is set in negative ion mode with a reverse-phase column (Targa C18, 5 µm, 2.1 × 40 mm) [7, 10].

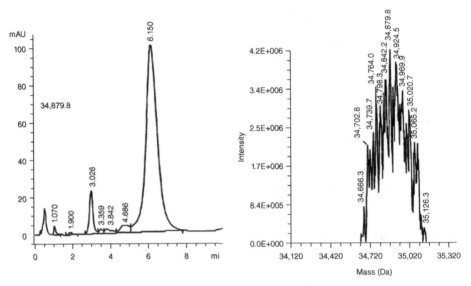

Figure 6.7. LC/MS analysis after a synthesis cycle. A chromatogram of a DEL after four cycles of synthesis and its corresponding mass spectrum. The DEL appears as a large single peak in the chromatogram, while the mass spectrum shows average masses [15].

Figure 6.7 shows a typical chromatogram of a DEL after four cycles of synthesis [15]. The chromatogram shows a single major peak with an average MW of $34,880\,m/z$. The accompanying deconvoluted ion signal shows a large number of MW within this peak, which contains over 800 million components. The deconvolution of such a massive collection of high-MW compounds can only be achieved with specialized software [18]. Quantifying the yield of each round of chemical synthesis is extremely difficult. At best, one can observe gross shifts in the average mass of the growing DEL after the addition of a collection of building blocks.

Although the large number of molecules within the DEL severely limits the usefulness of the LC/MS data, it is still important to run LC/MS samples after each chemistry cycle to follow the changing characteristics of the growing DEL. Subtle differences in polarity created by each building block family will cause noticeable changes in the LC/MS chromatogram that will give clues to whether the chemistry was successful or if an error has occurred.

6.6 RECORDKEEPING

It goes without saying that good recordkeeping is essential for successful DEL synthesis. The key to unlocking the chemical structure of any hit identified through the selection process is to match the encoding DNA tags with the correct building blocks. Equally important is keeping track of every step performed during the DEL synthesis. As you will see in Table 6.10, Table 6.11, Table 6.12, Table 6.13, and Table 6.14, synthesizing a DEL on a single 96-well plate involves more than 33 processes, many of

TABLE 6.10. P ligation (for color detail, please see color plate section)

Step	P ligation				
1	Add	Volume	Mols	Equiv.	Total volume (mL)
1.1	AOP-HP (50 µM in H₂O)	50 mL	50 µmol	1	50
1.2	10× ligation buffer[a]	20 mL	–	–	70
1.3	T4 DNA ligase	2 mL	–	–	72
1.4	H₂O	64 mL	–	–	136
1.5	DNA P (1.0 mM in H₂O)	64 mL	64 µmol	1.28	200
1.6	Stand overnight at room temperature				
1.7	Analyze by gel electrophoresis				
Step	Ethanol crash				
2	Add	Volume			Total volume
2.1	5 M NaCl(aq)	20 mL			220 mL
2.2	Cold EtOH	440 mL			660 mL
2.3	−78°C for ≥30 min or −20°C for 1 h				
2.4	Centrifuge and discard solvent				
2.5	Lyophilize DNA pellet				
2.6	Dissolve in approx. 50 mL of H₂O to make 1.0 mM solution				
2.7	Confirm conc. via OD measurement				

Table 6.10 outlines a step-by-step approach to attaching the P to the AOP-HP described in Table 6.4.
[a] 500 mM Tris pH 7.5, 500 mM NaCl, 100 mM $MgCl_2$, 100 mM DTT, 20 mM ATP.

which involve tedious yet precise additions of reagents. Any loss of concentration or missed step can have terrible consequences for the outcome of your DEL. Therefore, it is important to prepare well-thought-out checklists that will enable you to keep track of each step. It is also advisable to work in pairs so that your partner can double-check each addition.

6.7 TEST LIBRARIES

Once you have chosen a target scaffold and identified a set of high-yielding building blocks through a validation process, you are almost ready to start creating a DEL. The best advice we can offer is to start small and work your way up. By small, we mean a

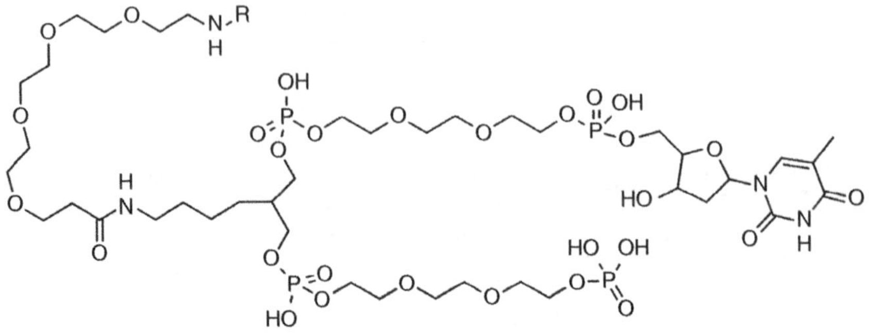

R = Synthesized molecule

Figure 6.8. Digestion of a DEL. Using enzymes such as DNase and S1 nuclease will allow the removal of the encoding DNA to give only the chemically synthesized molecule attached to a remnant of the AOP-HP [7].

single-component DEL. This will involve all of the necessary ligation and chemistry steps to produce a single molecule tagged to DNA and will give you the opportunity to become familiar with the analytical and purification techniques.

Once you have gained some confidence making a single DNA-tagged molecule, try making a small library of less than 20 components. This will require using the split-and-pool method and give you a feel for analyzing a mixture of compounds in a single sample. An advantage of creating a small DEL is that you can perform a "postmortem" analysis to understand where things might have gone wrong. A portion of the DEL can be digested using a combination of DNase and S1 nuclease to remove the encoding DNA and give only the chemistry products attached to the linker and a single nucleotide (Fig. 6.8).

By interrogating the mixture by LC/MS and NMR analyses, you will hopefully identify the desired molecules, but you may also observe truncations and side products that will give you clues to areas that still need optimization. Another portion of the DEL should be amplified by PCR to confirm that the encoding tags are in synchrony with the building blocks.

6.8 MAKING YOUR FIRST MILLION

Once you are comfortable making a small DEL, you are ready to move to a 96-well plate format and begin synthesizing a much more diverse library. GSK scientists recently published a paper [10] outlining the synthesis of a four-billion-component library that, when screened against the metalloprotease ADAMTS-5, identified a series of novel inhibitors. When the hits were resynthesized off-DNA, an ADAMTS-5 inhibitor with $IC_{50} = 30$ nM was obtained. This DEL provides an excellent example of what can be achieved through DNA-encoded synthesis.

The details provided in the supplemental section of Ref. [10] outline the many steps involved in DEL production. However, due to the immense size and molecular complexity of this DEL, which involved multiple 96-well plates and four different scaffolds, the inexperienced scientist may find the details of this synthesis difficult to follow. To make things easier for the reader, we have provided a single 96-well plate interpretation of this information in the form of Table 6.11, Table 6.12, Table 6.13, and Table 6.14. These tables provide color-coded step-by-step instructions that should allow you to understand the many steps, the reaction scale, and the volumes involved in performing a multicycle synthesis to build a multimillion-component DEL. The single-plate example provided in Table 6.11, Table 6.12, Table 6.13, and Table 6.14 outlines four synthesis cycles. Cycle 1 (violet) uses 96 Fmoc-amino acids (AAs); Cycle 2 (green) uses cyanuric chloride followed by 96 amines; Cycle 3 (blue) uses 96 diamines; and Cycle 4 (pink) uses 96 sulfonyl chlorides. This combination will generate 84,934,656 components. Scaling up the DEL synthesis to generate even greater diversity would simply involve adding additional plates and introducing additional building blocks at each cycle. However, if this is your first attempt at making a million-component DEL, sticking to a single 96-well plate is highly recommended. This will provide you valuable laboratory experience and also produce ample material for performing selections against an immobilized protein target.

6.8.1 Things You Will Need

The following list represents many of the components needed to make a DEL. Essential tools and equipment are also illustrated in Figure 6.9 and Figure 6.10.

- AOP-HP solution ($50\,\mu M$ in H_2O)
- Building block solutions ($200\,mM$ in either DMF, ACN, 1:1 ACN:H_2O, or DMA)
- Centrifuge with plate holders (ideally refrigerated)
- Cold room (optional)
- DMT-MM solution ($200\,mM$ in H_2O)
- DNA P and tags ($1.0\,mM$ in H_2O)
- Ethanol (99% pure)
- Gel electrophoresis kit
- HPLC instrument
- LC/MS instrument
- 10× Ligation buffer stock ($500\,mM$ Tris pH 7.5, $500\,mM$ NaCl, $100\,mM$ MgCl$_2$, $100\,mM$ DTT, and $25\,mM$ ATP)
- Lyophilizer
- Orbital shaker (to agitate plates)
- Pipettes (single and multichannel)
- Sodium borate buffer (pH = 9.4, $150\,mM$)
- Sodium chloride solution ($5\,M$)

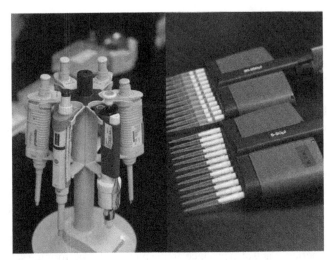

Figure 6.9. Essential tools for DEL synthesis. A collection of well-maintained single and multichannel pipettes will be used throughout the DEL production. Single pipettes will be needed for preparing building block solutions and reagents, while the multichannel pipettes will drive the split-and-pool plate syntheses. For color detail, please see color plate section.

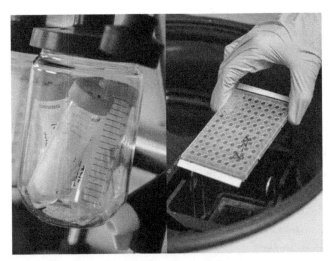

Figure 6.10. Essential equipment for DEL synthesis. Because the main solvent in DEL synthesis is water and purification is done by reverse-phase HPLC, a lyophilizer is essential for drying your product. A centrifuge with a dual plate holder is also crucial for isolating DNA after each ethanol precipitation step.

- T4 DNA ligase
- UV spectrophotometer for OD measurements
- Vortexer (for redissolving DNA pellets)
- 96-well plates

6.8.2 Step-by-Step

The first step for synthesizing your DEL is the attachment of DNA P to the AOP-HP, initially described in Table 6.4. The instructions in Table 6.10 start with 50 mL of a 50 μM stock solution of AOP-HP and outline the steps required to isolate 1.0 mM AOP-HP P (orange-coded product). This represents a substantial amount of starting material as only 6 mL is required to initiate a single-plate DEL synthesis. However, having excess material is always useful, and unused AOP-HP P can be stored at −80°C until needed again. The volumes in Table 6.11, Table 6.12, Table 6.13, and Table 6.14 represent single-well volumes, and the total volume per plate is 96 times greater. Also, the loss of material from each HPLC purification step is not factored into the volumes of starting material at the beginning of each cycle, so additional planning is needed to avoid running low on material.

The beginning of each synthesis cycle involves splitting the product from the previous cycle into a 96-well plate. The enzymatic ligation of the tags will always precede the chemical addition of the building blocks in each cycle. The first synthesis cycle is shown in Table 6.11. Fmoc-AAs are attached to the DEL moiety in two consecutive additions to generate 96 unique components (violet-coded product), which are pooled prior to HPLC purification.

TABLE 6.11. Cycle 1 (for color detail, please see color plate section)

Step	Split				
3	Split 6 mL of AOP-HP P (1.0 mM) into 96 wells				
Step	**Cycle 1— tag (T1) ligation**				
4	**Add**	**Volume**	**Mols**	**Equiv.**	**Total volume (per well)**
4.1	AOP-HP P (1.0 mM in H_2O)	62.5 μL	62.5 nmol	1	62.5 μL
4.2	10× ligation buffer	25 μL	–	–	87.5 μL
4.3	T4 DNA ligase	2.5 μL	–	–	90 μL
4.4	H_2O	35 μL	–	–	125 μL
4.5	DNA tags (T1) (1.0 mM in H_2O)	125 μL	125 nmol	2	250 μL
4.6	Stand overnight at room temperature				
4.7	Analyze by gel electrophoresis				

(Continued)

TABLE 6.11. (cont'd)

Step	Ethanol crash		
5	**Add**	**Volume**	**Total volume (per well)**
5.1	5 M NaCl(aq)	25 µL	275 µL
5.2	Cold EtOH	550 µL	825 µL
5.3	−78°C for ≥30 min or −20°C for 1 h		
5.4	Centrifuge and discard solvent		

Step	Cycle 1—building block addition				
6	**Add**	**Volume**	**Mols**	**Equiv.**	**Total volume (per well)**
6.1	Sodium borate buffer (150 mM, pH 9.4)	62.5 µL	62.5 nmol	1	62.5 µL
6.2	Fmoc-AAs (200 mM in DMF)	12.5 µL	2.5 µmol	40	75 µL
6.3	DMT-MM (200 mM in H_2O)	12.5 µL	2.5 µmol	40	87.5 µL
6.4	Agitate for 2 h at 4°C				
6.5	Fmoc-AAs (200 mM in DMF)	12.5 µL	2.5 µmol	40	90 µL
6.6	DMT-MM (200 mM in H_2O)	12.5 µL	2.5 µmol	40	92.5 µL
6.7	Agitate overnight at 4°C				
6.8	Analyze by LC/MS				

Step	Pool
7	Pool all 96 wells into a single vessel

Step	Ethanol crash		
8	**Add**	**Volume**	**Total volume (per well)**
8.1	5 M NaCl(aq)	0.89 mL	9.77 mL
8.2	Cold EtOH	19.5 mL	29.3 mL
8.3	−78°C for ≥30 min or −20°C for 1 h		
8.4	Centrifuge and discard solvent		
8.5	Dissolve in H_2O to make 1.0 mM solution		

Step	Purify
9	Purify by reverse-phase HPLC
9.1	Lyophilize pooled fractions

(Continued)

TABLE 6.11. (cont'd)

Step	Remove Fmoc		
10	**Add**	**Volume**	**Total volume**
10.1	10% piperidine in H$_2$O	36 mL	36 mL
10.2	Agitate for 1 h at room temperature		
10.3	Analyze by LC/MS		
Step	Ethanol crash		
11	**Add**	**Volume**	**Total volume**
11.1	5 M NaCl(aq)	3.6 mL	39.6 mL
11.2	Cold EtOH	80 mL	119.6 mL
11.3	−78°C for ≥30 min or −20°C for 1 h		
11.4	Centrifuge and discard solvent		
11.5	Lyophilize DNA pellet		
11.6	Dissolve in H$_2$O to make 1.0 mM solution		
11.7	Confirm conc. via OD measurement		

Table 6.11 outlines a step-by-step approach to performing the first cycle of a four-cycle DEL using a single 96-well plate.

The second synthesis cycle is shown in Table 6.12. After splitting the Cycle 1 product (violet code) into 96 wells and attaching the Cycle 2 tags (T2), two consecutive synthesis steps are used to install the triazine moiety followed by 96 unique amines to generate 9,216 pooled components (green-coded product). HPLC purification may be necessary at this point depending on the level of purity of the DEL mixture as determined by LCMS and gel electrophoresis analysis (Table 6.12).

The third synthesis cycle is shown in Table 6.13. The Cycle 2 product (green code) was first ligated with Cycle 3 tags (T3) followed by reaction with 96 diamines consisting of 60 Nvoc-protected diamines, 20 unprotected symmetrical diamines (e.g., piperazine), and 16 unprotected aromatic/aliphatic diamines (e.g., 4-aminobenzylamine). The pooled mixture is then irradiated at 365 nm to remove the Nvoc group and then purified by HPLC to give 884,736 components (blue-coded product).

The fourth synthesis cycle is shown in Table 6.14. The Cycle 3 product (blue code) is split into a 96-well plate, encoded with Cycle 4 tags (T4), and then reacted with 96 individual sulfonyl chlorides. This final diversity-creating step will generate 84,934,656 components (pink-coded product).

If a similar yield to the ELT library in Ref. [10] is encountered, you could expect to recover 500 nmol of DEL product using the single-plate example represented in Table 6.11, Table 6.12, Table 6.13, and Table 6.14. Although this does not sound like much, as little at 5 nmol is needed per selection.

TABLE 6.12. Cycle 2 (for color detail, please see color plate section)

(13) 96 ligations
(15) Cyanuric chloride/96 amines

(1 plate = 9,216 components)

Step	Split				
12	Split 5.76 mL of Cycle 1 product (1.0 mM) into 96 wells				
Step	**Cycle 2—tag (T2) ligation**				
13	Add	Volume	Mols	Equiv.	Total volume (per well)
13.1	Cycle 1 product (1.0 mM in H₂O)	60 μL	60 nmol	1	60 μL
13.2	10 × ligation buffer	24 μL	–	–	84 μL
13.3	T4 DNA ligase	2.4 μL	–	–	86.4 μL
13.4	H₂O	33.6 μL	–	–	120 μL
13.5	DNA tags (T2) (1.0 mM in H₂O)	110 μL	110 nmol	1.8	230 μL
13.6	Stand overnight at room temperature				
13.7	Analyze by gel electrophoresis				
Step	**Ethanol crash**				
14	Add	Volume		Total volume (per well)	
14.1	5 M NaCl(aq)	23 μL		253 μL	
14.2	Cold EtOH	506 μL		759 μL	
14.3	−78°C for ≥30 min or −20°C for 1 h				
14.4	Centrifuge and discard solvent				
Step	**Cycle 2—building block addition**				
15	Add	Volume	Mols	Equiv.	Total volume (per well)
15.1	Sodium borate buffer (150 mM, pH 9.4)	60 μL	60 nmol	1	60 μL
15.2	Cyanuric chloride (200 mM in ACN)	3 μL	0.6 μmol	10	63 μL
15.3	Agitate for 1 h				
15.4	Cool plate to 4°C				

(Continued)

TABLE 6.12. (cont'd)

15	Add	Volume	Mols	Equiv.	Total volume (per well)
15.5	Amines (200 mM in DMA or ACN:H$_2$O (1:1))	15 µL	3 µmol	50	78 µL
15.6	Agitate for 40 h at 4°C				
15.7	Analyze by LC/MS				
Step	**Pool**				
16	Pool all 96 wells into a single vessel				
Step	**Ethanol crash**				
17	Add	Volume		Total volume	
17.1	5 M NaCl(aq)	0.75 mL		8.24 mL	
17.2	Cold EtOH	16.5 mL		24.7 mL	
17.3	−78°C for ≥30 min or −20°C for 1 h				
17.4	Centrifuge and discard solvent				
17.5	Dissolve in approx. 5.76 mL H$_2$O to make 1.0 mM stock				
17.6	Confirm conc. via OD measurement				

Table 6.12 outlines a step-by-step approach to performing the second cycle of a four-cycle DEL using a single 96-well plate.

TABLE 6.13. Cycle 3 (for color detail, please see color plate section)

(19) 96 ligations
(21) 96 diamines (inc.60 Nvoc-diamines)
(24) 365 nm

(1 plate = 884,736 components)

Step	Split				
18	Split 5.76 mL of Cycle 2 product (1.0 mM) into 96 wells				
Step	**Cycle 3—tag (T3) ligation**				
19	Add	Volume	Mols	Equiv.	Total volume (per well)
19.1	Cycle 2 product (1.0 mM in H$_2$O)	60 µL	60 nmol	1	60 µL
19.2	10 × ligation buffer	24 µL	–	–	84 µL

(Continued)

TABLE 6.13. (cont'd)

19	Add	Volume	Mols	Equiv.	Total volume (per well)
19.3	T4 DNA ligase	2.4 μL	–	–	86.4 μL
19.4	H_2O	33.6 μL	–	–	120 μL
19.5	DNA tags (T3) (1.0 mM in H_2O)	180 μL	180 nmol	3	300 μL
19.6	Stand overnight at room temperature				
19.7	Analyze by gel electrophoresis				
Step	**Ethanol crash**				
20	Add	Volume		Total volume (per well)	
20.1	5 M NaCl(aq)	30 μL		330 μL	
20.2	Cold EtOH	660 μL		990 μL	
20.3	−78°C for ≥30 min or −20°C for 1 h				
20.4	Centrifuge and discard solvent				
Step	**Cycle 3—building block addition**				
21	Add	Volume	Mols	Equiv.	Total volume (per well)
21.1	Sodium borate buffer (150 mM, pH 9.4)	60 μL	60 nmol	1	60 μL
21.2	Nvoc-diamines (200 mM in DMA or ACN:H_2O (1:1))	15 μL	3 μmol	50	75 μL
21.3	Agitate for 8 h at 80°C				
21.4	Analyze by LC/MS				
Step	**Pool**				
22	Pool all 96 wells into a single vessel				
Step	**Ethanol crash**				
23	Add	Volume		Total volume	
23.1	5 M NaCl(aq)	0.72 mL		7.92 mL	
23.2	Cold EtOH	15.8 mL		23.7 mL	
23.3	−78°C for ≥30 min or −20°C for 1 h				
23.4	Centrifuge and discard solvent				
Step	**Nvoc deprotection**				
24	Dissolve in 3.6 mL AcOH aq. buffer (100 mM, pH 4.5)				
24.1	Transfer to flat-bottomed crystallizing dish				
24.2	Cool to 4°C and irradiate at 365 nm for 16 h				
24.3	Transfer to a centrifuge tube				

(Continued)

TABLE 6.13. (cont'd)

Step	Ethanol crash		
25	**Add**	**Volume**	**Total volume**
25.1	5 M NaCl(aq)	0.36 mL	3.96 mL
25.2	Cold EtOH	7.9 mL	11.86 mL
25.3	−78°C for ≥30 min or −20°C for 1 h		
25.4	Centrifuge and discard solvent		
25.5	Dissolve in H₂O		
Step	**Purify**		
26	Purify by reverse-phase HPLC		
26.1	Lyophilize pooled fractions		
26.2	Dissolve in H₂O to make 1.0 mM stock		
26.3	Confirm conc. via OD measurement		

Table 6.13 outlines a step-by-step approach to performing the third cycle of a four-cycle DEL using a single 96-well plate.

TABLE 6.14. Cycle 4 (for color detail, please see color plate section)

(28) 96 ligations
(30) 96 sulfonyl chlorides

(1 DEL = 84,934,656 components)

Step	Split				
27	Split 2.5 mL of Cycle 3 product (1.0 mM) into 96 wells				
Step	**Cycle 4—tag (T4) ligation**				
28	**Add**	**Volume**	**Mols**	**Equiv.**	**Total volume (per well)**
28.1	Cycle 3 product (1.0 mM in H₂O)	26 µL	26 nmol	1	26 µL
28.2	10 × ligation buffer	10.4 µL	–	–	36.4 µL

(Continued)

TABLE 6.14. (cont'd)

28	Add	Volume	Mols	Equiv.	Total volume (per well)
28.3	T4 DNA ligase	1.04 µL	–	–	37.4 µL
28.4	H₂O	14.6 µL	–	–	52.0 µL
28.5	DNA tag (T4) (1.0 mM in H₂O)	60 µL	60 nmol	2.3	112 µL
28.6	Stand overnight at room temperature				
28.7	Analyze by gel electrophoresis				

Step	Ethanol crash		
29	Add	Volume	Total volume (per well)
29.1	5 M NaCl(aq)	11.2 µL	123 µL
29.2	Cold EtOH	246 µL	369 µL
29.3	−78°C for ≥30 min or −20°C for 1 h		
29.4	Centrifuge and discard solvent		

Step	Cycle 4—building block addition				
30	Add	Volume	Mols	Equiv.	Total volume (per well)
30.1	Sodium borate buffer (150 mM, pH 9.4)	26 µL	26 nmol	1	26 µL
30.2	Sulfonyl chlorides (500 mM in ACN)	2.1 µL	1.05 µmol	40	28.1 µL
30.3	Agitate for 16 h at room temperature				
30.4	Analyze by LC/MS				

Step	Pool
31	Pool all 96 wells into a single vessel

Step	Ethanol crash		
32	Add	Volume	Total volume
32.1	5 M NaCl(aq)	270 µL	2.97 mL
32.2	Cold EtOH	5.93 mL	8.90 mL
32.3	−78°C for ≥30 min or −20°C for 1 h		
32.4	Centrifuge and discard solvent. Dissolve DNA pellet in H₂O		

Step	Purify
33	Purify by reverse-phase HPLC
33.1	Dissolve in H₂O to make 1.0 mM solution
33.2	Confirm conc. via OD measurement

Table 6.14 outlines a step-by-step approach to performing the last cycle of a four-cycle DEL using a single 96-well plate.

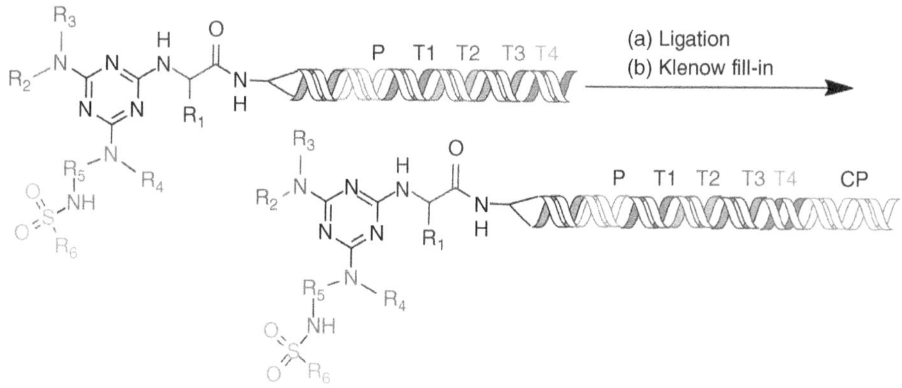

Figure 6.11. CP and Klenow fill-in. The final step in DEL synthesis is the ligation of the CP followed by Klenow fill-in. This can be done on a portion of the Cycle 4 product.

The Cycle 4 product (pink code) represents the completion of the DEL synthesis. All that remains is the ligation of the CP followed by a Klenow fill-in of the degenerate region using a mixture of dATP, dCTP, dGTP, and dTTP (Fig. 6.11).

Following the attachment of the CP, using similar conditions to those outlined in Table 6.10, the product is incubated at 37°C for 30 min in ligation buffer (50 mM Tris pH 7.5, 50 mM NaCl, 10 mM MgCl$_2$, 10 mM DTT, 2 mM ATP) supplemented with 10 mM dNTP and DNA polymerase I Klenow fragment. After completion of the Klenow reaction, the precipitated DEL is reconstituted in water to a final concentration of 0.5 mM and is ready for affinity selections.

6.9 FURTHER CONSIDERATIONS

Although DEL synthesis may appear intimidating at first glance, we hope that the instructions provided in this chapter will help alleviate any trepidation you may hold towards considering this powerful drug discovery tool.

Some have suggested that the large number of incremental steps makes the synthesis cumbersome and susceptible to significant "noise" as unknown reactions may create unwanted and/or unknown molecules attached to the DNA [17]. Furthermore, excess tags can result in cross-contamination during the ligation steps, resulting in mistagging, which may lower the efficiency of sequencing. Also, the HP that covalently links two DNA strands together may increase the melting point of the duplex and consequently hinder the accessibility of the P-binding region for polymerase amplification (e.g., PCR). To reduce the amount of library noise, you should always consider ways to reduce the overall number of steps that could lead to potential errors. For example, the number of ligation events could be reduced by incorporating encoding sequences in the P tags. This would require a unique P tag for each building block, but the result would be one less DEL ligation event. Also, deletions within molecules could be prevented by performing capping steps after each building block reaction to prevent unreacted material participating in the following synthesis cycle.

As mentioned in the introduction, just because many methods of creating DELS have have already been established, it does not mean there is little left to invent. The number of groups making use of this technology is still relatively small. As interest builds and more scientists become familiar with using DNA as a substrate for performing chemistry, we will undoubtedly see new and innovative ways of building DELs. Hopefully, through the knowledge presented in this chapter, you will be able to build, expand upon, and invent new and creative ways of building DELs that will benefit future drug discovery programs.

REFERENCES

1. Kemp, M. M., Weïwer, M., and Koehler, A. N. (2012) Unbiased binding assays for discovering small-molecule probes and drugs, *Bioorg. Med. Chem.*, *20*, 1979–1989.
2. Mannocci, L. (2009) DNA-encoded chemical libraries. Doctoral dissertation. Retrieved from e-collection, ETH Institutional Repository. http://dx.doi.org/10.3929/ethz-a-005783014. Accessed on November 26, 2013.
3. Kleiner, R. E., Dumelin, C. E., and Liu, D. R. (2011) Small-molecule discovery from DNA-encoded chemical libraries, *Chem. Soc. Rev.*, *40*, 5707–5717.
4. Scheuermann, J. and Neri, D. (2010) DNA-encoded chemical libraries: a tool for drug discovery and for chemical biology, *ChemBioChem*, *11*, 931–937.
5. Buller, F., Mannocci, L., Scheuermann, J., and Neri, D. (2010) Drug discovery with DNA-encoded chemical libraries, *Bioconjugate Chem.*, *21*, 1571–1580.
6. Hansen, M. H., Blakskjær, P., Petersen, L. K., Hansen, T. H., Højfeldt, J. W., Gothelf, K. V., and Hansen N. J. V. (2009) A yoctoliter-scale DNA reactor for small-molecule evolution, *J. Am. Chem. Soc.*, *131*, 1322–1327.
7. Clark, M. A., Acharya, R. A., Arico-Muendel, C. C., Belyanskaya, S. L., Benjamin, D. R., Carlson, N. R., Centrella, P. A., Chiu, C. H., Creaser, S. P., Cuozzo, J. W., Davie, C. P., Ding, Y., Franklin, G. J., Franzen, K. D., Gefter, M. L., Hale, S. P., Hansen, N. J. V., Israel, D. I., Jiang, J., Kavarana, M. J., Kelley, M. S., Kollmann, C. S., Li, F., Lind, K., Mataruse, S., Medeiros, P. F., Messer, J. A., Myers, P., O'Keefe, H., Oliff, M. C., Rise, C. E., Satz, A. L., Skinner, S. R., Svendsen, J. L., Tang, L., van Vloten, K., Wagner, R. W., Yao, G., Zhao, B., and Morgan, B. A. (2009) Design, synthesis and selection of DNA-encoded small-molecule libraries, *Nat. Chem. Biol.*, *5*, 647–654.
8. Merrifield, R. B. (1963) Solid phase peptide synthesis. I. The synthesis of a tetrapeptide, *J. Am. Chem. Soc.*, *85*, 2149–2154.
9. Merrifield, B. (1986) Solid phase synthesis, *Science*, *232*, 341–347.
10. Deng, H., O'Keefe, H., Davie, C. P., Lind, K. E., Acharya, R. A., Franklin, G. J., Larkin, J., Matico, R., Neeb, M., Thompson, M. M., Lohr, T., Gross, J. W., Centrella, P. A., O'Donovan, G. K., Bedard, K. L., van Vloten, K., Mataruse, S., Skinner, S. R., Belyanskaya, S. L., Carpenter, T. Y., Shearer, T. W., Clark, M. A., Cuozzo, J. W., Arico-Muendel, C. C., and Morgan, B. A. (2012) Discovery of highly potent and selective small molecule ADAMTS-5 inhibitors that inhibit human cartilage degradation via encoded library technology (ELT), *J. Med. Chem.*, *55*, 7061–7079.

11. Lipinski, C. A., Lombardo, F., Dominy, B. W., and Feeney, P. J. (2001) Experimental and computational approaches to estimate solubility and permeability in drug discovery and development settings, *Adv. Drug Deliver. Rev. 46*, 3–26.

12. Freskgard, P., Franch, T., Gouliaev, A. H., Lundorf, M. D., Felding, J., Olsen, E. K., Holtmann, A., Jakobsen, S. N., Sams, C. K., Glad, S. S., Jensen, K. B., and Pedersen, H. (2012) Method for the synthesis of a bifunctional complex, U.S. Patent 8,206,901, PCT filed October 30, 2003 and issued June 26, 2012.

13. Morgan, B., Hale, S., Arico-Muendel, C. C., Clark, M., Wagner, R., Israel, D. I., Gefter, M. L., Benjamin, D., Hansen, N. J. V., Kavarana, M. J., Creaser, S. P., Franklin, G. J., Centrella, P. A., and Acharya, R. A. (2011) Methods for synthesis of encoded libraries, U.S. Patent 7,935,658, filed October 4, 2007 and issued May 3, 2011.

14. Barbas, C. F. III, Burton, D. R., Scott, J. K., and Silverman, G. J. (2007) Quantitation of DNA and RNA, *Cold Spring Harb Protoc*. doi:10.1101/pdb.ip47.

15. Morgan, B., Hale, S., Arico-Muendel, C. C., Clark, M., Wagner, R., Israel, D. I., Gefter, M. L., Benjamin, D., Hansen, N. J. V., Kavarana, M. J., Creaser, S. P., Franklin, G. J., Centrella, P. A., and Acharya, R. A. (2011) Methods for identifying compounds of interest using encoded libraries, U.S. Patent 7,989,395, filed October 23, 2006 and issued August 2, 2011.

16. Leimbacher, M., Zhang, Y., Mannocci, L., Stravs, M., Geppert, T., Scheuermann, J., Schneider, G., and Neri, D. (2012) Discovery of small-molecule interleukin-2 inhibitors from a DNA-encoded chemical library, *Chem. Eur. J., 18*, 7729–7737.

17. Wagner, R. W. (2012) Methods of creating and screening DNA-encoded libraries, U.S. Patent Application 2012/0053091, PCT filed February 16, 2010 and published March 1, 2012.

18. ProMass software is available from Novatia, http://www.enovatia.com/products/promass/. Accessed on November 26, 2013.

7

THE DNA TAG: A CHEMICAL GENE DESIGNED FOR DNA-ENCODED LIBRARIES

Andrew W. Fraley

Moderna Therapeutics, Cambridge, MA, USA

7.1 INTRODUCTION

Drug discovery relies on technologies to efficiently generate and identify functional molecules that modulate the activity of a target protein. The rise of combinatorial chemistry and diversity-oriented synthesis has enabled screening (HTS) [1–6] of libraries with complexities up to 10^6 members. Library synthesis typically occurs on the solid phase, employing a split-and-pool methodology to generate a combinatorial set of compounds. The screening of such libraries has primarily been a one compound–one well approach requiring significant investments in robotic, material, and human resources [2]. Drug discovery's inherent need to cover more chemical diversity space has resulted in the need for ever-larger libraries and screening methods that can overcome the limits of classic HTS [7–10]. To improve throughput, many groups have investigated screening multiple compounds in a single well resulting with the need for a post-screening deconvolution event. Methods such as isotopic labeling for deconvolution by mass spec, deletion synthesis, or radiolabeling have been met with limited successes [11]. More recently, high-density compound arrays have approached this problem by spatially sequestering compounds in the 10,000s for deconvolution [12].

A Handbook for DNA-Encoded Chemistry: Theory and Applications for Exploring Chemical Space and Drug Discovery, First Edition. Edited by Robert A. Goodnow, Jr.
© 2014 John Wiley & Sons, Inc. Published 2014 by John Wiley & Sons, Inc.

In the early 1990s, multiple research groups began utilizing nucleic acids as an encoding scheme in the hunt for functional molecules. Akin to nature's biochemical relationship of genotype and phenotype, synthetic oligo-2'-deoxynucleotides were prepared to encode for the corresponding expressed RNA strand via *in vitro* transcription. Most notably, work by the Gold [13], Szostak [14], and Joyce [15] labs independently generated functional RNA aptamers and ribozymes. Encoding in this embodiment is the simplest form as the 2'-deoxynucleotide encodes for the corresponding nucleotide (e.g., 2'-deoxyadenosine for adenosine). A methodology proposal shortly followed by Brenner and Lerner [16] for the synthesis and subsequent tagging of peptides with oligonucleotides, where the DNA genotype is covalently attached to a linker molecule displaying the peptide phenotype. Based upon these initial ideas, numerous methodologies for generating DNA-tagged libraries have been realized over the past 20 years.

The aim of this chapter is to provide the reader with an insight into the thought processes behind genotype design for DNA-encoded libraries [17, 18]. The manner by which a given phenotype or molecule is encoded can vary significantly and depends on the technology utilized for synthesis. Yet, regardless of the technology employed, active molecules must be translated or decoded from their corresponding DNA strand. DNA sequencing readily provides an avenue to this result, requiring sequence design that is amenable to PCR amplification and primer incorporation for the chosen sequencing technology. Fortuitously, next-generation sequencing technologies developed at companies such as Illumina, Pacific Biosciences, Ion Torrent, and 454 Systems can sequence DNA on an unprecedented scale [19]. This advance now enables the determination of pre- and postselection sequence populations for small libraries (<1 million) and thus the corresponding compound populations and relative enrichments. For larger libraries (>1 million), this simply provides far greater sequence coverage and depth in regard to hits and structure–activity relationship data. The recent rise in productivity from DNA-tagged libraries has been thus facilitated by the development of high-throughput single-strand DNA sequencing. In one regard, if it were not for the rise in sequencing technology, DNA-encoded library technologies would have remained in their infancy.

The requirements to manipulate and decode the encoding DNA have to date resulted in overall architecture that, at a high level, is independent of the methodology used for library generation. As illustrated in Figure 7.1, the encoding oligo-2'-deoxynucleotide contains compound and/or building block identification via internal codons. Fixed sequences flank the encoding sequence regions for manipulation purposes such as primers for PCR and sequencing.

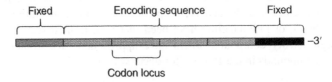

Figure 7.1. Overall template architecture for a DNA-encoded library. Fixed-sequence regions sit 5′ and 3′ to the internal coding sequence. The codon sequences are composed of codons that function as id tags for each building block. A series of loci for codons within an encoding sequence provide encoding for multiple components of a generated molecule.

7.2 PROGRAMMATIC DNA-ENCODED LIBRARIES

The majority of DNA-tagged library technologies can readily be classified into two distinct categories: (i) those that program or encode the compound prior to its construction and (ii) technologies that encode *in situ* or after a given reaction step. Programmatic encoded libraries can be prepared by biological or biochemical means, such as phage display [20] and mRNA display technologies [21]. Additionally, this category also includes library generation that utilizes various forms of DNA-templated chemistries [22–26]. In contrast, the second category covers platforms such as Lerner and Brenner's methodology [16] and YoctoReactor architecture [27] where DNA encoding occurs post reaction.

7.2.1 Phage Display

In a broad sense, the earliest and yet most complex DNA-encoded library is the cell. The classic dogma of biochemistry readily affords us the example of DNA transcribed into mRNA and then the mRNA translated into peptides and proteins. Similar to synthetic DNA-encoded libraries, cells employ flanking sequences around the exons to enable expression, replication, and manipulation. In prokaryotes, for example, upstream promoter sequences on the DNA duplex enable RNA polymerase binding and indicate the starting codon for transcription. Triplets of nucleotides within the DNA exon and mRNA function as codons for the eventual translation into single amino acids during protein synthesis.

Engineered cellular protein and peptide expression for library generation has been pioneered primarily using phage display, where an amino acid sequence is displayed on the N-terminus of a phage coat protein. In this library embodiment, a DNA construct must be cloned into the appropriate phage vector for the eventual expression of the desired specific and random amino acid sequences. Restriction and PCR primer sites must therefore be designed and incorporated into the flanking portions around the random coat protein exon construct to facilitate generation of the mature vector.

Examination of recent work by Heinis and colleagues [28] provides an interesting case study toward understanding considerations in phage library design. In their reported work, a 16-amino-acid polypeptide library was displayed on the N-terminus of the pIII phage coat protein. The peptide library contained three cysteines for postexpression modification with tris(bromomethyl)benzene to generate the library's bicyclic scaffold. The expressed protein–peptide library was expressed with the overall architecture shown in Figure 7.2. The pIII protein N-terminus is composed of a leader sequence for processing and eventual cleavage to the mature peptide. The adjacent library segment contains two random loops bookended with the fixed cysteines. This is followed by a linker segment positioning the build away from the D1 domain of the native pIII protein.

The corresponding DNA phage vector must then encode each element in the protein construct. While portions of the DNA sequence are fixed to express the leader peptide, the fixed cysteines, and the linker peptide, the random library portion requires a random trinucleotide codon cassette. To optimize library diversity, the cassette utilizes NNK degenerate codons. N represents an equal mixture of dG, dA, dC, and T with K

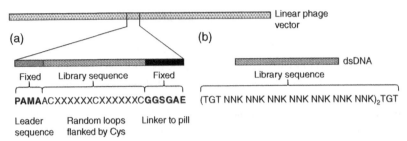

Figure 7.2. Template design for a phage display library. (a) A linear phage vector (dsDNA) is shown with a subportion highlighted that will generate the random peptide library adjacent to the pIII protein. The current example from Heinis [28] illustrates the amino acid leader sequence followed by the generated random loops between cysteines and connected via conserved linker sequence to pIII. (b) The DNA sequence used to generate the random portion of the library peptide employing the NNK degenerate codon.

NNN

	U	C	A	G	
U	UUU⎫Phe UUC⎭ UUA⎫Leu UUG⎭	UCU⎫ UCC⎪Ser UCA⎪ UCG⎭	UAU⎫Tyr UAC⎭ UAA stop UAG stop	UGU⎫Cys UGC⎭ UGA stop UGG Trp	U C A G
C	CUU⎫ CUC⎪Leu CUA⎪ CUG⎭	CCU⎫ CCC⎪Pro CCA⎪ CCG⎭	CAU⎫His CAC⎭ CAA⎫Gln CAG⎭	CGU⎫ CGC⎪Arg CGA⎪ CGG⎭	U C A G
A	AUU⎫ AUC⎪Ile AUA⎭ AUG Met	ACU⎫ ACC⎪Thr ACA⎪ ACG⎭	AAU⎫Asn AAC⎭ AAA⎫Lys AAG⎭	AGU⎫Ser AGC⎭ AGA⎫Arg AGG⎭	U C A G
G	GUU⎫ GUC⎪Val GUA⎪ GUG⎭	GCU⎫ GCC⎪Ala GCA⎪ GCG⎭	GAU⎫Asp GAC⎭ GAA⎫Glu GAG⎭	GGU⎫ GGC⎪Gly GGA⎪ GGG⎭	U C A G

NNT

	U	C	A	G	
U	UUU Phe	UCU Ser	UAU Tyr	UGU Cys	U
C	CUU Leu	CCU Pro	CAU His	CGU Arg	
A	AUU Ile	ACU Thr	AAU Asn	AGU Ser	
G	GUU Vall	GCU Ala	GAU Asp	GGU Gly	

NNK

	U	C	A	G	
U	UUU Phe⎫UCU⎫ UUG Leu⎭UCG⎭Ser	UAU Tyr UGU Cys UAG stop UGG Trp	U G
C	CUU⎫Leu CCU⎫Pro CUG⎭ CCG⎭	CAU His CGU⎫Arg CAG Gln CGG⎭	U G
A	AUU Ile ACU⎫Thr AUG Met ACG⎭	AAU Asn AGU Ser AAG Lys AGG Arg	U G
G	GUU⎫Val GCU⎫Ala GUG⎭ GCG⎭	GAU Asp GGU⎫Gly GAG Glu GGG⎭	U G

	NNN	NNK	NNT
Codons	64	32	16
Native AAs	20	20	16
Stop	3	1	0

Figure 7.3. Three examples of codon degeneracy. The wild-type NNN consisting of all 64 native codons, the NNK consisting of 32 codons, and the NNT consisting of 16.

representing an equal mix of dG and T. This reduces the standard genetic code from 61 to 32, while maintaining expression of all 20 amino acids, and removes two stop codons (Fig. 7.3). The removal of redundant codons results in a significant increase in diversity due to a more even amino acid distribution and a decrease in truncates via stop codon removal. As such, the diversity and composition of phage display libraries and all biologically translated libraries can be manipulated by the type of codon degeneracy employed [29].

7.2.2 RNA SELEX

The *in vitro* transcription of DNA into RNA and subsequent subjection of the resultant RNA library to a selection pressure for aptameric or catalytic activity have become commonplace for the discovery of functional RNA molecules [30]. The identification of active RNA molecules is accomplished by sequencing the corresponding reverse-transcribed DNA of the resultant active RNA sequences. Thus, decoding is quite straightforward with typically a direct one-to-one translation from ribose to $2'$-deoxyribose. Comparable to phage display, an initial DNA template construct must be generated composing the "exon" or RNA library portion and flanking sequences for manipulation. In the case of SELEX, a T7 promoter sequence is required to initiate RNA transcription and then flanking constant sequences for PCR primer sites that enable reverse transcription and PCR amplification. RNA selection from an encoding perspective is quite simple compared to phage display. Design aspects such as length or the design of constant regions can be taken into consideration.

7.2.3 mRNA Display

A logical progression of RNA selection technologies has been the development of mRNA display. mRNA display takes the SELEX technology one step further by the biochemical creation of protein and peptide libraries via *in vitro* mRNA translation. Though RNA is readily reverse transcribed into DNA for hit analysis, the resultant protein product is disconnected from the encoding mRNA during native protein translation. Hit identification thus becomes an extreme challenge, and the ability to regenerate an enriched template DNA strand pool becomes impossible. To bridge this divide, a short RNA oligonucleotide with a terminal puromycin is ligated to the nascent mRNA strand prior to translation. Puromycin stalls mRNA translation and functions as nucleophile to ligate the mRNA with the newly formed peptide [21].

The encoding DNA duplex design must contain the appropriate upstream sequences to initiate transcription (e.g., T7 promoter site), enable ribosome binding (e.g., Shine–Dalgarno or ΔTMV), and then encode the open reading frame for mRNA generation. The open reading frame can be composed of many features depending on the type of protein or peptide expressed. For example, Szostak and coworkers [31] have reported the incorporation of a His tag for affinity capture and a conserved gly-ser-gly-ser linker to position the peptide away from the His tag and mRNA. Figure 7.4 highlights these elements in the provided DNA and the resultant RNA–peptide conjugate.

Analogous to phage display, Guillen's reported work further illustrates the requirement to consider codon choice and degeneracy for nucleotide triplet cassettes. Here once again, the NNK degenerate codon is utilized. In addition to rebalancing the amino acid population and removing two stop codons, the NNK codon set removes redundant tRNAs from the system. The removal of redundant tRNAs has become key for the

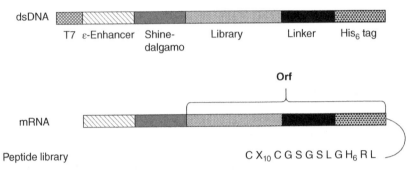

Figure 7.4. The dsDNA template architecture for an mRNA display library composed of the fixed upstream T7 promoter and enhancer sequences followed by a Shine–Dalgarno and random library sequence. Downstream from the library sequence are a conserved linker and His tag for construct manipulation. The generated mRNA–peptide library is shown in the following text highlighting the open reading frame for mRNA expression and the resultant peptide library.

incorporation of nonnative and N-methylated amino acids. Amino acid incorporation outside of the biological native 20 requires a significant investment of time and resource for the manual preparation of each charged tRNA. Thus, the removal of degenerate tRNAs can dramatically reduce the work required to generate nonnative libraries. Currently, methods such as the PURE system [32, 33] or the Flexizyme ribozyme [34] are used to charge tRNAs. For example, in Szostak's work, thirteen nonnative and seven naturally occurring amino acids were charged to 20 respective tRNAs compared to 61 via the PURE system.

The Suga research group [34] has further exploited the use of degenerate codons by preparing constructs that are composed of NNT, where the codon set now decreases to 16 codons and removes all three stop codons (Fig. 7.3). In this library design, 4 of the 16 amino acids were N-methylated, and the NNT set was utilized to ensure distribution. The assignment of codons TTT, TCT, TAT, and GCT for 4 for the N-methylated building blocks provides a 1 in 4 chance of N-methylation. As the NNT set does not provide a stop codon, an iodoacetic acid-modified tryptophan was employed to function as a terminator and encoded using a small population of the TAT codon. As these examples highlight, design aspects for protein and peptide libraries using biochemical machinery requires a multifaceted approach to generate the desired build.

7.2.4 DNA-Templated Chemistries

7.2.4.1 Solution Phase.
As discussed earlier, translation in biological and *in vitro* biochemical systems utilizes a trinucleotide codon to hybridize mRNA with tRNA. This interaction provides the next amino acid to the growing peptide, while sequestering peptide bond formation to the correct mRNA within the ribosome. The specificity provided by sequence-dependent oligonucleotide hybridization can be exploited in a similar manner to bring two chemically reactive groups together for nonribosomal translational systems. When two reactive building blocks are covalently tethered to

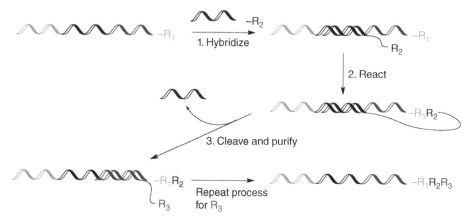

Figure 7.5. Pathway for the generation of DNA-programmed molecules developed by Liu and coworkers [25, 26].

respective complementary DNA oligonucleotides, hybridization increases their local and thereby effective concentration. This rise in concentration increases reaction rate and sequesters the building blocks for an intracomplex reaction.

DNA-templated reaction systems can readily utilize multiple hybridization events to program the construction of complex molecules. In a similar manner to biological translation, a template DNA strand is sequentially hybridized with reagent strands delivering payloads for the coupling of building blocks in a linear progression. This work, pioneered by the Liu group [25, 26], is illustrated in Figure 7.5. Directing the correct reagent to the correct loci and codon on the template strand is a sequence-driven process, and correct hybridization codon sequences must be designed appropriately to ensure fidelity.

In biological systems, the mRNA–tRNA 3-mer duplex forms in the ribosome A site, where the ribosomal machinery can stabilize and assist in discerning the correct tRNA. In contrast, DNA-templated chemical reactions are performed in the absence of the ribosome simply in buffered solution and typically require at least 10–12 nucleotides to ensure a stable duplex. Choice of sequence then becomes critical to ensure correct and stable hybridization. Thus, as a DNA-templating chemical system becomes more complex, sequence choice becomes a higher-order challenge.

As an example, the generation of a three-building-block molecule could employ a template with a single locus and a single codon to deliver each building block. The codon pair is used twice in sequential order to hybridize with a new building block-charged reagent strand (Fig. 7.6). As the molecule being generated is a single entity, a decoding event is not required. The codon in this case simply provides a sequence for hybridization. A template containing three loci with one codon at each position ($1 \times 1 \times 1$) could also be utilized in a sequential fashion, where the codon at each locus encodes for a given building block. The $1 \times 1 \times 1$ system has greater complexity by requiring an initial template codon to encode the first building block and two codon pairs to encode and deliver building blocks two and three. Now, sequence choice is required to ensure each reagent strand hybridizes to the correct codon sequence at a given locus.

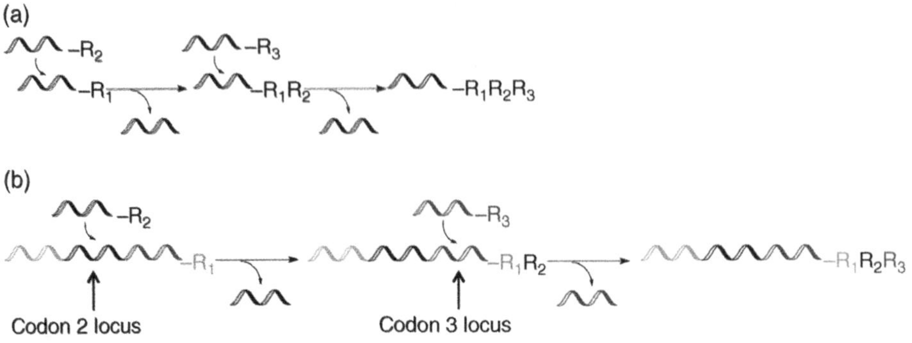

(a)

(b)

Codon 2 locus   Codon 3 locus

Figure 7.6. (a) Programmatic construction for one molecule composed of three building blocks utilizing one codon locus and two complementary codons for the entire process. (b) Programmatic construction for one molecule composed of three building blocks utilizing three codon loci with hybridization occurring at Loci 2 and 3 for sequential steps in the build.

The power of this platform clearly presents itself when utilizing multiple codons at a particular locus to enable multiple specific reactions in the same reaction vessel. Sequence-specific hybridization not only increases the effective concentration of two building blocks but now further directs multiple building block-oligo reagent constructs to their corresponding template strands. In addition to ensuring stable hybridization, a given codon-partner duplex must also outcompete incorrect duplex formation. The correct encoding of a library occurs only when each reagent strand hybridizes with the correct template strand. For example, a 1×10 template set will generate 10 uniquely tagged compounds, where the first building block position encodes the initial building block attached to the template strand (Fig. 7.7). The second locus then encodes for 10 building blocks for the reaction, resulting in a 10-member library where each member's identification can be ascertained if hybridization fidelity is high. The reagent strands in the reaction must therefore meet the following conditions:

1. Hybridize to the correct template codon.
2. Outcompete potential incorrect hybridization due to competition from other reagent strands.
3. Avoid cross-hybridization to an incorrect locus on the template oligo-2'-deoxynucleotide.
4. Avoid hybridization with other reagent strands.

The requirements outlined earlier hold true for any multicodon DNA-templated reaction. A jump in complexity from 10 compounds to greater numbers requires the same characteristics in codon set design. A library with a full combinatorial matrix of $10 \times 10 \times 10$ results in the generation of 1000 molecules each tagged with a unique template sequence. While only 3×10 templating codons are required to generate the 1000 unique template sequences, the complexity and potential cross-hybridization events increase exponentially. In this case, the build is a two-step reaction pathway where each reaction in the process requires 10 reagents (and template codon partners).

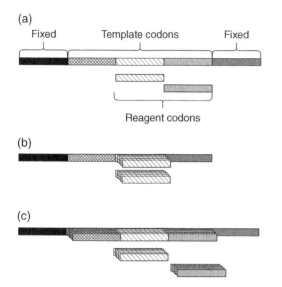

(d)

Diversity	Total codons	Codon type
(a) 1×1×1 (1 member)	5	3 Template 2 Reagent
(b) 1×10 (10 members)	21	11 Template 10 Reagent
(c) 10×10×10 (1000 members)	50	30 Template 20 Reagent

Figure 7.7. Template design and codon allocation for the following: (a) 1 x 1 x 1 three-building-block single-molecule template composed of 3 codon loci flanked by fixed sequences. (b) 1 x 10 library template set composed of 2 codon loci. The first one is fixed for the first building block and the second locus composed of 10 codons. (c) 10 x 10 x 10 library template set with 3 loci each composed of 10 codons resulting in the full combinatorial mix of 1000 unique template sequences. (d) Codon usage for each of the three constructs.

Figure 7.7 outlines this complexity where 30 codons are required for the templates and additional 20 reagent codons complementary to the template set are also required, giving 50 unique codon sequences in total.

The potential size of a DNA-programmed library is initially a question of sequence space—in other words, a choice of one of the four nucleotides per residue across a given codon length. For example, in the case of 3-nucleotide, 5-nucleotide, or 10-nucleotide codons, the resultant maximum number of codons is 64, 1024, and 1048576, respectively. The actual number of codons that can be employed in a combinatorial matrix is a small fraction, as the full combinatorial matrices outlined will contain codons that readily cross-hybridize. The determination of maximum diversity is then a clique of codons for the template and their reagent complements that conform to a chosen set of parameters.

The determination of a max clique of codons is driven by finding a set of codons that all meet the four conditions outlined earlier. Correct hybridization is the driving factor and must occur under the reaction conditions, for example, buffer choice, pH, salt concentrations, and temperature. Likewise, incorrect hybridization and competing template secondary structures must be minimized under the same conditions.

Codon choice then becomes an algorithmic logic puzzle to generate lower ΔG values for each correct duplex formation versus ΔG values for each incorrect duplex. The difference in ΔG for correct versus incorrect duplex formation enables specificity for the correct reagent strand, outcompetition of incorrect reagent strands, and hybridization to the correct template partner. The fidelity of the system relies simply on thermodynamics and one's ability to correctly calculate the energetics of duplex formation. Fortunately, the thermodynamics of DNA duplex formation is well understood and can be reasonably predicted with computer software packages. Upon generation of an *in silico* codon set, specificity testing for correct hybridization is performed in the wet lab to confirm a functional set.

Codon design is required to ensure fidelity not only for the generation of the library during the hybridization event but during the decoding process as well. Sequences that ensure fidelity during DNA sequencing are critical as mutations that can arise during PCR amplification or misreads during sequencing have the potential to provide incorrect data. Therefore, the choice of codon sequence and the Hamming distance between codons should preserve fidelity in the event of a mutation or misread. In more concrete terms, two codons that differ in two bases in contrast to differing by one base hold an advantage with regard to fidelity.

Similar to other DNA-encoded libraries, there is a design aspect to the overall template architecture, and the codon set must finally take into consideration the overall template design. Fixed sequences that flank the encoding portion can function as template tagging schemes and PCR primer sites. Therefore, codon sets must be chosen in the context of the overall sequence environment to ensure that the fixed and reaction template sequence regions do not interfere with reagent hybridization.

7.2.4.2 Solid-Phase DNA-Templated Chemistry.

Additional technology for the DNA-programmatic construction of functional molecules has been developed in the Harbury research group [22–24]. In this embodiment, the reagent oligo codon is tethered to a solid support, and the template strand is subsequently hybridized to the corresponding reagent bead. Each reagent codon bead is isolated in a respective column, and the template pool is effectively sorted via repeated passage over all of the potential columns. After the sorting process, the initial template pool has been separated by codon for a particular template codon (Fig. 7.8). Once the template pool is bound, the building block is introduced in solution as a free species.

Similar to Liu's work, Harbury's method utilizes hybridization to determine specificity, but it is maintained via spatial isolation in a reaction vessel. Thus, the incoming building block can be provided at a high concentration in solution similar to more traditional solid-phase methodologies.

The nature by which the incoming building block is introduced has very little influence on sequence and codon design. Both methods require codon sequences to hybridize correctly to their complement in the presence of competing DNA sequences. The main difference between the two hybridization events is that Harbury's method presents sequentially only one reagent at a time to the template set, while in Liu's method, all reagent strands are presented simultaneously in the same vessel. Therefore, in Harbury's method, there is no competition between reagent strands, only between template codons and the incoming reagent codon. Consequently, the system can readily be seen as less complex in regard to codon design. Yet, to ensure fidelity, the codons still must adhere to the guidelines outlined in the preceding section. The only exception is that reagent cross-hybridization does not have the potential to occur. This can be taken into account when assigning codon sequences and can potentially enable the generation of a large codon clique.

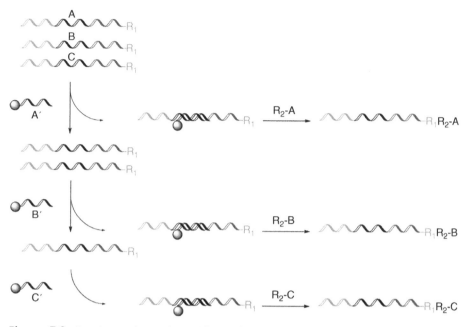

Figure 7.8. Routing and reaction pathway for DNA-programmed chemistry technology developed by Harbury and coworkers [22, 26].

7.3 POSTCHEMISTRY MANUAL ENCODING

The methodologies discussed in the preceding section highlight the connection between genotype and phenotype for biological and chemical systems—namely, the programmed expression of DNA into a chemical entity. While the DNA-programmatic construction of libraries can be an efficient means to generate a large collection of compounds, non-DNA-programmatic methodologies can also be employed to generate diverse libraries of potentially functional molecules. In the end, functional molecule discovery via DNA-encoded libraries simply needs to utilize the DNA as a tag to decode active compounds from a selection assay. In practice, non-DNA-programmatic methods typically involve a two-step process of chemical reaction and subsequent DNA tagging. Library design is accomplished by the manual choice of reagent to add to each split-and-mix reaction well. Encoding occurs in a similar manner, as each codon is manually added to a specific reaction well.

Lerner and Brenner's proposal of a chemical reaction event followed by DNA sequence tagging provided the first method ideation for postreaction encoding. The reduction of this method to practice was reported in 2012 by Praecis Pharmaceuticals [35]. Barry Morgan and coworkers developed methodology for building block encoding post reaction via double-stranded DNA ligation.

In practice, the tagging system developed at Praecis commences when an initial DNA headpiece is coupled to the starting building block and the first and subsequent

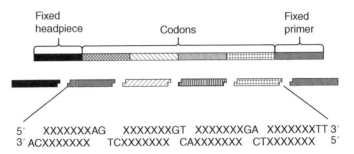

Figure 7.9. Template design and tagging scheme for postreaction encoding utilized by Morgan et al. [35].

building blocks are added via split-and-pool methodology. As each chemical reaction is spatially separate, the introduction of a respective DNA tag or codon, sequence choice is no longer dictated by hybridization.

This process actually requires little in the way of codon design. Codons are introduced as double-stranded DNA oligos with 2 Base-Pair (BP) overhangs, where the limit of codons for a given location is determined primarily by sequence length while ensuring a stable duplex is provided to the tagging mixture. The two-base overhang provides a means for hybridization to the growing DNA tag chain, where the overhangs are designed to ensure only the correct tag is ligated during enzymatic ligation (Fig. 7.9).

In the case of Morgan's initial publication [35], seven BP codons were employed, enabling potentially 8192 (4^7/2) codons and thus building blocks for a given step. Though not addressed in the respective publication, practical considerations for sequence design could actually limit this number. Brenner and Lerner highlighted this issue in their original work, as highly related sequences may prove to cloud deconvolution. For example, a frameshift or point mutation can result in incorrect sequencing.

In the context of overall sequence design, the construct employs a double-stranded headpiece connecting to the nascent molecule. The headpiece not only functions to receive the incoming cycle tag but further functions as a PCR primer for amplification prior to sequencing. Likewise, there is a conserved sequence tagged to the distal duplex terminus that also functions as a PCR primer.

7.3.1 Postchemistry Self-Encoding

One final methodology, developed by Hansen and coworkers at Vipergen [27], is a post-chemical encoding technology, where short DNA oligos charged with building blocks are brought into close proximity via duplex hybridization. Three oligos are assembled into a three-way junction pseudotrefoil structure (or YoctoReactor). The first step in a library construction anneals two oligos each supplying a building block and a third oligo functioning as a helper strand to complete the trefoil. The building blocks are reacted and the DNA ligated to link the encoding sequences. The third and final building block is provided by replacement of the third helper oligo with a charged one. Figure 7.10 outlines the structure formation and reaction pathway for this technology.

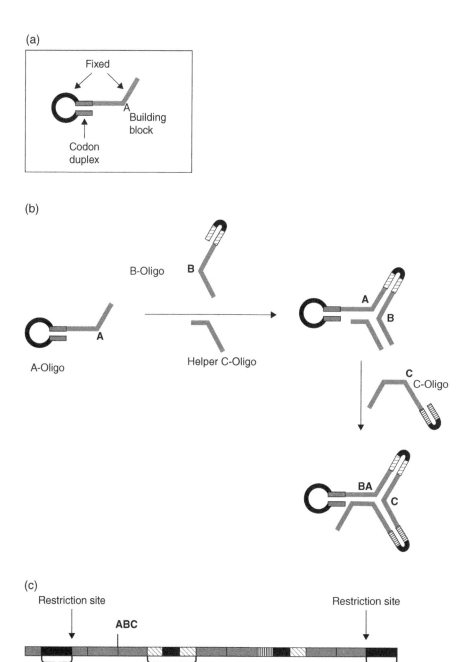

Figure 7.10. (a) A-oligo design. (b) Schematic for a two-step YoctoReactor build. (c) Linearized final product from a YoctoReactor build and schematic of fixed and variable sequences.

One unique aspect of the Vipergen's technology is that the library is self-encoded, whereby each incoming building block supplies its own encoding information. In contrast to programmatic methods, Vipergen's library construction relies on universal sequences to direct the assembly of the trefoil, enabling equal hybridization of all complementary oligos in the mixture. When the trefoil is used in a combinatorial format by introducing multiple building blocks for each position, the result is a combinatorial matrix.

Sequence design for the YoctoReactor is centered upon the formation of a stable and functional three-way junction in a stepwise fashion. As illustrated in Figure 7.10, initially, the A-2'-deoxynucleotide (A-oligo) hairpin is designed where the duplex of the hairpin encodes the first building block. The loop of the hairpin is fixed and composed of 18 bases functioning as a PCR site for eventual deconvolution. Downstream of the hairpin, two conserved sequences are designed to anneal to the incoming B-oligo tethering the second building block and the helper C-oligo.

The B-oligo akin to the A-oligo is designed in a similar manner. The hairpin duplex encodes for building block B, and though the hairpin is a conserved, it does not function as a PCR primer and is therefore shorter. The two conserved sequences hybridize to the A-oligo and helper C-oligo to complete the junction.

Sequences are designed so that upon completion of the first reaction (A and B), the A-oligo and B-oligo are spatially arranged for enzymatic ligation, thusly connecting the encoding information of the first two building blocks. As provided in Figure 7.10, the hairpins enable head-on adjacent alignment of the A- and B-oligos. The helper C-oligo lacks a hairpin and cannot function as a substrate for ligation to the B-oligo and is spaced apart from the A-oligo to avoid ligation.

After appropriate chemistry to generate the step-one product and a cleanup procedure, the charged C-oligo is presented to complete the final three-way junction. The C-oligo encodes its cargo again within a short hairpin, and downstream conserved sequences are designed to anneal to the ligated A/B-oligo. Here again, the hairpin provides correct alignment for ligation to the B portion of the construct, generating a DNA strand that fully encodes the three building blocks. The C-oligo additionally is composed of an overhang to function as a PCR primer site. This terminal conserved sequence additionally extends past the other terminus of the construct, which ensures the pseudotrefoil is not cyclized.

In regard to the overall template design, the mature pseudotrefoil construct is topologically straightened via primer extension using a terminal PCR site to the resultant duplex and is ready to be subjected to a selection assay. The construct additionally contains two restriction sites to shorten selective sequences for library propagation via rolling translation.

The conserved sequences and codons for Vipergen's technology, in one respect, function in the same role as other synthetic-encoded library approaches discussed, namely, the use of codons to encode building blocks and conserved sequences to facilitate library construction and deconvolution. However, sequences also function quite differently as they are designed to bring building blocks together in a combinatorial manner to generate the three-way junction. The guidelines for template design and codon choice are the same as other platforms, ensuring that the correct sequences efficiently find their partners and avoid cross-hybridization events. Thermodynamics again directs the sequence choices that result in hairpin, codons, and the conserved arms with the end result of a functional chemical system that ensures fidelity for its respective molecular products.

7.4 CONCLUDING REMARKS

The power of DNA-encoded libraries to rapidly investigate large areas of chemical space has matured extensively in the past number of years. Regardless of choice of technology used, template design is critical to ensure the build is produced effectively and generated with fidelity. Codon choice can make a significant impact in regard to library diversity observed in the usage of degenerate trinucleotide codons for mRNA and phage display libraries. Codon choice is also critical in programmatic technologies to enable a given reagent to find its template partner, thereby ensuring the correct compound is synthesized. From a wider perspective, template design has numerous roles to play from library expression for mRNA display to enabling the three-way junction formation of the YoctoReactor. Finally, after a selection, template constructs must enable facile and accurate retrieval for the active encoded library members. The study and employment of nucleic acids for library display readily reaffirms that they are a dynamic class of molecules and their full potential has yet to be realized.

REFERENCES

1. Kenny, B.A., Bushfield, M., Parry-Smith, D.J., Fogarty, S., Treherne, J.M. (1998). The application of high-throughput screening to novel lead discovery. *Prog. Drug Res.*, *51*, 245–269.

2. Posner, B.A. (2005). High-throughput screening-driven lead discovery: meeting the challenges of finding new therapeutics, *Curr. Opin. Drug Disc.*, *8*, 487–494.

3. Bleicher, K.H., Bohm, H.J., Muller, K., Alanine, A.I. (2003). Hit and lead generation: beyond high-throughput screening. *Nat. Rev. Drug. Disc.*, *2*, 369–378.

4. Huser, J. (2006). *High-throughput screening in drug discovery*, vol. *35*, Wiley-VCH, Weinheim.

5. Furka, A., Bennett, W.D. (1999). Combinatorial libraries by portioning and mixing. *Comb. Chem. High Throughout Scr.*, *2*, 105–122.

6. Spandl, R.J., Bender, A., Spring, D.R. (2008). Diversity-oriented synthesis; a spectrum of approaches and results. *Org. Biomol. Chem.*, *6*, 1149–1158.

7. Nicholls, A., McGaughey, G.B., Sheridan, R.P., Good, A.C., Warren, G., Mathieu, M., Muchmore, S.W., Brown, S.P., Grant, J.A., Haigh, J.A., Nevins, N., Jain, A.N., Kelley, B. (2010). Molecular shape and medicinal chemistry: a perspective. *J. Med. Chem.*, *53*, 3862–3886.

8. Keller, T.H., Pichota, A., Yin, Z. (2006). A practical view of "druggability". *Curr. Opin. Chem. Biol.*, *10*, 357–361.

9. Zhang, M.-Q., Wilkinson, B. (2007). Drug discovery beyond the "rule-of-five". *Curr. Opin. Biotechnol.*, *18*, 478–488.

10. Sauer, W.H., Schwarz, M.K. (2003). Molecular shape diversity of combinatorial libraries: a prerequisite for broad bioactivity. *J. Chem. Inf. Comp. Sci.*, *43*, 987l.

11. Barnes, C., Balasubramanian, S., (2000). Recent developments in the encoding and deconvolution of combinatorial libraries. *Curr. Opin. Chem. Biol.*, *4*, 346–350.

12. MacBeath, G., Koehler, A.N., Schrieber, S.L. (1999). Printing small molecules as microarrays and detecting protein-ligand interactions en masse. *J. Am. Chem. Soc.*, *121*, 7967–7968.

13. Tuerk, C., Gold, L. (1990). Systematic evolution of ligands by exponential enrichment: RNA ligands to bacteriophage T4 DNA polymerase. *Science*, *249*, 505–510.

14. Ellington, A.D., Szostak, J.W. (1990). In vitro selection of RNA molecules that bind specific ligands. *Nature*, *346*, 818–822.

15. Robertson, D.L., Joyce, G.F. (1990). Selection in vitro of an RNA enzyme that specifically cleaves single-stranded DNA. *Nature*, *344*, 467–468.

16. Brenner, S., Lerner R.A. (1992). Encoded combinatorial chemistry. *Proc. Natl. Acad. Sci. USA*, *89*, 5381–5383.

17. Melkko, S., Dumelin, C.E., Scheuermann, J., Neri, D. (2007). Lead discovery by DNA-encoded chemical libraries. *Drug Discov. Today*, *12*, 465–471.

18. Mannocci, L., Leimbacher, M., Wichert, M., Scheuermann, J., Neri, D. (2011). 20 years of DNA-encoded chemical libraries. *Chem. Commun.*, *47*, 12747–12753.

19. Tucker, T., Marra, M., Friedman, J.M. (2009). Massively parallel sequencing: the next big thing in genetic medicine. *Am. J. Hum. Genet.*, *85*, 142–154.

20. Miersch, S., Sidhu, S.S. (2012). Synthetic antibodies: concepts, potential and practical considerations. *Methods*, *57*, 48–498.

21. Roberts, R.W., Szostak, J.W. (1997). RNA-peptide fusions for the in vitro selection of peptides and proteins. *Proc. Natl. Acad. Sci. USA*, *94*, 12297–12302.

22. Halpin, D.R., Harbury, P.B. (2004). DNA display I. Sequence-encoded routing of DNA populations. *PLoS Biol.*, *2*, 1015–1021.

23. Halpin, D.R., Harbury, P.B. (2004). DNA display II. Genetic manipulation of combinatorial chemistry libraries for small-molecule evolution. *PLoS Biol.*, *2*, 1022–1030.

24. Weisinger, R.M., Wrenn, S.J., Harbury, P.B. (2012). *Highly parallel translation of DNA sequences into small molecules. PLoS ONE*, *7*, e28056.

25. Gartner, Z.J., Tse, B.N., Grubina, R., Doyon, J.B., Snyder, T.M., Liu, D.R. (2004). DNA-templated organic synthesis and selection of a library of macrocycles. *Science*, *305*, 1601–1605.

26. Tse, B.N., Snyder, T.M., Shen, Y., Liu, D.R. (2008). Translation of DNA into a library of 13000 synthetic small-molecule macrocycles suitable for in vitro selection. *J. Am. Chem. Soc.*, *130*, 15611–15621.

27. Hansen, M.H., Blackskjaer, P., Peterson, L.K., Hansen, T.H., Hojfeldt, J.W., Gothelf, K.V., Hansen, N.J.V. (2009). A yoctoliter-scale DNA reactor for small-molecule evolution. *J. Am. Chem. Soc.*, *131*, 1322–1327.

28. Heinis, C., Rutherford, T., Freund, S., Winter, G. (2009). Phage-encoded combinatorial chemical libraries based on bicyclic peptides. *Nat. Chem. Biol.*, *5*, 502–507.

29. Krumpe, L.R.H., Schumacher, K.M., McMahon, J.B., Makowski, L., Mori, T. (2007). Trinucleotide cassettes increase diversity of T7 phage-displayed peptide library. *BMC Biotechnol.*, *7*, 65.

30. Breaker, R. (2004). Natural and engineered nucleic acids as tools to explore biology. *Nature*, *432*, 838–845.

31. Guillen Schlippe, Y.V., Hartman, M.C.T., Josephson, K., Szostak, J.W. (2012). In vitro selection of highly modified cyclic peptides that act as tight binding inhibitors. *J. Am. Chem. Soc.*, *134*, 10469–10477.

32. Hartmann, M.C.T., Josephson, K., Lin, C.-W., Szostak, J.W. (2007). An expanded set of amino acid analogs for the ribosomal translation of unnatural peptides. *PLoS ONE*, *2*, e972.

33. Josephson, K., Hartman, M.C.T., Szostak, J.W. (2005). Ribosomal synthesis of unnatural peptides. *J. Am. Chem. Soc.*, *127*, 11727–11735.

34. Yamagishi, Y., Shoji, I., Miyagawa, S., Kawakami, T., Katoh, T., Goto, Y., Suga, H. (2011). Natural product-like macrocyclic N-methyl-peptide inhibitors against a ubiquitin ligase uncovered from a ribosome-expressed de novo library. *Chem. Biol.*, *18*, 1562–1570.

35. Deng, H., O'Keefe, H., Davie, C.P., Lind, K.E., Acharya, R.A., Franklin, G.J., Larkin, J., Matico, R., Neeb, M., Thompson, M.M., Lohr, T., Gross, J.W., Centrella, P.A., O'Donovan, G.K., Bedard, K.L., van Vloten, K., Mataruse, S., Skinner, S.R., Belyanskaya, S.L., Capenter, T.Y., Shearer, T.W., Clark, M.A., Cuozzo, J.W., Arico-Muendel, C.C., Morgan, B.A. (2012). Discovery of highly potent and selective small molecule ADAMTS-5 inhibitors that inhibit human cartilage degradation via encoded library technology (ELT). *J. Med. Chem.*, *55*, 7061–7079.

8

ANALYTICAL CHALLENGES FOR DNA-ENCODED LIBRARY SYSTEMS

George L. Perkins[1] and G. John Langley[2]

[1] *PerkinElmer, Inc, Branford, CT, USA*
[2] *Department of Chemistry, University of Southampton, Southampton, UK*

8.1 INTRODUCTION

The majority of biological compounds analyzed are referred to as "large" molecules, the most common examples of which are proteins, peptides, and oligonucleotides. The size of these compounds creates many challenges in analytical chemistry, and when this is combined with the need for fast, reliable, selective, and sensitive assays, the challenge becomes more complex.

With the elucidation of the molecular structure of DNA in 1953 by Watson and Crick, the so-called central dogma of molecular biology was born [1]. This proclaims that proteins are formed from amino acids that have been translated from the nucleobases of RNA, which is transcribed from DNA. Analytical techniques have been deployed to try to unravel the relationship between the genome and the proteome, the regulatory mechanism involved in this process, and further modifications that may occur.

Many analytical techniques have been used to analyze both single- and double-stranded oligonucleotides; separation techniques such as High-Performance Liquid Chromatography (HPLC) and latterly Ultra-High-Performance Liquid Chromatography (UHPLC) or Capillary Electrophoresis (CE) and Mass Spectrometry (MS) have become

A Handbook for DNA-Encoded Chemistry: Theory and Applications for Exploring Chemical Space and Drug Discovery, First Edition. Edited by Robert A. Goodnow, Jr.
© 2014 John Wiley & Sons, Inc. Published 2014 by John Wiley & Sons, Inc.

established approaches for the characterization of nucleic acids. These techniques are capable of analyzing oligonucleotides of less than 10 to many thousands of bases. Though separation techniques and MS can be used independently of each other to analyze nucleic acids, the online hyphenation of these two powerful techniques inherently leads to a more robust and automatable routine analysis. A number of reviews and books on these techniques have been published and are used as reference articles throughout this chapter [2–4].

Typically, DNA-encoded compounds and libraries are a complicated mix of DNA of varying sizes, often with only limited amounts of sample available. Due to the complex mixtures present in these samples, following reactions performed on DNA can be very difficult by traditional methods of analysis used in synthetic chemistry laboratories such as NMR spectroscopy. With its combination of separation and characterization, HPLC–MS is the most appropriate method for analyzing DNA-encoded compounds and library samples.

The following sections of this chapter will address the various methods of analysis available including separation techniques and mass spectrometric detection and will provide some methodology already published that would be applicable to DNA-encoded compounds and libraries.

8.2 CAPILLARY ELECTROPHORESIS

CE was developed in the 1980s as a high-resolution separation technique, often with the advantage of rapid analysis. Much of the early work in CE development was undertaken by Jorgenson et al. who demonstrated that separations could be achieved based on electrophoretic mobilities in an electric field within a small-diameter capillary (typically 50–100 μm id) [5]. With improvements in injection methods, detector sensitivity, capillary surface deactivation and coating technologies, buffer systems, and ease of hyphenation to MS, there has been a large increase in the number of areas where CE is applicable. CE typically uses injection volumes of 1–10 nL of 10^{-6} M analyte solutions. This very small sample consumption is advantageous, especially for biological samples, where only small amounts of sample may be available. This ability to handle very small injection volumes allows CE to be used in areas that are very challenging and problematic including the analysis of, or sampling from, single biological cells [6].

The majority of the current work in DNA analysis using CE has been to allow fast sequencing or forensic identification of DNA fragments. One review of this area, presented in 2002, clearly outlined the progress made with this technique and the possibilities for the future as the development of capillary array electrophoresis and microfabricated capillary arrays, including micromachined or microfabricated or microchip devices, continues to advance [7].

Even with the recent developments in CE, the authors feel that this is not the best technique for an open-access system that is being used by nonexperts in the analytical field, as sample preparation is very exacting, and concentration needs to be in a very narrow band; too high a concentration and separation resolution will be adversely affected; too low a concentration and detection becomes an issue. The presence of any particulates will cause the system to fail due to blockage of the capillary.

8.3 HIGH-PERFORMANCE LIQUID CHROMATOGRAPHY (HPLC)

There are a number of different HPLC approaches commonly used for the separation of DNA mixtures. The more common HPLC approaches are:

1. Size-exclusion chromatography—due to its low resolution, it is mainly used for prefractionating, desalting, or rebuffering of nucleic acids [8].
2. Anion-exchange HPLC—this has high chromatographic resolution, but the high ionic strength salts and the often used nonvolatile additives such as detergents make its use with MS difficult [9, 10].
3. Mixed-mode HPLC—this mode has the same issues as anion-exchange HPLC [11].
4. Hydrophilic Interaction Liquid Chromatography (HILIC) and the related Electrostatic Repulsion Hydrophilic Interaction Chromatography (ERLIC)— these have been used for analysis postdigestion of nucleotides/nucleosides [12, 13] and for oligonucleotides [14–16], including the analysis of oligonucleotides by HILIC coupled to inductively coupled plasma mass spectrometry [17].
5. Affinity chromatography—this is based on biospecific interactions of two complementary nucleic acid strands, one bound to the support and the other the intended analyte. The limitations of this method are the need to generate a new stationary phase for every new separation problem and its use with single-stranded DNA only [18, 19].
6. Reversed-phase HPLC.
7. Ion-pair reversed-phase HPLC.

Reversed-phase HPLC relies on the interaction between the hydrophobic nucleobases of the nucleic acids and the nonpolar surface of the stationary phase. Increasing the nonpolar nature of the mobile phase by increasing the concentration of organic solvent present, typically methanol or acetonitrile, interrupts this interaction causing the DNA to elute. Though buffers are often present in the aqueous phase, they are normally volatile and of a low ionic strength, ensuring compatibility with electrospray ionization (ESI). Reversed-phase HPLC is particularly relevant to single-stranded DNA that has been modified with hydrophobic groups, for example, trityl or dimethoxytrityl [20], or fluorescent dyes. In particular, double-stranded DNA is not separated under these conditions as the hydrophobic nucleobases are shielded by the hydrophilic sugar–phosphodiester backbone, leading to little if any retention. This can be an advantage if the separation of single-stranded from double-stranded DNA fragments is required. The need to use modified single-stranded DNA and the lack of suitability of this method to double-stranded DNA limit the number of applications that can use reversed-phase HPLC.

Ion-pair reversed-phase HPLC addresses the issues observed in reversed-phase HPLC and permits its use with double-stranded DNA. The ion-pair reagent is usually a quaternary amine having amphiphilic properties, that is, carrying a charge and having hydrophobic groups. It has been proposed that the positively charged, hydrophobic

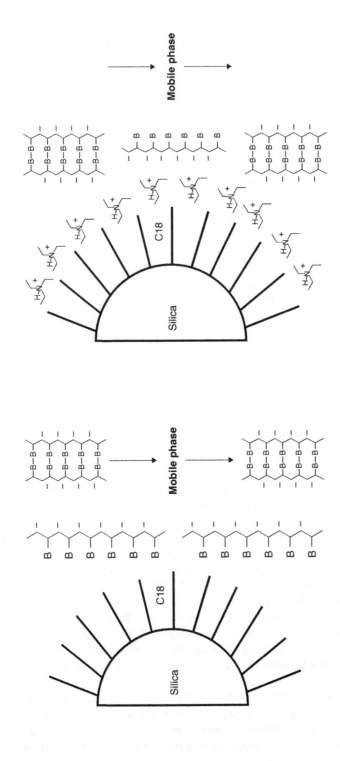

Partitioning between single-stranded DNA bases and stationary phase causes retention and selectivity. Double-stranded bases are shielded and elute without retention.

Both single-stranded DNA and double-stranded DNA can be retained with the addition of an ion pair, which causes electrostatic interactions.

Figure 8.1. Illustration of retention mechanism for both reversed-phase HPLC and ion-pair reversed-phase HPLC.

alkylammonium ions are adsorbed onto the nonpolar surface of the stationary phase, resulting in the formation of an electric double layer that has an excess of positive charges near the surface [21] (see Fig. 8.1). A number of different tetra-alkylammonium salts have been used as counterions in the analysis of DNA. The most commonly used counterion is Triethylamine (TEA) or one of its salts, such as Triethylammonium Acetate (TEAA). It is important to use a volatile salt if the HPLC is to be hyphenated with a mass spectrometer. The use of these mass spectrometric compatible mobile phases makes ion-pair reversed-phase HPLC the chromatographic mode of choice for DNA analysis. Retention and separation selectivity are controlled by numerous mobile-phase parameters including hydrophobicity of the ion pair and ion-pair concentration, pH, ionic strength, stationary-phase choice, organic-phase choice, and size and charge of the analyte; with this large choice of parameters, the separation of very complex samples is possible including mixtures of ionic and neutral solutes. Obviously, with such a large number of variables controlling separation selectivity, optimizing the separation involves the assessment of how each variable affects the separation and what the optimum value is for that variable. To assess this for all the variables will involve a large number of experiments and will take a considerable amount of time and effort. These variables have to be balanced to achieve the best separation without adversely affecting the ionization. This will be discussed in detail in Section 8.7 and has been reviewed in further detail by Huber et al. and by Banoub and Limbach [2, 3].

Typically, during the analysis, an increasing gradient of the organic phase causes a decrease in the concentration of the amphiphilic ion from the stationary phase; this decreases the surface potential of the stationary phase and leads to the elution of the nucleic acids. As the size of the nucleic acid increases, the number of charges on the nucleic acid also increases. For double-stranded DNA molecules, this charge and size increase is uniform and the molecules separate according to chain length [22, 23]. In the case of single-stranded DNA, RNA, and partially denatured double-stranded DNA, the unshielded bases will interact with the hydrophobic stationary phase and retention will depend on the exact sequence of the analyte involved. It has been demonstrated that temperature can have a large impact on the separation of DNA. Improvements in the resolution can be shown if the sample is subjected to a high temperature (>50°C) prior to analysis on column and high column temperatures (>50°C) have also been shown to improve the resolution. This is believed to be due to partial denaturing of the DNA and "melting" where the bases are no longer hydrogen bonded to one another or adjacent bases are no longer stacked. Both of these aspects give more degrees of freedom to the DNA, allowing more interactions with the hydrophobic stationary phase [24, 25].

8.4 MASS SPECTROMETRY OF NUCLEIC ACIDS

The underlying principles of MS were described by J.J. Thompson in 1913 in a paper in *Nature* [26] and further developed by Aston [27], Dempster [28], Stephens [29], Paul and Steinwedel [30], and Marshall [31] to name but a few. The basic composition of a mass spectrometer has changed little over the years; typically, it is comprised of an inlet, ionization source, mass analyzer, vacuum chamber, and detector. The use of computers to control the instrument and to acquire, record, and save the data is an essential part of modern mass spectrometers.

There are many ionization sources, and most have been used to analyze nucleic acids including electron ionization (EI) [32], chemical ionization (CI) [33], fast atom bombardment/liquid secondary ion mass spectrometry (FAB/LSIMS) [34, 35], matrix-assisted laser desorption/ionization (MALDI) [36], and ESI [37]. The two most commonly used sources today for nucleic acid analysis are MALDI and ESI, and these will be covered in more detail in the following.

8.5 IONIZATION

8.5.1 Matrix-Assisted Laser Desorption/Ionization (MALDI)

MALDI was developed from the work in which lasers were used both to desorb and ionize small organic compounds in the early 1970s [38, 39]. In these early experiments, it was necessary for the compounds to contain chromophores to absorb the energy from the laser; these initial experiments used high-power UV lasers that subject the sample to high thermal energy. In the late 1980s, both Hillenkamp et al. and Tanaka et al. [40] reported on the use of a matrix, which absorbed the laser energy and allowed desorption and ionization of both the matrix and the analyte of interest, without thermally stressing the compound. This development allowed the analysis of large biopolymers such as proteins, DNA, and RNA oligomers. Initially, the MALDI-MS method was predominantly applied to proteins and peptides as extending this technique to nucleic acid needed considerable work on the choice of matrix and the type of laser. The difficulties associated with MALDI and its application to nucleic acids have been extensively reviewed [41].

The use of delayed extraction by Vestel et al. [42] increased the resolution of the spectra, and this combined with simpler software interfaces promoted the move of this technique from the hands of specialists into the wider research community.

The underlying MALDI process is still unclear, and Gluckmann et al. proposed that differences in desorption/ionization processes occur depending on the specific application [43]. A number of publications have proposed mechanisms, which have been reviewed by Knochenmuss [44]. One suggested mechanism is that once in the gas phase, a proton is transferred from the matrix to the analyte of interest or a proton is transferred from the analyte of interest to the matrix forming positive and negative analyte ions, respectively (Fig. 8.2).

Both the matrix, selected because of its ability to absorb energy at the laser wavelength and its miscibility with the analyte in the solid state, and the analyte are spotted onto the MALDI plate and allowed to co-crystallize prior to irradiation. This produces both protonated (for positive ionization) or deprotonated (for negative ionization) analyte and matrix molecules that are commonly observed in the mass spectrum. The MALDI process tends to be more tolerant of impurities such as salts, but these can interfere with the crystallization process and therefore influence sensitivity, selectivity, and quality of the resulting spectra.

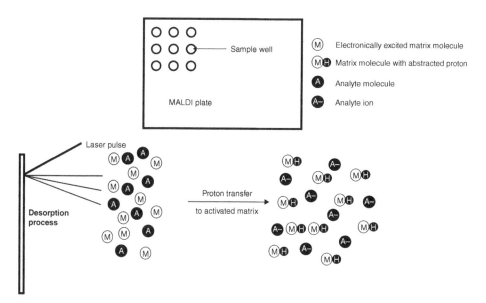

Figure 8.2. MALDI process: laser irradiation causes both the matrix and analyte to desorb followed by proton transfer ionization.

MALDI affords a fast method of analysis, particularly applicable to nucleic acids of less than 20 kDa, and could be used as a QC method for DNA-encoded single compounds or of small DNA-encoded libraries [45].

8.5.2 Electrospray Ionization (ESI)

ESI is a well-established ionization technique for a range of compound classes from low-mass polar compounds to whole ribosomes (>2 MDa) [46]. Yamashita and Fenn et al. [47] by extending the work of Dole et al. [48] described the initial experiments with electrospray in 1984, and Fenn's work with Mann et al. [49] showed that multiply charged species could be deconvoluted to yield the molecular mass of large biopolymers, typically with a mass accuracy between 0.01% and 0.1% [50]. ESI has a number of benefits:

1. It is compatible with the majority of commonly used mass analyzers: sector field, quadrupole, quadrupole ion trap (QIT), Fourier transform ion cyclotron resonance (FT-ICR) and time-of-flight (TOF) mass spectrometers [51].
2. ESI transfers only a low energy to the molecular species, and the ability for that ion to be "cooled" in the ionization process leads to few if any fragmentation products.
3. It can generate multiply charged ions of high-molecular-weight biopolymers in the gas phase.

It is this last point that has allowed ESI to have such a large impact in the biopolymer research arena. Most commercially available mass spectrometers have a mass range of less than 6000 m/z units, and in the case of quadrupoles and QIT, that mass range is usually less than 3000 m/z units. Though it is referred to as a mass range, it is in fact a mass-to-charge (m/z) range. This is important in biopolymers as they become multiply charged effectively lowering the mass at which the ion is observed. For proteins, it has been proposed that for approximately every thousand mass units, there will be one charge [52]. This allows analyzers with a mass range of only a couple of thousand m/z units to be used for biopolymer analysis, as a protein with a mass of 10,000 Da will have 10 charges, which would mean a peak would be observed at m/z 1001 ((10,000 + 10)/10) in the spectrum.

The ESI process consists of three important steps to produce a gas-phase ion from an electrolyte solution. Initially, the solution is passed through a capillary, and due to the high potential difference between the tip of the capillary and the counter electrode (typically 3–5 kV), droplets are produced. These droplets are exposed to a heated area, either a heated gas (usually nitrogen) or a heated capillary; this induces solvent evaporation of the droplet, and because of the electric field, the droplet distorts in shape. This distortion allows smaller droplets to form from the original droplet before the limit of the Rayleigh equation (see succeeding text) is reached:

$$q^2 = 8\pi^2 \varepsilon_0 \gamma D^3 \qquad (8.1)$$

where

q is the charge

ε_0 is the permittivity of the environment

γ is the surface tension

D is the diameter of a supposed spherical droplet

The first-generation droplet will carry approximately 50,000 elementary charges and be approximately 1.5 μm in diameter. The second-generation droplet will carry 300–400 elementary charges but have a diameter of 0.1 μm representing an increase of a factor of seven in the charge per unit volume. The second-generation droplets will continue to lose solvent, and when the electric field on their surface becomes large enough, desorption of ions from the surface occurs [53]. If a mixture of analytes is present in the droplet, the analytes that predominate at the surface of the droplet, usually the compound most easily protonated or deprotonated, will have a higher sensitivity than those within the bulk of the droplet. This explains why some compounds mask the presence of others when analyzed together; this is often referred to as ion suppression (Fig. 8.3).

As the molecular weight of the analytes increases above 5,000–10,000 Da, the ESI process is thought to change. As the droplets become smaller, they eventually have only one analyte present in each droplet. This one analyte carries many charges, if a suitable number of ionizable sites are present. The ion does not evaporate from the surface to form a gas-phase ion; instead, the solvent evaporates from the droplet until only the ion

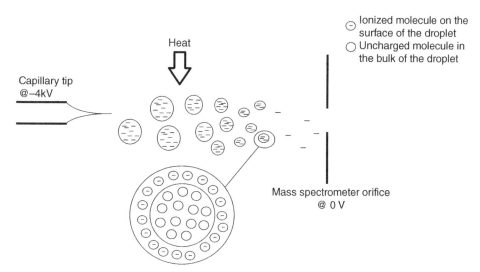

Figure 8.3. Schematic representation of negative ion electrospray process.

is left. These processes have been investigated in depth and were summarized in detail by Cole [54], Amad et al. [55], and Kebarle [56].

Oligonucleotides are typically analyzed using negative ion ESI as this takes advantage of the easily ionizable phosphodiester backbone. The latter has labile acidic protons that are easily removed during the ionization process. As oligonucleotides have a large number of these phosphodiester moieties, each oligonucleotide will have multiple charges. A typical spectrum will consist of a series of peaks in the spectrum increasing in m/z value as the number of charges decreases (see Fig. 8.4). It has been shown by Mann et al. that signals from a multiply charged series can be readily deconvoluted to yield the molecular mass of the original molecule [49], using the negative ion deconvolution equations of

$$z_1 = \frac{j\left(m_2 + m_p\right)}{\left(m_2 - m_1\right)} \quad \text{and} \quad M = z_1\left(m_1 + m_p\right) \tag{8.2}$$

where

z_1 is the charge/number of protons removed
m_1 and m_2 are two m/z values from peaks on the spectrum
j is the number of peaks in increasing m/z ratio separating the masses m_1 and m_2
m_p is the mass of a proton (1.007)
M is the mass of the uncharged compound

By selecting two peaks from the spectrum shown in Figure 8.4; one at m/z 1049.88 and one at m/z 1432.01, which are eight peaks apart in the spectrum, substituting these

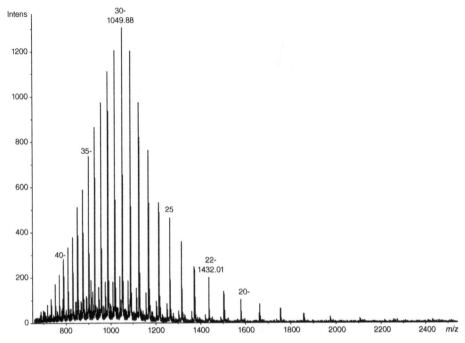

Figure 8.4. Negative ion ESI mass spectrum of a 105-mer oligonucleotide.

values into the equations, $z_1 = 8(1432.01 + 1.007)/(1432.01 - 1049.88)$; therefore, $z_1 = 30$. Substituting this value for z_1 into the second equation, $M = 30(1049.88 + 1.007)$ and therefore, $M = 31526.6$. This can be repeated for all the peaks in the spectrum and averaged to give a value for M.

Obviously, to repeat this manual deconvolution process for each peak and to average the result is a time-consuming exercise. This process becomes significantly more difficult when multiple compounds or cationic species are present. Fortunately, there are a number of computer programs capable of performing these calculations such as Promass [57] and maximum entropy [58].

One issue affecting ESI is the propensity of the sugar–phosphate backbone to partially substitute the protons for other cations typically monocations, for example, sodium and potassium, and dications, for example, calcium, magnesium, iron, and zinc. The presence of these cations means the ion currents are dispersed among several species of the same DNA composition each having a characteristic m/z ratio for its series of multiply charges. These additional series further complicate the spectrum and lead to an overall decrease in sensitivity [37, 59]. This can be a particular issue when analyzing large nucleic acids, because the various signals produced by the different cations for the higher-charge states merge into one unresolved peak, whose apex is shifted to higher mass than expected for the fully protonated moiety. The removal of the cations is therefore extremely important to ensure the mass spectrum is of high quality and the masses obtained from the spectra of nucleic acids can be confidently related to the compound analyzed. There are many different approaches to sample preparation, both off-line and online, to minimize the effect of the cations that may be present in a sample. These include ethanol precipitation [60], solid-phase extraction [61], microdialysis [62], affinity purification [63], cation exchange [64], and combinations of these [65]. The use of various organic cations has been explored to reduce the formation of the cation-adducted species of oligonucleotides. A strong base such as TEA or piperidine has been shown to provide the best suppression with up to a 100-fold decrease in the presence of cation-adducted species [66].

Electrospray is the most commonly used method for DNA analysis and has been used for the analysis of DNA-encoded compounds and libraries [67].

8.6 MASS ANALYZERS

The mass analyzer is the center of the mass spectrometer; it is used to separate and measure the ion generated by the ionization method of choice. The key parameter of a mass analyzer is its ability to separate ions of differing m/z values. There are many different types of mass analyzers, and they can be classified in a number of ways, but two very broad categories are (i) scanning mass analyzers (these allow ions of successive m/z ratio to traverse the analyzer, e.g., sector field analyzers or quadrupole mass analyzers) and (ii) simultaneous mass analyzers (these allow the transmission of all m/z ratios at the same time: examples of this include TOF analyzers and QIT analyzers).

All types of analyzers have undergone continuous development throughout the years; for example, conventional QITs based on the original Paul and Steinwedel [30] design have some limitations that have been addressed by the development of linear ion traps (LITs). A more recent advance is the development of the Orbitrap™ mass analyzer by Makarov [68], which is an electrostatic ion trap using Fourier transform to obtain mass spectra. A further development in mass analyzers has been to combine two or more analyzers to increase the types of experiments that can be performed. This was seen early on in mass analyzer development with the combination of electrostatic sectors and magnetic sectors, followed by the development of triple quadrupole mass spectrometers, and more recently the development of hybrid instruments such as the quadrupole TOF (QTOF) or the LIT/Orbitrap™.

Some important concepts for mass analyzers are as follows: (i) Mass range—this is the m/z limit of the analyzer and varies considerably depending on the type of analyzer; typically, quadrupole and QIT analyzers have the lower range of 2000–3000 m/z units, with TOF and FT-MS capable of 10,000 m/z units, or greater. (ii) Analysis speed—often referred to as scan speed, this is the time taken for the mass spectrometer to perform a single spectral acquisition. (iii) Ion transmission—this is a ratio of the number of ions exiting the source region into the analyzer region and the number of ions reaching the detector. (iv) Mass accuracy—there are two parameters to consider here. The initial one is the actual accuracy of the mass that the mass spectrometer provides compared to the theoretical value. Of equal importance is the precision of this result; this is the ability of the mass spectrometer to give the same result on each subsequent analysis. (v) Resolution—as mentioned, this is a key parameter in mass accuracy. The resolution of an instrument is defined as the ability of the analyzer to show distinct signals for two ions with a small m/z difference. If Δm is the smallest mass difference that shows two discernable signals for masses m and $m + \Delta m$, then the resolving power R can be represented by the equation $R = m/\Delta m$. The values for Δm can be defined by having two masses that are close in m/z value, but it can also be determined from a single peak. In this case, Δm is defined as the peak width at a given height; the exact height is much debated within the scientific community. Traditionally, the resolution was determined using the two-peak definition and a separation of a 10% above baseline for the valley. This value was based on a Gaussian-shaped peak, which was the case for sector field instruments. With the popularity of TOF instruments, which produce a Lorentzian peak shape, the calculation for Δm was changed to the single-peak calculation. This will be seen in the literature as FWHM, which stands for full width at half maximum and is the value of Δm at 50% peak height.

All the common mass analyzers have been used to analyze nucleic acids including sector field analyzers, quadrupole mass filters, ion trap analyzers (QIT, LIT, ion cyclotron, and Orbitrap™ analyzers), and TOF analyzers. Given that practitioners of DEL libraries are likely to be reliant on analytical systems involving details of mass spectrometers, the authors discuss in more detail quadrupole mass filters, QIT/LIT, TOF, FT-ICR, and LIT/Orbitrap™.

8.6.1 Quadrupole Mass Analyzers

The principle of the quadrupole was first described by Paul and Steinwedel [30], and commercialization of these instruments took place from the work carried out by a number of people including Post, Shoulders, Finnigan [69], and Story. Quadrupole mass analyzers are made of four rods that should ideally be hyperbolic in shape but are often circular for ease of manufacture. These rods are positioned equidistant from one another and are divided into two pairs connected electrically as shown in Figure 8.5.

At any given value of RF and DC, only one mass is stable, so the values of these are increased to allow the instrument to scan from one m/z to another. The stability of an ion

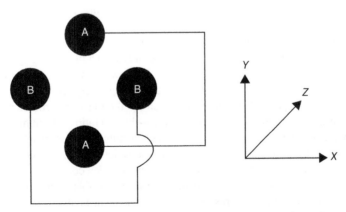

Figure 8.5. Schematic representation of a cross section of a quadrupole analyzer.

in an electric field, such as the one generated between the quadrupole rods, is given by the simplified Mathieu equation:

$$a_x = -a_y = \frac{8zU}{m\omega^2 r_0^2} \quad \text{and} \quad q_x = -q_y = \frac{4zV}{m\omega^2 r_0^2} \tag{8.3}$$

where

z is the number of charges on the ion
U is the amplitude of the applied DC voltage
V is the amplitude of the applied RF voltage
m is the mass of the ion in daltons
ω is the angular frequency of the RF voltage
r_0 is the distance from the center to the rod

For a given quadrupole, r_0 and ω are maintained constant, and the variables for any given m/z are the values of U and V. The quadrupole analyzer is a true m/z ratio analyzer and does not rely on the kinetic energy of the ion leaving the source. The kinetic energy has to be sufficient for the ion to traverse the analyzer but not so fast that the ion does not experience a number of oscillations. As the ions only require one to a few hundred electron volts of kinetic energy to leave the source area, the pressure in the source area can be relatively high, which makes quadrupole analyzers well suited to coupling to separation technology. This is aided by the linear scanning of the voltage and RF, which allows scans in excess of 10,000 m/z units/s.

Quadrupole analyzers are relatively inexpensive and are available in most large laboratories and could easily be applied to DNA-encoded library analysis. The low resolution would mean that the spectrum would show only the average of the isotopic pattern but this would increase sensitivity. The use of a triple quadrupole instrument is shown in the work by Hail [57].

8.6.2 Two-Dimensional and Three-Dimensional Quadrupole Ion Traps

The QIT uses the same principles described for the quadrupole analyzer. The ion trap uses an RF quadrupole field that traps ions in two (2D) or three (3D) dimensions. The first QITs were 3D ion traps or Paul traps, based on the original work by Paul and Steinwedel [30] and commercialized by Stafford et al. [70]. The 3D QIT uses a central ring electrode and two ellipsoid end caps as shown in Figure 8.6.

Ions of all masses are trapped in the analyzer when a direct potential and an alternating potential are applied. Ions can be ejected from the trap and detected by applying a resonance frequency along the z-axis. This frequency is directly related to the m/z, and therefore, the m/z can be measured. As the ions of all the m/z values are present in the trap at one time, the ions repel one another, and this would lead to their trajectories

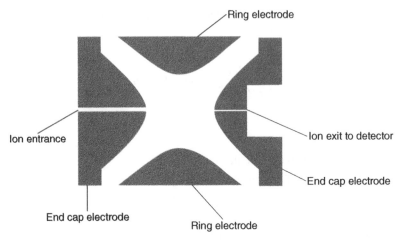

Ring electrode

Ion entrance

Ion exit to detector

End cap electrode

End cap electrode

Ring electrode

Figure 8.6. Schematic diagram of a 3D/Paul ion trap.

expanding over time. To lower the expansion of the trajectories, the trap is filled with helium gas at a relatively high pressure of 10^{-3} mbar, which ensures collisions occur between the ions and the helium gas, which reduces the trajectories; this is often referred to as cooling the ions.

The growth in the application of LIT in the last decade, which has been reviewed in detail by Douglas et al. [71], is due to two advantages that the LIT has over the Paul or 3D trap. The LIT has higher injection efficiencies and higher ion storage capacities. Injecting ions into the trap is obviously important in achieving good sensitivity, and the improved injection efficiency of the LIT gives it better overall sensitivity when compared to the 3D trap. The trapping efficiency for a 3D trap is only of the order of 5% of the ions injected into a 3D trap, whereas for an LIT, it can be in excess of 50%.

With the adoption of LITs and the sensitivity these afford, they are an ideal analyzer for DNA-encoded libraries. This type of analyzer has been used for the analysis of DNA-encoded libraries and compounds [67].

8.6.3 Fourier Transform Ion Cyclotron Resonance Mass Spectrometry (FT-ICR MS)

The FT-ICR MS, or as it is more commonly known the FT-MS, has two characteristic attributes: (i) the analyzer is under ultra-high vacuum, in the order of 1×10^{-10} mbar, and (ii) the analyzer cell is housed within a superconducting magnet, requiring the use of cryogens to obtain low enough temperatures to maintain superconductivity of the magnet. These two characteristics contribute to the ability of the FT-MS to provide high spectral resolution and high mass accuracy.

The first application of ion cyclotron resonance to MS was by Sommer et al. [72], and the FT-MS was first described by Comisarow and Marshall in 1974 [31, 73] and has been thoroughly reviewed by Amster [74].

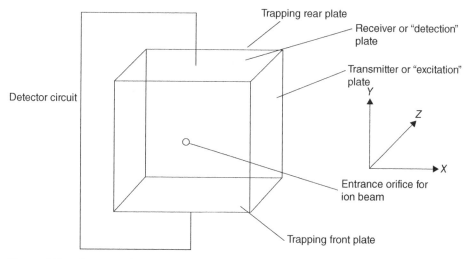

Figure 8.7. Schematic of a cubical or Penning trap.

When an ion enters a strong magnetic field, it will achieve a circular motion that is perpendicular to the magnetic field; this is known as cyclotron motion. To be able to contain this ion in a reasonably sized trap, the static magnetic field has to be homogenous; this can be achieved with a permanent magnet or electromagnets, but these types of magnets are restricted to a magnetic field strength of 2 Tesla (T). More commonly, superconducting magnets are used that provide magnetic field strengths of 3–12 T. This cyclotron motion is directly related to the m/z ratio of the ion from the idealized equation:

$$f = \frac{qB}{2\pi m} \tag{8.4}$$

where

f is the cyclotron frequency
q is the charge on ion
B is the magnetic field strength
m is the mass of ion

For any given instrument, the magnetic field strength is a constant; therefore, it can be seen that cyclotron frequency is inversely related to the m/z ratio.

Of the many geometrical types of traps used in FT-MS, the two most commonly used are cylindrical and cubical in geometry. To better understand the operation of the FT-MS, consider a cubical or Penning trap as shown in Figure 8.7.

If a single m/z ion is present in the cell, it would be relatively easy to calculate the frequency of this waveform and therefore convert it to an m/z value. In reality, like all traps, all m/z ions produced by the source are present in the cell at the same time, and therefore, the detector signal is a representation of all the corresponding sinusoidal waveforms. A single recording of the signal is known as a transient, and normally, multiple transients are collected to form one spectrum. If one accumulates n transients, the signal-to-noise ratio will improve by $n^{1/2}$ [75]; therefore, there are diminishing returns in increasing the number of transients.

The FT-MS analyzer has the resolution necessary to resolve isotopic peaks or a mixture of compounds that are not separated prior to ionization. These are relatively expensive instruments and need an expert to get the most from them. Thus, for application to the analysis of DNA-encoded libraries, this type of mass analyzer is appealing for its resolving power; however, from the point of view of walk-up use, it presents certain issues.

8.6.4 The Orbitrap™ or Electrostatic Trap

This is a recent addition to mass analyzers that was developed and reported by Makarov [76] in 2000. This uses an electrostatic trap comprised of an outer electrode and an inner electrode, both of which have a specific cylindrical geometry. The inner electrode is shaped like a spindle and is essentially at ground potential, whereas the outer electrode, which has a curved geometry, has a DC voltage of several kilovolts (positive if analyzing negative ions) applied to it. An off-axis orifice allows packets of ions to enter the trap. These ions are injected with several kilo-electron volts, inducing a circular or oval trajectory around the inner electrode:

$$\omega = \left(\left(\frac{q}{m} \right) k \right)^{1/2} \tag{8.5}$$

where

ω is the oscillation frequency
q is the total number of charges
m is the mass of the ions
k is the field curvature

From the equation earlier, it can be seen that the oscillation frequency is directly related to the m/z ratio of the ion.

The oscillating frequency of the ions in the Orbitrap™ is independent of the ion's initial velocity, and therefore the ion's initial kinetic energy, as is the case for FT-MS. By contrast, other mass analyzers are affected by the spread of the ion's kinetic energy and the ability of these analyzers to minimize this energy spread to achieve resolution. Because the frequency is independent of the kinetic energy and the frequency can be measured accurately, resolving powers in excess of 100,000 have been routinely reported [77].

The Orbitrap™ has comparable resolution to the FT-MS analyzer and therefore offers the same performance as described in Section 8.6.3. The advantage of the Orbitrap™

is that it does not require cryogens and therefore requires less regular maintenance than the FT-ICR MS but it would still require an expert operator to get the most from this analyzer.

8.6.5 Time of Flight

The first experiments with TOF analyzers go back to Stephens in 1946 [29], and the first commercial instrument was described in 1955 by Wiley and McLaren [78]. The original instruments were linear TOF analyzers where the ions are separated according to their velocities in an electrostatic drift field. It was shown that an ion's drift time to traverse the drift-free region was directly related to its m/z value. The following equation shows the relationship between the m/z value and the drift time:

$$(m/z) = \frac{2eVt^2}{L^2} \tag{8.6}$$

where

V is the acceleration potential

t is the time to traverse the drift-free region

L is the length of the drift-free region

This equation can be further simplified as V, e, and L are constants for any given instrument; therefore,

$$(m/z) = \frac{2eVt^2}{L^2}$$

$$(m/z)^{1/2} = \left(\frac{(2eV)^{1/2}}{L}\right)t \tag{8.7}$$

$$(m/z)^{1/2} = At$$

It can be seen from this expression that the longer the drift time, the higher the m/z of the ion that is being studied. In reality, there are other parameters that need to be taken into account, such as time delays in cables and detector circuitry. This of course should be a constant for any given instrument and can be represented by the term B; therefore,

$$(m/z)^{1/2} = At + B \tag{8.8}$$

Clearly, this is a linear equation and shows that by using two known compounds, a straight-line response for a given TOF can be determined and thus any unknown m/z determined. Obviously, there are certain caveats to this, the most important of which is that the known compounds should not be too close together in m/z value and preferably bracket the mass of the unknown.

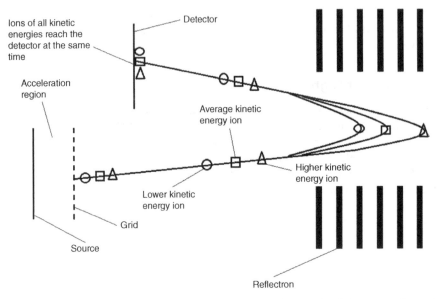

Figure 8.8. Schematic of a TOF analyzer with reflectron and how this helps correct for variations in an ion's kinetic energy.

It is critical to the resolution of the instrument that the ions have the same kinetic energy when they enter the drift tube. An accelerating voltage in the order of 20 kV is applied as a short pulse, which minimizes the difference between each ion's kinetic energy. Vestel et al. reported on the benefits of delayed extraction in increasing resolution [42]. The resolution of a TOF analyzer is also related to the length of the flight tube, as can be seen from the following equation:

$$R = \frac{L}{2\Delta x} \tag{8.9}$$

where

    R is the resolution
    L is the length of the flight tube
    x is the "thickness" of the ion packet

So, if the length of the flight tube is increased, the resolution will increase for the ions that are being analyzed. The downside to a longer flight tube is that sensitivity will decrease because of the loss of ions due to scatter, and housing the instrument in a laboratory becomes more difficult as the flight tube becomes longer. Both are addressed to a degree by using an electrostatic reflector more commonly known as a reflectron.

The use of a reflectron was first proposed by Mamyrin et al. [79] in 1973; this "ion mirror" effectively doubles the flight path of any given drift tube, which if all other parameters are kept the same doubles the resolution of the instrument (Fig. 8.8). An added benefit of the reflectron is it actually decreases the ion packet's "thickness," which also increases the resolution of the instrument. The reflectron in its simplest design is a series of rings that have a potential gradient applied to them.

As discussed, the MALDI source uses a pulsed laser to volatilize and ionize molecules and is ideally suited to large biological molecules as is the TOF analyzer. In fact, molecules in excess of a megadalton (1×10^6 Da) have been analyzed using MALDI-TOF MS [80] and even intact virus head pieces with a mass in excess of 13 MDa [81].

Even with the averaging of microscans, the TOF analyzer has one of the fastest spectral acquisitions. This makes it the ideal analyzer to work with UHPLC that produces narrow chromatographic peaks and therefore needs a fast "scanning" analyzer to ensure the spectra of each component are acquired. UHPLC offers higher separation resolution that is often needed in DNA-encoded samples due to their complexity.

8.7 HYPHENATION OF LC AND MS

So far, we have seen the benefits of Liquid Chromatography (LC) to separate complex mixtures or to desalt oligonucleotides and the power of MS to provide structural information. The combination of these two techniques provides us with a powerful technique that can separate mixtures of oligonucleotides and provide structural information on each one of the components.

Having considered the use of MALDI-MS as a means to ionize oligonucleotides, it has been seen that it is not a routine technique for the analysis of large, >20 kDa, nucleic acids. Normally, MALDI-MS is coupled to chromatographic techniques off-line after fractionation of the sample [82]. There have, however, been a number of attempts to couple separation techniques and MALDI-MS online, and this has been thoroughly reviewed by Murray [83].

ESI uses a liquid flow to generate a spray from which gas-phase ions are produced, making it the obvious choice for coupling online to LC systems. ESI has been shown to be adaptable to various flow rates from nL/min to mL/min and to a variety of mobile phases varying in pH, organic solvent choice, and ionic strength. This versatility combined with the ability to multiply charged ions has meant that electrospray is used for the majority of biological LC–MS applications.

The previous discussion showed that to analyze single- and double-stranded DNA by HPLC, an ion-pair method is preferred. When using ESI to ionize the eluted compounds, the choice of ion pair and other additives is restricted to those that will volatilize and not adversely affect the ionization process. The mobile phase has to be able to suppress the formation of adduct ions as nucleic acids form cation adducts with metals and other cations present, as previously discussed. One approach to reduce the formation of adduct ions is to use volatile bases in the mobile phase. Ammonium ions can be used effectively to replace other cations bound to the DNA in solution and will be released from the DNA during the ionization process [84]. As discussed previously, the use of stronger bases, such as TEA or piperidine, has been shown to be more effective than ammonium at replacing the cations particularly for larger nucleic acid moieties [66]. The addition of ethylenediaminetetraacetic acid at low concentrations (*ca* 10 μM) has also been used to reduce the number of cations present [57].

Obviously, the retention of the nucleic acid on the column of an HPLC system gives the opportunity to wash the salts from the nucleic acid prior to elution. A decrease in sensitivity of nucleic acids is seen at high aqueous concentration and with the use of ion-pair compounds, which is typical of the elution conditions used. These conditions increase the surface tension and the conductivity of the eluent used to form the droplet in the ESI process, leading to a poor response [2]. A number of studies have been carried out to find the ideal conditions for LC–MS of nucleic acids. A balanced approach is required, which neither biases the analysis toward the LC separation nor toward ESI sensitivity. The use of 1,1,1,3,3,3-hexafluro-2-propanol (HFIP) has been used in mobile phases and has a number of functions: (i) being an organic solvent, it helps lower the viscosity and therefore the surface tension of the droplet, and (ii) it has an acidic proton, which can be used to replace cations on the surface of the nucleic acid phosphodiester backbone, thus eliminating some of the cations that may be present. The one stipulation with using HFIP is that it is insoluble in acetonitrile, so the method has to use methanol as the organic modifier, if it is desired to dissolve the HFIP in the organic modifier or to use high organic modifier concentrations. By titrating the HFIP with a triethylammonium base, usually TEAA, the pH of the mobile phase can be adjusted to pH 7. This is ideal for chromatographic conditions, but a higher pH would be more suitable for the negative ion ESI. It has been postulated that the HFIP is volatile and in the droplet is lost preferentially, at least at the droplet surface, and this causes the pH to rise, leading to the dissociation of the ion pair between the nucleic acid and triethylammonium to give good sensitivity for the nucleic acid [85]. Various alkylammonium and their corresponding salts have been investigated [86, 87]. Chromatographically, it was shown that using acetate, bicarbonate, formate, or chloride anion made very little difference, with only a slight change in retention time. To MS, the anion makes a difference to sensitivity; sensitivity increases in the order of chloride < formate < bicarbonate < acetate, and it has been proposed that this is due to the competition for ionization between the anion and the nucleic acid. The most common ion pair used is triethylammonium, which has given many good results. By using a more lipophilic alkyl chain such as butyldimethylammonium as an ion pair, greater retention can be achieved. This means more organic modifier is necessary to elute the nucleic acid, which in turn leads to better sensitivity in the mass spectrometer due to lower viscosity of the droplets.

8.8 REPORTED ANALYTICAL METHOD CONDITIONS

Examples of oligonucleotide chromatographic and mass spectroscopic characterizations are shown in Figure 8.9 and Figure 8.10.

Figure 6.7 shows the chromatographic and mass spectroscopic characterization of a small-molecule DNA-encoded library.

The following four analytical methods have been shown to work well for oligonucleotides and would therefore be appropriate for the analysis of DNA-encoded libraries.

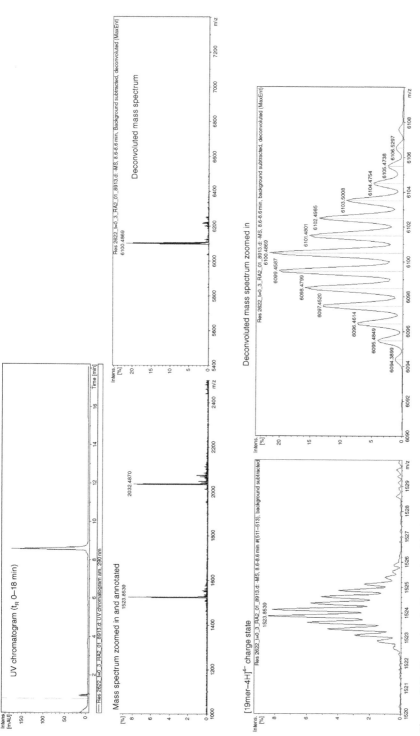

Figure 8.9. Example of oligonucleotide characterization. Sample: 19-mer DNA/RNA chimera TGcAuBrTGcAuBrTGcA (upper case = DNA, lower case = RNA, uBr = 5-bromouridine); MW(av) 6100.37, MW(mono) 6094.64. Chromatographic conditions: Dionex UltiMate 3000 LC system; Acquity UHPLC BEH C18 column (1.7 mm, 1 × 100 mm, Waters, Milford, MA, USA); column temperature 40°C; UV absorbance measured at 290 nm; binary gradient solvent system—mobile phase A, 10 mM TEAA and 100 mM hexafluoroisopropanol (HFIP) in water, and mobile phase B, 20 mM TEAA in acetonitrile, time (min) (%B) 0 (5), 1 (5), 15 (40), 5 (15.1), 18 (5); flow rate 100 µL/min; injection volume 2 mL. Mass spectroscopy conditions: ESI MS conditions—Bruker MicrOTOF, negative ion electrospray, capillary 0 V, capillary cap +4 kV, nebulizer gas 2.0 bar, drying gas 6 L/min.

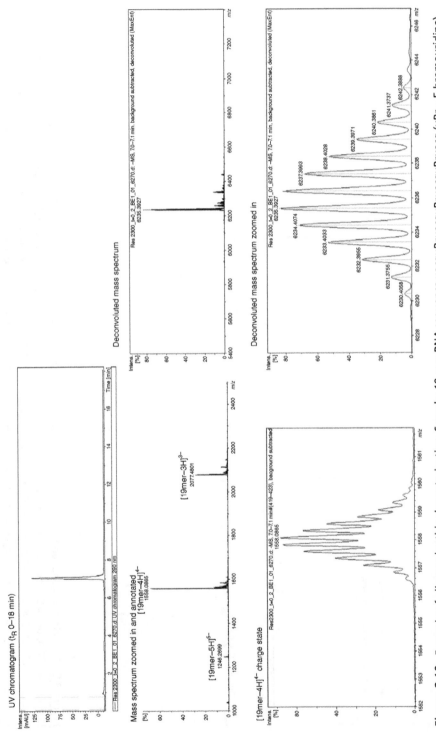

Figure 8.10. Example of oligonucleotide characterization. Sample: 19-mer RNA sequence uacguBruacguBruacguBruacg (uBr = 5-bromouridine). MW(av) 6236.23, MW(mono) 6230.51. Chromatographic and mass spectroscopy conditions are the same as in Figure 8.9.

8.8.1 Method 1

This method was discussed earlier and combines the use of HFIP, TEA, and EDTA with a water/methanol mobile phase and switching valves for optimum throughput [57].

The system uses two switching valves, one of which has two small (1×10 mm) columns (Novatia Oligo HTCS trap columns) to provide desalting of the samples only: the mobile phase is 60/40 water/methanol with 0.75% HFIP, 0.0375% TEA, and 10 μM EDTA. The other valve has the option of a longer column (ACE 3 C18-300 50×2.1 mm) for desalting and separation of the components present in the sample before ionization: the mobile phase is a water/methanol gradient with 0.75% HFIP, 0.0375% TEA, and 10 μM EDTA.

The mass spectrometer used was a Finnigan TSQ 7000 and negative ion ESI data were acquired.

8.8.2 Method 2

This method is a longer method, which could be used for high-quality chromatographic separation for complex mixtures [88].

The mobile phase is a water/acetonitrile gradient. The aqueous mobile phase has 10 mM TEAA and 100 mM HFIP. The organic modifier is 20 mM TEAA dissolved in acetonitrile. The column was a UHPLC BEH C18 (1.7 μm 1×100 mm, Waters, Milford, MA, USA) that was held at 40°C and with a flow rate of 100 μL/min. The gradient begins at 5% organic modifier that is held for 1 min, followed by a gradient from 5% to 40% over 14 min before returning to 5% of organic modifier.

The mass spectrometer used was a Bruker MicrOTOF and negative ion ESI data were acquired over the *m/z* range 250–3500.

8.8.3 Method 3

This method uses capillary columns and low flow rates and the addition of butyldimethylammonium bicarbonate instead of the more usual TEA or TEAA. Also, this method does not use HFIP in the mobile phase [89].

The mobile phase is a water/acetonitrile gradient. The aqueous mobile phase contains 25 mM butyldimethylammonium bicarbonate, and the organic modifier is acetonitrile. The gradient is from 4% to 22% acetonitrile in 15 min with a flow rate of 2.5 μL/min. The column is made in house following a published protocol [90]. The injection volume is only 500 nL to match the low flow rate.

The mass spectrometers used were either a Bruker Esquire 3000 or a Thermo Electron Corp. LCQ Deca; both are QITs. Negative ion ESI data were acquired on both instruments.

8.8.4 Method 4

This is a UHPLC method. The column is a Waters Acquity UHPLC OST 2.1×50 mm, 1.7 μm particle size held at 60°C [91]. The mobile phase is a water/methanol gradient from 31% to 47% mobile phase B. Mobile phase A consists of 15 mM TEA and 400 mM

HFIP in water, and mobile phase B is 50% (v/v) methanol in mobile phase A. The flow rate is 0.2 mL/min.

The mass spectrometer used was a Waters Corp. LCT Premier XE and negative ion ESI data were acquired.

Other examples of the chromatographic and mass spectroscopic characterization of DNA-encoded libraries have been reported [67, supplemental material].

8.9 CONCLUSIONS

As should be clear from the preceding discussion, there are aspects that require special consideration when approaching the analysis of DNA oligonucleotides. Such large and highly charged molecules present different analytical challenges compared with the analysis of small molecules for drug discovery research. The size and charged nature of DNA require modification of chromatography conditions for the separation of oligonucleotide mixtures. An understanding of the different mass detection systems is helpful when selecting and adapting methods for the analysis of DNA and small-molecule DNA constructs. The development of new chemistry for DELs as well as the production of DEL libraries requires reliable and robust analytical methods, the majority of which will be provided by hyphenating LC and MS technologies.

REFERENCES

1. Watson, J.D., Crick, F.H.C. (1953). Molecular structure of nucleic acids: a structure for deoxyribose nucleic acid. *Nature*, *171* 737–738.
2. Huber, C.G., Oberacher, H. (2001). Analysis of nucleic acids by on-line liquid chromatography-mass spectrometry. *Mass Spectrom. Rev.*, *20*, 310–343.
3. Banoub, J.H., Limbach, P.A. (2010). *Mass spectrometry of nucleosides and nucleic acids*, CRC Press, Boca Raton
4. Smith, R.D., Wahl, J.H., Goodlett, D.R., Hofstadler, S.A. (1993). Capillary electrophoresis/ mass spectrometry. *Anal. Chem.*, *65*, 574–584.
5. Jorgenson, J.W., Lukacs, K.D. (1983). Capillary zone electrophoresis. *Science*, *222*, 266–272.
6. Yeung E.S. (1999). Study of single cells by using capillary electrophoresis and native fluorescence detection. *J. Chromatogr. A*, *830*, 243–262.
7. Righetti, P.G., Gelfi, C., D'Acunto, M.R. (2002). Recent progress in DNA analysis by capillary electrophoresis. *Electrophoresis*, *23*, 1361–1374.
8. Ellegren, H., Låås, T. (1989). Size-exclusion chromatography of DNA restriction fragments. Fragment length determinations and a comparison with the behaviour of proteins in size-exclusion chromatography *J. Chromatogr.*, *426*, 217–226.
9. Kato, Y., Sasaki, M., Hashimoto, T., Murotsu, T., Fukushige, S., Matsubara, K. (1983). Separation of DNA restriction fragments by high-performance ion-exchange chromatography on a non-porous ion exchanger. *J. Chromatogr.*, *265*, 342–346.
10. Westman, E., Eriksson, S., Laas, T., Pernemalm, P.-A., Skold, S.-E. (1987). Separation of DNA restriction fragments by ion-exchange chromatography on FPLC columns Mono P and Mono Q. *Anal. Biochem.*, *166*, 158–171.

11. McLaughlin, L.W. (1989). Mixed-mode chromatography of nucleic acids. *Chem. Rev.*, *89*, 309–319.

12. Koivisto, P., Peltonen, K. (2010). Analytical methods in DNA and protein adduct analysis. *Anal. Bioanal. Chem.*, *398*, 2563–2572.

13. Zhang, J.-J., Zhang, L., Zhou, K., Ye, X., Liu, C., Zhang, L., Kang, J., Cai, C. (2011). Analysis of global DNA methylation by hydrophilic interaction ultra high-pressure liquid chromatography tandem mass spectrometry. *Anal. Biochem.*, *413*, 164–170.

14. Alpert, A.J. (1990). Hydrophilic-interaction chromatography for the separation of peptides, nucleic acids and other polar compounds. *J. Chromatogr. A*, *499*, 177–196.

15. Alpert, A.J. (2008). Electrostatic repulsion hydrophilic interaction chromatography for isocratic separation of charged solutes and selective isolation of phosphopeptides. *Anal. Chem.*, *80*, 62–76.

16. Holdšvendová, P., Suchánková, J., Bunček, M., Bačkovská, Coufal, P. (2007). Hydroxymethyl methacrylate-based monolithic columns designed for separation of oligonucleotides in hydrophilic-interaction capillary liquid chromatography. *J. Biochem. Biophys. Methods*, *70*, 23–29.

17. Easter, R.N., Kröning, K.K., Caruso J.A., Limbach, P.A. (2010). Separation and identification of oligonucleotides by hydrophilic interaction liquid chromatography (HILIC)-inductively coupled plasma mass spectrometry (ICPMS). *Analyst*, *135*, 2560–2565.

18. Goss, T.A., Bard, M., Jarrett, H.W. (1990). High performance affinity chromatography of DNA. *J. Chromatogr.*, *508*, 279–287.

19. Goss, T.A., Bard, M., Jarrett, H.W., (1991). High performance affinity chromatography of messenger RNA. *J. Chromatogr.*, *588*, 157–164.

20. Huang, G.J., Krugh, T.R. (1990). Large-scale purification of synthetic oligonucleotides and carcinogen-modified oligodeoxynucleotides on a reverse-phase polystyrene (PRP-1) column. *Anal. Biochem.*, *190*, 21–25.

21. Bartha, A., Stahlberg, J. (1994). Electrostatic retention model of reversed-phase ion-pair chromatography. *J. Chromatogr. A*, *668*, 255–284.

22. Huber, C.G., Oefner, P.J., Bonn, G.K. (1995). Rapid and accurate sizing of DNA fragments by ion-pair chromatography on alkylated nonporous poly(styrene-divinylbenzene) particles. *Anal. Chem.*, *67*, 578–585.

23. Huber, C.G., Oefner, P.J., Bonn, G.K. (1993). High resolution liquid chromatography of oligonucleotides on nonporous alkylated styrene-divinylbenzene copolymers. *Anal. Biochem.*, *212*, 351–358.

24. Huber, C.G., Berti, G.N. (1996). Detection of partial denaturation in AT-rich DNA fragments by ion-pair reversed-phase chromatography. *Anal. Chem.*, *68*, 2959–2965.

25. Oefner, P.J. (2000). Allelic discrimination by denaturing high-performance liquid chromatography. *J. Chromatogr. B.*, *739*, 345–355.

26. Thomson, J.J. (1913). On the appearance of helium and neon in vacuum tubes. *Nature*, *90*, 645–647.

27. Aston, F.W. (1919). A positive ray spectrograph. *Philos. Mag.*, *38*, 707–714.

28. Dempster, A.J. (1918). A new method of positive ray analysis. *Phys. Rev.*, *11*, 316–325.

29. Stephens, W.E. (1946). A pulsed mass spectrometer with time dispersion. *Phys. Rev.*, *69*, 691–691.

30. Paul, W., Steinwedel, H. (1953). Ein neues Massenspektrometer ohne Magnetfeld. *Z. Naturforsch. A*, *8*, 448–450.

31. Comisarow, M.B., Marshall, A.G. (1974). Fourier transform ion cyclotron resonance spectroscopy. *Chem. Phys. Lett.*, *25*, 282–283.

32. Biemann, K., McCloskey, J.A. (1962). Applications of mass spectrometry to structure problems VI. Nucleosides. *J. Am. Chem. Soc.*, *84*, 2005–2007.

33. Wilson, M.S., McCloskey, J.A. (1975). Chemical ionization mass spectrometry of nucleosides mechanisms of ion formation and estimations of proton affinity. *J. Am. Chem. Soc.*, *97*, 3436–3444.

34. Toren, P.C., Betsch, D.F., Weith, H.L., Coull, J.M. (1986). Determination of impurities in nucleoside 3′-phosphoramidites by fast atom bombardment mass spectrometry. *Anal. Biochem.*, *152*, 291–294.

35. Fujitake, M., Harusawa, S., Araki, L., Yamaguchi, M., Lilley, D.M.J., Zhao, Z.Y., Kurihara, T. (2005). Accurate molecular weight measurements of nucleoside phosphoramidites: a suitable matrix of mass spectrometry. *Tetrahedron*, *61*, 4689–4699.

36. Karas, M., Bachmann, D., Bahr, U., Hillenkamp, F. (1987). Matrix-assisted ultraviolet laser desorption of non-volatile compounds. *Int. J. Mass Spectrom. Ion Processes*, *78*, 53–68.

37. Smith, R.D., Loo, J.A., Edmonds, G.G., Barinaga, C.J., Udseth, H.R. (1990). New developments in biochemical mass spectrometry: electrospray ionization. *Anal. Chem.*, *62*, 882–899.

38. Cotter, R.J. (1987). Laser mass spectrometry: an overview of techniques, instruments and applications. *Anal. Chim. Acta*, *195*, 45–59.

39. Ball, R.J.(2005). A mass spectrometry based hybridization assay for single nucleotide polymorphism analysis. Doctor of Philosophy thesis, University of Southampton.

40. Tanaka, K., Waki, H., Ido, Y., Akita, S., Yoshida, Y., Yoshida, T. (1988). Protein and polymer analyses up to m/z 100,000 by laser ionization time-of-flight mass spectrometry. *Rapid Commun. Mass Spectrom.*, *2*, 151–153.

41. Nordhoff, E., Kirpekar, F., Roepstorff, P. (1996). Mass spectrometry of nucleic acids. *Mass Spectrom. Rev.*, *15*, 67–138.

42. Vestel, M.L., Juhasz, P., Martin, S.A. (1995). Delayed extraction matrix-assisted laser desorption time-of-flight mass spectrometry. *Rapid Commun. Mass Spectrom.*, *9*, 1044–1050.

43. Gluckmann, M., Pfenninger, A., Kruger, R., Thierolf, M., Karasa, M., Horneffer, V., Hillenkamp, F., Strupat, K. (2001). Mechanisms in MALDI analysis: surface interaction or incorporation of analytes? *Int. J. Mass Spectrom.*, *210/211*, 121–132.

44. Knochenmuss, R. (2006). Ion formation mechanisms in UV MALDI. *Analyst*, *131*, 966–986.

45. Chen, C.-H., Taranenko, N.I., Zhu, Y., Allman, S.L. (1996). MALDI for fast DNA analysis and sequencing. *Lab. Robot. Autom.*, *8*, 87–99.

46. Videler, H., Ilag, L.L., McKay, A.R.C., Hanson, C.L., Robinson, C.V. (2005). Mass spectrometry of intact ribosomes. *Fed. Eur. Biochem. Soc. Lett.*, *579*, 943–947.

47. Yamashita, M., Fenn, J.B. (1984). Electrospray ion source. Another variation on the free-jet theme. *J. Phys. Chem.*, *88*, 4451–4459.

48. Dole, M., Mack, L.L., Hines, R.L., Mobley, R.C., Ferguson, L.D., Alice, M.B. (1968). Molecular beams of macroions. *J. Chem. Phys.*, *49*, 2240–2249.

49. Mann, M., Meng, C.K., Fenn, J.B. (1989). Interpreting mass spectra of multiply charged ions. *Anal. Chem.*, *61*, 1702–1708.

50. Potier, N., Van Dorsselaer, A., Cordier, Y., Roch, O., Bischoff, R. (1994). Negative electrospray ionization mass spectrometry of synthetic and chemically modified oligonucleotides. *Nucleic Acids Res.*, *22*, 3895–3903.

51. Cole, R.B. (1997). *Electrospray mass spectrometry: fundamentals, instrumentation & applications*, John Wiley & Sons Inc., New York.

52. De Hoffmann, E., Stroobant, V. (2007). *Mass spectrometry: principles and applications*, John Wiley & Sons Ltd, Chichester.

53. Gomez, A., Tang, K. (1994). Charge and fission of droplets in electrostatic sprays. *Phys. Fluids*, 6, 404–414.

54. Cole, R.B. (2000). Some tenets pertaining to electrospray ionization mass spectrometry. *J. Mass Spectrom.*, 35, 763–772.

55. Amad, M.H., Cech, N.B., Jackson, G.S., Enke, C.G. (2000). Importance of gas-phase proton affinities in determining the electrospray ionization response for analytes and solvents. *J. Mass Spectrom.*, 35, 784–789.

56. Kebarle, P. (2000). A brief overview of the present status of the mechanisms involved in electrospray mass spectrometry. *J. Mass Spectrom.*, 35, 804–817.

57. Hail, M.E., Elliott, B., Nugent, K., Whitney, J.L., Detlefsen, D.J. (2003). High-throughput analysis of oligonucleotides using automated electrospray ionization mass spectrometry. *Am. Biotechnol. Lab.*, 22, 12–14.

58. Ferrige, A.G., Seddon, M.J., Jarvis, S. (1991). Maximum entropy deconvolution in electrospray mass spectrometry. *Rapid Commun. Mass Spectrom.*, 5, 374–379.

59. Bleicher, K., Bayer, E. (1994). Various factors influencing the signal intensity of oligonucleotides in electrospray mass spectrometry. *Biol. Mass Spectrom.*, 23, 320–322.

60. Stults, J.T., Marsters, J.C. (1991). Improved electrospray ionization of synthetic oligodeoxynucleotides. *Rapid Commun. Mass Spectrom.*, 5, 359–363.

61. Brezinschek, H.P., Brezinschek, R.I., Lipsky, P.E. (1995). Analysis of the heavy chain repertoire of human peripheral B cells using single-cell polymerase chain reaction. *J. Immunol.*, 155, 190–202.

62. Xu, N., Lin, Y., Hofstadler, S.A., Matson, D., Call, C.J., Smith, R.D. (1998). A microfabricated dialysis device for sample cleanup in electrospray ionization mass spectrometry. *Anal. Chem.*, 70, 3553–3556.

63. Ross, P.L., Davis, P.A., Belgrader, P. (1998). Analysis of DNA fragments from conventional and microfabricated PCR devices using delayed extraction MALDI-TOF mass spectrometry. *Anal. Chem.*, 70, 2067–2073.

64. Huber, C.G., Buchmeiser, M.R. (1998). On-line cation exchange for suppression of adduct formation in negative-ion electrospray mass spectrometry of nucleic acids. *Anal. Chem.*, 70, 5288–5295.

65. Wunschel, D.S., Tolić, L.P., Feng, B., Smith, R.D. (2000). Electrospray ionization Fourier transform ion cyclotron resonance analysis of large polymerase chain reaction products. *J. Am. Soc. Mass Spectrom.*, 11, 333–337.

66. Grieg, M., Griffey, R.H. (1995). Utility of organic bases for improved electrospray mass spectrometry of oligonucleotides. *Rapid Commun. Mass Spectrom.*, 9, 97–102.

67. Clark, M.A., Acharya, R.A., Arico-Muendel, C.C., Belyanskaya, S.L., Benjamin, D.R., Carlson, N.R., Centrella, P.A., Chiu, C.H., Creaser, S.P., Cuozzo, J.W., Davie, C.P., Ding, Y., Franklin, G.J., Franzen, K.D., Gefter, M.L., Hale, S.P., Hansen, N.J.V., Israel, D.I., Jiang, J., Kavarana, M.J., Kelley, M.S., Kollmann, C.S., Li, F., Lind, K., Mataruse, S., Medeiros, P.F., Messer, J.A., Myers, P., O'Keefe, H., Oliff, M.C., Rise, C.E., Satz, A.L., Skinner, S.R., Svendsen, J.L., Tang, L., van Vloten, K., Wagner, R.W., Yao, G., Zhao, B., Morgan, B.A. (2009). Design, synthesis and selection of DNA-encoded small-molecule libraries. *Nat. Chem. Biol.*, 5, 647–654.

68. Makarov, A.A. (1999). Mass spectrometer. US Patent, 5886346.

69. Finnigan, R.E. (1994). Quadrupole mass spectrometers. *Anal. Chem.*, *66*, 969A–975A.

70. Stafford, G.C., Kelley, P.E., Syka, J.E.P., Reynolds, W.E., Todd, J.F.J (1984). Recent Improvements in and analytical applications of advanced ion trap technology. *Int. J. Mass Spectrom. Ion Processes*, *60*, 85–98.

71. Douglas, D.J., Frank, A.J., Mao, D. (2005). Linear ion traps in mass spectrometry. *Mass Spectrom. Rev.*, *24*, 1–29.

72. Hipple, J.A., Sommer, H., Thomas, H.A. (1949). A precise method of determining the Faraday by magnetic resonance. *Phys. Rev.*, *76*, 1877–1878.

73. Comisarow, M.B., Marshall, A.G. (1974). Frequency-sweep Fourier transform ion cyclotron resonance spectroscopy. *Chem. Phys. Lett.*, *26*, 489–490.

74. Amster, I.J. (1996). Fourier transform mass spectrometry. *J. Mass Spectrom.*, *31*, 1325–1337.

75. Wronska, L.V.M. (2009). The factors and protocols that influence accuracy, precision and uncertainty of accurate mass measurements by Fourier transform ion cyclotron resonance mass spectrometry to validate the assignment of elemental composition. University of Southampton. Doctor of Philosophy thesis.

76. Makarov, A. (2000). Electrostatic axially harmonic orbital trapping: a high-performance technique of mass analysis. *Anal. Chem.*, *72*, 1156–1162.

77. Hu, Q., Noll, R.J., Li, H., Makarov, A., Hardman, M., Cooks, R.G. (2005). The Orbitrap: a new mass spectrometer. *J. Mass Spectrom.*, *40*, 430–443.

78. Wiley, W.C., McLaren, I.H. (1955). Time of flight mass spectrometer with improved resolution. *Rev. Sci. Instrum.*, *26*, 1150–1157.

79. Mamyrin, B.A., Karataev, V.I., Shmikk, D.V., Zagulin, V.A. (1973). Mass Reflectron. New nonmagnetic time-of-flight high-resolution mass spectrometer. *Zhurnal Eksperimental'noi I Toereticheskoi Fizki*, *64*, 82–9.

80. Wenzel, R.J., Matter, U., Schultheis, L., Zenobi, R. (2005). Analysis of megadalton ions using cryodetection MALDI time-of-flight mass spectrometry. *Anal. Chem.*, *77*, 4329–4337.

81. Bier, M.E., Aksenov, A.A., Ozdemir, A., Sipe, D., Hendrix, R.W., Firek, B. (2007). MALDI TOF MS of biomacromolecular complexes using a superconducting tunnel junction (STJ) cryodetector. Abstracts of papers, *234th ACS National Meeting*, Boston, MA.

82. Vallone, P.M., Devaney, J.M., Marino, M.A., Butler, J.M. (2002). A strategy for examining complex mixtures of deoxyoligonucleotides using ion-pair-reverse-phase high-performance liquid chromatography, matrix-assisted laser desorption ionization time-of-flight mass spectrometry and informatics. *Anal. Biochem.*, *304*, 257–265.

83. Murray, K.K. (1997). Coupling matrix-assisted laser desorption/ionization to liquid separations. *Mass Spectrom. Rev.*, *16*, 283–299.

84. Oberacher, H. (2008). On the use of different mass spectrometric techniques for characterization of sequence variability in genomic DNA. *Anal. Bioanal. Chem.*, *391*, 135–149.

85. Apffel, A., Chakel, J.A., Fischer, S., Lichtenwalter, K., Hancock, W.S. (1997). Analysis of oligonucleotides by HPLC-electrospray ionization mass spectrometry. *Anal. Chem.*, *69*, 1320–1325.

86. Huber, C.G., Krajete, A. (1999). Analysis of nucleic acids by capillary ion-pair reversed-phase HPLC coupled to negative-ion electrospray ionization mass spectrometry. *Anal. Chem.*, *71*, 3730–3739.

87. Oberacher, H., Parson, W., Mühlmann, R., Huber, C.G. (2001). Analysis of polymerase chain reaction products by on-line liquid chromatography-mass spectrometry for genotyping of polymorphic short tandem repeat loci. *Anal. Chem.*, *73*, 5109–5115.

88. Riley, J.-A., Brown, T., Gale, N., Herniman, J., Langley, G.J. (2012). Self reporting RNA probes as an alternative to cleavable small molecule mass tags. *Analyst*, *137*, 5817–5822.

89. Mayr, B.M., Kobold, U., Moczko, M., Nyeki, A., Koch, T., Huber, C.G. (2005). Identification of bacteria by polymerase chain reaction followed by liquid chromatography-mass spectrometry. *Anal. Chem.*, *77*, 4563–4570.

90. Premstaller, A., Oberacher, H., Huber, C.G. (2000). High-performance liquid chromatography-electrospray ionization mass spectrometry of single- and double-stranded nucleic acids using monolithic capillary columns. *Anal. Chem.*, *72*, 4386–4393.

91. McCarthy, S.M., Ivleva, V., Fujimoto, G., Gilar, M. (2008). Optimization of LCT Premier XE MS settings for oligonucleotide analysis. Application note, Waters Corp.

INFORMATICS: FUNCTIONALITY AND ARCHITECTURE FOR DNA-ENCODED LIBRARY PRODUCTION AND SCREENING

John A. Feinberg[1,2] and Zhengwei Peng[1,3]

[1] *Formerly of Roche, Nutley, NJ, USA*
[2] *Currently of Accelrys, Inc., Bedminster, NJ, USA*
[3] *Currently of Merck & Co., Rahway, NJ, USA*

9.1 INTRODUCTION

High-Throughput Screening (HTS) of large compound collections has been established as one of the proven strategies for hit/lead generation in drug discovery [1]. Those corporate compound collections used by pharmaceutical companies generally contain about a few million compounds (10^6–10^7), which come from either single-compound synthesis or combinatorial library synthesis. Though the standard combinatorial chemistry approach has the potential to generate a huge number of compounds (10^{14}–10^{18}), only a very small portion (~10^6) of those virtual compounds has actually been synthesized [2]. The main reason is the cost [3] associated with the synthesis and screening of so many compounds using traditional methods, which require that each compound be stored in separate containers (e.g., wells and plates) and screened individually against selected protein targets.

During the last few years, DNA-encoded combinatorial libraries have been recognized by the pharmaceutical industry as a tool for overcoming the prohibitive cost barrier associated with traditional methods [4]. In 2009, Clark et al. from GlaxoSmithKline (GSK) reported a four-component combinatorial library containing

0.8×10^9 compounds inside a single tube [5]. Each of those library compounds was covalently tagged by a unique DNA sequence that encoded the identifiers for the reaction building blocks used to synthesize that compound [5]. The GSK group went on to demonstrate that the full library could be screened simultaneously against a target protein via an enrichment process based on physical binding between individual library compounds and the immobilized target proteins. By amplifying and sequencing the DNA tags remaining after the enrichment step, they were able to identify a subset of compounds that were potent inhibitors of the target protein p38 MAPK kinase. Here, it is important to keep in mind that the goal of the sequencing step within the context of DNA-encoded library screenings is not to elucidate the sequences (since they are known already) but rather to detect and count them as individual sequences in order to derive an assay readout that is related to the binding affinity of the DNA-encoded library compounds with the target proteins. This goal is different from the goal of genomic sequencing, where the results of multiple reads are pieced together as a consensus to elucidate the sequence of an unknown DNA segment.

Compared with the hit generation process utilizing traditional methods of singleton and combinatorial library synthesis and HTS assay screening, the hit generation process based on DNA-encoded libraries creates unique demands on the informatics system that supports it. In this chapter, an informatics perspective on the workflow associated with DNA-encoded library production and utilization will be given. It will be followed by a high-level discussion of the business needs, requirements, and architecture of a hypothetical informatics system. Naturally, this discussion is heavily shaped by the published work [4, 5]. Next is a more detailed discussion of the DNA encoding of identifiers for the reaction building blocks and the need for sequence error detection in the encoding strategy. Finally, this chapter will conclude with a summary of the highlighted points and a short outlook to the future.

9.2 AN INFORMATICS PERSPECTIVE ON THE WORKFLOW ASSOCIATED WITH DNA-ENCODED LIBRARIES

Based on the published report by the GSK group [5], the main business workflow may be summarized as library production, library screening, and hit analysis and triage (see Fig. 9.1). The first step, library production, may be further refined as idea evaluation, reaction development, validation of reaction building blocks, and the actual DNA-encoded library synthesis. Within the library production step, the innovative feature introduced by the use of DNA encoding is the one-to-one association of a unique DNA tag with a unique reaction building block identifier. The second step, library screening, also has several features unique to the use of DNA-encoded libraries. It may be further refined to include affinity-based enrichment against immobilized target proteins, DNA tag amplification and sequencing, and the extraction of the assay readout from the sequence data. Unlike standard DNA sequencing used to elucidate the sequence of an unknown DNA fragment, the objective of DNA sequencing and assay readout extraction steps in this context is to count the occurrences of expected DNA sequences and to use such frequently occurring sequences to infer the identity of library compounds that bind to a target protein with higher affinity relative to average

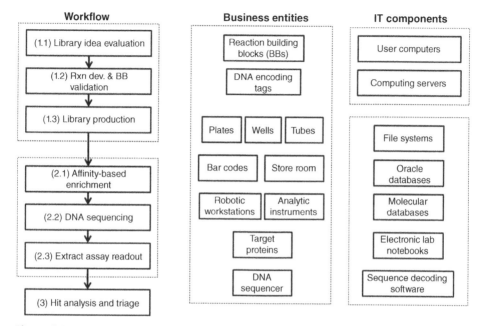

Figure 9.1. Workflow, business entities, and informatics components to support the production and utilization of DNA-encoded libraries. "BB" stands for reaction Building Block. "Rxn dev. & BB validation" stands for Reaction Development and Reaction Building Block Validation.

background binding frequencies. The final step, hit analysis and triage, is not unique to the use of DNA-encoded libraries, and a commercial tool Spotfire [6a] has been reported for this purpose [5].

Based on an anticipated workflow, a list of business entities participating in the implementation of a DNA-encoded library process is given in Figure 9.1. The required ingredients of the process are the reaction building blocks and DNA-encoding tags. For automation purposes, those reaction building blocks and DNA tags have been most likely prepared in solution phase, stored in plate format, barcoded, and housed in special storerooms before the actual library synthesis. To streamline the library synthesis, various robotic workstations are likely to be used.

9.3 INFORMATICS NEEDS, FUNCTIONALITY, AND ARCHITECTURE

The high-level summary of informatics components necessary to support the production and utilization of DNA-encoded libraries is also depicted in Figure 9.1. These informatics components combine support for three types of users (chemists, biologists, and computation/informatics data experts) involved in the production, screening, and analysis of DNA-encoded libraries (see Fig. 9.2). Web tools and molecular database access tools (such as ISIS/Base [7] depicted in Fig. 9.2) can be used by the chemists to interact with the informatics system (to search, access, register, upload, etc.). The molecular

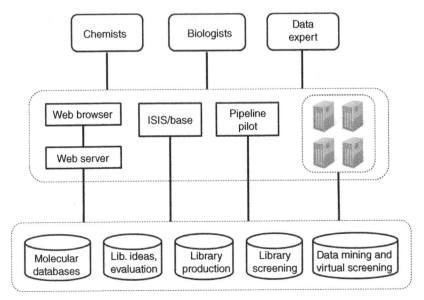

Figure 9.2. A high-level architecture of a hypothetical informatics system to support the production and utilization of DNA-encoded libraries. Here, three types of user groups are assumed: chemists for reaction development and library production; biologists for target protein production, library screening, and DNA tag sequencing; and data analysis experts who perform the postsequence data analysis, hit analysis and triage, and data mining. Depending on the nature of a user's work, he/she could interact with the informatics system via web, stand-alone GUI tools (e.g., ISIS/Base and Pipeline Pilot rich client), or even Unix/Linux command line interface.

biologist and the computational data expert may directly access the computational cluster via the Unix command line interface and perform various data analyses and mining runs via flexible and powerful analysis tools such as Pipeline Pilot [8]. In the following sections, the requirements of and possible informatics solutions for each of the steps inside the organizational workflow will be discussed.

9.3.1 Library Idea Evaluation

In this step, the library team members propose reactions, search for suitable reaction building blocks available from both commercial vendors and in-house stockrooms, and assess the potential additional value offered by a new library over the existing libraries already produced. Molecular databases of reaction building blocks are likely available in most pharmaceutical companies. To evaluate and prioritize proposed libraries, one might want to enumerate a portion of the virtual compound space and compute the physical properties and molecular descriptors and then assess the value added by the new library. Interested readers are encouraged to read Chapter 12 in this volume for more on this subject.

9.3.2 Reaction Development and Reaction Building Block Validation

During this step, an electronic chemistry laboratory notebook can provide sufficient support for experiment design and result capture. Also, existing informatics systems for reagent ordering can be used. Since many reaction building blocks need to be validated (e.g., 192 Fmoc-protected amino acids + 32 bifunctional acids + 340 amines + 384 amines = 948 building blocks used in the second library reported by Clark et al. [5]), material management, plate formatting, barcoding, and handling all need to work seamlessly together to maximize efficiency. Furthermore, the validation results for each reaction building block must be captured for later use by the library team in evaluating newly proposed libraries.

9.3.3 Library Production

During this step, the library to be produced must be defined and registered. All associations of reaction building block IDs with DNA tags must be recorded precisely so that extraction of the assay readout from DNA sequences will be possible and error free. It is unlikely that any existing electronic chemistry laboratory notebook offers this functionality. Rather, this functionality must be custom-built. However, software components to manage the storeroom and drive the robotic workstations and analytical instruments are likely available directly from vendors. Even so, there are still some choices to be made regarding registration of the synthesized DNA-encoded libraries. For example, a DNA-encoded library may be registered as a single entity, or the individual compounds within that library may be fully enumerated and registered as individual molecules within a molecular database. Given that current informatics systems can handle at most 10^8 explicit molecules [9], it would seem that the size of the reported 0.8×10^9 GSK library approaches the limits of capability. This challenge will be visited again in Section 9.5.

9.3.4 Affinity-Based Enrichment

A single test tube containing all compounds of a DNA-encoded library is used as the input. At the end of the affinity-based enrichment step, another test tube is obtained as the output. This output test tube contains compounds with DNA tags still remaining after the enrichment step, and the relative concentrations of those DNA tags inside the test tube reflect the binding affinities of their corresponding library compounds against the target proteins. This simple workflow does not pose any special challenges beyond the requirement that explicit experimental conditions used for enrichment must be captured. Yet it is also possible that the informatics needs may grow significantly in the future when more elaborate workflows are implemented for this step.

9.3.5 DNA Sequencing

The input for this step is a test tube containing the DNA tags isolated from the selection experiment. Each DNA segment is about 50–100 bases long (estimate based on the Clark report [5]). The output of this step is a file containing the sequences of those

DNA segments in either FASTA [10] or FASTQ [11] format. With the advent of next-generation sequencing, it is likely that all software modules required for this step are available directly from the vendor of the DNA-sequencing hardware. Furthermore, this step may be fully outsourced to external service providers to leverage their scale and expertise while eliminating the cost associated with building up the required capabilities in-house.

9.3.6 Extraction of Assay Readout

Once the DNA sequence data are obtained, a software component is required to translate the data into binding affinities (relative to background) between the library compounds and the target protein (see Fig. 9.3). As the following discussion makes clear, this component is unique to the DNA-encoded library technology and therefore must be custom-built.

The basic principle was expressed simply by Mannocci and coworkers [4c] as follows: "The counts for individual library codes… indicate the abundance of the corresponding oligonucleotide-compound conjugates." Yet, the actual practice seems to be much more involved, as the same authors have shown in their work [4c]. This is due in part to the background noise arising from the existence of nontarget-specific selections, which must be properly considered and evaluated (see both Fig. 9.4 and Figure S1 of Ref. [4c]). Thus, in Table S1 of Ref. [4c], Mannocci and coworkers compared the counts of the 11 DNA sequences for their library compounds after the enrichment step against immobilized streptavidin and also the K_d values of the associated small molecules against streptavidin. If the count of a given DNA sequence is indeed a good descriptor for the binding affinity of the encoded library compound against the target protein streptavidin, then a high correlation is expected between the sequence counts and the corresponding K_d values of the small molecules. To test this hypothesis, the data from Table S1 in Ref. [4c] is analyzed using a linear fit of $-\log(K_d)$ as a

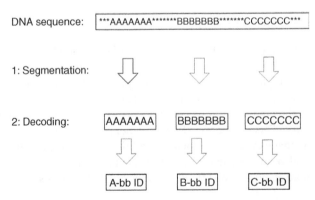

Figure 9.3. Decoding of a library compound's DNA tags to its synthetic origin. Here, a three-component library compound is used for illustration. The first step (segmentation) is to extract the appropriate sections from the full input DNA sequence to individual DNA tags for the A, B, and C reaction building blocks. The second step is a straightforward lookup process to convert a DNA tag to its corresponding reaction building block ID.

TABLE 9.1. Correlation between sequence counts and K_d values reported in Table S1 of Ref. [4c] by Mannocci and coworkers (2008)

Compound	Count[a]	K_d (μM)[b, c]	$-\log(K_d/1.0\ \mu M)$[d]
15–117	0	99	−2.00
15–78	7	79	−1.90
13–40	2	54	−1.73
02–107	0	50	−1.70
07–78	73	11	−1.04
11–78	48	3.5	−0.544
16–78	55	1.1	−0.0414
02–49	41	0.8	0.097
02–78	108	0.38	0.420
17–78	70	0.37	0.432
17–49	32	0.35	0.456

[a] The original column label in Table S1 of Ref. [4c] is "Count after DEL4000 selection."
[b] The original column label in Table S1 of Ref. [4c] is "Streptavidin K_d, μM."
[c] The table is sorted in descending order of K_d.
[d] Linear fitting: $-\log(K_d/1.0\ \mu M) = -1.535(\pm0.325) + 0.0214\ (\pm0.0062) \times$ count with $R^2 = 0.57$.

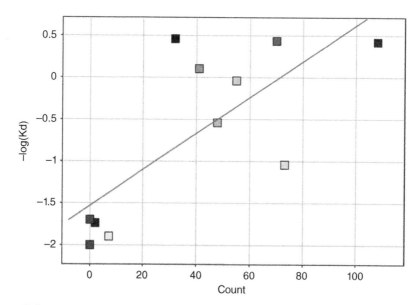

Figure 9.4. Correlation between the sequence count and the K_d value. The line represents the linear fit of $-\log(K_d)$ as a function of count. The legend in Table 9.1 for the source and annotation of the data plotted inside this figure.

function of the observed sequence count (see Table 9.1 and Fig. 9.4). The observed correlation turned out not to be very high, with an R^2 of only 0.57. It is possible that the large statistical uncertainties associated with the low count values (0–108; see Table 9.1) might account, to some extent, for the low correlation observed. Nevertheless, their work suggests that the experimental descriptor (e.g., such as count after enrichment) chosen to aid prioritization of library hits for hit follow-up must be validated first using K_d values of resynthesized small molecules against the same target protein.

Another consideration is experimental error in the PCR amplification and sequencing step, which may degrade the quality of the assay data. A more detailed discussion of sequence error and its potential impact and possible detection is given in Section 9.4.

9.3.7 Hit Analysis and Triage

Many data analysis tools may be used for this step, which involves SAR analysis, trend discovery, and clustering (e.g., Spotfire and Vortex; see Ref. [6]). The GSK group reported the utilization of Spotfire [6a] for this step. As long as it is assumed that one DNA sequence corresponds to a unique molecular structure, then the analysis of hits from DNA-encoded libraries should be similar to that for hits from regular compound collections. However, a unique DNA sequence actually represents a family of related molecular structures (see Fig. 5.6). Therefore, an analysis feature may be custom-built to relate any one DNA sequence to the family of molecular structures with which it is associated.

9.4 EXTRACTION OF ASSAY READOUT AND IMPACT OF SEQUENCE ERRORS ON ENCODING STRATEGY

Given a full DNA sequence associated with a specific library compound, one can extract those sections corresponding to the coding regions for the reaction building block IDs as the precise locations for their starting and ending positions are known. If there is no sequence error in a given DNA tag, then a simple DNA tag to reaction building block ID lookup will translate a given full DNA sequence into a library compound identity (a combination of reaction building block IDs). These two steps are depicted in Figure 9.3 as segmentation and decoding, and their implementation in software is also expected to be straightforward.

In reality, DNA sequences experimentally obtained in the sequencing step do contain errors. For example, in a 2011 paper, Minoche and coworkers reported that the expected error rate for each of the bases inside raw sequence reads obtained from the Illumina HiSeq system may be as high as 10^{-1} (7–16%) [12]. In comparison, the error rate at the base level for genomic sequencing has been reported to be around 10^{-3} to 10^{-2} [13]. This apparent discrepancy in error rate could be considered in light of the fact that base calling for genomic sequencing is in fact a result of the consensus among many individual reads (multiple coverage). On the other hand, each raw sequence tag used in DNA-encoded libraries has to be treated individually since the observed frequency of each unique sequence tag is related to the binding affinity of a library compound encoded by that tag with the target protein.

The expected error rate reported by Illumina is approximately 0.01%. However, it is important to keep in mind that this error rate is calculated from an Illumina analysis of a control sample (PhiX) and only uses a percentage of the data (typically 85–90%) that has a quality score at Q30 or better. Q30 is the probability of any single base containing a 1 in 10,000 chance of containing an error. This is what gives rise to the 0.01% error rate. This is usually not what one sees in practice because all reads generated are typically not "perfect" like the PhiX control. There are many things that can affect error rates, including GC content, sequence diversity, and cluster density on the flow cell. Therefore, it is not surprising that one may see a slightly higher error rate. For sequences that contain common base pair sequences depending on the tagging architecture, there will be a higher error rate. This is expected given that every sequence on the flow cell may contain the exact same nucleotide, and therefore, they are all lighting up at once when their fluorescent tag is excited. This makes it more difficult for the instrument to determine where exactly the light is coming from, thereby decreasing the quality score (or confidence of the base call) and increasing the error rate.

Such a high error rate in the raw sequences would have a significant impact on the assay signal extraction. Assuming that the length of DNA tags used to encode reaction building blocks is seven (as reported by Clark et al. [5]), the chance of obtaining four perfect DNA tags used to encode four reaction building blocks of a library compound would be $(1 - 0.1)^{7 \times 4} = 0.05$, given the error rate of 0.1 (10^{-1}). This means that more than 95% of the tags present in the raw sequence data will be corrupted. Even if the error rate is reduced to 10^{-2}, the chance of obtaining four perfect DNA tags would be $(1 - 0.01)^{7 \times 4} = 0.75$. This still would result in 25% of the tags having sequence data corruption. Fortunately, the research field on information encoding and decoding and error detection is very mature with broad applications known in the communication and information industries. Interested readers are encouraged to find more general discussions on this topic. A simplified example will be given in the succeeding text to illustrate this point.

Let us assume that there are four reaction building blocks (bb_1, bb_2, bb_3, and bb_4). How would one encode their identities using DNA sequences? The simplest one is to use a DNA tag of length 1 according to the following: A → bb_1, G → bb_2, C → bb_3, and T → bb_4. In a world without error, this encoding scheme would be sufficient.

In the aforementioned example, there are only four reaction building blocks for a single reaction component to be encoded. In real DNA-encoded library production, however, the number of reaction building blocks could be in the hundreds or even thousands. Given the numbers of building blocks involved, it is clear that there must be a software tool to systematically and efficiently design DNA tags. One can envision two types of use case for this software tool. In the initial use case, a set of DNA tags is generated with user-specified constraints required for efficient reaction development and library production. For example, the constructed set of DNA tags must be sufficient in number to code for the given number of reaction building blocks (e.g., 1000). Once the DNA tags are designed in accordance with these parameters, they may be ordered from external vendors that are specialized in oligonucleotide synthesis. Thus, for example, for the 32 bifunctional acids used by Clark et al. [5] in the second library, 32 tags were selected from the existing DNA tag library of 7 bases long that are 5 or more mutations apart.

9.5 SUMMARY AND OUTLOOK

In this chapter, an overview of the workflow related to the production and utilization of DNA-encoded libraries is given from the informatics perspective. The informatics needs for the steps within the workflow and their possible solutions are identified, discussed, and combined into an informatics system with a hypothetical three-tiered architecture having the following characteristics:

- It is intended to be used by a small team specialized in DNA-encoded library production and utilization for hit generation;
- It should be built in a modular way for flexibility, robustness, and adaptability;
- It enables many existing systems and/or components to be leveraged and reused to support DNA-encoded library activities (a molecular structure database for reaction building blocks, a building block inventory system, electronic laboratory notebooks, etc.);
- It enables the intelligent design and utilization of DNA tags to maximize signal and minimize the impact of sequencing errors.

The highest number of total sequencing reads reported by Clark et al. [5] is slightly less than 200,000 (or 0.2 million reads). Modern HTS machines could produce far higher amounts of data. For example, in 2011, 246 million read pairs (2×95-nucleotides) were reported to have been generated in a single experiment by the Illumina HiSeq 2000 [12]. This 1000-fold increase in sequencing throughput translates into significantly enhanced sensitivity of the assay as well as increased demands for computer hardware (e.g., CPUs, file space, and Oracle table space) and software to decode the DNA sequences and extract the assay signal related to physical binding affinity between the encoded library compound and the target protein. This increased demand needs to be taken into consideration at the planning stage of implementing an informatics system to support DNA-encoded library production and utilization.

As the number of DNA-encoded libraries produced increases, interest in the following three areas are expected to grow:

1. Development of a more systematic evaluation of new library proposals (see Chapter 12 for more discussion);
2. Development of the capability for Virtual Screening (VS) and data mining against those libraries to determine which libraries should (or should not) be screened against a given protein target where the screening of all produced libraries becomes impractical;
3. Development of the capacity for VS against those libraries as a means of idea generation for compound design such that any compound obtained may be followed up by project medicinal chemists via actual synthesis.

All three areas could benefit from a potential database containing DNA-encoded library compounds in which one can perform various searches and data analyses. Even though the size of DNA-encoded libraries is generally too large if their explicit member compounds are fully enumerated [9], there are now implicit ways to store those combinatorial libraries without the need for full enumeration while still allowing end users to perform various searches (exact, substructure, and similarity) (see Ref. [14] and references cited within).

ACKNOWLEDGMENTS

Jade Carter is thanked for his helpful comments regarding error rates in HTS.

REFERENCES

1. (a) Broach, J.R., Thorner, J. (1996) High-throughput screening for drug discovery. *Nature*, 384, Suppl, 14–16. (b) Bleicher, K.H., Boehm, H.-J., Muller, K., Alanine, A.I. (2003) Hit and lead generation: Beyond high-throughput screening. *Nat. Rev Drug Discov.*, 2, 369–378. (c) Goodnow Jr., R.A. (2006) Hit and lead identification: Integrated technology-based approaches. *Drug Discov. Today*, 3, 367–375.

2. For example, only 2×10^6 combinatorial compounds were synthesized at Pfizer out of a virtual compound space of 1.3×10^{18} in size. See Hu, Q., Peng, Z., Sutton, S.C., Na, J., Kostrowicki, J., Yang, B., Thacher, T., Kong, X., Mattaparti, S., Zhou, J.Z., Gonzalez, J., Ramirez-Weinhouse, M., Kuki, A. (2012) Pfizer global virtual library (PGVL): A chemistry design tool powered by experimentally validated parallel synthesis information. *ACS Comb. Sci.* 14, 579–589.

3. References for cost associated with synthesis and screening of traditional compounds synthesized by traditional combinatorial chemistry methods. HTS cost per well is ~$0.07 in 2009. (a) Glaser, V. (2009) High throughput screening retools for the future, BioIT World, Jan. 20, 2009. (http://www.bio-itworld.com/BioIT_Content.aspx?id=86836, accessed on November 27, 2013). (b) See discussion in Chapter 18.

4. (a) Clark, M.A. (2010) Selecting chemicals: The emerging utility of DNA-encoded libraries. *Curr. Opin. Chem. Biol.*, 14, 396–403. (b) Mannocci, L., Leimbacher, M., Wichert, M., Scheuermann, J., Neri, D. (2011) 20 years of DNA-encoded chemical libraries. *Chem. Commun.*, 47, 12747–12753. (c) Mannocci, L. Zhang, Y., Scheuermann, J., Leimbacher, M., De Bellis, G., Rizzi, E., Dumelin, C., Melkko, S., Neri, D. (2008) High-throughput sequencing allows the identification of binding molecules isolated from DNA-encoded chemical libraries. *Proc. Natl. Acad. Sci. U.S.A.*, 105, 17670–17675.

5. Clark, M.A., Acharya, R.A., Arico-Muendel, C.C., Belyanskaya, S.L., Benjamin, D.R., Carlson, N.R., Centrella, P.A., Chiu, C.H., Creaser, S.P., Cuozzo, J.W., Davie, C.P., Ding, Y., Franklin, G.J., Franzen, K.D., Gefter, M.L., Hale, S.P., Hansen, N.J.V., Israel, D.I., Jiang, J., Kavarana, M.J., Kelley, M.S., Kollmann, C.S., Li, F., Lind, K., Mataruse, S., Medeiros, P.F., Messer, J.A., Myers, P., O'Keefe, H.,Oliff, M.C., Rise, C.E., Satz, A.L., Skinner, S.R., Svendsen, J.L., Tang, L., van Vloten, K., Wagner, R.W., Yao, G., Zhao, B., Morgan, B.A. (2009) Design, synthesis and selection of DNA-encoded small-molecule libraries. *Nat. Chem. Biol.*, 5, 647–654.

6. (a) Spotfire from Tibco (https://silverspotfire.tibco.com, accessed on November 27, 2013). (b) Vortex from Dotmatics (http://www.dotmatics.com/products/vortex/, accessed on November 27, 2013).

7. ISIS/Base from Accelrys (http://accelrys.com/products/informatics/decision-support/isis. html, accessed on November 27, 2013).

8. Pipeline Pilot from Accelrys (http://accelrys.com/products/pipeline-pilot/, accessed on November 27, 2013).

9. The Accelrys data sheet on its Direct chemistry cartridge (2011) "Tested and proven with databases containing over 17 million reactions and 100 million structures." (http://accelrys. com/products/datasheets/accelrys-direct.pdf, accessed on November 27, 2013). Similar search capability is also provided by ChemAxon's JChem Base/Cartridge (http://www. chemaxon.com/jchem/doc/admin/Performance.html#max_structures, accessed on November 27, 2013).

10. Lipman, D. J, Pearson, W. R (1985) Rapid and sensitive protein similarity searches. *Science*, 227, 1435–1441.

11. Cock, P.J.; Fields, C.J.; Goto, N.; Heuer, M.L.; Rice, P.M. (2010) The Sanger FASTQ file format for sequences with quality scores, and the Solexa/Illumina FASTQ variants. *Nucleic Acids Res.*, 38, 1767–1771.

12. Minoche, A.E., Dohm, J.C., Himmelbauer, H. (2011) Evaluation of genomic high-throughput sequencing data generated on Illumina HiSeq and Genome Analyzer systems. *Genome Biol.*, 12, R112.

13. Reported error rate in genomic DNA sequencing: (a) Fabret, C., Quentin, Y., Guiseppi, A., Busuttil, J., Haiech, J., Denizot, F. (1995) Analysis of errors in finished DNA sequences: The surfactin operon of *Bacillus subtilis* as an example. *Microbiology*, 141, 345–350. (b) Wesche, P. L., Gaffney, D. J., Keightley, P. D. (2004) DNA sequence error rates in Genbank records estimated using the mouse genome as a reference. *DNA Seq.*, 15, 362–364.

14. (a) Hu, Q.Y., Peng, Z., Kostrowicki, J., Kuki, A. (2011) LEAP into the Pfizer Global Virtual Library (PGVL) Space: Creation of readily synthesizable design ideas automatically. Chemical Library Design, Editor(s): Zhou, J.Z.X. *Methods in Molecular Biology*, 685, 253–276. (b) Peng, Z. (2013) Very large virtual compound spaces: Construction, storage, and utility in drug discovery. *Drug Discov. Today*, 10, e387–e394.

10

THEORETICAL CONSIDERATIONS OF THE APPLICATION OF DNA-ENCODED LIBRARIES TO DRUG DISCOVERY

Charles Wartchow

Formerly of Hoffmann-La Roche Inc.
Roche Discovery Technologies
Nutley, NJ, USA
Currently of Novartis Institute for
Biomedical Research, Emeryville, CA, USA

10.1 INTRODUCTION

In order to understand the potential of the application of DNA-Encoded Libraries (DELs) in the drug discovery process, an understanding of the current discovery methods is important. In this chapter, the various components of successful drug discovery are discussed, including the challenges of emerging target classes, the chemical space used to interrogate targets, the analytical methods used to identify drugs, and the intrinsic challenges associated with these methods. These aspects of drug discovery are discussed in the context of screening targets with DELs, which have immense potential for the identification of drug-like inhibitors, allosteric modulators, and tool compounds for probing biochemical and cellular activity. The factors that impact hit identification are also discussed, including target behavior, chemical space, compound behavior, and aspects of the DEL screening or "selection" method.

A Handbook for DNA-Encoded Chemistry: Theory and Applications for Exploring Chemical Space and Drug Discovery, First Edition. Edited by Robert A. Goodnow, Jr.
© 2014 John Wiley & Sons, Inc. Published 2014 by John Wiley & Sons, Inc.

10.2 THE DRUG DISCOVERY PROCESS

Despite immense progress in the fundamentals of disease biology, target validation, analytical methods, and advances in medicinal chemistry [1], drug discovery is very challenging, with significant opportunities in many disease areas including infectious disease [2] and oncology [3]. An early step in the small-molecule discovery process focuses on the identification of a hit, defined as a molecule with some confirmed activity against the target in question. There are several ways to identify hits, including High-Throughput Screening (HTS) of random collections, combinatorial libraries, and natural products; studies of peptides and peptidomimetics; fragment screening and assembly; structure-based design; chemogenomics; virtual HTS; and information-based approaches. With respect to screening, discovery typically involves screening a library of compounds with a protein target of interest in a biochemical or cell-based assay, followed by confirmation of hits and the elimination of false-positive results with one or more secondary assay(s) that utilize different readouts, such as orthogonal biochemical assays, biophysical assays, or NMR assays (Fig. 10.1). These assays generate "validated hits" that are often examined in X-ray crystallographic studies with the protein of interest. This process is highly labor and resource intensive, and success is limited by a number of factors, including the resources required for the discovery process, the nature of the target, the composition of the library, the library size, and the inherent limitations, both known and undiscovered, in any given analytical method. While many discovery campaigns are fruitful, others are often unproductive. There is therefore a constant need for more efficient discovery tools to address these limitations. Reports of DEL screens (also known as "selections,") which are built on the successes of combinatorial chemistry, molecular biology, and DNA sequencing, are emerging as a provocative and influential option.

The primary goal of the drug discovery process is to identify drugs that can be tested in the clinic in order to provide new medicines for patients. The focus of most successful "first-in-class" drug development campaigns involves strategies including target-based biochemical screening and phenotypic screening [4]. Assessment of the potential for success in a discovery campaign with a given target and library is highly desirable. In attempting to classify targets, some have coined the term "druggability" [5], or perhaps more precisely "ligandability" [6], which is assessed based on the outcome

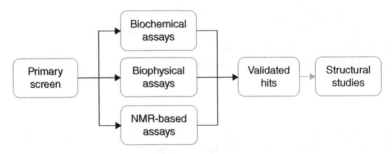

Figure 10.1. An example of the drug discovery process.

of screening metrics including hit rates and the number of unique chemical series [5]. Hit rates are dependent on the target itself, the analytical method utilized, and the composition of the library, which may be biased during library design with compounds that prefer tractable classes of targets such as kinases. The application of DEL technology to the discovery process potentially streamlines this process by reducing the scope of the early discovery process, allowing for more efficient identification of true hits, and assessments of the ligandability of new targets may be possible.

10.3 THE CHALLENGES OF EMERGING TARGETS

The hit discovery process is typically highly productive for well-behaved enzyme targets like kinases and proteases, but it can be very challenging for more complex targets like membrane-bound targets and targets that modulate Protein–Protein Interactions (PPI). Success in these screens is also highly dependent on the chemical content in the library, as well as the purity of the compounds therein. Less tractable targets include those that modulate PPI because they contain large contact areas and they often lack the well-defined binding pockets of the more ligandable or tractable protein classes [6–8]. Additionally, PPI targets may modulate interactions through less conventional mechanisms including conformational changes and the formation of transient binding pockets [6]. Despite these difficulties, some PPI targets contain "hot spots," which are capable of binding small ligands [7]. If these small binding pockets are in close proximity, potency can be improved with a larger molecule that binds to multiple pockets. This concept was demonstrated for an inhibitor of the IL-2/IL-2 receptor interaction where potency was improved by >100-fold [7]. In this case, the original inhibitor was designed to bind to the receptor but was eventually shown to bind to IL-2, which demonstrates the complexity of rational inhibitor design based on PPI.

The inhibition of the GTP/GDP binding protein ras is an excellent example of discovery efforts with a very challenging therapeutic target. Disruption of the ras–raf interaction has tremendous implications for the treatment of numerous forms of cancer [9], but the contact interface for these two proteins is relatively flat, and the identification of a potent disrupter of this interaction remains elusive despite decades of effort. More recent studies describe cell-based activity with a purported allosteric modulator identified in a virtual screen [10] and the STD NMR study for the identification of fragments that bind to a small pocket in the SOS binding site of ras [11]. In this study, fragment binding to other sites on ras, including the nucleotide binding site and corresponding loop and the ras–raf interface, was not observed, and the affinity of the SOS-site binders on ras was in the millimolar range. In another approach, NMR studies showed improvement of affinity to 190 μM [12], demonstrating the potential for improvements in affinity, but clearly there is significant opportunity for further improvement in potency. The application of DEL screening to more challenging target classes, like those described above, has the potential to improve the outcomes of screening campaigns by sampling a larger number of compounds, in a more efficient process, which could be as brief as 6 weeks [13].

The successful discovery efforts for inhibitors of MDM2 and Bcl-2, Nutlin-3, and ABT-737, respectively [14, 15], exemplify the immense potential for drug discovery

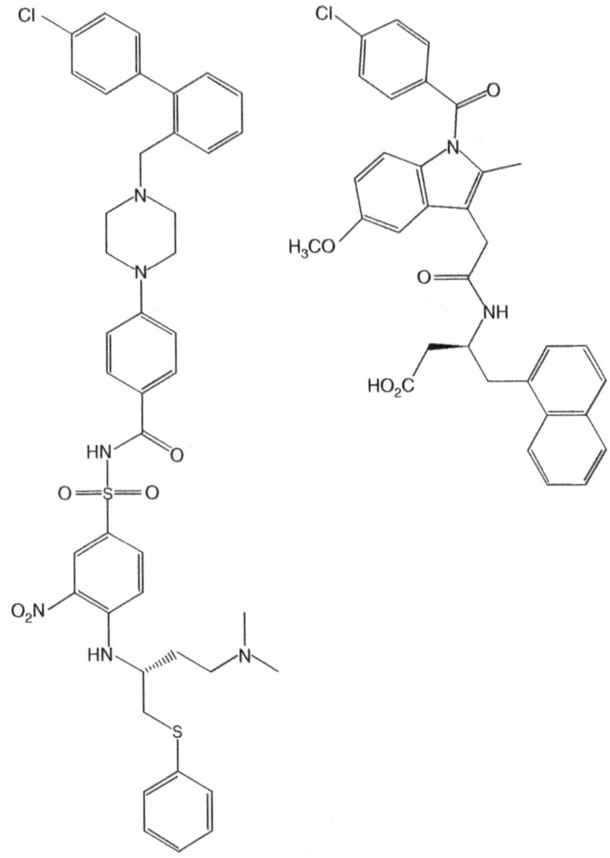

ABT-737 (K_d 0.6 nM)

19/93 (K_d 0.93 μM)

Figure 10.2. Structures of an optimized Bcl-xL inhibitor ABT-737 [18] and inhibitor identified from an encoded library campaign [19].

with difficult and seemingly intractable targets, and DEL screening has the potential to have an impact on the discovery process for these types of targets. Both of these targets contain a binding pocket that is defined by adjacent alpha-helices, and they bind a helical motif in the corresponding protein partner (MDM2/p53 and Bcl-xL/BH3), suggesting that the inhibition of helix/helix interactions is an attractive strategy for disrupting PPI [16]. With respect to Bcl-xL, a member of the Bcl-2 family of proteins that is also a target for ABT-737 (K_d < 1 nM), a small-molecule inhibitor with a K_d of 0.93 μM was identified with a DEL consisting of a mere 4000 members [17]. The structures of ABT-737 and the hit from the DNA-encoded screen are shown in Figure 10.2. For other targets, including p38 MAP kinase and trypsin, inhibitors related to hits from encoded library screens have potencies in the low nanomolar range, and other targets generated hits in the micromolar IC_{50} range [19].

10.4 CHEMICAL SPACE

The chemical space occupied by organic compounds is "for all practical purposes infinite" [20]. Chemical space is dependent on the content available through chemical synthesis or by isolation of compounds from natural sources. In any given compound library, chemical space consists of optimized compounds, precursors, and permutations from legacy programs, and many libraries have been supplemented with commercially available libraries. Choosing a subset of compounds in drug discovery campaigns depends on many factors, including, in the simplest of terms, an optimized arrangement of atoms that form suitable interactions with a specific protein target. For example, chemotypes for target classes such as kinases can be modeled on ATP, a known component of the phosphorylation process. In general, inhibitors that modulate PPI do not share a common chemotype [8]. For these targets, peptide mimics have been proposed including alpha-helix mimetics based on terphenyl, terephthalamide, and pyrrolopyrimidine cores [21]; universal mimetics based on oxadiazole, triazole, and diacetylene cores [22]; branched peptides [23]; triaryl amides [24]; and beta-turn mimetics based on a bicyclic core [25].

The process of identifying ligands for therapeutic targets involves an intensive campaign that samples a small fraction of chemical space [20]. With respect to drug development, Lipinski's well-known "Rule of Five (Ro5)" guidelines focus this space on compounds with properties consistent with known drugs, including limits on molecular weight (<500 daltons (Da)), calculated octanol–water partitioning ($clogP \leq 5$), hydrogen bond donors (≤ 5), and hydrogen bond acceptors (≤ 10), and compounds that violate two or more of these rules have unfavorable *in vivo* absorption and cell permeability [26]. Many drug discovery scientists prefer to associate target classes with unique chemotypes that occupy a subset of chemical space [20]. Some drugs are known to contain exceptions to these rules. Additional algorithms for characterizing "drug-likeness" have emerged that consider, in addition to Lipinski's parameters, polar surface area, the number of rotatable bonds, the number of aromatic rings, and the number of chemical structure alerts [27].

Fragment-Based Screening (FBS) has more recently become an attractive drug discovery strategy [28, 29], relying essentially on the identification of weak-binding compounds generally less than 300 Da that can be synthetically modified to increase potency and selectivity based on guidance from X-ray crystallography studies [29]. An additional advantage of FBS is that lower-molecular-weight compounds are "well behaved" in screening assays, which results in higher success rates for lead identification, due in part to greatly reduced false-positive rates, due to the enhanced solubility of low-molecular-weight compounds relative to larger molecules [29]. With respect to chemical space, relatively small libraries are required to sample chemical space; a recent calculation estimates that "a library of 1000 compounds, each with 12 or fewer heavy atoms (MW <170 Da), samples ~0.001% of the theoretical total number of compounds of this size, whereas the corresponding estimate for compounds with 25 or fewer heavy atoms (<350) is $10^{-14}\%$" of possible larger molecules [29]. Although compounds have binding constants in the high micromolar range, the ligand–target contacts that are present are generally retained throughout the optimization process [29]. With respect to DELs, the compound molecular weight is typically higher (300–1000 Da) [19], but they are attached to a soluble carrier (DNA), and individual members are screened at very low concentrations. As a result, false-positive results due to compound aggregation or solubility issues may be more limited than for traditional biochemical screens, including FBS campaigns.

An additional opportunity for the application of DEL-based screening methods is the identification of tool compounds for probing biological processes. These compounds have unique criteria that are separate from those identified for drugs [30]. These criteria include *in vitro* characteristics including potency and selectivity for a given target and related targets (desired and undesired), proof of mechanism, Structure–Activity Relationships (SAR) for related compounds, and structure data from X-ray crystallography studies. Cell-based studies with tool compounds should be designed so that the readout is consistent with the proposed mechanism and hypothesis, and the compounds should not be cytotoxic. Lastly, the compounds should be stable, should be readily available, and should not be chemically reactive (unless desired).

With respect to challenging targets, chemical diversity is particularly important since each target may require a new structural strategy. For PPI targets, for example, common chemotypes for a given class may be uncommon [8]. PPI inhibitors that block the p53–HDM2 and Bcl-xL/BH3 interactions, Nutlin-3, and A-385358, respectively, represent two unrelated chemotypes, even though both of these PPI involve helix/helix interactions between the two binding partners [16]. In this context, numerous scaffolds that mimic protein secondary structural elements like the alpha-helix and the beta-sheet have been proposed, as mentioned earlier, and may be suitable for DEL chemistry.

The chemical space of DELs has the potential to vastly surpass the content of conventional libraries. DELs consisting 100 million to 1 billion members are typically designed with defined chemical strategies, and each library is typically 100- to 1000-fold larger than any given conventional compound repository, since each building block varies by hundreds to thousands of members, and libraries consist of three or more building blocks in a given library. Because of the large library size, if a preferred chemotype for a given target is present in a DEL, numerous structurally related compounds are present within a given library, and many of these derivatives will be identified in a screen, which increases the likelihood that an authentic chemotype will be identified [31]. Expansion of chemical space within the framework of DELs is limited merely by the identification of novel combinatorial chemistry strategies that are compatible with the DNA-based chemistry, the number of available building blocks compatible with the synthetic scheme, and the synthetic yield for a given building block. The "dual-pharmacophore" strategy where each DNA strand contains a small molecule [19] is interesting in that it uses a strategy in common with fragment screening, where two separate molecules that bind to the binding pocket in separate locations can later be attached chemically to generate a more potent inhibitor.

10.5 BIOCHEMICAL SCREENING METHODS AND CAMPAIGNS

The ideal screening campaign will include compounds from chemical space relevant to the target and to lead optimization, compounds that are "well behaved" in aqueous solution, suitable assay characteristics including a favorable Z'-factor [32], acceptable throughput, reasonable cost, and a method that is relatively insensitive to artifacts resulting from the physical and chemical properties of the compounds. Screening methods typically involve biochemical events including substrate modification by a target enzyme (e.g., kinase or protease assay), binding assays with a known binding partner (e.g., protein–peptide interaction or PPI), or a direct binding assay that does not require prior knowledge of the binding partner or mechanism. The discovery methods may be classified as *a priori*, where compound binding to any pocket, known or unknown, can be identified, or *a posteriori*, where knowledge of a substrate or binding partner is required for assay design and implementation. *A priori* methods are valuable in the early discovery process where binding partners or substrates are unknown, since they can detect hits to previously unknown pockets that may be involved in known and unknown biological pathways. Examples of targets with previously undiscovered binding pockets include FKBP, E2–31, survivin, and CMPK, which were identified by heteronuclear NMR-based screening [5], and compounds that bind to the inactive form of JNK-1, which were identified using a filtration assay that removes unbound ligands from those bound to JNK-1, followed by identification of the bound ligands with analysis by mass spectrometry [33]. Likewise, DEL screens have the potential to identify hits that bind to known pockets and undiscovered but relevant pockets.

Numerous analytical techniques for the identification of lead compounds in drug discovery are known [34–37]. *A priori* discovery methods include crystallography [37], Surface Plasmon Resonance (SPR) [38, 39], Bio-Layer Interferometry (BLI) [40], 1D and 2D NMR [5, 41], thermal shift assays [42], Isothermal Calorimetry (ITC) [43], mass spectrometry assays, and DEL technology [19, 31]. A posteriori discovery methods include biochemical assays, including those that utilize TR-FRET, or substrate hydrolysis to generate fluorescent products. In practice, all screening technologies have their strengths and weaknesses. The advantages and disadvantages of these methods are summarized in Table 10.1, and the choice of method used for a given campaign depends on many factors including resources, cost, level of difficulty, robustness, and user expertise. In general, hits are identified with one or more of the aforementioned assays, and they are confirmed in a secondary assay with an orthogonal detection mechanism to rule out false positives. Structural evidence of binding is also highly valued, including 2D NMR studies and, ultimately, data from X-ray crystallography that show the binding location of the compound in question.

Structure-based screening methods based on X-ray crystallography and 2D NMR are favored because these methods provide unambiguous proof of interaction with the target, and they are less susceptible to false-positive results due to compound interferences. However, they are limited by throughput and are typically used for the confirmation of hits from high-throughput methods or for screens with a limited number of compounds. Biosensor-based methods and thermal shift assays are also favored for screening, but these techniques have moderate throughput, and like HTS methods, they are susceptible to interferences. DEL screens, in contrast, are much less susceptible to interferences and are capable of screening millions to billions of compounds in a single day.

Because biochemical assays involve the interaction between the target protein and a known binding partner or substrate, biochemical assays are typically specific for a

TABLE 10.1. Summary of screening methods: Low, medium, and high throughput are 100–1000, 1001–50,000, 50,001–1,000,000, respectively

Screening method	Description	Features	Challenges
DEL technology	Combinatorial libraries of compounds are encoded with a unique DNA tag, and the DEL is incubated with a protein target. Compounds that bind to the target are isolated, sequenced, synthesized, and tested biochemical assays	Large libraries (10^8–10^9 members), potential for identifying known and novel binding pockets, immune to issues in HTS such as compound aggregation and optical interferences	Encoding represents synthetic pathway, not actual compound; linker location may impact binding of hits, and other challenges are emerging
X-ray crystallography	Compound density is observed in X-ray studies with crystals prepared in the presence of compound or where compound is soaked into the crystal	Provides structural information for optimization of potency, low false-positive rate	Limited to crystallizable targets, low throughput
SPR	Biosensor-based method where compound is injected into a microfluidic cartridge containing multiple immobilized targets, depending on platform	Provides information on binding quality as well as affinity and kinetic information, validated extensively for fragment screening, low throughput	Nonspecific binding results in false positives and complicates interpretation
BLI	Biosensor-based method where a biosensor containing immobilized target is dipped into a 96- or 384-well plate containing compound	Provides information on binding quality as well as affinity and kinetic information	Same as SPR
ITC	The enthalpy released upon the binding of a compound to a target is measured for a dilution series	Provides thermodynamic binding information including changes in enthalpy and entropy, affinity, and binding stoichiometry, low throughput	Analysis of compounds with affinities in the mM range is difficult, binding event must release a measurable amount of heat

given interaction, such as a known protein–ligand interaction, PPI, or enzymatic modification of a natural substrate or derivative thereof. These assays are generally amenable to high-throughput formats for screening 100,000–1,000,000 compounds. Inhibition of these assays in screening campaigns with a ligandable target provides true hits that specifically inhibit the interaction of interest. These assays also generate a significant number of false positives, which can have an impact on the progress of the discovery campaign, since significant resources may be required to eliminate them. The number of false positives is dependent on the physical properties of the compound and target, the concentration of the compound used for the screen, the type of compound, the purity of the compound, the interferences from impurities and nonbinding compounds, the buffer, the detection method, and the quality of the assay. Some false positives can be eliminated by the use of a second biochemical or biophysical assay to confirm the hits. For proteases, a second biochemical assay was sufficient to improve the identification of true hits [44]. When choosing a second assay for hit confirmation, an orthogonal detection method is desirable, to rule out bias due to optical interferences, for example. False-negative results are also an important component of biochemical assays, and these occur as a result of systematic errors, a lack of assay robustness, and suboptimal assay quality. However, they are of less concern in a successful campaign that produces hits that can be validated in NMR and X-ray studies, are of sufficient chemical novelty, and can be optimized during the discovery process.

The greatest challenge in any screening campaign is the identification of true hits among a significant number of false-positive results. 2D NMR methods appear to be the least sensitive to false-positive results, and this method empowered successful lead identification studies with ras, one of the most challenging targets [11, 12]. In these studies, numerous hits were identified, and many yielded structural information in X-ray studies. In one NMR study with ras, 3300 compounds were screened with STD NMR resulting in the identification of 240 hits, only 25 of which were confirmed in subsequent 2D NMR studies [11]. In this study, the compound may be interacting with the target protein through nonspecific mechanisms. Even with this high rate of attrition, this campaign was highly successful since hits with affinity constants in the millimolar range were successfully crystallized with the target.

False-positive results in biochemical assays are generated through many mechanisms including compound aggregation, chemical reactivity, and redox activity [45, 46]. HTS campaigns employ robotics and low volume plates (384 well and 1536 well), and false positives may also arise due to issues with delivery of reagents, solubilization of compounds from DMSO stock solutions, and nonspecific binding of the compound to the target [47–49]. With respect to identification of false-positive results, false positives that are unique to the detection method can be eliminated with a secondary assay using an orthogonal method, but if both assays are sensitive to the mechanism that generates the false result, such as compound aggregation, then these results will not be eliminated. Aggregators can be identified through many secondary assays including, examining inhibition in the presence and absence of detergent with a standardized beta-lactamase assay [49]. Inactivation of the target by chemical reactivity is generally easy to identify from examination of compound structure, and these compounds should be eliminated from libraries or excluded during library design [45].

Interferences are not limited to biochemical assays and are known in other assay formats, including biosensor-based assays, assays performed with cell extracts, ligand-observed NMR assays, and cell-based assays. Biosensor-based detection methods including SPR and BLI are valuable because in addition to quantitative kinetic information including affinity constants (K_d), off-rates (k_{off}), and on-rates (k_{on}), these methods provide qualitative information through visualization of responses for the association and dissociation of the compound with the target protein. These data provide important information for well-behaved and poorly behaved interactions of compounds with immobilized targets, including estimation of binding stoichiometry relative to a positive control compound and assessments of ideal and nonideal binding profiles [50]. Pull-down assays, where a protein is typically immobilized on a bead and is incubated with a cell extract containing a target protein of interest, are often used to show that a ligand can inhibit a PPI. These assays, although performed with cell-based components, are susceptible to the same interferences as biochemical assays. Ligand-observed NMR studies monitor changes in the resonances of the ligand, and therefore, any interaction, specific or nonspecific, will result in changes in intensities of NMR signals. Cell-based assays are a very important component of drug discovery, and phenotypic screening in particular is important for successful drug development but can be limited by off-target interactions of the compound with undesired targets that are part of the pathway that causes the change in phenotype. Cytotoxicity can also be a source of false-positive results, but can be monitored. Nonetheless, interpretation of results from cell-based assays can be complicated by the high variability that is intrinsic to these methods.

10.5.1 Factors Impacting Hit Identification in DNA-Encoded Library Screens

A successful DNA-encoded library screening campaign that yields validated hits is dependent on many factors, including those related to the target protein, protein and library concentration, the structure of the compound attached to the DNA strand, the immobilization method, the washing protocol, the affinity of the DNA-tagged compound for the target, the amplification of the bound library members via Polymerase Chain Reaction (PCR), and the sensitivity and resolution of sequencing, which is capable of capturing and amplifying individual DNA molecules that correspond to bound library members. And most importantly, the outcome of the screen is informatics, not actual binding data for a discreet compound. For confirmation, hits are resynthesized in the absence of DNA for studies in one or more confirmation assays. Depending on the goal, verification of binding may be sufficient, or if modulation of a particular enzymatic or binding event is desired, then an activity assay is performed. The reported screening process is summarized generally in Figure 10.3.

An important attribute of screening DELs is the sensitivity of the method, which allows for screening very large libraries (0.1–1 billion members) at high copy number (~0.3–3 million copies of each compound) at low concentrations for each member. For libraries of this size, the concentration of any given library member is 100 fM, assuming that the library contains 100 million members and a total library concentration of

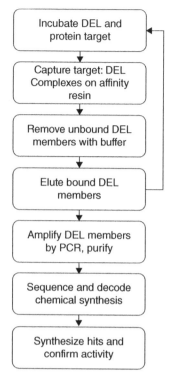

Figure 10.3. Summary of the generally reported DEL screening process.

10 μM. The femtomolar concentrations of compound–DNA library members in a typical selection campaign are well below the low micromolar affinities for unoptimized hits identified from traditional screening campaigns, yet hits have been identified in numerous campaigns [19, 31]. While mathematical models that enumerate the concentrations of eluted members have not been reported, the identification of hits is most likely complex and impacted by numerous variables including protein concentration, which is significantly higher for DEL screens (~1 μM, [31]), and because the DNA that encodes the hits can be amplified by the PCR prior to sequencing, the identification of specific binding molecules is possible.

An additional factor that may result in decreases in off-rates for DEL members is the phenomenon of "rebinding," which is known for biosensor-based methods where dissociation rates of binding partners decrease as a result of high local concentration of protein [51]. For example, a fivefold increase in ligand loading density on a solid support resulted in a ~1.7-fold increase in retention of a nuclease in a column chromatography experiment [52]. The increase in retention time is due to an increase in partitioning of the nuclease onto the solid support, which was derivatized with a thymidine derivative that binds to the nuclease. In an experiment where a protein–DEL member complex is present on a solid support, dissociation occurs, and the elution of the DEL member will be dependent on the density of the captured protein on the solid support, where higher densities of protein will result in longer retention times for the bound DEL members.

Since screening campaigns are performed in solution by incubation of the target with the library, the equilibrium concentration of a given complex between the target protein and specific library members is represented in simplest terms by steady-state kinetic expressions from the interaction of a ligand "L" (e.g., small molecule or small molecule attached to DNA) with a protein "R" [53, 54]. The concentration of the protein–ligand complex "RL" is determined by Equation 10.1, where k_{on} and k_{off} are the rate constants for association and dissociation of a library member (L) with a target protein (R) for the dissociation of the protein–ligand complex (RL) (Equation 10.1). In this equation, R and L are the unoccupied and unbound forms, respectively. For this interaction, increasing protein concentration results in an increase in the concentration of RL complex. In a DEL screen, protein concentrations are on the order of 0.5–1 µM [31], which is many orders of magnitude higher than the concentration of any given DEL member. For comparison, this protein concentration is up to 1000-fold higher than that typically used in common HTS campaigns, where the compound is screened in excess relative to the protein. Thus, a distinction in DEL screens is that the protein concentration is many orders of magnitude greater than the concentration of individual DEL members, which may help drive the equilibrium toward formation of the RL complex in concentrations sufficient for subsequent amplification of the DNA tags that encode for the synthetic scheme of the ligand that binds to the target. The sensitivity of quantitative PCR is reported to be better than 0.1 attomolar for the amplification of DNA containing 200–250 base pairs for sequencing on the Illumina platform [55]. PCR sensitivity can vary significantly depending on the sequence that is amplified, the amplification protocol, and the primers used for amplification. Sequencing is also highly sensitive, requiring 2–8 pM DNA [55], with the ability to read hundreds of millions of sequences in a single run.

$$R + L \underset{k_{off}}{\overset{k_{on}}{\rightleftharpoons}} RL \tag{10.1}$$

In comparison to HTS campaigns, each library member in a DEL may be present at subpicomolar concentrations, and many of the challenges of traditional biochemical and biophysical screening, including interferences due to compound aggregation and limitations intrinsic to the analytical readout of a given technology, do not apply to screens with DELs, since screening campaigns are performed with very low concentrations of individual compounds (~10–100 fM) that are well below the concentrations known to cause physical interferences. Furthermore, the readout of a DEL screen, unlike many other HTS methods, is not optical, and the results are unaffected by spectral interferences. However, when compounds are resynthesized in the absence of DNA for hit confirmation, assays may require concentrations in the ranges where interferences are known, and interpretation of the data is subject to the boundaries of the strengths and weaknesses of a given assay format. Thus, hit confirmation strategies and caveats for hits from DEL screens are the same as those for biochemical screens.

With respect to the target protein, several factors impact the potential success of a given campaign. First, the target must contain one or more binding pockets or surfaces that are suitable for the binding of the small molecules in the library; it must be ligand-able. The protein must be compatible with the immobilization method, and the binding pockets should be located distal to the affinity tag used to capture the protein onto the solid support. If the binding pockets are near the location of the affinity tag, then the DNA molecule may interfere with capture of the target complex with a bound DEL member, or the library members may dissociate resulting in capture of target only.

The properties of the compound attached to DNA in a DEL may also impact the results of a DEL screen. With respect to the location of the linker on the compound portion of the DEL member, the linker location will preclude the binding of molecules that may bind to the target in the absence of the DNA tag, resulting in a false-negative result. The physical properties of the compound may also play a role in the outcome of a screen. Since the DEL screen uses affinity chromatography [13] or filtration of immobilized proteins [17] to separate bound DEL members from unbound DEL members, nonspecific partitioning of the compound to the protein and/or to the resin is also possible and may play a role in the outcome of the selection by generating false-positive results. And lastly, some compounds, such as the agent doxorubicin, are known to intercalate DNA [56], which may generate false-positive results by binding to DNA attached to true binders or false-negative results if the intercalator is also a true binder to the protein target.

The method of capture of the complex of protein and DEL member may also have an impact on the outcome of a DEL screen. Key factors to consider when capturing the target protein include the complexity and efficiency of the capture method, the impact on the protein structure and the protein–DEL member complex, the impact of the capture method on activity, the construction of an appropriately tagged protein, and the potential for nonspecific binding of DEL members to the solid support. In a screen with metalloprotease ADAMTS-5, the protein was captured with a streptavidin resin, and a control experiment without target protein was performed to identify and remove compounds that bind to the resin or to streptavidin [13]. The conjugation of proteins to solid supports is a mature science, and there are numerous options for both covalent and noncovalent immobilization [57, 58]. Covalent strategies typically involve the conjugation of surface lysines to activated groups on the solid support or resin, including NHS esters or epoxides. With these methods, there is the potential for attachment of multiple lysines on the same protein, which may impact the function of the immobilized protein. A more recent example of site-specific covalent immobilization is the generation of an aldehyde on a specific peptide sequence [59]. Noncovalent methods include the use of AviTag, a 15-amino-acid peptide that is incorporated into the gene encoding the protein of interest [60], and other methods including FLAG peptide, histidine tag, and glutathione S-transferase. The advantage of these tags is that they can be attached specifically to the N-terminus or C-terminus of the protein, and the protein is therefore oriented in a uniform manner on the solid support.

The activity of the target is also an important consideration. The protein must be stable under the conditions of the DEL screening process, and stability can be determined if a suitable binding partner is known. Stability is assessed with a binding assay with a positive control or with an enzymatic activity assay. Binding assays can provide stoichiometry between the target and the binding partner, depending on the affinity, whereas enzyme assays do not. Thus, enzymes can be largely inactive but show sufficient response in an assay, which can be misleading, since binding assays with DELs will be performed with a mixture of active and inactive protein. Characterizing the activity of the protein is desirable in any assay and requires some prior knowledge of the protein structure and function. Method development is an important component of any successful experiment, and qualification of the activity of the protein can have a significant impact on the quality and outcome of the selection campaign. Most assays with protein targets require optimization of buffer conditions, including detergents and additives that minimize background, the addition of salts and/or cofactors that are necessary for activity, and the addition of stabilizers like dithiothreitol. Optimization can be a very important component for the success of a screening campaign. For example, in the reported DEL screen with ADAMTS-5, an osteoarthritis target, the selection buffer contains DNA and the detergent Brij35 [13], presumably to reduce background binding of DEL members to the solid support or to the protein target.

10.6 SUMMARY

In summary, DEL screening is a promising method for drug discovery. It has several advantages over existing screening methods, including HTS. DEL screens are performed at compound concentrations that are well below those used for traditional screens where compounds are known to cause interferences that impede the discovery process and outcome. DELs contain very high numbers of compounds, from hundreds of millions to billions, and lastly, because DEL screens use PCR amplification, the detection of compounds in the subpicomolar concentration range is possible when incubating DELs with protein targets.

REFERENCES

1. Drews, J. (2000). Drug discovery: a historical perspective. *Science*, 287, 1960–1964.
2. Jabes, D. (2011). The antibiotic R&D pipeline: an update. *Curr. Opin. Microbiol.*, 14, 564–569.
3. Popovic, R., Licht, J. D. (2012). Emerging epigenetic targets and therapies in cancer medicine. *Cancer Discov.*, 2, 405–413.
4. Swinney, D. C., Anthony, J. (2011). How were new medicines discovered? *Nat. Rev. Drug Discov.*, 10, 507–519.
5. Hajduk, P. J., Huth, J. R., Fesik, S. W. (2005). Druggability indices for protein targets derived from NMR-based screening data. *J. Med. Chem. J. Med. Chem.*, 48, 2518–2525.
6. Surade, S., Blundell, T. L. (2012). Structural biology and drug discovery of difficult targets: the limits of ligandability. *Chem. Biol.*, 19, 42–50.
7. Arkin, M. R., Wells, J. A. (2004). Small-molecule inhibitors of protein-protein interactions: progressing towards the dream. *Nat. Rev. Drug Discov.*, 3, 301–317; Tilley, J. W., Chen, L., Fry, D. C., Emerson, S. D., Powers, G. D., Biondi, D., Varnell, T., Trilles, R., Guthrie, R., Mennona, F., Kaplan, G., LeMahieu, R. A., Carson, M., Han, R. J., Liu, C. M., Palermo, R.,

Ju, G. (1997). Identification of a small molecule inhibitor of the IL-2/IL-2Rα receptor interaction which binds to IL-2. *J. Am. Chem. Soc.* 119, 7589–7590.

8. Mullard, A. (2012). Protein-protein interaction inhibitors get into the groove. *Nat. Rev. Drug Discov.*, 11, 173–175.

9. Gysin, S., Salt, M., Young, A., McCormick, F. (2011). Therapeutic strategies for targeting Ras proteins. *Genes Cancer*, 2, 359–372.

10. Grant, B. J., Lukman, S., Hocker, H. J., Sayyah, J., Brown, J. H., McCammon, J. A., Gorfe, A. A. (2011). Novel allosteric sites on Ras for lead generation. *PloS ONE*, 6, 1–10.

11. Maurer, T., Garrenton, L. S., Oh, A., Pitts, K., Anderson, D. J., Skelton, N. J., Fauber, B. P., Pan, B., Malek, S., Stokoe, D., Ludlam, M. J. C., Bowman, K. K., Wu, J., Giannetti, A. M., Starovasnik, M. A., Mellman, I., Jackson, P. K., Rudolph, J., Wang, W., Fang, G. (2012). Small-molecule ligands bind to a distinct pocket in Ras and inhibit SOS-mediated nucleotide exchange activity. *Proc. Natl. Acad. Sci. USA*, 109, 5299–5304.

12. Sun, Q., Burke, J. P., Phan, J., Burns, M. C., Olejniczak, E. T., Waterson, A. G., Lee, T., Rossanese, O. W., Fesik, S. W. (2012). Discovery of small molecules that bind to K-Ras and inhibit SOS-mediated activation. *Angew. Chem. Intl. Ed.*, 51, 6140–6143.

13. Deng, H., O'Keefe, H., Davie, C. P., Lind, K. E., Acharya, R. A., Franklin, G. J., Larkin, J., Matico, R., Neeb, M., Thompson, M. M., Lohr, T., Gross, J. W., Centrella, P. A., O'Donovan, G. K., Bedard, K. L., van Vloten, K., Mataruse, S., Skinner, S. R., Belyanskaya, S. L., Carpenter, T. Y., Shearer, T. W., Clark, M. A., Cuozzo, J. W., Arico-Muendel, C. C., Morgan, B. A. (2012). Discovery of highly potent and selective small molecule ADAMTS-5 inhibitors that inhibit human cartilage degradation via encoded library technology (ELT). *J. Med. Chem.*, 55, 7061–7079.

14. Vassilev, L. T., Vu, B. T., Graves, B., Carvajal, D., Podlaski, F., Filipovic, Z., Kong, N., Kammlott, U., Lukacs, C., Klein, C., Fotouhi, N., Liu, E. A. (2004). In vivo activation of the p53 pathway by small-molecule antagonists of MDM2. *Science*, 303, 844–848.

15. Oltersdorf, T., Elmore, S. W., Shoemaker, A. R., Armstrong, R. C., Augeri, D. J., Belli, B. A., Bruncko, M., Deckwerth, T. L., Dinges, J., Hajduk, P. J., Joseph, M. K., Kitada, S., Korsmeyer, S. J., Kunzer, A. R., Letai, A., Li, C., Mitten, M. J., Gettesheim, D. G., Ng, S., Nimmer, P. M., O'Connor, J. M., Oleksijew, A., Petros, A. M., Reed, J. C., Shen, W., Tahir, S. K., Thompson, C. B., Tomaselli, K. J., Wang, B., Wendt, M. D., Zhang, H., Fesik, S. W., Rosenberg, S. H. (2005). An inhibitor of Bcl-2 family proteins induces regression of solid tumours. *Nature*, 435, 677–681.

16. Jochim, A. L., Arora, P. S. (2010). Systematic analysis of helical protein interfaces reveals targets for synthetic inhibitors. *ACS Chem. Biol.*, 5, 919–923.

17. Melkko, S., Mannocci, L., Dumelin, C. E., Villa, A., Sommavilla, R., Zhang, Y., Grütter, M. G., Keller, N., Jermutus, L., Jackson, R. H., Scheuermann, J., Neri, D. (2010). Isolation of a small-molecule inhibitor of the antiapoptotic protein Bcl-xL from a DNA-encoded chemical library. *ChemMedChem*, 5, 584–590.

18. Lee, E. F., Czabotar, P. E., Yang, H., Sleebs, B. E., Lessene, G., Colman, P. M., Smith, B. J., Fairlie, W. D. (2009). Conformational changes in Bcl-2 pro-survival proteins determine their capacity to bind ligands. *J. Biol. Chem.*, 284, 30508–30517.

19. Mannocci, L., Leimbacher, M., Wichert, M., Scheuermann, J., Neri, D. (2011). 20 years of DNA-encoded chemical libraries. *Chem. Commun.*, 47, 12747–12753.

20. Lipinski, C. Hopkins, A. (2004). Navigating chemical space for biology and medicine. *Nature*, 432, 855–861.

21. Lee, J.-H., Zhang, Q., Jo, S., Chai, S. C., Oh, M., Im, W., Lu, J., Lim, H.-S. (2011). Novel pyrrolopyrimidine-based alpha-helix mimetics: cell-permeable inhibitors of protein-protein interactions. *J. Am. Chem. Soc.*, 133, 676–679.

22. Ko, E., Liu, J., Perez, L. M., Lu, G., Schaefer, A., Burgess, K. (2011). Universal peptidomimetics. *J. Am. Chem. Soc.*, 133, 462–477.

23. Ruvo, M., Sandomenico, A., Tudisco, L., De Falco, S. (2011). Branched peptides for the modulation of protein-protein interactions: more arms are better than one? *Curr. Med. Chem.*, 18, 2429–2437.

24. Shaginian, A., Whitby, L. R., Hong, S., Hwang, I., Farooqi, B., Searcey, M., Chen, J., Vogt, P. K., Boger, D. L. (2009). Design, synthesis, and evaluation of an alpha-helix mimetic library targeting protein-protein interactions. *J. Am. Chem. Soc.*, 131, 5564–5572.

25. Eguchi, M., Lee, M. S., Nakanishi, H., Stasiak, M., Lovell, S., Kahn, M. (1999). Solid-phase synthesis and structural analysis of bicyclic beta-turn mimetics incorporating functionality at the i to i + 3 positions. *J. Am. Chem. Soc.*, 121, 12204–12205.

26. Lipinski, C. A., Lombardo, F., Dominy, B. W., Feeney, P. J. (2001). Experimental and computational approaches to estimate solubility and permeability in drug discovery and development settings. *Adv. Drug Deliv. Rev.*, 46, 3–26.

27. Bickerton, G. R., Paolini, G. V., Besnard, J., Muresan, S., Hopkins, A. L. (2012). Quantifying the chemical beauty of drugs. *Nat. Chem.*, 4, 90–98.

28. Erlanson, D. A. (2012). Introduction to fragment-based drug discovery. *Top. Curr. Chem.*, 317, 1–32.

29. Murray, C. W., Verdonk, M. L., Rees, D. C. (2012). Experiences in fragment-based drug discovery. *Trends Pharmacol. Sci.*, 33, 224–232.

30. Frye, S. V. (2010). The art of the chemical probe. *Nat. Chem. Biol.*, 6, 159–161.

31. Clark, M. A., Acharya, R. A., Arico-Muendel, C. C., Belyanskaya, S. L., Benjamin, D. R., Carlson, N. R., Centrella, P. A., Chiu, C. H., Creaser, S. P., Cuozzo, J. W., Davie, C. P., Ding, Y., Franklin, G. J., Franzen, K. D., Gefter, M. L., Hale, S. P., Hansen, N. J. V., Israel, D. I., Jiang, J., Kavarana, M. J., Kelley, M. S., Kollmann, C. S., Li, F., Lind, K., Mataruse, S., Medeiros, P. F., Messer, J. A., Myers, P., O'Keefe, H., Oliff, M. C., Rise, C. E., Satz, A. L., Skinner, S. R., Svendsen, J. L., Tang, L., van Vloten, K., Wagner, R. W., Yao, G., Zhao, B., Morgan, B. A. (2009). Design, synthesis, and selection of DNA-encoded small-molecule libraries. *Nat. Chem. Biol.*, 5, 647–654.

32. Zhang, J.-H., Chung, T. D. Y., Oldenburg, K. R. (1999). A simple statistical parameter for use in evaluation and validation of high throughput screening assays. *J. Biomol. Scr.*, 4, 67–73.

33. Corness, K. M., Sun, C., Abad-Zapatero, C., Goedken, E. R., Gum, R. J., Borhani, D. W., Argiriadi, M., Groebe, D. R., Jia, Y., Clampit, J. E., Haasch, D. L., Smith, H. T., Wang, S., Song, D., Coen, M. L., Cloutier, T. E., Tang, H., Cheng, X., Quinn, C., Liu, B., Xin, Z., Liu, G., Fry, E. H., Stoll, V., Ng, T. I., Banach, D., Marcotte, D., Burns, D. J., Calderwood, D. J., Hajduk, P. J. (2011). Discovery and characterization of non-ATP site inhibitors of the mitogen activated protein (MAP) kinases. *ACS Chem. Biol.*, 6, 234–244.

34. Kemp, M. M., Weïwer, M., Koehler, A. N. (2012). Unbiased binding assays for discovering small-molecule probes and drugs. *Bioorg. Med. Chem.*, 20, 1979–1989.

35. Zhu, Z., Cuozzo, J. (2009). High-throughput affinity-based technologies for small-molecule drug discovery. *J. Biomol. Scr.*, 14, 1157–1164.

36. Ma, H., Deacon, S., Horiuchi, K. (2008). The challenge of selecting protein kinase assays for lead discovery optimization. *Expert Opin. Drug Discov.*, 3, 607–621.

37. Blundell, T. L., Jhoti, H., Abell, C. (2002). High-throughput crystallography for lead discovery in drug design. *Nat. Rev. Drug Discov.*, 1, 45–54.

38. Perspicace, S., Banner, D., Benz, J., Müller, F., Schlatter, D., Huber, W. (2009). Fragment-based screening using surface plasmon resonance technology. *J. Biomol. Scr.*, 14, 337–349.

39. Giannetti, A. M. (2011). From experimental design to validated hits: a comprehensive walk-through of fragment lead identification using surface plasmon resonance. *Method Enzymol.*, 493, 169–218.

40. Wartchow, C. A., Podlaski, F., Li, S., Rowan, K., Zhang, X., Mark, D., Huang, K. S. (2011). Biosensor-based small molecule fragment screening with biolayer interferometry. *J. Comput.-Aid. Mol. Des.*, 25, 669–676.

41. Barelier, S., Pons, J., Gehring, K., Lancelin, J.-M., Krimm, I. (2010). Ligand specificity in fragment-based drug design. *J. Med. Chem.*, 53, 5256–5266.

42. Lo, M.-C., Aulabaugh, A., Jin, G., Cowling, R., Bard, J., Malamas, M., Ellestad, G. (2004). Evaluation of fluorescence-based thermal shift assays for hit identification in drug discovery. *Anal. Biochem.*, 332, 153–159.

43. Ladbury, J. E., Klebe, G., Freire, E. (2010). Adding calorimetric data to decision making: a hot tip. *Nat. Rev. Drug Discov.*, 9, 23–27.

44. Boettcher, A., Ruediser, S., Erbel, P., Vinzenz, D., Schiering, N., Hassiepen, U., Rigollier, P., Mayr, L. M., Woelcke, J. (2010). Fragment-based screening by biochemical assays: systematic feasibility studies with trypsin and MMP12. *J. Biomol. Scr.*, 15, 1029–1041.

45. Baell, J. B., Holloway, G. A. (2010). New substructure filters for removal of pan assay interference compounds (PAINS) from screening libraries and for their exclusion in bioassays. *J. Med. Chem.*, 53, 2719–2740.

46. Šink, R., Gobec, S., Pečar, S., Zega, A. (2010). False positives in the early stages of drug discovery. *Curr. Med. Chem.*, 17, 4231–4255.

47. Liu, Y., Beresini, M. H., Johnson, A., Mintzer, R., Shah, K., Clark, K., Schmidt, S., Lewis, C., Liimatta, M., Elliot, L. O., Gustafson, A., Heise, C. E (2012). Case studies of minimizing nonspecific inhibitors in HTS campaigns that use assay-ready plates. *J. Biomol. Scr.*, 17, 225–236.

48. Jadhav, A., Ferreira, R. S., Klumpp, C., Mott, B. T., Austin, C. P., Inglese, J., Thomas, C. J., Maloney, D. J., Shoichet, B. K., Simeonov, A. (2010). Quantitative analyses of aggregation, autofluorescence, and reactivity artifacts in a screen for inhibitors of a thiol protease. *J. Med. Chem.*, 53, 37–51.

49. Feng, B. Y., Simeonov, A., Jadhav, A., Babaoglu, K., Inglese, J., Shoichet, B. K., Austin, C. P. (2007). A high-throughput screen for aggregation-based inhibition in a large compound library. *J. Med. Chem.*, 50, 2385–2390.

50. Giannetti, A. M., Koch, B. D., Browner, M. F. (2008). Surface plasmon resonance based assay for the detection and characterization of promiscuous inhibitors. *J. Med. Chem.*, 51, 574–580.

51. Myszka, D., He, X., Dembo, M., Goldstein, B. (1998). Extending the range of rate constants available from BIACORE: interpreting mass transport-influenced binding data. *Biophys. J.*, 75, 583–594.

52. Dunn, B. M., Chaiken, I. M. (1974). Quantitative affinity chromatography. Determination of binding constants by elution with competitive inhibitors. *Proc. Natl. Acad. Sci. USA*, 71, 2382–2385.

53. Krohn, K. A., Link, J. M. (2003). Interpreting enzyme and receptor kinetics: keeping it simple, but not too simple. *Nucl. Med. Biol.*, 30, 819–826.

54. Hulme, E. C., Trevethick, M. A. (2010). Ligand binding assays at equilibrium: validation and interpretation. *Br. J. Pharmacol.*, 161, 1219–1237.

55. Buehler, B., Hogrefe, H. H., Scott, G., Ravi, H., Pabón-Peña, C., O'Brien, S., Formosa, R., Happe, S. (2010). Rapid quantification of DNA libraries for next-generation sequencing. *Methods*, 50, S15–S18.

56. Chen, N.-T., Wu, C.-Y., Chung, C.-Y., Hwu, Y., Cheng, S.-H., Mou, C.-Y., Lo, L.-W. (2012). Probing the dynamics of doxorubicin-DNA intercalation during the initial activation of apoptosis by fluorescence lifetime imaging microscopy (FLIM). *PloS ONE*, 7, e44947.

57. Rusmini, F., Zhong, Z., Feijen, J. (2007). Protein immobilization strategies for protein biochips. *Biomacromolecules*, 8, 1775–1789.

58. Wong, L. S., Khan, F., Micklefield, J. (2009). Selective covalent protein immobilization: strategies and applications. *Chem. Rev.*, 109, 4025–4053.

59. Rabuka, D., Rush, J. S., deHart, G. W., Wu, P., Bertozzi, C. R. (2012). Site-specific chemical protein conjugation using genetically encoded aldehyde tags. *Nat. Protoc.*, 7, 1052–1067.

60. Li, Y., Rousa, R. (2012). Expression and purification of E. coli BirA biotin ligase for in vitro biotinylation. *Protein Express. Purif.*, 82, 162–167.

11

BEGIN WITH THE END IN MIND: THE HIT-TO-LEAD PROCESS

John Proudfoot

Boehringer Ingelheim Pharmaceuticals Inc.,
Ridgefield, CT, USA

11.1 INTRODUCTION

Over the past decade, DNA-encoded combinatorial chemistry has emerged as a promising component of lead generation. The screening of DNA-encoded combinatorial libraries provides hits against the target of interest, and the process that follows, the Hit-to-Lead (HtL) process, guides the transformation of these hits into lead series suitable for the optimization to clinical candidates. The HtL process has grown in sophistication and scope since it was initially described, and an understanding of the challenges that are commonly addressed provides the opportunity to guide library design toward more readily exploitable drug-like and lead-like space.

The terms "hit" and "lead" may be applied with different definitions across organizations. For the purpose of this chapter, a "hit" is defined as a sample that produces activity above the hit threshold in an assay [1] and a "validated hit" as a compound of verified structure and purity that has confirmed concentration-responsive potency in the assay system of interest [2]. A "lead" (compound or compound series) is one that satisfies organizationally defined criteria for further structure and activity optimization. These will typically include demonstrated activity at a targeted level in a relevant

A Handbook for DNA-Encoded Chemistry: Theory and Applications for Exploring Chemical Space and Drug Discovery, First Edition. Edited by Robert A. Goodnow, Jr.
© 2014 John Wiley & Sons, Inc. Published 2014 by John Wiley & Sons, Inc.

biochemical assay, appropriate selectivity, tractable Structure–Activity Relationship (SAR), and activity in a relevant cell-based assay and *in vivo* model.

Historically, the number of structures that have been exploited as leads for drug discovery is rather limited. In 1996, Sneader identified 244 drug prototypes, fewer than 140 of which would be categorized as drug-like by current standards [3]. Over the past 30 years or so, Lead Identification (LI) has progressed from a process focused on endogenous ligands and natural products or competitor drugs as starting points to a point where a variety of approaches may be available depending on the target type. Technologies and processes such as structure-based design, High-Throughput Screening (HTS), combinatorial chemistry, virtual screening, encoded libraries, diversity-oriented synthesis, and Fragment-Based Screening (FBS) have emerged as viable options to support LI. Some of these, for example, FBS, are of such recent origin that a valid assessment of their utility toward the delivery of approved drugs would be premature [4, 5]. For others, structure-based design and HTS in particular, a clear positive impact is readily apparent. For example, in contrast to the situation in 1985 where all the new drugs approved by the FDA originated from the exploitation of already known drugs as leads [6], in each year since 2001, at least one and, in some years, up to three approvals worldwide trace the lead structure to an HTS origin. The discovery of HIV-1 protease and integrase inhibitors approved for the treatment of AIDS relied heavily on structure-based methods. For other approaches, a positive impact is not so clear. Although combinatorial chemistry [7] was enthusiastically adopted by many in the early 1990s and rapidly progressed to application toward drug-like molecules [8, 9], there are few published examples to date of a positive impact on the design of an approved drug. The acknowledgment of a combinatorial or high-throughput chemistry impact on the development of sorafenib appears to be the sole published instance to date [10] (Fig. 11.1).

As new processes have been applied to LI, our understanding of the advantages and limitations of each new practice has gradually become more sophisticated. The early obsession with a "numbers only" approach in the context of random screening and library synthesis and design has been replaced with a sharper focus on the properties of screening decks and designed libraries. Particularly influential in regard to categorizing the molecular properties desirable in drug candidates (and by extension influencing desirable properties of leads and exploitable hits) are the contributions of Lipinski through the Rule of Five (Ro5) [12], Hann in assessing the impact of molecular complexity on lead-finding activities [13], Teague in developing the concept of lead-like libraries [14], and Hopkins in introducing the concept of Ligand Efficiency (LE) [15]. With regard to screening decks, the assessment of quality in relation to size, diversity, and the nature of frequent hitter and promiscuous chemotypes has also evolved [16].

One likely outcome of any HTS approach to LI is the identification of hundreds or even thousands of "hits" per campaign. The artifacts due to signal interference, compound aggregation, or compound reactivity might confound the recognition of validated hits, and even among the validated hits, there will likely be a ranking based on potency and other attributes. Therefore, as important as the categorizations highlighted in the previous paragraph was the introduction in 1996 of a specialized process focused on identifying the worthwhile hits and leads from HTS campaigns.

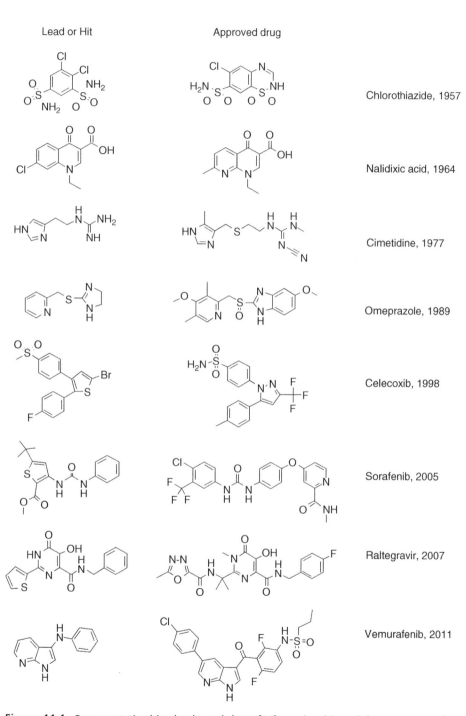

Figure 11.1. Representative hits, leads, and drugs [11]. Lead or hit and drug structures for selected approved drugs from 1950s to present, with approval dates.

11.2 HISTORICAL HIT-TO-LEAD PROCESS

In 1996, Michne [17] introduced the concept of an HtL phase that followed the completion of a screening campaign that had the goal of selecting those hits with the highest probability of successful progression toward a lead or lead series. The concept was introduced in the context of a screening deck of around 150,000 samples and prosecution of 16 campaigns/year. Although screening capacity has increased tremendously in the intervening time period, the guiding principles are still relevant today and are the foundation for most current HtL processes.

The process as originally designed had the following goals in mind:

a. Assess the purity of the hit sample(s)
b. Identify the minimum active fragment (develop a pharmacophore hypothesis)
c. Enhance selectivity versus closely related molecular targets
d. Separate structure-based from property-based mechanisms of action
e. Increase potency

11.2.1 Assess the Purity of the Hit Sample

The priority assigned to assessing the purity (and implicitly the verification of the assigned structure) of the hit sample was critical at the time and is still important. In the past, purity requirements for samples submitted to screening decks were often inconsistent, and samples from combinatorial or high-throughput synthesis methods were often made available for screening without a purification step. Additionally, there is the possibility that certain dissolved stock samples, subjected to multiple freeze–thaw cycles, might degrade over time. It is good practice to verify that the observed activity is reproduced with a sample of defined purity and confirmed structure, giving a "validated hit" and removing the chance that the observed activity is due to a reactive impurity. In some instances, this data can be obtained from a purified sample of the initial hit; in other cases, a resynthesis is required.

11.2.2 Identify the Minimum Active Fragment

The identification of the minimum active fragment related to the hit lays the foundation for subsequent determination of the opportunities offered to improve potency along with addressing any liabilities, whether structural or associated with activity against closely related targets or off-targets. In many instances, the clusters of related active compounds may provide an SAR pointing to a minimum active core and an essential substitution pattern. In the case of singleton hits, where no relevant analogs are present in the hitset, the synthesis of key analogs is required to provide this information. At this stage, the necessity of any hydrogen bond features in the hit, particularly hydrogen bond donors, should be clarified. This activity also provides an opportunity to replace any problematic structural features, such as aromatic nitro groups or activated olefins.

11.2.3 Enhance Selectivity versus Closely Related Molecular Targets

Assessing selectivity versus closely related targets and identification of opportunities to achieve a targeted selectivity profile is an important consideration for the prioritization of hits. This is generally a component of activity (b) earlier (see Section 11.2.2).

11.2.4 Separate Structure-Based from Property-Based Mechanisms of Action

In the original HtL process, this was described in the context of ion channel blockers where membrane disruption by lipophilic amines might unselectively block a channel. It was articulated as a requirement to demonstrate that an SAR rather than a structure–property relationship exists. More generally, this process defines the mechanism of action that gives rise to the observed activity and demonstrates that it is due to direct interaction with the target of interest and exploitable based on the desired lead profile. For example, hits might inhibit enzyme activity through competitive or allosteric mechanisms of action or covalent modification, and not all might be acceptable for progression.

11.2.5 Increase Potency

Potency improvement, although last on the list, was categorized in the original process as arguably the most important objective. In the context of current practice, illustrated in the following, where several additional considerations are important, this is no longer the most important consideration. At the early stages of the HtL process, understanding the opportunities and liabilities of the available hits is more critical than a focus on improving potency.

11.3 MODERN HIT-TO-LEAD PROCESS AND IMPLICATIONS FOR LIBRARY FOLLOW-UP

The concepts underlying this early HtL process have been applied broadly as a standard for the identification of leads from screening campaigns. As originally described, the process captured activities ranging from data analysis on the initial hitset through the generation of SAR with the aim of initiating lead optimization. Some of these activities, such as assessing the purity of the hit samples and separating structure-based from property-based mechanisms of action, are proximal to the initial HTS data delivery. Others, such as enhancing potency and selectivity, are ongoing activities that, while initially described as occurring within the space of a few months, nowadays typically consume a longer time frame. The modern process is often fragmented to a hit-triage phase, which is concerned with the identification of "validated hits," a hit-evaluation stage where the SAR tractability is ascertained and a broader profiling to assess selectivity and off-target liabilities is completed, followed by the HtL phase where the focus is on improving potency and addressing the liabilities that emerge in hit evaluation.

The original process predated issues and concepts that are an embedded part of the drug discovery process today. The discussions around drug-likeness had not yet appeared in the literature; the appreciation of hERG inhibition as a feature to be incorporated into lead optimization was emerging in relation to drug withdrawals (e.g., terfenadine); and the analyses that led to the identification of PK issues as a major factor in clinical attrition were ongoing. An understanding of the importance of these concepts and issues has led to the incorporation of many additional data points in decision making during the HtL process. For example, the following are usually now taken into account:

a. Compound physicochemical properties, particularly Molecular Weight (MW) and lipophilicity
b. Solubility
c. Permeability
d. Metabolism rates
e. Cytochrome P450 (CYP) enzyme inhibition and induction
f. Activity on key off-targets (e.g., hERG)
g. Intellectual Property (IP)
h. Promiscuity
i. LE

11.3.1 (a), (b), (c) Compound Physicochemical Properties, Solubility, and Permeability

A number of these additional parameters are strongly influenced by a more sophisticated perspective on drug-likeness and "lead-likeness" than was available in 1996. While recognizing that the Ro5 was constructed with the particular goal of categorizing compounds by their likelihood to be orally bioavailable, it is often applied in assessing the quality of screening collections and hitsets. Compounds that are Ro5 compliant are more likely to display better solubility and cell permeability than those that are not. This is particularly so if two or more of the Ro5 parameters are violated. A number of modifications of the Ro5 metrics have appeared (Table 11.1) ranging

TABLE 11.1. List of properties for molecules classified according to drug-like/Ro5, lead-like, and fragment-like

	Drug-like, Ro5	Lead-like	Fragment-like (Ro3)
MW	≤500	≤460	≤300
H-donors	≤5	≤5	≤3
H-acceptors	≤10	≤9	
Lipophilicity	≤5	≤4.2	≤3
Rotatable bonds	≤10	≤10	≤3

TABLE 11.2. List of properties for tiering approach that prioritizes hits based on the ranges of computed properties including Ro4 and Ro5 and the fraction of sp3 centers (Fsp3)

Tier 1	Tier 2	Tier 3	Tier 4	Tier 5	Tier 6
Ro4	Ro4	Ro4	Ro5	Ro5	Ro5
NAR<4	NAR<4		NAR<4	NAR<4	
Fsp3 >0.5			Fsp3 >0.5		

from the proposals of Hann and Oprea [18] that categorize "lead-likeness" in relation to library design to the fragment-like criteria [19], which have also been proposed as lead-like criteria [2].

Table 11.1 also incorporates guidance on the number of rotatable bonds, a parameter that was also found to influence oral bioavailability, independent of MW [20]. As applied to assessing the quality of hits and leads, the Ro5 parameters are considered rather generous, especially since the progression of hits to leads to drug candidates more often than not involves increases in MW and lipophilicity. It is prudent to begin with hits that allow for some growth in MW and lipophilicity in order to arrive at an endpoint that is drug-like.

Beyond the molecular properties in the table earlier, the analysis by MacDonald on the impact of aromatic ring count (NAR) on compound developability [21] is also beginning to influence the hit-evaluation process. Briefly, the presence of more than three aromatic rings in a molecule correlates with poorer compound developability and an increased risk of attrition in development. This parameter has been incorporated by Abbott (Table 11.2) into a tiering approach [22] that prioritizes hits based on ranges of computed properties that also include Ro4 and Ro5 and the fraction of sp3 centers (Fsp3). While the Ro5 is quite well known, the Ro4 is less so. These rules are as follows: Ro4 compliance-C log P < 4, MW < 400, HBD < 5, HBA <10, and tPSA <140 and Ro5 compliance-C log P < 5, MW <500, HBD <5, HBA <10, and tPSA <140.

From correlations of these properties with in-house *in vitro* data, it was concluded that optimal physicochemical space may be defined by the following criteria: Ro4 compliance, NAR<4, and Fsp3 >0.5. Compounds within this chemical space will likely have superior drug-like properties. It is instructive that this tiering approach is applied at the very early stages of hit evaluation and that tier 1 is a significantly more stringent categorization than the Ro5.

11.3.2 Metabolism Rates, Cytochrome P450 Enzyme Inhibition and Induction, and hERG Inhibition

The measurement and assessment of metabolic susceptibility, CYP inhibition and induction, and hERG inhibition have been brought forward early into the LI process in order to address key issues that led to significant clinical and market failures during the 1990s. In general, while any of these parameters independently might not constitute a no-go for a particular hit or hit series, knowing earlier rather than later that a liability related to any of these parameters exists allows for an informed prioritization of the available hits. The generation of this relevant data at the earlier stage also allows an assessment of whether the SAR associated with target engagement is compatible with addressing the specific liability at hand. If a hit carries multiple liabilities, it is an indication that the path to lead optimization will be particularly challenging.

11.3.3 Intellectual Property

Although not explicitly stated among the objectives in the original HtL process, it is clear from the conclusion in the article that the assessment of IP was embedded in the original process. It is no less important at present, especially given the very broad patenting strategies that are commonplace. Some structural types are so heavily patented, especially among kinase targets, that it is critical to provide an early assessment on whether any realistic opportunities exist to generate a defensible patent position.

11.3.4 Ligand Efficiency

The concept of LE, although introduced quite recently, has been especially influential in providing a path to a balanced perspective on the opportunities offered by high-MW hits with superior potency compared to lower-MW hits with weaker potency. As presented by Hopkins, who built on the earlier work of Kuntz, LE is defined as ΔG/HA (where ΔG is the free energy of binding and HA is the number of nonhydrogen atoms in the ligand). The variations of this equation are also commonly used (e.g., pIC_{50}/HA or pIC_{50}/MW), but all retain the concept of normalizing target engagement relative to molecular size. While LE was adopted initially in the context of fragment-based LI, where a metric to prioritize very small molecules with weak affinity was a necessity, the value is such that it is now a standard parameter in assessing hit quality. One particularly beneficial outcome of the application of LE to decision making is movement from a pathological affinity to hit potency regardless of molecular properties and the addition of transparency to situations where a modest gain in potency is achieved through the addition of disproportionately large substitution.

11.3.5 Compound Promiscuity

The assessment of compound promiscuity [23], which can emerge in a number of contexts, is also important at the early stages of LI. There may be apparent promiscuity due to interference with a particular detection method in an assay, essentially resulting in artifactual activity [24]. Some structural features may result in activity across a number of target types. The extent and impact of promiscuity are subject to interpretation, and it may be more or less important depending on the particular observation and the number and quality of hits already available. If several hit candidates are available, then broad profiling at an early stage allows for the prosecution of the scaffolds or series with the fewest initial liabilities.

11.4 THE IMPORTANCE OF A HIGH-QUALITY SCREENING DECK

While it is important to recognize that the focus of the HtL phase is to develop the validated hits into new leads rather than discover a drug directly, one should be mindful of the likely relationship between any validated hit and a subsequent drug candidate. For example, a final candidate will likely be of higher MW and increased lipophilicity compared to the initial hit [25, 26]. One previous criticism of HTS-derived hits, a tendency to select hits (leads) with increased lipophilicity and MW [27], is no longer true based on the evolved HtL processes. In the past, when analog-based approaches were prevalent, drug candidates typically bore strong structural resemblances to the lead starting point [28]. This is no longer so apparent, and recent examples show significant structural evolution between hit and drug structures [29]. From its very nature, a hit carries some liabilities that prevent it from being designated a lead, and the nature and significance of these liabilities will strongly influence the rate of progression toward a high-quality lead. Some liabilities are easier to address than others, and not all should be weighted equally. For example, medicinal chemists will often find that potency and selectivity are relatively straightforward parameters to improve, especially if structural information on the target is available, compared to solving challenges related to metabolic stability or hERG inhibition. Most often, the significant challenge is in finding the modifications that satisfy more than one required optimization parameter.

Screening decks and libraries, however assembled, provide opportunities and liabilities related to their origins. Corporate screening decks, which were typically assembled over long time periods, contain compounds from many sources including legacy projects, acquisition programs (academic and commercial sources), and combinatorial and focused libraries.

Compounds from legacy projects are likely to be drug-like rather than lead-like. They are also likely, especially if generated later in a project, to be more rather than less complex and, based on the perspective provided by Hann, less likely to hit in an assay. Compounds added through acquisition are likely to be rather more lead-like than drug-like, especially if the appropriate computational filters were applied during the selection process, but on the other hand are unlikely to be IP friendly. Combinatorial libraries made before 2000 are likely to be of little value for LI [30], but more recent libraries, particularly those designed using more modern perspectives on "lead-likeness" and diversity, are of distinct value [31] (Fig. 11.2). Screening decks are also likely to be enriched in "privileged structures" [32]—those structures that appear to contain features that facilitate binding to various target surfaces. These structures are capable of providing useful ligands for more than one target through judicious modification.

The original HtL process was constructed to provide guidance on the analysis of HTS hitsets such as those originating from screening collections described earlier. The HtL principles are also applicable to any screening collection, including those provided by presynthesized libraries, encoded or not.

Encoded libraries provide some significant advantages over standard screening decks. The combinatorial nature of the syntheses provides a far greater number of

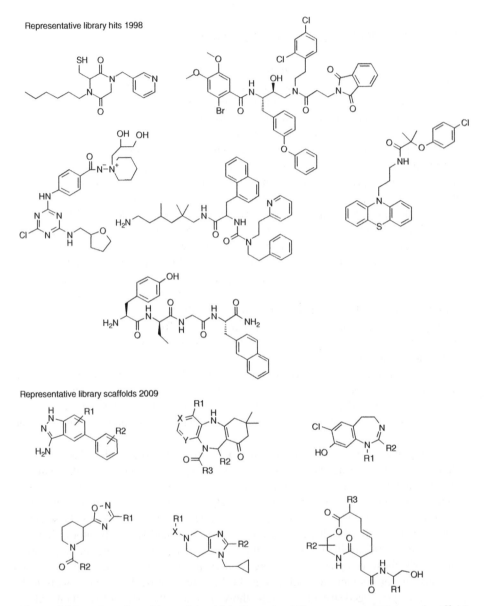

Figure 11.2. Selected combinatorial or high-throughput library representatives and scaffolds from 1998 to 2009 [31].

samples than is present in any standard screening deck (10^7–10^9 vs. 10^6). The additional numbers should give a greater probability of identifying "hit" samples. Well-designed libraries can be biased toward drug-like structures, and structural alerts can be avoided through the appropriate selection or deselection of building blocks. The growing variety of synthetic methods that are enabled for combinatorial synthesis and that are compatible with the tagging methods should allow a decreased reliance on reductive amination and amide and sulfonamide formation reactions as the vehicles for library diversity [33]. Multiple use of these specific reactions is highlighted in particular because of the implicit generation of multiple hydrogen-bonding features in the product molecules. The analysis by Hann [13] teaches that the likelihood of finding productive directional interactions with a target decreases with increasing molecular complexity, and the very elegant study by Bartlett [34] is convincing in highlighting the penalties associated with mismatched H-bonding interactions between ligand and target. H-donors or H-acceptors projected into an inappropriate environment are likely to abrogate binding affinity.

Following the identification of the hit samples, the process of generating validated hits is analogous to that defined for HTS campaigns. In fact, because the screening samples are covalently tagged, the synthesis of the hit in untagged form is a requirement to confirm the observed activity. For well-designed libraries, where systematically substituted building blocks are used, the potential for an SAR to emerge rapidly from the initial hits is a distinct possibility.

The hit-identification process employed for encoded libraries, usually some variant of an affinity selection method, is analogous to a binding assay employed in an HTS campaign. The advantages are similar (generally a straightforward categorization of actives and inactives), as are the disadvantages (no immediate indication of relevant function associated with the binding event). The possibility of employing negative selection for counter targets to provide hits with a predefined selectivity profile is a unique feature. As indicated earlier, the generation of data in a functional assay (especially if the function relies on interaction with an intracellular target) follows the synthesis of the hit in untagged form. It is this expectation that the hits will be useful beyond a biochemical tool that makes the library design so important. If the appropriate physicochemical properties (e.g., those compatible with permeation across cell membranes) are embedded in the design process, then there is increased likelihood that functional activity will be observed with the initial validated hits.

One feature that remains unclear at present is whether the increased numbers and diversity of the screened libraries will address the druggability of "difficult" target types, particularly those that involve the disruption of Protein–Protein Interactions (PPIs). For these presently intractable targets, it is sensible to cast a broad net in the search for leads, and the large numbers of compounds in these libraries might provide the tipping point toward identifying tractable lead structures. However, success in this challenging area is likely to emerge not just from the application of any particular new technology or process in isolation but rather from the intelligent use of any available information and technologies. In this regard, just as high-throughput crystallography enabled and facilitated the fragment approach to LI, it is just as likely that crystallography will be the key enabling technology for the successful prosecution of an encoded library approach to disrupting PPIs. If available, cocrystal structures provide direction on minimum pharmacophores and guidance on regions of the hit structure where improvements in physical properties and/or interactions with the target could be attained.

11.5 CURRENT PRACTICE

In 2003, Kubinyi espoused the position that the lack of productivity in the pharma industry reflected a shortage of new lead structures that can be optimized into therapeutically useful drugs [35]. However, even in just the past 10 years, major advances that should enable LI have occurred, including FBS approaches, high-throughput crystallography, and encoded libraries. The generation of viable lead series requires (i) a clear and rational definition of the requirements of the lead in terms of physical, chemical, biological, and pharmacological properties and (ii) a strategy to achieve the target profile [36]. A clear understanding of the final target profile should inform the prioritization of candidate hits: if the final candidate is to be codosed with multiple medications, then a clean DDI (e.g., CYP) profile is important; if CNS penetration is required, then consideration of tPSA should be included in the decision making; and if the targeted mechanism requires extended coverage at high plasma levels, then metabolism rates and perhaps protein binding become important considerations. The HtL process provides a framework for the identification and progression of hits regardless of their origin. The process can be equally applied to HTS campaigns, FBS campaigns, or even analog-based approaches. The process is an integrated one in that that requires contributions beyond medicinal chemistry alone for success but has been clearly demonstrated to be successful in the HTS world and will likely also contribute to successful LI from other approaches.

DNA-encoded combinatorial chemistry conducted for the purposes of lead generation is likely to contribute significantly to drug discovery in the future. It is important to consider the process that must be conducted after the identification of active compounds from such libraries and take advantage of the lessons that have been learned over the past 15 years regarding the HtL process. An understanding of the common challenges that are addressed in the HtL process should facilitate the design of libraries with superior drug-like and lead-like characteristics.

REFERENCES

1. Proudfoot, J., Nosjean, O., Blanchard, J., Wang, J., Besson, D., Crankshaw, D., Gauglitz, G., Hertzberg, R., Homon, C., Llewellyn, L., Neubig, R., Walker, L., Villa, P. (2011). Glossary of terms used in biomolecular screening (IUPAC Recommendations 2011). *Pure Appl. Chem.*, 83, 1129–1158.

2. Wunberg, T., Hendrix, M., Hillisch, A., Lobell, M., Meier, H., Schmeck, C., Wild, H., Hinzen, B. (2006). Improving the hit to lead process. *Drug Discov. Today*, 11, 175–180.

3. Sneader, W. S., *Drug prototypes and their exploitation*, John Wiley & Sons: Chichester, 1996, 3–10. Although the prototypes are not directly comparable to the definition of leads, this was the first attempt to systematically categorize drug discovery starting points.

4. The identification and prosecution of the lead structure for the approved drug Tipranavir could be regarded as an early example of the process. Thaisrivongs, S., Tomich, P. K., Watenpaugh, K. D., Chong, K.-T., Howe, W. J., Yang, C.-P., Strohbach, J. W., Turner, S. R., McGrath, J. P., Bohanon, M. J., Lynn, J. C., Mulichak, A. M., Spinelli, P. A., Hinshaw, R. R., Pagano, P. J.,

Moon, J. B., Ruwart, M. J., Wilkinson, K. F., Rush, B. D., Zipp, G. L., Dalga, R. J., Schwende, F. J., Howard, G. M., Padbury, G. E., Toth, L. N., Zhao, Z., Koeplinger, K. A., Kakuk, T. J., Serena, S. L., Cole, L., Zaya, R. M., Piper, R. C., Jeffrey, P. (1994). Structure-based design of HIV protease inhibitors: 4-hydroxycoumarins and 4-hydroxy-2-pyrones as non-peptidic inhibitors. *J. Med. Chem.*, 37, 3200–3204.

5. Bollag, G., Tsai, J., Zhang, J., Zhang, C., Ibrahim, P., Nolop, K., Hirth, P. (2012). Vemurafenib: the first drug approved for BRAF-mutant cancer. *Nat. Rev. Drug Discov.*, 11, 873–886.

6. Freter, K. R. (1988). Drug discovery—today and tomorrow: the role of medicinal chemistry. *Pharm. Res.*, 5, 397–400.

7. (a) Lowe, G. (1995). Combinatorial chemistry. *Chem. Soc. Rev.*, 24, 309–317. (b). Patel, D. V., Gordon, E. M. (1996). Applications of small-molecule combinatorial chemistry to drug discovery. *Drug Discov. Today*, 1, 134–144.

8. (a) Bunin, B. A., Ellman, J. A. (1992). A general and expedient method for the solid-phase synthesis of 1,4-benzodiazepine derivatives. *J. Am. Chem. Soc.*, 114, 10997–8. (b) Bunin, B. A., Plunkett, M. J., Ellman, J. A. (1994). The combinatorial synthesis and chemical and biological evaluation of a 1,4-benzodiazepine library. *Proc. Natl. Acad. Sci. USA*, 91, 4708–4712.

9. Bunin, B. A., Plunkett, M. J., Ellman, J. A., Bray, A. M. (1997). The synthesis of a 1680 member 1,4-benzodiazepine library. *N. J. Chem.*, 21, 125–130.

10. Lowinger, T. B., Riedl, B., Dumas, J., Smith, R. A. (2002). Design and discovery of small molecules targeting Raf-1 kinase. *Curr. Pharm. Des.*, 8, 2269–2278.

11. (a) Chlorothiazide: Novello, F. C., Sprague, J. M. (1957). Benzothiadiazine dioxides as novel diuretics. *J. Am. Chem. Soc.*, 79, 2028–2029. (b) Nalidixic acid: Lesher, G. Y., Froelich, E. J., Gruett, M. D., Bailey, J. H., Brundage, R. P. (1962). 1,8-Naphthyridine derivatives. A new class of chemotherapeutic agents. *J. Med. Pharm. Chem.*, 5, 1063–1065. (c) Cimetidine: Ganellin, C. R. (1982). Cimetidine. *Chronicles of drug discovery*, vol. 1, pp. 1–39. Bindra, J. S., Lednicer, D. eds. Wiley, New York, NY. (d) Omeprazole: Lindberg, P., Brändström, A., Wallmark, B., Mattsson, H., Rikner, L., Hoffmann, K.-J. (1990). Omeprazole: the first proton pump inhibitor. *Med. Res. Rev.* 10, 1–54. (f) Celecoxib: Penning, T. D., Talley, J. J., Bertenshaw, S. R., Carter, J. S., Collins, P. W., Docter, S., Graneto, M. J., Lee, L. F., Malecha, J. W., Miyashiro, J. M., Rogers, R. S., Rogier, D. J., Yu, S. S., Anderson, G. D., Burton, E. G., Cogburn, J. N., Gregory, S. A., Koboldt, C. M., Perkins, W. E., Seibert, K., Veenhuizen, A. W., Zhang, Y. Y., Isakson, P. C. (1997). Synthesis and biological evaluation of the 1,5-diaryl-pyrazole class of cyclooxygenase-2 inhibitors: identification of 4-[5-(4-methylphenyl)-3-(trifluoromethyl)-1H-pyrazol-1-yl]benzenesulfonamide (SC-58635, Celecoxib). *J. Med. Chem.*, 40, 1347–1365. (g) Sorafenib: see Ref. [10] above. (h) Raltegravir: Summa, V., Petrocchi, A., Bonelli, F., Crescenzi, B., Donghi, M., Ferrara, M., Fiore, F., Gardelli, C., Gonzalez Paz, O., Hazuda, D. J., Jones, P., Kinzel, O., Laufer, R., Monteagudo, E., Muraglia, E., Nizi, E., Orvieto, F., Pace, P., Pescatore, G., Scarpelli, R., Stillmock, K., Witmer, M. V., Rowley, M. (2008). Discovery of raltegravir, a potent, selective orally bioavailable HIV-integrase inhibitor for the treatment of HIV-AIDS infection. *J. Med. Chem.*, 51, 5843–5855. (i) Vemurafenib: see Ref. [5] above.

12. Lipinski, C. A., Lombardo, F., Dominy, B. W., Feeney, P. J. (1997). Experimental and computational approaches to estimate solubility and permeability in drug discovery and development settings. *Adv. Drug Deliv. Rev.*, 23, 3–25.

13. Hann, M. M., Leach, A. R., Harper, G. (2001). Molecular complexity and its impact on the probability of finding leads for drug discovery. *J. Chem. Inf. Comput. Sci.*, 41, 856–864.

14. Teague, S. J., Davis, A. M., Leeson, P. D., Oprea, T. (1999). The design of leadlike combinatorial libraries. *Angew. Chem. Int. Ed.*, 38, 3743–3748.

15. Hopkins, A.L. Groom, C. R., Alex, A. (2004). Ligand efficiency: a useful metric for lead selection. *Drug Discov. Today*, 9, 430–431.

16. (a) Davis, A. M., Keeling, D. J., Steele, J., Tomkinson, N. P., Tinker, A. C. (2005). Components of Successful Lead Generation. *Curr. Top. Med. Chem.*, 5, 421–439. (b) Rishton, G. M. (2008). Molecular diversity in the context of leadlikeness: compound properties that enable effective biochemical screening. *Curr. Opin. Chem. Biol.*, 12, 340–351. (c) Pearce, B. C., Sofia, M. J., Good, A. C., Drexler, D. M., Stock, D. A. (2006). An empirical process for the design of high-throughput screening deck filters. *J. Chem. Inf. Model.*, 46, 1060–1068.

17. Michne, W. F. (1996). Hit-to-Lead chemistry: a key element in new lead generation. *Pharm. News*, 3, 19–21.

18. Hann, M. M., Oprea, T. I. (2004). Pursuing the "lead-likeness" concept in pharmaceutical research. *Curr. Opin. Chem. Biol.*, 8, 255–263.

19. Congreve, M., Carr, R., Murray, C., Jhoti, H. (2003). A 'Rule of Three' for fragment-based lead discovery? *Drug Discov. Today*, 8, 876–877.

20. Veber, D. F., Johnson, S. R., Cheng, H.-Y., Smith, B. R., Ward, K. W., Kopple, K. D. (2002). Molecular properties that influence the oral bioavailability of drug candidates. *J. Med. Chem.*, 45, 2615–2623.

21. (a) Ritchie, T. J., Macdonald, S. J. F (2009). The impact of aromatic ring count on compound developability—are too many aromatic rings a liability in drug design? *Drug Discov. Today*, 14, 1011–1020. (b) Ritchie, T. R., Macdonald, S. J. F., Young, R. J., Pickett, S. D. (2011). The impact of aromatic ring count on compound developability: further insights by examining carbo- and hetero-aromatic and –aliphatic ring types. *Drug Discov. Today*, 16, 164–171.

22. Cox, P. B., Gregg, R. J., Vasudevan, A. (2012). Abbott physicochemical tiering (APT)—a unified approach to HTS triage. *Bioorg. Med. Chem.*, 20, 4564–4573.

23. Dimova, D., Hu, Y., Bajorath, J. (2012). Matched molecular pair analysis of small molecule microarray data identifies promiscuity cliffs and reveals molecular origins of extreme compound promiscuity. *J. Med. Chem.*, 55, 10220–10228.

24. (a) Baell, J. B., Holloway, G. A. (2010). New substructure filters for removal of pan assay interference compounds (PAINS) from screening libraries and for their exclusion in bioassays. *J. Med. Chem.*, 53, 2719–2740. (b) Johnston, P. A., Soares, K. M., Shinde, S. N., Foster, C. A., Shun, T. Y., Takyi, H. K., Wipf, P., Lazo, J. S. (2008). Development of a 384-well colorimetric assay to quantify hydrogen peroxide generated by the redox cycling of compounds in the presence of reducing agents. *Assay Drug Dev. Technol.*, 6, 505–518.

25. Perola, E. (2010). An analysis of the binding efficiencies of drugs and their leads in successful drug discovery Programs. *J. Med. Chem.*, 53, 2986–2997.

26. Oprea, T. I., Davis, A. M, Teague, S. J., Leeson, P. D. (2001). Is there a difference between leads and drugs? A historical perspective. *J. Chem. Inf. Comput. Sci.*, 41, 1308–1315.

27. Lipinski, C. A. (2000). Drug-like properties and the causes of poor solubility and poor permeability. *J. Pharm. Toxicol.*, 44, 235–249.

28. Proudfoot, J. R. (2002). Drugs, leads, and drug-likeness: an analysis of some recently launched drugs. *Bioorg. Med. Chem. Lett.*, 12, 1647–1650.

29. (a) Roehrig, S., Straub, A., Pohlmann, J., Lampe, T., Pernerstorfer, J., Schlemmer, K.-H., Reinemer, P., Perzborn, E. (2005). Discovery of the novel antithrombotic agent 5-chloro-N-({(5S)-2-oxo-3-[4-(3-oxomorpholin-4-yl)phenyl]-1,3-oxazolidin-5-yl}methyl)thiophene-

2-carboxamide (BAY 59-7939): an oral, direct factor Xa inhibitor. *J. Med. Chem.*, 48, 5900–5908. (b) Cui, J. J., Tran-Dube, M., Shen, H., Nambu, M., Kung, P.-P., Pairish, M., Jia, L., Meng, J., Funk, L., Botrous, I., McTigue, M., Grodsky, N., Ryan, K., Padrique, E., Alton, G., Timofeevski, S., Yamazaki, S., Li, Q., Zou, H., Christensen, J., Mroczkowski, B., Bender, S., Kania, R. S., Edwards, M. P. (2011). Structure based drug design of crizotinib (PF-02341066), a potent and selective dual inhibitor of mesenchymal-epithelial transition factor (c-MET) kinase and anaplastic lymphoma kinase (ALK). *J. Med. Chem.*, 54, 6342–6363.

30. Dolle, R. E., Nelson, K. H. (1999). Comprehensive survey of combinatorial library synthesis: 1998. *J. Comb. Chem.*, 1, 235–282.

31. Dolle, R. E., Le Bourdonnec, B., Worm, K., Morales, G. A., Thomas, C. J., Zhang, W. (2010). Comprehensive survey of chemical libraries for drug discovery and chemical biology: 2009. *J. Comb. Chem.*, 12, 765–806.

32. Evans, B. E., Rittle, K. E., Bock, M. G., DiPardo, R. M., Freidinger, R. M., Whitter, W. L., Lundell, G. F., Veber, D. F., Anderson, P. S., Chang, R. S. L., Lotti, V. J., Cerino, D. J., Chen, T. B., Kling, P. J., Kunkel, K. A., Springer, J. P., Hirshfield, J. (1988). Methods for drug discovery: development of potent, selective, orally effective cholecystokinin antagonists. *J. Med. Chem.*, 31, 2235–2246.

33. (a) Dolle, R. E., Le Bourdonnec, B., Goodman, A. J., Morales, G. A., Thomas, C. J., Zhang, W. (2008). Comprehensive survey of chemical libraries for drug discovery and chemical biology: 2007. *J. Comb. Chem.*, 10, 753–802.

34. Morgan, B. P., Scholtz, J. M., Ballinger, M. D., Zipkin, I. D., Bartlett, P. A. (1991). Differential binding energy: a detailed evaluation of the influence of hydrogen-bonding and hydrophobic groups on the inhibition of thermolysin by phosphorus-containing inhibitors. *J. Am. Chem. Soc.*, 113, 297–307.

35. Kubinyi, H. (2003). In search for new leads. EFMC yearbook, pp. 14–28. http://www.kubinyi. de/EFMC-Pub-2003.pdf. Accessed on December 14, 2013.

36. Keseru, G. M., Makara, G. M. (2006). Hit discovery and hit-to-lead approaches. *Drug Discov. Today*, 11, 741–748.

12

ENUMERATION AND VISUALIZATION OF LARGE COMBINATORIAL CHEMICAL LIBRARIES

Sung-Sau So

Formerly of Hoffmann-La Roche Inc., Nutley, NJ, USA
Currently of Merck & Co., Kenilworth, NJ, USA

12.1 INTRODUCTION

The advent of combinatorial chemical library synthesis and, more recently, DNA-encoded chemical library synthesis offers drug discovery researchers access to billions of unique chemical entities that can be useful candidates for screening against therapeutic drug targets [1]. The ability to rapidly access new chemical space has stimulated considerable interest in the area of chemical diversity analysis [2–6] and also in the development of efficient methods to sample and compare large chemical libraries [7–11]. From the early days of combinatorial chemistry, chemists have long recognized that an astonishingly large number of library products can be synthesized from just several thousands of common blocking blocks. Yet, typical combinatorial libraries that are actually produced and screened for drug discovery programs nowadays seldom exceed 10^3 molecules. Synthesis, purification, analysis, and storage for a huge number of individual compound arrays are not economical or practical, and compound screening can also become very costly. In contrast, the use of mixture-based combinatorial libraries offers substantial saving in logistics, but this benefit is largely offset by the tremendous effort required for iterative deconvolution in hit identification and validation. So, the emphasis of computational library design in the past 20 years has primarily been the selection of optimal building blocks to produce chemical libraries at a modest scale [12–17]. In the case of exploratory or probe libraries that are not target dependent, the goal is often to make a

A Handbook for DNA-Encoded Chemistry: Theory and Applications for Exploring Chemical Space and Drug Discovery, First Edition. Edited by Robert A. Goodnow, Jr.

minimal subset of diverse library compounds that best represent the entire library. For focused or biased libraries, one would optimize building block selection to produce compounds that exhibit the necessary pharmacophores to target specific protein families such as kinases, GPCRs, or proteases or to bias the synthesis of products with a desirable physicochemical property profile, such as compliance with the rule of five [18].

The recently developed DNA-encoded library technology is a disruptive technology that also changes the landscape of combinatorial library design. Not only is there now an effective way to produce and handle massive numbers of combinatorial products, but there are means to detect hits from such libraries with unparalleled sensitivity. A unique DNA tag attached to each library member serves as its identification bar code, from which a signal is subsequently amplified by PCR and analyzed by next-generation high-throughput sequencing technology. Computational library design of billions or even trillions of molecules is now of practical use and also a reality.

The purpose of this chapter is to provide a brief overview of methods and tools that are currently available to enumerate and analyze very large combinatorial chemical libraries of more than one billion members. Here, a combinatorial library produced from a four-component Ugi reaction serves to illustrate some practical considerations for enumeration and analysis [19]. In this reaction, isocyanides, aldehydes, amines, and carboxylic acids are mixed in one pot, and a very large array of combinatorial products is formed. The application of cheminformatics tools to select building blocks, enumerate library products, and also calculate several key molecular properties to profile the library is discussed. At the end of the chapter, several methods to compare chemical space for four compound libraries, either qualitatively through visualization or quantitatively, with molecular similarity calculations have been suggested.

12.2 ENUMERATION

In this section, the enumeration of a combinatorial chemical library for the four-component Ugi library is used as an illustrative example. A schematic representation of the reaction is shown in Figure 12.1.

12.2.1 Reagent Identification

The initial step in library enumeration is the selection of appropriate sets of reagents for the library. In this case, four classes of reagents—isocyanides, aldehydes, amines, and

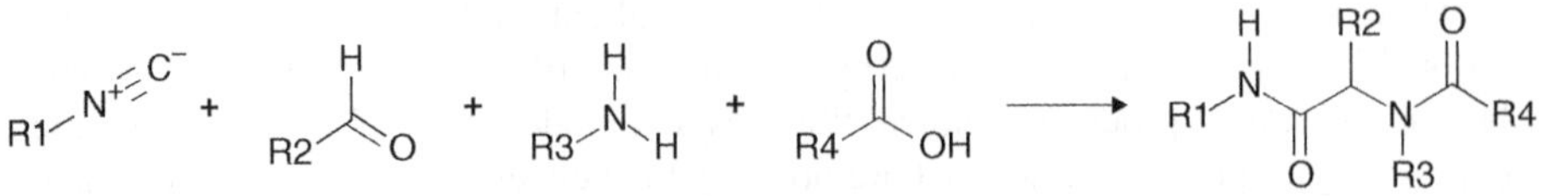

Figure 12.1. Four-component Ugi reaction. The four diversity sites (R1–R4) of the final products are introduced by isocyanides, aldehydes, amines, and carboxylic acids building blocks. The reaction introduces an uncontrolled stereocenter (the tetrahedral carbon attached to R2), and an epimeric mixture is produced.

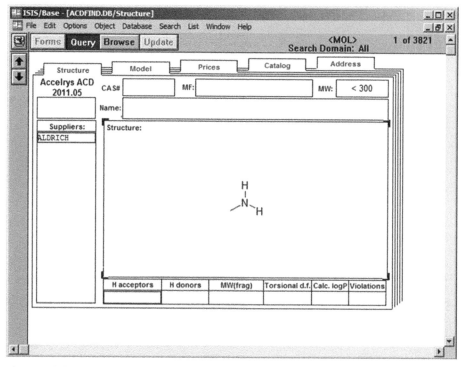

Figure 12.2. Example of a substructure query to search for low MW and commercially available primary amines from the ACD database through an ISIS database user interface.

carboxylic acids—are needed to build this library. One can start to assemble a short list of reagents from either an in-house repository or commercial suppliers. As shown in Figure 12.2, a substructure query was made in an ISIS/Base Available Chemicals Directory (ACD) [20] application to search for low molecular weight (MW < 300) primary amines that are commercially available from Aldrich. A major limitation of using a substructure-based query such as the one shown here is that the structure representation often lacks sufficient specificity. For example, primary carboxamides, which contain an amino group but cannot undergo the Ugi reaction, are also retrieved from this substructure query and become part of an initial selection of 3821 building blocks. It is therefore important that the list of building blocks is further refined so that only those that are chemically compatible with the reaction are selected prior to enumeration.

12.2.2 Reagent Filtering

The refining of the initial list of building blocks can be easily accomplished using tools from a cheminformatics platform such as Pipeline Pilot [21] or through scripting using Daylight [22] or OEChem toolkits [23]. In the current example, Pipeline Pilot was used for postprocessing the hits obtained from this primary amine ISIS query on the ACD [20].

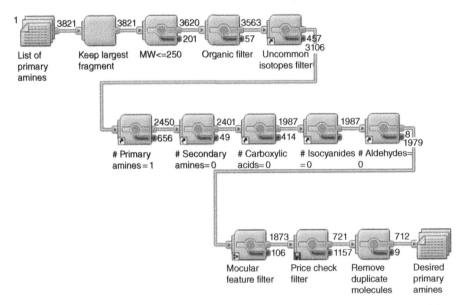

Figure 12.3. Pipeline Pilot workflow used to filter a list of primary amines obtained from the ACD.

In the workflow shown in Figure 12.3, the initial list of building blocks was processed through a chain of molecular filter and manipulator components to remove those that were incompatible or seemed less desirable. Here, the largest molecular fragment in each building block was kept (e.g., removal of any counterion). It was decided to apply an upper MW threshold, and only the building blocks with MW ≤ 250 were retained. An "organic" filter and an "uncommon isotope" filter eliminate organic molecules with unusual elements (i.e., molecules with H, C, N, O, P, S, F, Cl, Br, or I atoms are acceptable, but not those with B or Si), and also those without unusual isotopes were kept. During this initial stage of filtering, more than 700 building blocks were removed. The goal for the next stage of filtering is to remove reagents that contain incompatible or potentially problematic functional groups. This can be achieved through the use of a SMARTS expression, which is a powerful and versatile text-based cheminformatics language. The significant advantage of SMARTS over an ISIS-based query (such as the one shown in Figure 12.2) is that the substructure specification can be very precise. For example, a recursive SMARTS pattern such as

$$[NX3; H2; +0; !\$(NC = [O, S, N]); !\$(NS = O); !\$(N - N = C)]$$

identifies only primary amines, but does not identify secondary amines, primary carboxamides, or sulfonamides. This type of SMARTS-based molecular filter can be readily incorporated as a customized Pipeline Pilot component to detect the presence of specific chemical substructures. In this example, there are 2450 building blocks in the reagent list that contain exactly one primary amine in the molecule. The other 656 building blocks, which either did not match the SMARTS pattern or matched the query more than once, were removed. Likewise, one can further remove other building blocks with chemical features that are considered problematic using a chain of SMARTS-based filter components. For example, one might prefer to exclude amino acids from this list of primary amine building blocks. The SMARTS patterns to identify aldehyde, carboxylic acid, primary amine, and isocyanide substructures are listed below:

Aldehyde	[CX3H1](=O)[#6]
Carboxylic acid	[CX3](=[OX1])[OX2H1,O-]
Primary amine	[NX3;H2;+0;!$(NC=[O,S,N]);!$(NS=O);!$(N-N=C)]
Isocyanide	[NX2;+1]#[CX1;-1]

It is beyond the scope of this chapter to discuss the SMARTS syntax in detail. Interested readers are referred to a manual and examples from Daylight for additional information [24].

At this point, 1979 building blocks have been identified from the original list of 3821 containing exactly one primary amine (but no functional group from other deselected reagent classes) after the second stage of filtering. In the final stage of filtering, additional molecular feature filters were applied to remove other undesirable or reactive chemical groups (e.g., alkyl halides, sulfonyl halides), again using SMARTS patterns to flag such chemical substructures. So far, all filters being deployed were related to either chemical structures or molecular properties. It should be noted that other types of filters relating to logistical consideration (e.g., price of reagent) could also be used. In the current example, a customized "price check" filter was incorporated to positively select building blocks that satisfied a set of predetermined price limits (e.g., maximum of $50/g or $250/5 g). The more expensive reagents or those without pricing information were removed from further consideration. Finally, any duplicate building blocks from the list were dropped, resulting in a filtered list of 712 primary amines that are suitable for library enumeration. Shown in Figure 12.4 are some 40 examples from the list that have survived the three-stage filtering process. This procedure was repeated for the other three reagent classes by running substructure queries against the ACD and then filtering the substructure hits accordingly. In addition to the 712 primary amines, 687 carboxylic acids, 320 aldehydes, and 10 isocyanides were identified as the building blocks for this Ugi combinatorial chemistry library. In the subsequent sections of this chapter, the enumeration and property calculation of a 1.6 billion compound library derived from this reagent set are reported.

12.2.3 Building Block Selection

In some circumstances, there may be logistical considerations or budgetary concerns that prevent the sourcing and use of all available building blocks. From a practical point of view, smaller focused libraries produced from only a subset of building blocks could offer substantial cost saving in term of synthesis, storage, and screening against therapeutic targets. Indeed, depending on the purpose and the format of library synthesis,

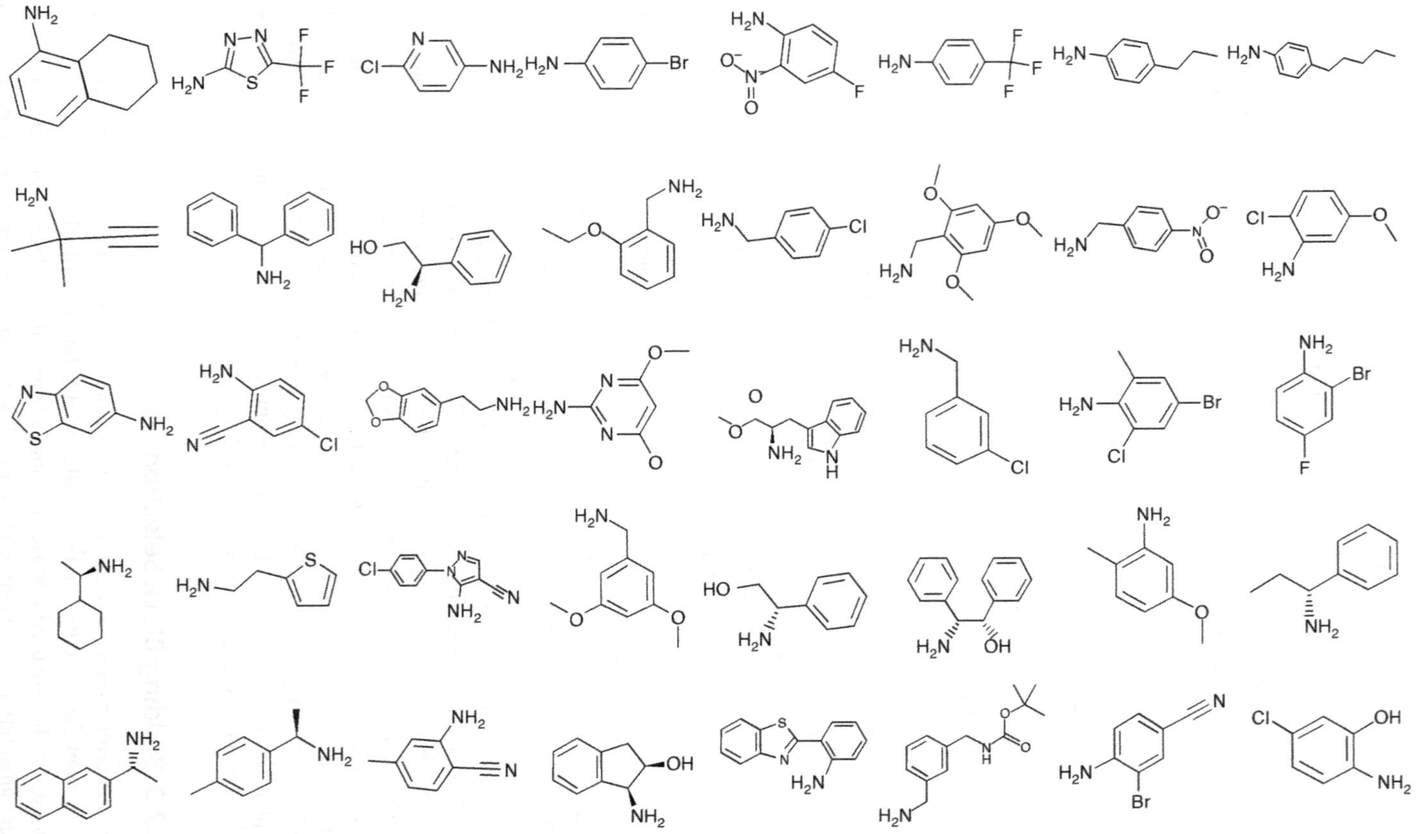

Figure 12.4. Examples of primary amine building blocks.

not all available building blocks should be included. For example, one may wish to eliminate building blocks that do not result in compounds containing a required pharmacophore feature (e.g., hydrogen donor/acceptor motif targeting protein kinases). If the goal is to produce general screening libraries that are rule-of-five compliant, then it would be beneficial to limit the number of highly lipophilic building blocks. Many powerful optimization algorithms originating from the field of computer science have also been adopted and deployed for optimal selection of building blocks. These included simulated annealing, Monte Carlo simulations, genetic algorithms, and Pareto sorting. The objective design functions vary according to the purpose of the focused library, with some examples including predicted activity from a QSAR model, molecular similarity against an active therapeutic ligand, or a docking score against a protein receptor.

One approach to select building blocks is the lockdown method proposed by Spellmeyer and coworkers [25]. The beauty of the lockdown method is that it is conceptually simple, easy to implement, and extremely efficient in execution. In this method, building blocks from different reagent classes are iteratively trimmed by strategically starting from the reagent class with the fewest candidates to ensure more efficient sampling. Through a series of random sampling of enumerated products, the subset of building blocks in the class that are associated with a minimum threshold of desirable compounds (e.g., matching a target molecular property profile or pharmacophore constraints) is retained. This subset of building blocks will be locked down as the preferred list in the class for the subsequent stages of sampling when the same procedure is applied to select another subset from other reagent classes. In the original publication, the authors demonstrated the use of this method to optimize for monomer selection of an Ugi library that maximized pharmacophore matching from a known thrombin inhibitor. Here, the use of this method to select a building block subset to optimally produce library products with no more than one Lipinski violations was demonstrated [18].

The result of the lockdown experiment on this current list of Ugi building blocks is summarized in Table 12.1. As discussed, the first lockdown stage was used to select building blocks from the reagent class with fewest candidates, which were the 10 isocyanides in this case. Random library products, 100,000 in number, were enumerated to sample how the 10 different diversity elements introduced by isocyanides could impact the overall physicochemical parameters of the final products. Only 19% of this initial sample of combinatorial library products passed with no more than one Lipinski violation. Four isocyanide building blocks produced at least 20%—the initial filter threshold—of their enumerated products, and they passed the desirability criteria. The four building blocks were subsequently locked down during the next stages of sampling. The procedure was reiterated to identify a subset of aldehydes (the class with second fewest number of building blocks) now using a reduced reagent set. Higher filter threshold values were set at the later stages so that the proportion of products passing the filtering criteria would steadily increase. After 10 lockdown stages, the number of building blocks for each reagent class was substantially reduced. The final selection of 290 amines, 201 acids, 41 aldehydes, and 4 isocyanides yielded a biased sublibrary containing approximately 10 million products, or about 0.6% of the size of the original library. This sublibrary was enumerated, and indeed, nearly all products $(9,481,172/9,559,560 = 99.2\%)$ had no more than one Lipinski violation. The lockdown

TABLE 12.1. Building blocks selection for the Ugi library using the lockdown approach

Stage	Class sampled	Amines	Acids	Aldehydes	Isocyanides	Possible products	Product sampled	Product passed	Filter threshold (%)
1	Isocyanides	712	687	320	10	1,565,260,800	100,000	18,848	20
2	Aldehydes	712	687	320	4	626,104,320	100,000	29,400	30
3	Acids	712	687	140	4	273,920,640	100,000	45,057	40
4	Amines	712	386	140	4	153,905,920	100,000	59,982	50
5	Aldehydes	468	386	140	4	101,162,880	100,000	73,778	60
6	Acids	468	386	127	4	91,769,184	100,000	75,636	70
7	Amines	468	234	127	4	55,632,096	100,000	84,125	80
8	Aldehydes	307	234	127	4	36,493,704	100,000	91,153	95
9	Acids	307	234	41	4	11,781,432	100,000	97,777	95
10	Amines	307	201	41	4	10,119,948	100,000	98,858	95
Final		290	201	41	4	9,559,560	9,559,560	9,481,172	

Successive stages of the lockdown process where the number of building blocks in each reagent class being sampled is shown. Product passed denotes the number of random enumerated products that have no more than one Lipinski violations. Initially, only 19% of the products from the full library would pass. One hundred thousand random samples were used at every stage to obtain statistics determining which building blocks (in the reagent class being sampled) have passed the minimum filter threshold set at that stage. Subsequently, this subset of building blocks in the reagent class would survive to the next lockdown stage. Note that the filter threshold criteria become increasingly more stringent at later stages to enforce more optimal subset selection. After 10 stages, a significantly reduced focused library was obtained, and more than 99% of its products have no more than one Lipinski violations.

procedure was computationally very efficient and the calculation required less than 10 min of CPU time for the current example.

As demonstrated, the lockdown procedure proves very useful for biased library design when an objective function is known. However, such knowledge is often not available, particularly when designing exploratory chemical libraries for hit identification purposes. To reduce the number of building blocks used in such a library, an alternative approach called molecular clustering can be applied to remove some building blocks that are already richly represented. In clustering, building blocks with similar chemical structures are grouped together to form a cluster. To exemplify each compound cluster, a representative structure (sometimes referred to as the cluster center) is picked from the group, while the remaining cluster members are discarded. A number of different clustering methods and fingerprinting types have been developed and reported, and interested readers should refer to several excellent reviews in this area [26–29].

To illustrate how clustering is applied to trim a list of building blocks, a Pharmacophore Graph (PG)-based method developed by DISCNGINE [30] was used to analyze and cluster the initial 712 primary amine data set. First, each amine was converted to an object called a PG using a decomposition algorithm that assigns a specific pharmacophoric feature to each of its constituent multiatom fragment units. Since multiple types of fragment units (which are structurally similar) can map to same pharmacophore feature, it follows that similar molecules can also be reduced to an identical PG. 167 unique PGs were found in this set of 712 amines, with the group of molecules that were converted to the same graph forming a compound cluster. One such cluster containing 11 amines is shown in Figure 12.5. The high pharmacophoric similarity shared by this set of compounds seems obvious here—all molecules have an amine functional group directly attached to a saturated atom of either an indane or tetrahydronaphthalene ring. In particular, the building block MFCD00003799 seems a good choice as a general representative for this cluster. This compound is a racemic mixture of two single enantiomer amines (MFCD00216669 and MFCD00216670) in the set, and it also shares close (sub)structural similarity with other group members. By selecting a single structure to represent each cluster, the size of the amine data set was significantly reduced from 712 to 167, leading to a fourfold reduction of the overall size of the combinatorial

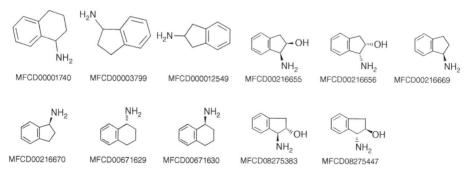

Figure 12.5. Structure of 11 amines grouped into the same cluster using a PG-based clustering method.

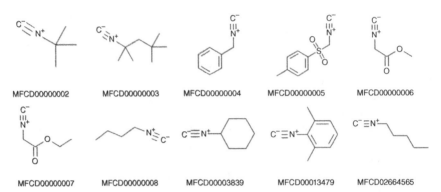

Figure 12.6. Structures of the 10 isocyanide building blocks used for this study. The PG-based algorithm groups MFCD00000008 and MFCD02664565 into the same cluster and perceives other eight building blocks as singletons. On the other hand, a human expert would likely expect an additional group containing the two isocyano-acetic acid esters (MFCD00000006 and MFCD00000007).

library. A similar factor of size reduction was also observed for the carboxylic acids (from 687 to 171) and also for the aldehydes (from 320 to 78) when this clustering method was applied. Interestingly, the majority of isocyanides (8 out of 10) in the set exist as singletons (i.e., they have a unique PG). Only two building blocks—1-isocyano-butane (MFCD00000008) and 1-isocyano-pentane (MFCD02664565)—were grouped into the same cluster (Fig. 12.6). This is an indication of either a greater level of structural diversity or paucity within this class of reagents. The isocyanide example also served as a useful demonstration that a fully automated clustering method does not always yield the type of result that is intuitive to a human expert. In addition to the 1-isocyano-butane/pentane pair, most medicinal chemists would also group isocyano-acetic acid methyl ester (MFCD00000006) and the corresponding ethyl ester (MFCD00000007) to the same cluster. However, the perception of the two molecules by computer is, in fact, different due to the algorithmic implementation for how PG are generated. The methyl ester is by itself a distinct pharmacophore unit, and any additional aliphatic group (i.e., the extra carbon in ethyl ester) attached is assigned to an additional aliphatic pharmacophore feature. This example is a reminder of the technical challenge in developing a tool that can produce a fully satisfactory clustering scheme.

In a similar manner to the use of the lockdown procedure, a molecular clustering method was also used for significantly trimming of the number of building blocks in a combinatorial library. The PG-based method used here led to a diversity-based selection of 167 amines, 171 acids, 78 aldehydes, and 9 isocyanides. An enumeration of this subset of building blocks yielded approximately 20 million products, which is approximately 1% of the original library. An interesting question arises: would this subset of compounds, which is derived from reduction of chemical space at reagent level, be able to approximate the great diversity of the overall product space? For example, can this sublibrary accurately reproduce the molecular property profile of the original library? In the next section, the enumeration of both the full 1.6 billion member library and this sublibrary will be described to determine if there is a significant difference in the distributions of some key molecular properties.

12.2.4 Enumeration and Property Profiling

Enumeration and analysis of massive combinatorial libraries has been an active area of research in combinatorial library design. In spite of the rapid advancement in computational technologies, a library size of about one trillion (10^{12}) molecules remains a practical upper limit for explicit structure enumeration and property calculation today. In addition to the need for computing processing resources, storage requirements can quickly become an issue. A molecular data set containing 1 trillion molecules requires more than 20 terabytes of storage space even at a modest cost of 25 bytes per structure. To circumvent explicit enumeration, alternative approaches to evaluate the property profile of large combinatorial libraries have been developed by leveraging statistical sampling methods or even machine learning algorithms. Spellmeyer and coworkers published a binomial formula derived from statistical theory to estimate errors from quantities obtained from random sampling of a small number of library products [25]. Agrafiotis and Lobanov developed a novel approach called combinatorial neural networks, which is based upon the finding that most molecular descriptors that are commonly used in library design can be accurately predicted from the properties of their respective building blocks [31]. They were able to train a computational neural network to predict the coordinates of the enumerated products on a chemical space directly from properties of building blocks as inputs, circumventing expensive explicit structure enumeration steps.

In this section, the Ugi reaction has been used as an example for library enumeration and property calculation. A very simple method—random sampling on combinatorial products—was applied to estimate the statistical mean of several molecular properties. These properties include MW, calculated partition coefficient between octanol and water ($AlogP$), and Polar Surface Area (PSA). It should be noted that the average value of a "decomposable" property such as MW is readily calculable without enumeration—it is simply the average MW of reactants subtracted by the MW of the elements lost in the reaction [25]. However, properties such as $AlogP$ or PSA are not easily derived from simple addition of fragment contributions from reagents alone. The value of the property often depends on molecular connectivity, which may be modified by formation of new bonds during a reaction.

With currently available computing power, it is feasible to enumerate and to compute molecular properties of a virtual combinatorial library with the size of 10^9. Several standard cheminformatics components available in the Pipeline Pilot program were used to enumerate the Ugi library products and calculate the properties. The structure enumeration was performed using an "Enumerate Combinatorial Reaction" component, and a combinatorial reaction scheme was encoded using Daylight SMIRKS reaction transform language [32]. For each molecule, a canonical SMILES string was obtained and then used to calculate four molecular properties: MW, $AlogP$, PSA, and the number of Lipinski violations (NumLipinski). The data generated were stored as a text file in a compressed data format. In total, 1,565,260,800 library products were enumerated using the full set of filtered reagents: 712 amines, 10 isocyanides, 320 aldehydes, and 687 acids. The calculation, including both the enumeration and property calculation, required approximately 8 days on a single core of an Intel Xeon CPU (E7340 @ 2.4 GHz); approximately 30 gigabytes of data were generated. This translated to a resource requirement of 7 min of CPU time and 19 megabytes of disk storage for every 10^6 molecules. It should also be pointed out that library enumeration is a class of computational problem that can be handled with an "embarrassingly parallel" workload, where very little effort is required to distribute the calculation to multiple computer processors to gain substantial speedup. In the present case, the calculation could be readily split to 10 parallel threads, with each thread handling the enumeration of 1 of the 10 isocyanides, and the calculation could be done in less than a day in real time.

From the enumeration and property calculation of the full 1.6 billion compound library, an exact solution was obtained for the mean value of each molecular property. For the entire library, the means are as follows: MW = 612.72, $AlogP$ = 6.119, and PSA = 118.24. With this information, it is possible to determine how the accuracy of mean estimation from random sampling is affected by parameters such as sampling size. This experiment was begun with a small sample size of 16 random enumerated products, which corresponds to 10^{-6}% of the library. The mean of MW, $AlogP$, and PSA were estimated from these samples; this procedure was repeated 10 times to understand the variance within individual estimations. Sampling sizes belonging to 7 categories of sampling size, ranging from 10^{-6}% to 1% (i.e., 16–16 million samples), were investigated. The result of this sampling calculation is summarized in Table 12.2 and Figure 12.7a–c. The results from different percentage categories are grouped along the horizontal axis. The property mean computed from individual runs is plotted as black circles within each percentage group. The horizontal line in the middle of the "mean diamond" depicts the group mean, and the vertical apex provides the 95% confidence interval for the 10 runs in each group. The statistical mean for each molecular property, which was computed from the full 1.6 billion compounds library, is also shown on the plot as a dotted line. As expected, sampling accuracy steadily increases with higher sampling sizes along with lower variance among individual estimates. Consistent with the earlier reports [25, 33], this simulation suggests that statistical parameters such as mean value of a molecular property can be accurately determined using a small fraction of random enumeration products. For example, the result of sampling the 0.001% group (i.e., only 15,600 random samples) yielded mean estimates of MW = 612.86±0.71, $AlogP$ = 6.119±0.015, and PSA = 118.14±0.30. These estimates are already accurate to three significant figures when compared to the exact solutions. Clearly, enumeration of all library products for this purpose would be unnecessary and wasteful in this case.

In addition to estimation of a single measure such as statistical means, it was also interesting to see whether the sampling of a small portion of the library would accurately reproduce property distributions such as a histogram profile. In the lower graph of Figure 12.8a, a histogram shows the percentage of compounds with various numbers of Lipinski violations in the full library (100% group, shown with black vertical bars) and compared with the corresponding percentages that were derived from just 15,600 random samples (0.001% group, shown with light grey bars). The two histogram profiles are virtually indistinguishable from each other. In the upper graph, the histogram profile derived from a partial library enumeration (subset group, shown with dark grey bars) based on a subset of building blocks previously selected by PG-based clustering method is shown. Interestingly, the profile of this library subset is noticeably different from the full library, despite the fact that this sublibrary contains nearly 20

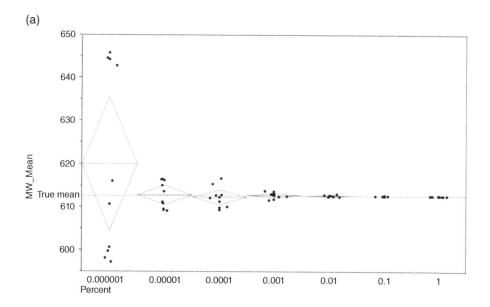

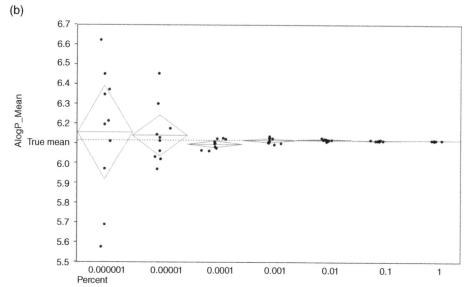

Figure 12.7. (a) Estimation of MW mean from random sampling of Ugi library products. (b) Estimation of AlogP mean from random sampling of Ugi library products.

(c)

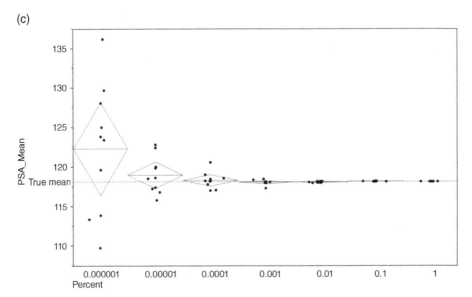

Figure 12.7. (*Continued*) (c) Estimation of PSA mean from random sampling of Ugi library products.

TABLE 12.2. Summary of the sampling runs using a small percentage (0.000001% to 1%) of enumerated library products to estimate various molecular properties

Run	MW	AlogP	PSA
0.000001	620.14±21.77	6.159±0.330	122.38±0.49
0.00001	612.91±3.11	6.144±0.144	119.04±0.17
0.0001	612.37±2.35	6.101±0.024	118.41±0.03
0.001	612.86±0.71	6.119±0.015	118.14±0.02
0.01	612.75±0.19	6.122±0.005	118.19±0.01
0.1	612.74±0.05	6.119±0.002	118.26±0.00
1	612.72±0.01	6.119±0.001	118.24±0.00
Full library	612.72	6.119	118.24
PG subset	639.01	5.927	132.89

The mean and standard deviation (from 10 sampling calculations) of MW, AlogP, and PSA is reported. Full library indicates the exact solution obtained from full enumeration of the 1.6 billion compound library. The PG subset indicates the property average derived from a 20 million compound library subset enumerated using a diverse set of reagents obtained from the PG-based clustering method.

million samples, which are already more than 1% of the full library. A similar result is also found for the histograms for MW (Fig. 12.8b). The MW profile from 0.001% random sampling in product space (light grey bars) is very similar to the full library (black bars), whereas the library subset derived from reagent clustering (dark grey bars) has a distribution that is right shifted relative to the full library. This is not a surprising result. Reagent-based diversity library design is a useful strategy to simplify the logistics of library production and to minimize cost. However, there is no reason to expect that a diverse set of reagents would necessarily lead to an optimally diverse set of products. There is already overwhelming evidence suggesting that a random selection based on products themselves can be substantially more diverse, perhaps by as much as 35–50% comparing with reagent-based selection [33–36]. In the current case, the estimated property means for MW, $A\log P$, and PSA are 639.01, 5.927, and 132.89,

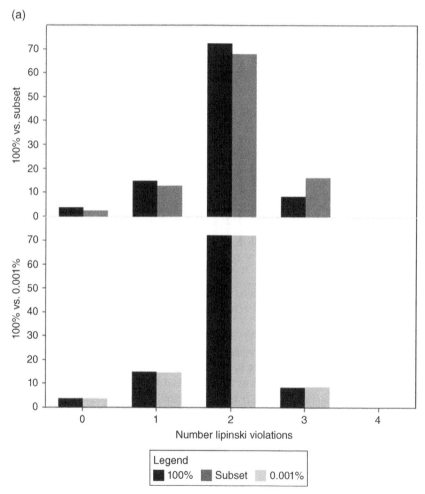

Figure 12.8. (a) Histogram showing distribution of Ugi library products according to the number of Lipinski violations. The lower plot shows a comparison of histograms derived from the full 1.6 billion compound data set (100%, black) versus 15,600 random samples (0.001%, light grey). The upper plot shows a comparison of histograms derived from the full data set (100%, black) versus a partial enumeration of the library (subset, dark grey) of 20 million compounds based on a subset of building blocks selected from a PG-based clustering method.

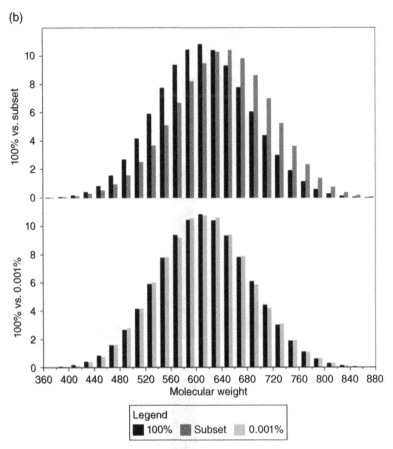

Figure 12.8. (*Continued*) (b) Histogram showing distribution of Ugi library products according to the MW. The lower plot shows a comparison of histograms derived from the full 1.6 billion compound data set (100%, black) versus 15,600 random samples (0.001%, light grey). The upper plot shows a comparison of histograms derived from the full data set (100%, black) versus a partial enumeration of the library (subset, dark grey) of 20 million compounds based on a subset of building blocks selected from a PG-based clustering method.

which deviate significantly from the grand means (612.72, 6.119, and 118.24) of the full library (Table 12.2).

12.3 CHEMICAL SPACE COMPARISON

The notion of chemical space is a concept in which chemists have a long-standing interest [37–42]. Different chemistry-space metrics have been developed to better characterize the broad range of chemical diversity in this extremely vast space [37, 40]. One central question is whether it is possible to compare the chemical space occupied by

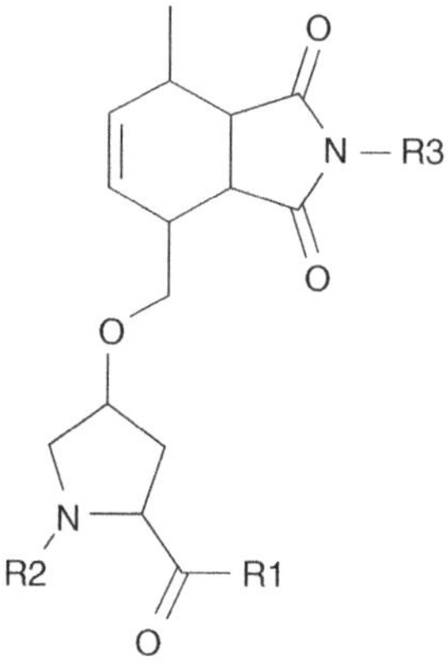

Figure 12.9. General scaffold for the DNA-encoded chemical library from Neri and coworkers. R1, R2, and R3 are the three diversity elements. A DNA tag is attached as part of R1 group.

different chemical libraries. To investigate this, three large compound libraries were used here and compared with the Ugi library enumerated in the previous section. The first library is the PubChem database [43], which is maintained by the National Center for Biotechnology Information (NCBI) and is a comprehensive source of chemical structures from the scientific literature. Currently, the database contains more than 30 million unique and chemically diverse compounds. The second library is the GDB-11 database, a (mostly) virtual library of small fragment structures generated by Reymond and coworkers [44]. This database contains about 26 million compounds that represent an exhaustive enumeration of small organic molecules containing five element types: C, N, O, F, and H, up to a total of 11 nonhydrogen atoms. A few simple chemical stability and synthetic feasibility rules were enforced during enumeration so that legitimate and stable organic molecules were produced. In essence, this data set represents the chemical space of small fragments, and similar to the PubChem database, it is very chemically diverse. The third library is a one million compound DNA-encoded chemical library published by Neri and coworkers [45]. In this third library, the key reaction step is a Diels–Alder cycloaddition between building blocks containing a hexadiene group and maleimides as dienophiles, leading to a 5,6-*cis*-fused bicyclic general scaffold with diversity elements shown in Figure 12.9. For the purpose of structure enumeration, the DNA tag that is attached to the R1 diversity element was replaced by a methyl group (i.e., *N*-methyl amide is the terminating group).

12.3.1 Comparison of Exact Structures

One way to compare chemical space is to identify the overlap of exact chemical entities between two libraries. Interestingly, comparison of exact structures is a relatively easy and computationally efficient process even between two extremely large libraries. For example, one can simply create a relational database with two data tables, each containing a text field of canonical SMILES strings generated from compounds in standardized

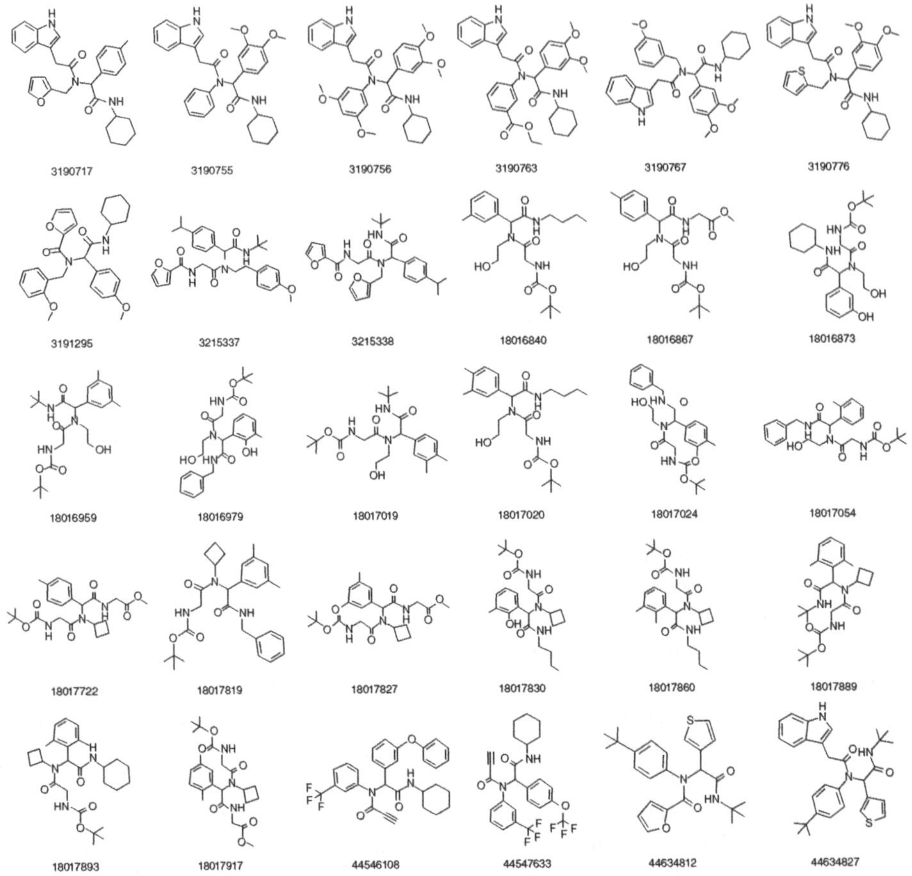

Figure 12.10. Examples of Ugi products from the current library that are found in the PubChem database. The label below each structure is the PubChem identifier.

tautomeric form from the library. Molecules that are common to both libraries can be quickly retrieved with an SQL join operation using the SMILES string as a key. Even for large data tables, this type of operation is very fast because the search can be indexed. The comparison of exact structures was done between the 1.6 billion compound Ugi library and the 30 million compound PubChem database. It is surprising that, given that the one-pot Ugi chemistry was published decades ago and the reaction has also been a benchmark for computational combinatorial library analysis, only 323 compounds from the current Ugi enumeration are known in PubChem. Selected examples of 30 such compounds are shown in Figure 12.10. It is intriguing to note, at least from this set of examples, the presence of some functional groups that are considered less common. For example, about one-quarter of compounds in this set contain the cyclobutane functionality (compared to the presence in less than 0.3% of compounds in PubChem overall). The prevalence of the *tert*-butyl functional group in this set of compounds is easy to comprehend. Many of them are synthetic intermediates derived from commercial Boc-protected amino acids.

The overlap of content between the GDB-11 and the PubChem databases was also analyzed. Compared to the Ugi library, the extent of exact structure matches for GDB-11 with PubChem is significantly higher. Approximately 100,000 GDB-11 molecules—but still only 0.4% of the library—have been synthesized and are known in the PubChem database. This fact gives evidence to the vastness and diversity of chemistry space. Even after more than a century of organic synthesis, only a small fraction of compounds belonging to this seemingly restricted chemical space of fragment-size molecules has been created and characterized. Finally, when comparing the structures from the Neri library and the PubChem database, it is not surprising to see that there was no structure match between the two as this library contains an uncommon general scaffold that is unlikely to be found in PubChem.

12.3.2 Chemical Space Heat Map

Clearly, chemical space is so huge that exact matching of chemical structures between even two very large libraries is rare. Different strategies to reduce molecular representation have been proposed, an interesting example of which is the 42-dimensional chemical space defined by Molecular Quantum Number (MQN) descriptors by Reymond and coworkers [40]. These are a set of integer-value descriptors encoding the counts of specific atomic, bond-type, or topological features (e.g., presence of three-membered rings) of a molecule. Molecules with identical MQN values would be regarded as "molecular isomers" in analogy to atoms with the same atomic number (isotopes). However, because of its high dimensionality, the MQN space itself is likely to be sparsely occupied, which makes comparison of chemical libraries challenging. In contrast to a chemical space encoded by intrinsic structural features, a small set of (ideally orthogonal) physicochemical properties (e.g., PSA) or pharmacophoric features (e.g., the number of aromatic hydrophobes) can be computed and used as coordinates of a low-dimensional property-based chemical space. Obviously, this type of chemical space is quite coarse and lacks any detailed description of the molecules. In addition, the space is likely to be highly degenerate as many different chemotypes can map to the same space by virtue of the similarities in molecular properties or pharmacophoric features that they share. To address such shortcomings, the BCUT metric, a set of descriptors that have been widely used to generate low-dimensional chemical space, was developed. The BCUT descriptors, which are related to the Burden index [46], were proposed and put into practical application by Pearlman and coworkers [37]. In essence, BCUT descriptors are eigenvalues of a modified adjacency matrix primarily encoding molecular connectivity information. Different types of BCUT descriptors result when atomic properties such as atomic charge, hydrogen bonding propensity, or polarizability are incorporated as the diagonal matrix elements. As a result, both chemical structural features and molecular property attributes of a molecule can be captured by just a few BCUT descriptors. One can then create an approximately orthogonal and low-dimensional chemical space by optimally choosing two or three descriptors that best distinguish the structural (or activity) differences between the compounds for the purpose of visualization and analysis. Alternatively, one can deploy dimensionality reduction methods to transform the full descriptor space to a lower dimension. Most commonly, this is done using Principal Component Analysis (PCA), multidimensional scaling, nonlinear mapping, or Kohonen self-organizing maps [47]. This way, the areas of space that are occupied by compound sets with different characteristics (e.g., drug vs. nondrug, fragments vs. peptides) can be visualized and examined on a two-dimensional plot. An example of this type of approach to visualize and navigate chemical space is the Chemical Global Positioning System (ChemGPS) method proposed by Oprea and Gottfries [48, 49]. In their method, 423 small molecules are selected as spatial reference of a drug-like universe, and "chemographic" map coordinates are extracted from PCA of a standard fixed set of molecular descriptors. Because of the invariant nature of this reference system, this method is well suited for comparing multiple chemical libraries.

As an illustration, the BCUT metric was applied and followed by PCA to compare the chemical space occupied by the Ugi, Neri, PubChem, and GDB-11 data sets. For this analysis, the 30 million PubChem compounds were chosen as a reference chemical space due to their great chemical diversity. For each molecule, six BCUT descriptors, which were the highest and the lowest eigenvalues of the three adjacency matrices modified by atomic charge, hydrogen bond, and polarizability, were computed. A PCA was performed on this large data matrix (30 million by 6). The first (PC1) and second principal component (PC2) explained 27% and 20% of the variance of the BCUT metrics. The coordinates projected from the two principal components were used as a new low-dimensional chemical space for all data sets. For this calculation, the BCUT descriptors were obtained using a ChemAxon Calculator component in Pipeline Pilot. The PCA and the generation of heat maps were performed using standard components.

To further simplify visualization, a cell-based data aggregation approach was adopted. First, a two-dimensional grid was created with 500 equally divided bins along each principal component axis. For each principal component, a range of ±5 standard deviations from the mean was defined as relevant chemical space since nearly all compounds have descriptor values within this range. Each compound in the PubChem database was assigned to one of $500 \times 500 = 250,000$ individual cells in the grid based on the binning of PC1 and PC2 values. Finally, the total number of compounds that were mapped to each cell on the grid was counted. A key benefit of such a data aggregation procedure is the transformation of a large array (30 million by 6) of floating-point numbers to a more compact matrix containing only integer elements. More importantly, the size of the matrix depends only on a user-defined grid solution and is entirely independent of library size. Because of this attribute, the chemical space of any library can be summarized as a heat map in a consistent format and visually compared. In the heat map, the two principal components lie along the horizontal and vertical axes, and the population of each individual cell is depicted using color shading, where a darker color indicates a higher compound population in the cell. Figure 12.11a shows a heat map of the 30 million PubChem compounds in the BCUT-derived PC1/PC2 space. The highest compound density is found in the cells close to the center (i.e., within ±1 standard deviation from the mean) as evident by dark color shading in the middle of the plot. Figure 12.11b is a histogram showing the distribution of cells according to different population ranges (in log order of magnitude). Despite a fine grid resolution, fewer than one-third of the cells are empty, and nearly half of the cells contain at least 10 compounds. Only a small percentage (0.7%) of the cells have very high compound density with greater than 10^4 compounds.

(a)

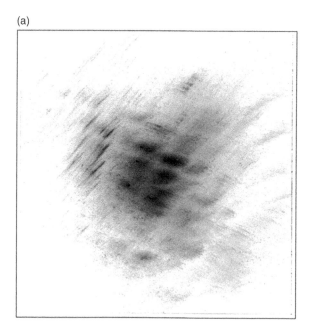

(b)

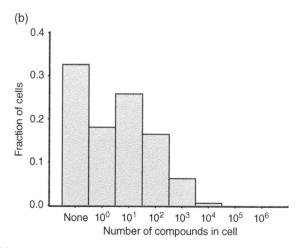

Figure 12.11. (a) Heat map showing a representation of the chemical space of the PubChem library. (b) Distribution of cells according to different population ranges for the PubChem database.

Overall, the excellent coverage of compounds across the cells along the two principal axes makes this a good reference chemical space for the purpose of library comparison.

Based on the principal component parameters obtained from the PubChem data set, another heat map (that has the same horizontal and vertical scales as Fig. 12.11a) was generated using a random selection of 30 million Ugi products (Fig. 12.12a). Not surprisingly, the chemical space of the Ugi library compounds is considerably more well defined when compared to the structurally diverse PubChem database. This also reflects a markedly different population distribution (shown in Fig. 12.12b). More than 80% of

(a)

(b)

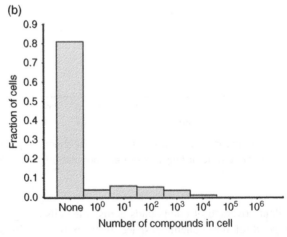

Figure 12.12. (a) Heat map showing a representation of the chemical space of the Ugi library. (b) Distribution of cells according to different population ranges for the Ugi library.

the cells are unoccupied, approximately 5% of the cells contain compounds in the 10^3 range, and a few cells contain more than 10^5 compounds.

Projection of the 23 million GDB-11 compounds onto the PubChem reference chemical space shows that these small molecular fragments occupy primarily the upper right quadrant (Fig. 12.13a), an area distinctly different from the location of the Ugi library. In fact, a significant portion of compounds from GDB-11 occupy a region of space that is sparsely represented by the PubChem compounds. Figure 12.13b shows a different heat map where this "virtual" chemical space (i.e., upper middle part of the plot) unique to the GDB-11 compounds is characterized. About 60% of the grid cells are

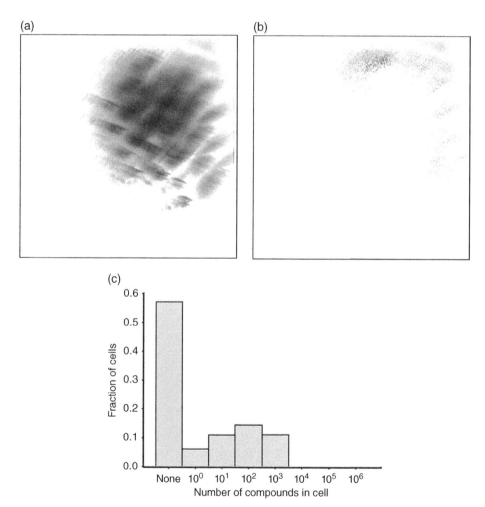

Figure 12.13. (a) Heat map showing a representation of the chemical space of the GDB-11 database. (b) Heat map highlighting the areas of chemical space that are represented by molecules in the GDB-11 database but not by any compound in the PubChem database. (c) Distribution of cells according to different population ranges for the PubChem database.

empty, but unlike either PubChem or Ugi libraries, the GDB-11 population is distributed quite evenly across the rest of the chemical space, and none of the cells have population density greater than 10^3 molecules (Fig. 12.13c).

Figure 12.14a shows the heat map that corresponds to the chemical space of the Neri library. In general, the location of chemical space occupied by the Neri library seems quite similar to the Ugi library (Fig. 12.14a). But due to very high structural similarity among its members, the associated chemical space seems quite confined. This

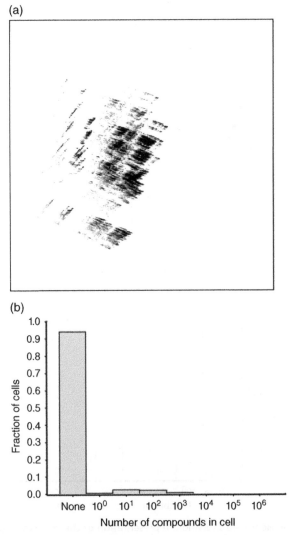

Figure 12.14. (a) Heat map showing a representation of the chemical space of the Neri DNA-encoded chemical library. (b) Distribution of cells according to different population ranges for the Neri DNA-encoded chemical library.

finding is also consistent with the fact that only 6% of the grids are occupied, and also a significant proportion of its population resides inside just a handful of cells.

12.3.3 Library Similarity Calculation

The heat maps shown in the previous section represent a useful qualitative tool that provides comparative visualization of how different compound libraries occupy distinct regions of chemical space. A method that yields a quantitative measure of library overlap using molecular similarity concept can provide complementary information. The Tanimoto coefficient is arguably the most commonly used formula (Eq. 12.1) for molecular similarity calculation, and these coefficients are readily computed based on the comparison of molecular fingerprints between a pair of molecules:

$$\text{Tanimoto}\,(A,B) = \frac{N_{AB}}{N_A + N_B - N_{AB}} \tag{12.1}$$

where

N_A and N_B are the number of bits in the fingerprint set in molecule A and molecule B
N_{AB} is the number of bits set common to both molecules

Tanimoto coefficient ranges between 0 and 1, where a higher value indicates a molecular pair that is structurally more similar. Here, the use of this concept is extended, and a mean Tanimoto coefficient value is derived from similarity calculations of all molecule pairs between two compound libraries. This new parameter is referred to as "library similarity," and the following simple worked example is used to illustrate the principle (Fig. 12.15).

In this example, there are three "libraries" of compounds, and the goal is to obtain a quantitative measure to determine whether Library 2 or Library 3 is structurally more

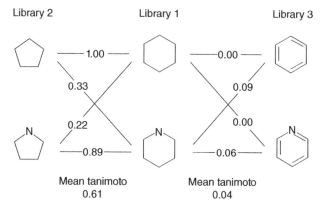

Figure 12.15. A simple example showing the calculation of mean Tanimoto coefficient (or library similarity) between library sets. The result indicates that Library 1 is structurally more similar to Library 2 according to library similarity calculation.

similar to Library 1. Library 1 contains cyclohexane and piperidine, Library 2 contains cyclopentane and pyrrolidine, and Library 3 contains benzene and pyridine. To compare Library 1 and Library 2, the Tanimoto coefficients of all compound combinations between the two library sets are computed based on a user-selected molecular fingerprint (FCFP_4 fingerprints in this example). This resulted in a 2×2 similarity matrix. The four individual Tanimoto coefficients in the matrix were averaged to give a library similarity value of 0.61 that indicates a relatively high degree of similarity between these two libraries. It seems reasonable as both set of compounds are cyclic aliphatic molecules. Repeating the procedure to compare Library 1 and Library 3 yielded a much lower library similarity (0.04), a reflection of the obvious difference in aromaticity between the two sets of compounds. This example illustrates that this type of calculation is quite easy to set up but can quickly become prohibitively expensive when comparing large numbers of input compounds because calculation of pairwise similarity is an $O(N^2)$ process. For example, it is estimated that the computation of a full similarity matrix between the 1.6 billion compound Ugi library and the 30 million compound PubChem database would require 30,000 years on a Xeon CPU (E7340 @ 2.4 GHz). So, a strategy to obtain an exact result of the full similarity matrix by using a brute-force approach (e.g., massive parallelization with a large array of computer processors) seems intractable and, most likely, wasteful. Instead, one would attempt to estimate the library similarity value between two compound libraries using small, multiple sets of random samples taken from each library. To test the effectiveness of this sampling approach, a subset of 50,000 compounds from the Ugi library was randomly selected and along with a subset of 50,000 compounds from the PubChem database. It was possible to determine an exact solution (0.1822) for the library similarity between the two library subsets by performing 2.5 billion similarity calculations, which took 13 CPU hours. Next, a series of sampling calculations was run on the subsets by varying both (i) the sample size (N_S) used for the similarity calculations and (ii) the number of repeated sampling runs (N_R) so that different sets of random samples can be drawn from the two 50,000 compound subsets. Table 12.3 shows the estimates of library similarity for a range of N_S and N_R parameters. First, it is remarkable that nearly all of the runs yielded very good estimates with the exception of those involving very few samples (i.e., N_S or $N_R \leq 5$). The N_S or N_R combinations shown in grey cells in the table produced estimates that were accurate to three significant figures to the exact result (0.1822). It is also interesting to compare the following pairs of result that are derived from same number of individual similarity calculations. For example, a 10×10 matrix repeated 100 times (library similarity estimate = 0.1832) and a 100×100 matrix done just once (0.1849) both took 10,000 similarity calculations, a 10×10 matrix repeated 1,000 times (0.1824) and a 100×100 matrix repeated 10 times (0.1830) both took 100,000 calculations, and a 50×50 matrix repeated 1,000 times (0.1824) and a 500×500 matrix repeated 10 times (0.1832) both took 2,500,000 calculations. In all cases, the observed trend strongly suggests that, given the same number of similarity calculations, a more rapid convergence of library similarity can be achieved by a higher number of repeated runs at the expense of smaller sample size drawn from the libraries.

To investigate the convergence behavior, the sampling calculation for the 10×10 matrix set was continued in order to draw additional random samples until the simulation reached 100,000 iterations. The value for the library similarity estimate obtained at each iteration step was tracked plotted against the number of iterations taken (Fig. 12.16).

TABLE 12.3. Variation of sampling accuracy according to sampling size of the similarity matrix ($N_S \times N_S$) and also the number of repeated sampling runs (N_R)

Matrix size ($N_S \times N_S$)	Number of sampling runs (N_R)								
	1000	500	250	100	50	25	10	5	1
1000×1000	0.1822	0.1822	0.1823	0.1824	0.1823	0.1822	0.1824	0.1832	0.1830
500×500	0.1822	0.1823	0.1823	0.1823	0.1823	0.1824	0.1832	0.1830	0.1850
250×250	0.1823	0.1823	0.1823	0.1822	0.1824	0.1828	0.1830	0.1840	0.1850
100×100	0.1825	0.1824	0.1823	0.1825	0.1832	0.1830	0.1830	0.1850	0.1849
50×50	0.1824	0.1823	0.1825	0.1834	0.1832	0.1842	0.1852	0.1849	0.1820
25×25	0.1823	0.1824	0.1829	0.1832	0.1842	0.1840	0.1851	0.1854	0.1656
10×10	0.1824	0.1833	0.1832	0.1832	0.1847	0.1841	0.1839	0.1810	0.1795
5×5	0.1835	0.1835	0.1843	0.1854	0.1844	0.1855	0.1826	0.1658	0.1858
1×1	0.1838	0.1872	0.1865	0.1871	0.1807	0.1622	0.1787	0.1679	0.1343

The exact solution of library similarity between the two library subsets is 0.1822.

273

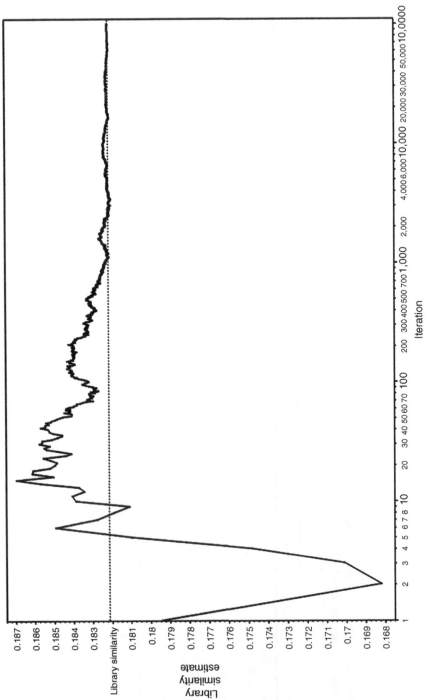

Figure 12.16. Estimate of library similarity as a function of iteration steps between the Ugi and the PubChem library subsets using a 10×10 sampling matrix. The exact solution for library similarity (0.1822) is represented by the horizontal dotted line. The horizontal axis of the figure is shown in log scale to highlight convergence behavior at the early stages of the calculation. A sampling accuracy to two, three, and four significant figures was reached after 100, 1,000, and 50,000 iterations, respectively.

The dotted line shows the exact solution of library similarity calculated from the full matrix. As expected, the estimate for library similarity fluctuated quite significantly at the early phase due to insufficient sampling of data. It was very encouraging that after only 100 iterations, the sampling calculation has already produced an estimate that is already accurate to two significant figures. At the 1000th iteration, sampling accuracy to three significant figures—a level of accuracy that is probably not needed for most practical purposes—was reached. In an actual application, one can define their desired level of sampling accuracy and terminate the simulation once this condition of interest is met. For this current case, the total number of similarity calculations that had been performed at this stage is $10 \times 10 \times 1,000 = 100,000$, a very small number (about 0.0004%) comparing to the 2.5 billion individual calculations needed for the full matrix calculation.

After this initial validation on smaller subsets, the same sampling protocol was used to estimate library similarity values between the Ugi, Neri, PubChem, and GDB-11 libraries. For the full runs, 100 random samples were taken from each library at a time, and the similarity matrix calculation was repeated 50,000 times (i.e., $100 \times 100 \times 50,000 = 500$ million similarity pairs). Based on this test run, this should provide sufficient sampling to generate reasonable library similarity estimates, and further sampling would yield only marginal improvement. Each library comparison calculation took approximately 4 CPU hours to complete.

The library similarity values from these pairwise library comparisons are shown in Table 12.4. The low values obtained from the comparisons between the various libraries against GDB-11 confirm earlier qualitative assessment from the PCA-based heat maps. The GDB-11 seems to be the most unique of the four libraries as these molecular fragments were occupying a distinct region of chemical space. Not surprisingly, the Neri library has the highest intralibrary similarity (0.457) of the four libraries. This is due to a high level of structure homogeneity originating from a unique scaffold that is present in every compound. The Neri library seems less structurally similar (library similarity = 0.170) to PubChem in general than the Ugi library to PubChem (library similarity = 0.182). This difference in library similarity value is significant from a computational library design point of view. For example, if the purpose of library building is to produce chemical moieties that are generally more novel than compounds that currently exist, one may want to prioritize the production of the Neri library over the Ugi library since most of its library members are considered more unique and chemically differentiated than those found in PubChem.

TABLE 12.4. The library similarity values from the comparison of four large chemical libraries (Ugi, Neri, PubChem, and GDB-11) used in this study

	Ugi	Neri	PubChem	GDB-11
Ugi	0.289 ± 0.006	0.216 ± 0.003	0.182 ± 0.004	0.099 ± 0.002
Neri		0.457 ± 0.007	0.170 ± 0.004	0.108 ± 0.003
PubChem			0.176 ± 0.006	0.099 ± 0.003
GDB-11				0.147 ± 0.005

The diagonal values are the intralibrary similarity values that reflect compound heterogeneity within a given library. The off-diagonal elements are the interlibrary similarity values between two different compound libraries.

12.4 SUMMARY

A major lesson learned from early days of combinatorial library synthesis and screening is that without proper library design, many chemical libraries either failed to improve hit rate against therapeutic targets or the poor drug-like characteristics of resulting hits had discouraged further medicinal chemistry follow-up [12, 50, 51]. Clearly, the ability to make millions or even billions of compounds is not sufficient to ensure success; a good library design is still a critical factor. In this chapter, a practical overview for reagent selection, enumeration, property profiling, and library comparison of some very large combinatorial libraries was shown. Cheminformatics workflow packages offer many useful tools for both the selection and filtering of building blocks. SMARTS-based queries are simple to implement and can quickly identify very specific substructures that are reaction compatible. Enumeration of combinatorial library with 10^9 members can now be done on a routine basis, although libraries with greater than 10^{12} members still pose technical challenges. Also demonstrated was that the mean and statistical distribution of several key properties commonly used in molecular design can be accurately estimated using a small fraction of random library products. These simulations suggested that random sampling on product space is deemed much more effective than on partially enumerated library products based on reagent-based sampling.

Heat maps derived from the first two principal components of six BCUT metrics were used to compare the chemical space of different libraries, using the PubChem data set as a spatial reference. The molecular similarity concept can be extended to provide a quantitative parameter, library similarity, to estimate overlap between two chemical libraries. This author suggests that library similarity with other libraries is a factor to be considered in the prioritization and production of large combinatorial chemical libraries.

ACKNOWLEDGMENTS

The author gratefully acknowledges the careful review of the manuscript and helpful comments by Drs. Robert Goodnow and Paul Gillespie.

REFERENCES

1. Kennedy, J. P., Williams, L., Bridges, T. M., Daniels, R. N., Weaver, D., Lindsley, C. W. (2008). Application of combinatorial chemistry science on modern drug discovery. *J. Comb. Chem.*, 10, 345–354.
2. Martin, E. J., Blaney, J. M., Siani, M. A., Spellmeyer, D. C., Wong, A. K., Moos, W. H. (1995). Measuring diversity: experimental design of combinatorial libraries for drug discovery. *J. Med. Chem.*, 38, 1431–1436.
3. Holliday, J. D., Ranade, S. S., Willett, P. (1995). A fast algorithm for selecting sets of dissimilar molecules from large chemical databases. *Quant. Struct.–Activity Relationships*, 14, 501–506.
4. Bender, A., Glen, R. C. (2004). Molecular similarity: a key technique in molecular informatics. *Org. Biomol. Chem.*, 2, 3204–3218.

5. Maldonado, A. G., Doucet, J. P., Petitjean, M., Fan, B.-T. (2006). Molecular similarity and diversity in chemoinformatics: from theory to applications. *Mol. Diversity*, 10, 39–79.

6. Gillet, V. J. (2011). Diversity selection algorithms. *WIREs Comput. Mol. Sci.*, 1, 580–589.

7. Ashton, M. J., Jaye, M. C., Mason, J. S. (1996). New perspectives in lead generation. II Evaluating molecular diversity. *Drug Discov. Today*, 1, 71–78.

8. Lewis, R. A., Mason, J. S., McLay, I. M. (1997). Similarity measures for rational set selection and analysis of combinatorial libraries: the diverse property-derived (DPD) approach. *J. Chem. Inf. Comput. Sci.*, 37, 599–614.

9. Flower, D. R. (1998). DISSIM: a program for the analysis of chemical diversity. *J. Mol. Graph. Model.*, 16, 239–253.

10. Lobanov, V. S., Agrafiotis, D. K. (2002). Scalable methods for the construction and analysis of virtual combinatorial libraries. *Comb. Chem. High Throughput Screen*, 5, 167–178.

11. Gillet, V. J. (2008). New directions in library design and analysis. *Curr. Opin. Chem. Biol.*, 12, 372–378.

12. Gillet, V. J., Khatib, W., Willett, P., Fleming, P. J., Green, D. V. (2002). Combinatorial library design using a multiobjective genetic algorithm. *J. Chem. Inf. Comput. Sci.*, 42, 375–385.

13. Jamois, E. A., Lin, C. T., Waldman, M. (2003). Design of focused and restrained subsets from extremely large virtual libraries. *J. Mol. Graph. Model.*, 22, 141–149.

14. Schneider, G., Schueller, A. (2009). Adaptive combinatorial design of focused compound libraries. *Methods Mol. Biol.*, 572, 135–147.

15. Yu, N., Bakken, G. A. (2009). Efficient exploration of large combinatorial chemistry spaces by monomer-based similarity searching. *J. Chem. Inf. Model.*, 49, 745–755.

16. Sciabola, S., Stanton, R. V., Johnson, T. L., Xi, H. (2011). Application of Free-Wilson selectivity analysis for combinatorial library design. *Methods Mol. Biol.*, 685, 91–109.

17. Truchon, J.-F. (2011). GLARE: a tool for product-oriented design of combinatorial libraries. *Methods Mol. Biol.*, 685, 337–346.

18. Lipinski, C. A., Lombardo, F., Dominy, B. W., Feeney, P. J. (1997). Experimental and computational approaches to estimate solubility and permeability in drug discovery and development settings. *Adv. Drug Deliv. Rev.*, 23, 3–25.

19. Ugi, I., Lohberger, S., Karl, R. The Passerini and Ugi reactions. In: *Comprehensive organic synthesis: selectivity for synthetic efficiency*, Vol. 2, Trost, B., Fleming, I., Eds. Pergamon Press, Oxford, 1991, pp. 1083–1109.

20. Accelrys Available Chemicals Directory (ACD). Accelrys, Inc. 10188 Telesis Court, Suite 100, San Diego, CA 92121, USA.

21. Pipeline Pilot, version 8.5. Accelrys, Inc. 10188 Telesis Court, Suite 100, San Diego, CA 92121, USA.

22. Daylight toolkit. Daylight Chemical Information Systems, Inc., PO Box 7737, Laguna Niguel, CA 92677, USA.

23. OEChem TK. OpenEye Scientific Software, 9 Bisbee Court, Suite D, Santa Fe, NM 87508, USA.

24. A tutorial and examples of SMARTS can be found following this URL: http://www.daylight.com/dayhtml_tutorials/languages/smarts/index.html. Accessed on November 28, 2013.

25. Beroza, P., Bradley, E. K., Eksterowicz, J. E., Feinstein, R., Greene, J., Grootenhuis, P. D. J., Henne, R. M., Mount, J., Shirley, W. A., Smellie, A., Stanton, R. V., Spellmeyer, D. C. (2000). Applications of random sampling to virtual screening of combinatorial libraries. *J. Mol. Graph. Model.*, 18, 335–342.

26. Hert, J., Willett, P., Wilton, D. J., Acklin, P., Azzaoui, K., Jacoby, E., Shuffenhauer, A. (2004). Comparison of fingerprint-based methods for virtual screening using multiple bioactive reference structures. *J. Chem. Inf. Comput. Sci.*, 44, 1177–1185.

27. Schuffenhauer, A., Brown, N., Ertl, P., Jenkins, J. L., Selzer, P., Hamon, J. (2007). Clustering and rule-based classifications of chemical structures evaluated in the biological activity space. *J. Chem. Inf. Model.*, 47, 325–336.

28. Steffen, A., Kogej, T., Tyrchan, C., Engkvist, O. (2009). Comparison of molecular fingerprint methods on the basis of biological profile data. *J. Chem. Inf. Model.*, 49, 338–347.

29. Willett, P. (2011). Similarity searching using 2D structural fingerprints. *Methods Mol. Biol.*, 672, 133–158.

30. DiscNgine. DISCNGINE S.A.S., Parc Biocitech, 102 route de Noisy, 93230, Romainville, France.

31. Agrafiotis, D. K., Lobanov, V. S. (2001). Multidimensional scaling of combinatorial libraries without explicit enumeration. *J. Comp. Chem.*, 22, 1712–1722.

32. A tutorial and examples of SMIRKS can be found following this URL: http://www.daylight. com/dayhtml_tutorials/languages/smirks/index.html. Accessed on November 28, 2013.

33. Lobanov, V. S., Agfrafiotis, D. K. (2000). Stochastic similarity selections from large combinatorial libraries. *J. Chem. Inf. Comput. Sci.*, 40, 460–470.

34. Gillet, V. J., Willet, P., Bradshaw, J. (1997). The effectiveness of reactant pools for generating structurally-diverse combinatorial libraries. *J. Chem. Inf. Comput. Sci.*, 37, 731–740.

35. Jamois, E. A., Hassan, M., Waldman, M. (2000). Evaluation of reagent-based and product-based strategies in the design of combinatorial library subsets. *J. Chem. Inf. Comput. Sci.*, 40, 63–70.

36. Gillet, V. J. (2002). Reactant- and product-based approaches to the design of combinatorial libraries. *J. Comput.-Aided Mol. Des.*, 16, 371–380.

37. Pearlman, R. S., Smith, K. M. (1998). Novel software tools for chemical diversity. *Perspect. Drug Discov.*, 9–11, 339–353.

38. Fink, T., Bruggesser, H., Reymond, J.-L. (2005). Virtual exploration of the small-molecule chemical universe below 160 daltons. *Angew. Chem. Int. Ed.*, 44, 1504–1508.

39. Medina-Franco, J. L., Martinez-Mayorga, K., Giulianotti, M. A., Houghten, R. A., Pinilla, C. (2008). Visualization of the chemical space in drug discovery. *Curr. Comput. Aided Drug Des.*, 4, 322–333.

40. Nguyen, K. T., Blum L. C., van Deursen, R., Reymond, J.-L. (2009). Classification of organic molecules by molecular quantum numbers. *ChemMedChem*, 4, 1803–1805.

41. Reymond, J.-L., Ruddigkeit, L., Blum, L., van Deursen, R. (2012). The enumeration of chemical space. *WIREs Comput. Mol. Sci.*, 2, 717–733.

42. Lopez-Vallejo, F., Giulianotti, M. A., Houghten, R. A., Medina-Franco, J. L. (2012). Expanding the medicinally relevant chemical space with compound libraries. *Drug Discov. Today*, 17, 718–726.

43. Bolton, E. E., Wang, Y., Thiessen, P. A., Bryant, S. H. (2008) PubChem: integrated platform of small molecules and biological activities. In: Annual reports in computational chemistry, Vol. 4, pp. 217–241. American Chemical Society, Washington, DC.

44. Fink, T., Reymond, J.-L. (2007). Virtual exploration of the chemical universe up to 11 atoms of C, N, O, F: assembly of 26.4 million structures (110.9 million stereoisomers) and analysis for new ring systems, stereochemistry, physico-chemical properties, compound classes and drug discovery. *J. Chem. Inf. Model.*, 47, 342–353.

45. Buller, F., Steiner, M., Frey, K., Mircsof, D., Scheuermann, J., Kalisch, M., Buehlmann, P., Supuran, C. T., Neri, D. (2011). Selection of carbonic anhydrase IX inhibitors from one million DNA-encoded compounds. *ACS Chem. Biol.*, 6, 336–344.

46. Burden, F. R. (1997). A chemically intuitive molecular index based on the eigenvalues of a modified adjacency matrix. *Quant. Struct.–Activity Relationships*, 16, 309–314.

47. Reutlinger, M., Schneider, G. (2012). Nonlinear dimensionality reduction and mapping of compound libraries for drug discovery. *J. Mol. Graph. Model.*, 34, 108–117.

48. Oprea, T. I., Gottfries, J. (2001). Chemography: the art of navigating in chemical space. *J. Comb. Chem.*, 3, 157–166.

49. Oprea, T. I. (2002). Chemical space navigation in lead discovery. *Curr. Opin. Chem. Biol.*, 6, 384–389.

50. Martin, E. J., Crichlow, R. E. (1999). Beyond mere diversity: tailoring combinatorial libraries for drug discovery. *J. Comb. Chem.*, 1, 32–45.

51. Valler, M. J., Green, D. (2000). Diversity screening versus focussed screening in drug discovery. *Drug Discov. Today*, 5, 286–293.

13

SCREENING LARGE COMPOUND COLLECTIONS

Stephen P. Hale

Ensemble Therapeutics, Cambridge, MA, USA

Two fundamentally different but complementary screening approaches are available to modern-day drug hunters, enabling the discovery of compounds possessing unique, therapeutically relevant functional activity. The two approaches are both the product of a desire to screen greater numbers of chemically and structurally diverse compounds against defined, well-characterized molecular targets. The primary discovery workhorse approach applied across the pharmaceutical industry is referred to as High-Throughput Screening (i.e., HTS) and essentially evaluates, in a highly parallel format, single compounds each in a single isolated assay. The process is the product of an effective miniaturization and automation of biochemical or cell-based functional assays with readouts tracking the effect of high test-compound concentrations. A complementary method for creating and screening large collections of chemical compounds has evolved from the fundamental concepts of biologically based techniques like phage and mRNA display techniques [1, 2] and was pioneered by Lerner and Brenner [3]. The critical concept is that if a compound's function, or phenotype, is linked to a coding feature, or genotype, then large collections of molecules can be synthesized as complex mixtures and screened for a particular function because the rare "active" molecules can be isolated and subsequently identified through reading of the genotypic codes. Nucleic acid polymers and specifically DNA are ideal as a coding genotype for synthetic libraries of

A Handbook for DNA-Encoded Chemistry: Theory and Applications for Exploring Chemical Space and Drug Discovery, First Edition. Edited by Robert A. Goodnow, Jr.
© 2014 John Wiley & Sons, Inc. Published 2014 by John Wiley & Sons, Inc.

compounds because of their high information-content density, their stability under a range of conditions, and the availability of many advanced tools for the high-fidelity synthesis, assembly, and decoding. High-sensitivity PCR and next-generation sequencing are available tools to provide a rapid and precise readout of a screening process that reduces the members in a complex mixture to a small set that possess the desired functional properties.

The development of techniques that allow for the creation of mixtures of DNA-encoded molecules as a source of diverse chemical structures has been paralleled by a development of methods to isolate and identify individual molecules possessing specific functional traits from such complex mixtures. A screen against collections of many individual single compounds, routinely performed in HTS campaigns, typically detects some readout signal from a target-specific assay that quantifies the individual compound's effect on the target's function. These types of assays, which depend on a readout signal, are practical for individual compounds that can be evaluated at high (i.e., μM) concentrations but not compatible with a screen that looks to identify a rare active compound present at low (i.e., pM) concentrations in a complex mixture of compounds. The ability to identify an active compound out of a complex mixture through the linkage of a readout signal, such as a fluorescent signal, to the function of an individual compound is not possible because the readout would effectively be an average of all of the signals from all of the compounds in that mixture. Rather than apply a direct measureable readout to compounds within the mixture upon functional modulation of the target, the individual compounds in "library" mixtures are simultaneously evaluated for their ability to associate with the target, based on affinity. The ability to detect molecules present in complex mixtures is possible because individual members of the library will each have a unique affinity, or phenotype, for the target of interest, and this physical property defines the differentiating characteristics of target-binding and nontarget-binding compounds. These affinity-based screens are well suited to interrogate complex mixtures, or libraries, of compounds as the process results in an association of compounds and target that can be subsequently exploited. The ability to track and identify these rare target-binding molecules is possible because of the one-to-one linkage between the coding DNA genotype and its covalently linked functional compound.

13.1 AFFINITY-BASED SCREENING: A CHROMATOGRAPHIC APPROACH

The creation of technologies to synthesize encoded compound collections allows access to very large numbers of compounds representing heretofore unexplored chemical diversity. Vast diversity can be represented in encoded libraries because a combinatorial approach is taken to react all potential combinations of the diversity elements available at each synthetic step. The encoding with DNA allows for the application of an array of molecular biological tools for the amplification, quantification, and decoding of the synthetic history associated with each member of the library. The synthesis of new chemical diversity is of practical significance if such diversity can be interrogated and

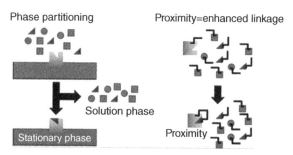

Figure 13.1. Identification of target-associated (■) compounds from libraries of encoded compounds (░░) can be achieved by a physical phase separation or signal generation linked to a target-proximity-dependent event. For color detail, please see color plate section.

individual compounds of interest can be identified and characterized. Affinity-based screening is well suited to interrogate encoded library-based compounds and allows each compound within the mixture to be simultaneously evaluated based on the fundamental property of target affinity. The interest of the investigator for compounds within an encoded collection is predicated on the fact that there is an affinity-driven functionally relevant interaction between each compound and the target that is of value.

Affinity-based screening as a process to isolate specific functional molecules from a large collection of mostly nonfunctional molecules can be accomplished by two operationally very different methods. Association and separation are the hallmarks of the first and primary approach, whereas colocalization is the key concept applied in the second approach (Fig. 13.1). Although these two methods approach the challenge of how to distinguish compounds having affinity for a target from compounds that do not have affinity for a target differently, the ability to identify with high-fidelity compounds that associate with the target of interest is the objective of both methods. The second approach applies the concept of colocalization as a process to screen libraries and has origins in assays based on the basic principles of a proximity effect (FRET or proximity ligation assay). These types of proximity assays take advantage of "wash-free homogeneous" methods and proximity-dependent signal creation. In one embodiment, the target and the library members are both tagged with DNA fragments, and when the two are brought into proximity through the affinity-based interaction of the compound with the target, they associate to produce an amplifiable DNA product. The DNA product is then amplified and decoded by DNA sequencing to generate the linkage between the encoded library compound and the screening target [4, 5].

Screening techniques that depend upon colocalization of the target and target-binding library compounds exploit the affinity certain compounds have for the target and are based on fundamental biochemical techniques but are not widely applied. Separation techniques such as affinity chromatography, size-exclusion chromatography, or ultracentrifugation are also fundamental to biochemistry and are the most widely applied techniques modified for library screening applications, with far and away the most common methods being based on affinity chromatography. Affinity-based chromatography is a powerful and versatile method for the screening of encoded compound libraries and is based on fundamental methods developed historically for the isolation or characterization of biological molecules. Phase-separation-based methods for affinity-based chromatography are ubiquitous, and a variety of commercially available tools are available to support the application and implementation of a screening campaign.

Affinity-based chromatography, as a form of adsorption chromatography, is a fundamental process that had its genesis over 100 years ago [6, 7]. The first biospecific adsorption of a protein onto a solid-phase matrix spawned an expansion of the affinity-based concept for the isolation of low-abundance compounds from complex biologically based sources such as tissue homogenates or cellular extracts. The methods of affinity chromatography have evolved significantly from their origins and are widely used in modern molecular biology and biochemistry as separation tools. Affinity chromatography as modified for screening of libraries is robust, reproducible, and elegant in its simplicity. The application of high-resolution affinity-driven phase-partitioning methods to the screening of DNA-encoded chemical compound collections is at its core a purification process but differs from a more traditional compound purification in that for a screening application the molecules of interest are ultimately identified through decoding and not collected for a subsequent physical use. The goal in screening encoded libraries of molecules as in affinity chromatography is to "purify" the compounds of interest having affinity for a target away from other members of the libraries that do not have affinity for the target. The significance of the DNA encoding in this process of affinity-based screening is that the DNA allows for the application of exquisitely sensitive molecular biology tools such as PCR and DNA sequencing to amplify and identify the products of affinity-driven separation processes even when the product concentrations would be too low to be identified directly using biophysical tools. Although the compounds recovered from an affinity-based screening campaign are isolated at very low concentrations, the "reading" of the encoding "DNA barcode" allows for the determination of frequency levels for each library member ultimately allowing for the identification of target-associated compounds out of complex, low-abundance mixtures.

Affinity-based chromatography methods have been adapted to the specific demands of compound-library screening but remain true to the fundamentals of the process. Modifications to basic methods take into consideration small volumes and the very low concentrations of compounds recovered from the screen. A reduced scale conserves the encoded library and allows for the screening of very large libraries to take place in a general laboratory. PCR and next-generation sequencing allow for the decoding of the screening output but also raise the potential for low levels of contamination to affect the results of the screen. To limit the potential for cross-contamination from screen to screen, it is preferred to use disposable labware throughout the entire process from target and buffer preparation, running of the screen, and processing of the screening output.

Core to the fundamental method is creating a benign linkage between the target and a solid-phase matrix to enable a phase separation between the solid-phase target and the solution-phase library of compounds. With an appropriately modified target enabling the partitioning of the target, the affinity-based partitioning process is defined by three events (Fig. 13.2): first, there is a binding event that allows for the compounds being screened to occupy binding sites on the target as a function of their affinity; second,

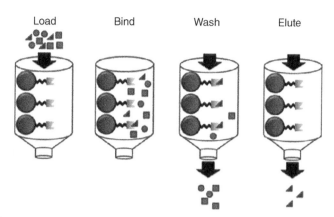

Figure 13.2. Affinity-based library screening applying a chromatographic phase-separation process is defined by the adsorption of target-binding library compounds (█▲●) to the solid-phase (●) target (▮) in the bind step, separation of weakly target-associated compounds in the wash step, and recovery of target-associated compounds in the elute step. For color detail, please see color plate section.

there is a washing event, usually multiple independent steps, to partition compounds having no or low affinity to the solution phase away from the solid-phase target and associated compounds; and third, there is the recovery or elution of target-associated compounds off of the solid phase and back into solution. The elution is designed to effect dissociation of compounds from the solid-phase target, resulting in the recovery of the target-associated compounds for subsequent rounds of screening or compound identification through reading of the compounds' encoding DNA barcodes.

There are several advantages to synthesizing and screening DNA-encoded collections of compounds to identify therapeutic lead compounds or "tool" compounds over the more traditional single-compound single-screen HTS paradigm. The numbers of DNA-encoded compounds far exceeding those available in all individual compound collections (i.e., HTS ready) can be created rapidly and simultaneously evaluated against a molecular target, enabling efficient exploration of novel chemical space not covered in available screening collections. The speed and simplicity of affinity-based screening allows for the rapid and efficient interrogation of new chemical space toward identifying functional and potentially novel chemical matter for targets that have not yielded to other discovery approaches.

Affinity-chromatography-based screening is at its most basic a method that allows the partitioning of target-associated molecules from free nontarget-associated molecules. The technique is in its design an efficient method for interrogating large collections of compounds because in a single screen all library compounds are simultaneously evaluated under identical conditions. Depending on the numbers of compounds in a particular library, an affinity-based screen could evaluate and generate target-binding data simultaneously on thousands, millions, or even billions of compounds.

The generalized nature of compound identification from an affinity-based screening event, in that compounds are identified that bind to the target as opposed to affecting a specific functional process, allows for the development of methods that can be applied to a wide range of targets. A core set of methods can be developed that can be routinely applied to a wide range of targets, usually with some minor optimization. The modification and evolution of fundamental biochemical techniques used to isolate compounds of interest from complex mixtures has produced some of the most versatile and robust affinity-based screening methodologies. Separation technologies such as affinity chromatography routinely depend on a phase partitioning, between a stationary and a mobile phase, to enable the separation of compounds of interest from undesirable compounds present in complex mixtures. DNA encoding of compound libraries allows for large highly diverse soluble mixtures of compounds to be synthesized that are well suited to be interrogated by affinity-based screens having their origins in chromatographic methods. The library compounds are presented in the solution phase to a target presented on a stationary phase, thereby setting up the ability to partition compounds based on their association level with the target. The encoding of library compounds with DNA barcodes enables the screening of very high-diversity libraries, where the concentration of each member can be in the pM range, with the ability to track the frequency of each compound throughout the process by sensitive and high-fidelity methods such as quantitative PCR and DNA sequencing.

13.2 SCREENING PREPARATION

The most commonly practiced affinity-based screening method applied to encoded libraries is a derivative of affinity chromatography and is fundamentally a separation process. The target protein is immobilized onto a stationary phase, and a library of soluble compounds, associated with the mobile phase, is exposed to the target. Compounds that have affinity for the target preferentially partition to the stationary phase and will be separated from the bulk of the library that remains in solution. DNA-encoded compound libraries are well suited to be interrogated by affinity-based screening methods as the enrichment and identification of target-binding members is essentially, at its most basic level, a "purification" of rare low-frequency compounds out of a complex mixture. The DNA barcode allows for amplification, exquisitely sensitive detection, and DNA-sequencing-based decoding throughout the process to monitor the fate of each library member.

Although the types of biological targets that are compatible with affinity-based screening approaches are diverse, protein targets dominate because of scientific interest and potential commercial value. Proteins are critical components in therapeutically relevant biological processes and therefore the focus of most drug discovery efforts. Protein targets are well suited to screening approaches that apply affinity-based methods not only because of their important structural and functional roles *in vivo* but there is also a large repertoire of chemical and molecular biological tools available for their expression, purification, and modification that support the affinity-based screening process.

The entire affinity-chromatography-based screening process that begins at target nomination and is complete upon hit identification is defined generally by what can be referred to as the initial preparation (Fig. 13.3) and the subsequent implementation (Fig. 13.4). The preparation phase of the screening process is an evaluation and

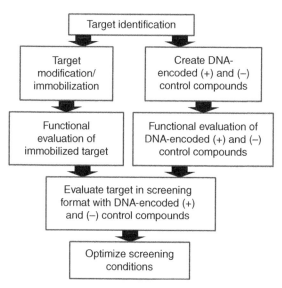

Figure 13.3. The preparation phase of the screening process sets a solid foundation for the subsequent affinity-based encoded library screening process.

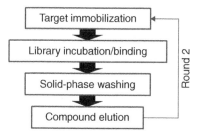

Figure 13.4. The implementation phase applies the process developed and optimized in the preparation phase to the screening of DNA-encoded libraries.

optimization process and occurs prior to the screening of compound libraries. The pre-screening preparation ensures that the compatible target form available is biologically relevant, that the target can be partitioned to the stationary phase in a manner compatible with the screening process, and that optimized methods are in place to ensure that compounds in the encoded library having affinity for the target do associate with the target under the defined conditions. To be compatible with chromatography-derived screening processes, the screening target needs to be of high function and in a physical form that maintains its function when immobilized onto a stationary phase.

In addition to identifying a form of the target and conditions ensuring maximum functionality, the creation and validation of control compounds is carried out simultaneously. Appropriate control molecules, DNA-tagged versions of known target-binding compounds, are valuable tools and serve a dual role in ensuring an optimized screening process and will be discussed in Section 13.2.5.

Buffer conditions need to be optimized to ensure target function and to minimize nonspecific interactions between the library compounds and both the target and solid-phase matrix. The preparation phase provides a foundation to build upon in the implementation phase that applies the optimized methods in the actual screening process of bringing target and library compounds together toward the identification of compounds having affinity for the target.

13.2.1 Screening Preparation: Target Considerations

Targets of interest for screening of DNA-encoded collections most frequently are complex biological molecules including proteins, nucleic acids, and carbohydrates. Proteins dominate the therapeutically relevant targets of most modern drug discovery programs and will be the focus here, but the principles and methods could be applied to a broader range of biological and nonbiological macromolecules. Consideration must be given to understanding the form and function of the target to be used in the screen. As the target-based screening model applied currently in the pharmaceutical industry is dependent upon the modulation of a specific and defined function to affect a therapeutic response, a well-understood and well-characterized target is critical for success. Translation of a biologically relevant target structure to a form that is amenable to screening will increase the potential for successfully identifying compounds having the desired effect on the target function. To recapitulate the target in a screening format requires an understanding of how it fits into a complex biological process and what role is played by such variables as buffer conditions, interactions with macromolecules such as proteins or membranes, and posttranslational modifications of the target itself. The objective is to produce a screening-compatible functional target that is biologically and therapeutically relevant.

Bcl-x_L exemplifies a molecular target that is well characterized and therapeutically relevant (Fig. 13.5). Bcl-x_L is an antiapoptotic protein that through a protein–protein interaction mediated by a BH3 peptide is a prosurvival factor for cancer cells [8]. A simple BH3-binding fluorescence polarization assay is available to monitor the

---

Target ID: Bcl-x_L

Functional assay: Inhibition of caspase 8-cleaved-BID mediated release of cytochrome c from isolated mouse liver mitochondria.

Functional binding: Direct binding to Bcl-x_L of a synthetic BH3 peptide derived from a pro-apoptotic Bcl-2 family member

Screening-hit criteria: Compounds must be competitive with Bcl-x_L binding to a synthetic BH3 peptide

---

Figure 13.5. An appropriate target for screening is characterized by a well-defined function, appropriate functional assays, and therapeutic relevance.

function of Bcl-x$_L$ and would be used throughout the initial preparation phase to ensure that the function of the target is maintained [9].

Protein targets can be sourced in many different ways. Targets can be isolated and purified from natural biological sources, or as is the case more routinely, targets can be produced in recombinant systems and purified to near homogeneity. Using a high-purity target preparation in a screen ensures maximum recovery of library compounds that have affinity for the desired target. Nontarget contaminants in the target preparation constitute screening targets in their own right, and library compounds that have affinity for the nontarget components will be recovered, effectively reducing the recovery of target-specific compounds. The purity of the target is also an important consideration from the perspective of library stability. Although the DNA component of the encoded library is quite stable, it could be susceptible to degradation by nucleases present in the target preparation. Prior to the screening of libraries, all targets should be evaluated for the presence of nucleases. A simple evaluation of the integrity of the encoding DNA before and after exposure of the library to the target, under the conditions of a typical screen, would ensure that a particular target preparation is not an issue.

If a target is deemed valuable from a therapeutic perspective but cannot be purified to high levels, there are limitations but screening can ultimately be productive. Membrane-associated cell surface receptors can in some cases be isolated and purified, but in other cases, the biologically relevant form cannot be isolated and can be screened in the context of the intact cell. The cell surface presents a complex menagerie of which even an overexpressed receptor represents a small percentage. This does not preclude the target receptor from being screened against encoded libraries [10–12], but the low-purity nonhomogeneous presentation of the target will increase the recovery of off-target library compounds and reduce the enrichment potential of target-specific compounds during each round of screening.

13.2.2 Native Targets

For native targets isolated and purified from biological sources, consideration must be given as to how the target will be prepared for immobilization onto a solid-phase matrix and subsequent screening of encoded libraries. The association of the target with the solid-phase matrix is fundamental to the chromatography-based screening process and can be enabled by specific chemical modification of the target. The modification of the target protein to introduce a so-called molecular handle considers the availability of a complementary partner that binds to the handle and can be coupled to a stationary-phase matrix. These association mediators allow for an indirect, noncovalent, and stable linkage to be formed between the target and the stationary-phase matrix. A set of association mediators well suited for a chromatography-based screening process are biotin and biotin-binding proteins such as streptavidin. Biotin is used widely as a modification agent enabling the capture of proteins because it is a small (molecular mass = 244.31 Daltons) stable molecule that has very high affinity ($K_d = 10^{-15}$ M) for streptavidin. The small size reduces the potential for an undesirable effect on target functionality upon modification, and the high affinity for streptavidin ensures rapid and efficient capture by the appropriately modified stationary-phase matrix [13]. A biotin tag can readily be introduced onto

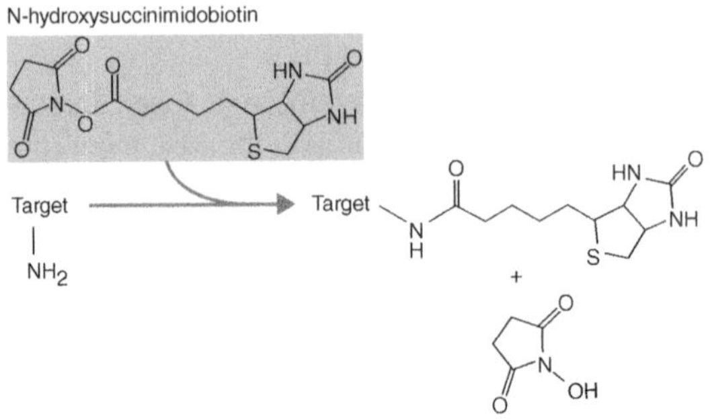

Figure 13.6. The "biotinylation" reaction between a primary amine of the target and an N-hydroxysuccinimide (NHS) ester of biotin.

the surface of a protein target through a chemical reaction between an appropriately reactive biotin reagent and specific amino acid side chain functional groups (e.g., primary amine, carboxylic acid, or thiol) present on the target (Fig. 13.6).

Modification of a target protein through chemical conjugation with reactive biotin analogs is rapid, simple, and efficient and enables subsequent immobilization of the target protein onto solid-surface-linked streptavidin.

The chemical conjugation of an appropriate association mediator like biotin to an isolated and purified target is an effective method to introduce a tag into the structure of an untagged native target protein.

13.2.3 Recombinant Targets

Many protein targets can be overexpressed in recombinant systems, allowing for the incorporation of gene-coded molecular tags into a specific functionally inert region of the target's protein sequence. Biotin can be introduced at a specific site in an expressed target by introducing a peptide sequence referred to as an AviTag™. The AviTag™ is a specific 15-amino-acid peptide sequence that is incorporated into the target protein and directs the posttranslational addition of a single biotin molecule to a specific lysine residue within the sequence. The AviTag is preferred to install a biotin over chemical modification as it is controlled and site specific, alleviating the potential for the random introduction of multiple biotin molecules that could compromise the target's function.

Short peptide sequences can readily be introduced into a target gene and expressed within a functionally inert region of a target. A His_6 tag (i.e., six consecutive histidine residues) is a small tag that is well tolerated in many expressed proteins and binds efficiently to certain metal ions tightly associated with a stationary-phase matrix [14]. The His_6 tag is typically introduced into the amino acid sequence of the target at either the C-terminus or the N-terminus and enables the efficient capture of the target onto a solid phase that has been functionalized with Nitrilotriacetic Acid (NTA) and loaded with Ni^{2+}. The His_6 tag of the target protein and the solid-phase NTA–Ni^{2+} reagents represent association mediators that are appropriate in many cases and enable a chromatography-based screening campaign.

A wide range of commercial products are available to introduce gene-coded peptide tags or protein fusions into a target and to capture the appropriately modified target onto a solid-phase matrix (Table 13.1).

TABLE 13.1. Some fusion protein or peptide tags available for recombinant incorporation into a target protein and their corresponding association partners

Tag designation	Protein modification	Incorporation method	Association partner	Source/examples
AviTag	Single biotin	Recombinant expression	Avidin/streptavidin	Avidity, LLC
Polyhistidine	His_6 or His_{10}	Recombinant expression	Ni–NTA/Co-NTA	Life Technologies, Inc.
GST fusion	Glutathione S-transferase	Recombinant expression	Glutathione	GE Healthcare Biosciences
Fc fusion	IgG Fc fusion	Recombinant expression	Protein A/G	InvivoGen, Inc.
MBP fusion	Maltose-binding protein fusion	Recombinant expression	Maltose	New England Biolabs, Inc.

$His_6 \rightarrow$ [15].
GST $\rightarrow$ [16].
MBP $\rightarrow$ [17].
Additional methods $\rightarrow$ [18].
Fc fusions $\rightarrow$ [19].

Although the introduction of a small tag such as biotin or His_6 can have limited to no impact on the functionality of the target protein, it is critical to ensure that the modification did not have a deleterious effect. A screening event against a modified target will have a greater chance of identifying therapeutically relevant compounds if the target structure recapitulates that of the native protein. If a screening target presents surfaces that are not components of the native target, then there is a potential for a screen to provide hit compounds against these new surfaces that could turn out to be functionally irrelevant. Targets can be very complex, and it is important to understand the function to be affected by the compounds that come out of the screening process, whether that is a catalytic activity, a protein–protein interaction signaling event, or other processes. The evaluation of the modified target should take into account the relevant function, and appropriate confirmation criteria should be applied prior to screening to ensure that the relevant function is appropriately intact. The functionality of the target protein should be evaluated throughout the screening preparation process.

The oncology target human Bcl-x_L is a 233-amino-acid protein that maintains function when expressed in a bacterial system as a truncated form with a His_6 tag introduced at the C-terminus. The functionality of the modified recombinant protein should be confirmed to ensure that the target that moves into screening represents an appropriate (i.e., therapeutically relevant) form (Fig. 13.7).

13.2.4 Screening Preparation: The Solid Phase

The partitioning of the target and its associated library compounds with the solid-phase matrix is fundamental to a chromatography-based screening process. The linkage between a screening target and the solid-phase matrix can be through a direct or an indirect

> **Target ID:** Bcl-x$_L$
>
> **Function:** Neutralization of pro-apoptotic Bcl-2 family members through a BH3-peptide-mediated protein:protein interaction
> **Protein sequence:** Ser2-Arg212
> **Protein form:** *E. coli* expressed with a C-terminal His$_6$ tag
> **Purity:** >95% by SDS-page

Figure 13.7. The function of a tagged recombinant protein target is confirmed with appropriate assays to ensure maintenance of a therapeutically relevant activity.

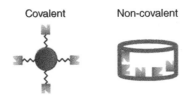

Covalent   Non-covalent

Figure 13.8. Direct linkage of the target (■) to the solid-phase matrix can be through the covalent coupling to an activated resin bead (blue sphere) or through noncovalent interactions with the plastic surface blue well of a polystyrene plate. For color detail, please see color plate section.

linkage. Direct linkage (Fig. 13.8) of the target to a solid-phase matrix is the default method if the target does not have an introduced tag or modifications that enable subsequent capture by an appropriately modified resin.

A direct linkage, as it implies, refers to a situation where the target plays a central role and is directly linked to the solid support. Direct coupling of a protein target to a stationary-phase matrix can be covalent or noncovalent and is an effective method to associate the target with a stationary phase. Direct covalent conjugation of a target protein to a solid-phase matrix is effective but limits a range of options such as the ability to vary the solid phase or carrying out the binding phase of the library and target in solution. Sugar-based chromatography beads (e.g., agarose or sepharose) or plastic (e.g., polystyrene) surfaces can be modified to present functionalities that react directly with groups that are typically on the surface of biological target molecules. These "functionalized" solid-phase matrices allow for the direct immobilization of biological targets like proteins to the matrix in preparation for an affinity-chromatography-based screening campaign.

Direct target linkage can be through a covalent or noncovalent interaction but typically is irreversible. The covalent coupling of a target typically occurs between a reactive function amino acid (e.g., lysine, aspartic acid, glutamic acid, or cysteine) and an appropriately activated solid-phase matrix (Fig. 13.9; AminoLink coupling resin, Thermo Scientific Inc.), while noncovalent interactions are routinely through adsorption-mediated capture onto 96-well polystyrene plates (high binding plates, Thermo Scientific Inc.). Both of these direct methods are effective and do not require a preemptive modification

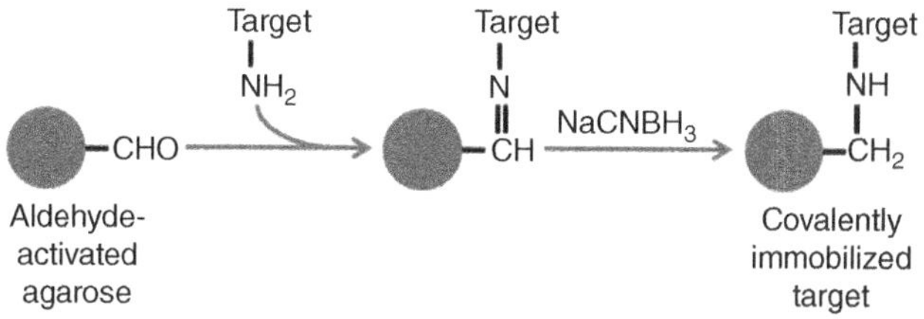

Figure 13.9. An aldehyde-activated agarose matrix can react with primary amines on the target to directly link the target to the solid phase.

of the target but are random in their modification of the target's functional groups and do not orient the target predictably or consistently on the solid surface.

A simple but effective method for the direct noncovalent capturing of a target protein is adsorption onto the polystyrene surface of a multiwell plate. The method is versatile as it requires no modification of the target protein, but orientation of the protein is random and the capacity of the surface is low compared to a chromatography matrix such as agarose beads. The relatively low protein capacity of the plate surface can limit the recovery of low-affinity library compounds that may be of interest. The target concentration is the variable factor influencing the occupancy of library compounds on the target, so to ensure the recovery of low-affinity compounds, the target concentration must be high. In a primary screen out of a naïve library, it might be desirable to recover even low-affinity library compounds against the target, so a bead-based system would be preferred.

The most versatile and ubiquitous method for enabling the association of a protein target onto a solid phase is one that applies a pair of association mediators, one on the target and one on the solid phase. Modifications to the target protein to introduce one of the association mediators, in the form of a molecular handle, are made, bearing in mind the availability of a complementary partner that can be coupled to a solid-phase matrix. The association between the molecular handle and the complementary partner ensures an indirect stable noncovalent capture of the target onto the solid phase. The primary considerations when evaluating a pair of association mediators for utility in the screening of DNA-encoded libraries are that the modification of the target with one of the mediators does not alter its functionality and that the linking of the other mediator to a solid-phase matrix maintains its ability to associate with its partner. Three of the most widely used solid-phase formats that allow for the straightforward linking of one of the association mediators to a solid-phase matrix are (i) agarose-based chromatography beads, (ii) polystyrene or silica paramagnetic beads, and (iii) plastic multiwell plates (commonly 96 well). The beads or the plate surface provides the insoluble solid-phase structure to which one of the association mediators will be linked. The tag incorporated into the target structure and the "capture moiety" on the solid phase represents a pair of association mediators, which enable the phase transfer of the target from solution to solid, critical to the screening process.

An example of a pair of effective association mediators that are widely available and applied is biotin and streptavidin (Fig. 13.10). Biotin, being a small molecule, is typically the target-modifying reagent so as to minimize the potential for a negative

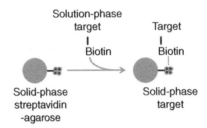

Figure 13.10. Solution-phase biotin-tagged targets are effectively transferred to the solid phase by association with immobilized streptavidin.

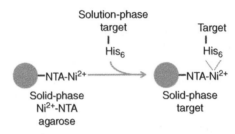

Figure 13.11. Solution-phase His_6-tagged targets are effectively transferred to the solid phase by association with immobilized Ni^{2+}.

effect on target function. Streptavidin is a very stable protein that maintains its ability to bind to biotin when directly immobilized to a solid-phase matrix. Streptavidin is preferred for screening of DNA-encoded libraries over the functionally equivalent avidin because of its acidic pI and lack of glycosylation, resulting in lower nonspecific interactions with the library. Streptavidin is linked to the solid-phase matrix covalently, or in the multiwell plate format, noncovalent adsorption of the streptavidin to the polystyrene plastic is also effective. The incorporation of biotin into the target structure and the linking of the streptavidin with a solid-phase matrix is an exemplary application of association mediators that do not dramatically affect the target function but do allow for a rapid and efficient solid-phase capture of the screening target.

An additional set of association mediators takes advantage of a short amino acid sequence and an immobilized metal ion (Fig. 13.11). Polyhistidine stretches inserted into the protein target's primary structure using recombinant methods enable noncovalent capturing by a solid-phase chemically chelated metal ion. This pair of association mediators is routinely used in screening campaigns as His_6 amino acid tags are commonly incorporated into recombinant proteins to facilitate the purification process and are therefore available to be exploited during screening.

Prior to initiating a screen using a set of an appropriately modified target and solid-phase matrix, an assessment of the extent of target capture and the evaluation of target functionality is appropriate. A given quantity of solid-phase matrix should be evaluated for its capacity to capture the solution-phase target onto the solid phase. A confirmation that the pair of association mediators is operating as designed and an estimate of the level of target loading onto the solid phase is necessary prior to moving on to library

1. Load 50 pmol of Bcl-x$_L$-His$_6$ onto a 5µL Ni^{2+}-NTA-Agarose column
2. Apply various amounts of a Fluor-BIM peptide to the solid-phase Bcl-x$_L$

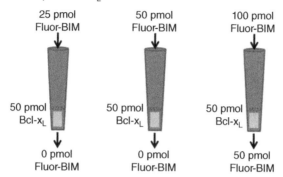

25 pmol Fluor-BIM	50 pmol Fluor-BIM	100 pmol Fluor-BIM
50 pmol Bcl-x$_L$	50 pmol Bcl-x$_L$	50 pmol Bcl-x$_L$
0 pmol Fluor-BIM	0 pmol Fluor-BIM	50 pmol Fluor-BIM

3. Quantify the Fluor-BIM peptide that can not be captured by the solid-phase Bcl-x$_L$
4. The capture of 50 pmol of Fluor-BIM indicates fully active solid-phase Bcl-x$_L$

Figure 13.12. The function of solid-phase Bcl-x$_L$ can be determined by evaluating the level of fluorescent BIM capture. Stoichiometric capture of the BIM ligand implies a fully active Bcl-x$_L$.

screening. Determining the level of target loading will ensure that the effective concentration and target numbers are appropriate for the subsequent library screens.

As the loading density of the target onto the solid phase may influence the function of the target, it is appropriate to evaluate several loading densities that increase systematically up to a level where saturation of all capture sites on the solid phase has been achieved. With an understanding of the capacity of the solid-phase matrix, the next step is to determine the functionality of the solid-phase target. The physical proximity of the target to the solid phase or the loading density of the target onto the solid phase could have a negative impact on the functionality of the target. At each step in preparing the target for screening, it is vital to apply the available tools to ensure that the target maintains its therapeutically relevant function. The function of an enzymatically active target can be evaluated with a soluble substrate, or a protein–protein interaction target can be evaluated for its ability to bind and capture its protein partner.

A convenient format for holding small volumes of a modified agarose matrix is in a pipet-tip-derived microcolumn (PhyNexus, Inc.). 5 µL of Ni^{2+}–NTA agarose packed into a microcolumn facilitates the capture of a His$_6$-tagged target and has appropriate target capacity for screening encoded libraries. The column format also enables the facile exposure of the matrix to solutions (i.e., binding, washing, and eluting) using a standard laboratory pipet. The level of functionality for a solid-phase target like Bcl-x$_L$ can be determined (Fig. 13.12) by its ability to bind to a fluorescently labeled BIM-derived BH3 peptide (e.g., 6-FAM-IWIAQELRRIGDEFNAYY). In the microcolumn, at a fixed loading density of target, the binding capacity of the immobilized target for the fluorescent peptide can be determined to ensure a highly functional target.

Critical to maintaining and allowing for maximum target functionality is providing a compatible buffer system. The identification of a buffer system that ensures target functionality is a requirement that must also be balanced with a need to minimize any nonspecific interactions between all of the components of the screen and the DNA-encoded library compounds. To reduce potential electrostatic interactions between the target and the DNA component of the libraries, it is preferred to include 150 mM NaCl as a source of counterion. There may be a balance to be struck between target functionality and reduction of nonspecific DNA interactions, but physiologically relevant levels of NaCl are a practical starting point. Additionally, the inclusion of biologically compatible levels of a nonionic detergent (e.g., 0.05% Tween-20) will function to reduce nonspecific interactions between the library components, the target, and the solid-phase components.

13.2.5 Screening Preparation: Encoded Control Compounds

To ensure that the target protein under the conditions of the screen is functionally relevant, its biochemical activity is monitored throughout the preparation process. In addition to confirming that the target maintains functionality under screening conditions and prior to the actual library screen, it is appropriate to confirm that there are not target-specific limitations to the screening of specifically DNA-encoded libraries. To ensure that there is no interference by the encoding DNA tag, appropriate control compounds are generated to confirm that the compound component of the DNA-encoded molecule has the capacity to interact specifically with the solid-phase target in a predictable fashion. The synthesis of DNA-encoded versions of molecules that are known to interact with the target (i.e., positive controls) is extremely valuable in the screening preparation phase to confirm that tagged versions of well-characterized target-binding compounds retain the ability to interact with the solid-phase target. A positive control candidate will have affinity for the target in a range that would predispose its recovery under the conditions of the screen. In a situation where a variety of affinity reagents are available, it is beneficial to create several positive control compounds that cover a range of affinities for the target. Additionally, DNA-encoded versions of compounds that do not have affinity for the target (i.e., negative controls) should not bind to the solid-phase target and ensure that a molecule's behavior is determined by the properties of the compound and not by the encoding DNA.

13.2.6 Encoded Positive Control Compounds

Small molecules, peptides, and even small proteins can function as control compounds. Control compound candidates must have well-characterized affinity for the target and a site amenable to chemical modification. The modification of the candidate through the introduction of an inert spacer and ultimately conjugation to an encoding DNA should ideally have a minor impact on the affinity of the candidate compound for the target. The design of the DNA-encoded control compounds should recapitulate the design of the DNA-encoded libraries to be subsequently screened, but the compound-specific DNA encoding sequences for the controls must be unique from any library members. The

Fluorescently labeled BIM BH3 peptide

(6-FAM)-IWIAQELRRIGDEFNAYY

DNA-conjugated BIM BH3 peptide

DNA—Linker—IWIAQELRRIGDEFNAYY

Figure 13.13. The label position on a Bcl-x$_L$-binding fluorescent BH3 peptide informs the DNA-conjugation site for the creation of an encoded positive control.

unique coding of control compounds enables their subsequent introduction into libraries for use as internal controls whose behavior in the screen demonstrates the behavior of a well-characterized compound of known affinity for the target. The unique DNA sequence encoding the control compounds allows for the application of quantitative PCR [20] to monitor the enrichment of the control (i.e., defined as the concentration of the control relative to the behavior of the bulk library) in optimization experiments and in the actual library screens.

Ideal candidates for use as DNA-encoded control compounds are those where there is precedence for modification of the compound (e.g., fluorescent-label conjugation) that would inform the DNA-conjugation efforts (Fig. 13.13). A fluorescently labeled BIM-derived BH3 peptide used in a fluorescent polarization activity assay for Bcl-x$_L$ [21] is functional as a conjugated derivative and thus represents a potential candidate for use as a DNA-encoded positive control for Bcl-x$_L$. The linkage point of the fluorophore to the BIM peptide has been demonstrated to be a region of the peptide that can be modified without loss of function and is a position available for DNA conjugation.

The design of the DNA-encoded control compound should incorporate a long flexible linker to minimize the potentially negative effect of DNA conjugation on the compound's functionality. To ensure that the DNA-conjugation process does not dramatically reduce the affinity of the compound for the target, it should be confirmed by a direct method such as Surface Plasmon Resonance. If the control compound affects a target function, then in addition to determining the affinity of the DNA conjugate for the target, an appropriate biochemical assay evaluating function should be applied. With the independent confirmation that the DNA-encoded version of the control compound has retained affinity/function for the target, the evaluation of the control in the screening format should then be investigated.

13.2.7 Encoded Negative Control Compounds

Additional confidence in the optimization and the ultimate performance of a screen can be achieved through monitoring the behavior of not only positive control compounds but also negative control compounds. A DNA-encoded negative control, in contrast to a positive control, does not have affinity for the target. When the negative control is included in a library at the concentration of a single library member, its behavior would predict the behavior of a library compound that does not bind to the target under the conditions of the screen.

Fundamentally, there are two types of negative controls. The first type serves as a matched set with the positive control and should be a similar structural class (i.e., peptide or small molecule) but lacking any significant affinity for the target. When conjugated to a unique DNA and incorporated into the library, the matched set negative control represents a nonbinding compound in the screen and will provide data to determine the level of "enrichment noise" in the screening process. The second type of negative control molecule is a DNA-only control and is a unique DNA of the same design as the encoding DNA within the library without a conjugated compound attached. The DNA-only control is added to the library at the concentration of a single library compound and, in conjunction with the compound–DNA negative control, would provide insight as to which component of the set of library compounds is contributing to the "enrichment noise" in the system, compound, or DNA.

13.2.8 Encoded Control Compound Evaluation

The control compounds allow for the conditions of the screen to be optimized to maximize the recovery of encoded compounds binding to the target and to minimize the recovery of encoded compounds that do not bind to the target. Conditions compatible with target functionality and the minimization of any nonspecific interactions between the DNA portion of the control and the target should be determined empirically in this initial evaluation. Buffered saline solutions that contain nonionic detergents are compatible with many protein targets and contain components to reduce nonspecific interactions. These physiologically compatible buffers are readily available and represent a good initial starting point to begin the optimization of the screening conditions. Phosphate-Buffered Saline (PBST) (i.e., PBS w/0.05% Tween-20) or Tris-Buffered Saline (TBST) (i.e., TBS w/0.05% Tween-20) contains physiological levels of NaCl and low levels of a nonionic detergent that effectively reduce nonspecific interactions between the DNA-conjugated compounds and the target or solid-phase components of the system. Additional suppression of DNA-mediated nonspecific interactions between the library members and various components of the screen can be achieved through the inclusion of PCR-silent blocking nucleic acids. Effective nucleic acid blockers are yeast tRNAs, sheared salmon sperm DNA, or synthetic DNAs that are similar in design to the encoded template design but lack the capacity to be PCR amplified.

The scale of the target and control evaluation should consider the scale appropriate for the screening of libraries at the bench in a modern research or discovery laboratory. A significant advantage of screening DNA-encoded libraries is the elegant simplicity and modest scale of the process that allows for rapid access to large numbers of diverse chemical structures. The use of modified pipet-tip solid-phase columns enables the binding, washing, and eluting steps (detailed method discussions in Sections 13.3.3, 13.3.4, and 13.3.5, respectively) using common laboratory instrumentation. A 5 µL column bed volume that can immobilize 50 pmol of target is appropriate for screening a library in a 50–100 µL volume. The evaluation of the control compounds in the screening format allows for the determination of their baseline behavior and for further optimization of the system prior to the actual library screen (Fig. 13.14).

Target
Recombinant Human Bcl-x$_L$
Ser2-Arg212 with a C-terminal His$_6$ tag
Molecular Mass: 24 kDa

Stationary-phase matrix
Ni^{2+} charged NTA-agarose
beads loaded in a 5 μL bed
volume pipet-tip column

1x TBST/tRNA buffer
50 mM Tris pH7.6
150 mM NaCl
0.05% Tween-20
1 mg/mL yeast tRNA

Control-DNA conjugate
BIM BH3 peptide-DNA

1. **Bind:** In a volume of 50 μL combine Bcl-x$_L$ (at 10 μM) and control-DNA conjugate (at 20 pM) in 1× TBST/tRNA buffer
2. Incubate at room temperature for 1 h
3. **Capture:** Apply sample for 10 min to a column packed with 5 μL of Ni^{2+} charged NTA-agarose
4. Remove the "flow through" sample from the column
5. **Wash:** Wash the column five times with 100 μL 1× TBST/tRNA
6. **Elute:** Elute the column in 25 μL 72°C 1× TBST/tRNA for 10 min
7. Add Bcl-x$_L$ to a final concentration of 10 μM in a final volume of 50 μL in 1× TBST/tRNA
8. Repeat steps 2–6

Figure 13.14. The appropriate functional evaluation of an encoded positive control ensures binding to the target and the ability to recover a PCR-competent DNA in a process that recapitulates the screening process.

An additional factor influencing the ultimate recovery of the positive control and library compounds associated with the solid-phase target is the stability of the linkage between the target and the solid phase. The linkage can be noncovalent but must be sufficiently stable under the conditions of the wash steps to ensure minimal loss of target and target-associated compounds. The stability of the linkage can vary dependent upon the capture format and should be determined empirically for each system prior to initiating the screen.

The detection limits of PCR should be considered and will influence the quantity of library to interrogate in each individual affinity-based screening event. The absolute number of each library member that is exposed to the target is critical, and that number will be influenced by the recovery efficiency of each round of the affinity screen. The lower limit for observing signal in the screen is defined by the efficiency of the PCR and should be empirically determined and optimized to ensure that both the positive control and the negative control are appropriately recovered.

DNA-encoded control compounds are valuable tools that allow for the optimization of screening conditions, resulting in increased recovery of target-binding positive controls and a decrease of background signal generated by the nontarget-directed recovery of PCR-competent library members. The evaluation and optimization of the encoded controls can be carried out in the context of an encoded library, or if library conservation is a priority, a PCR-competent DNA that recapitulates the design of the library DNA can

Pre screen compound frequency

• Control 1
• RT-PCR-determined relative library concentration = 1x
• RT-PCR-determined relative encoded-control concentration = 0.0001x
• Frequency is 1 in 10,000 (0.0001)

Post screen compound frequency

• Control 1
• RT-PCR-determined relative library concentration = 1x
• RT-PCR-determined relative encoded-control concentration = 0.01x
• Frequency is 1 in 100 (0.01)

Calculation of enrichment

Post screen frequency/pre screen frequency = 0.01/0.0001 = 100-fold

Figure 13.15. RT-PCR is a tool that can determine the relative concentrations of individual known DNA sequences, allowing for the calculation of an enrichment value for each compound relative to other members in the encoded library.

be used as a library substitute. DNA-encoded control compounds allow for the evaluation of the enrichment potential for compounds that have known affinity for the target. To achieve this, each encoded control is introduced into a library at the average concentration of a single library member. If a library is to be screened in which each library member is at a concentration of 20 pM, then the encoded controls should be introduced into that library to a final concentration of 20 pM. A 50 μL library aliquot at a concentration of 20 μM that contains 1,000,000 unique members is calculated to have 1 fmol of each library member. 1 fmol of each of the encoded controls would be introduced into the library and would represent unique library members having known affinity for the target.

If an encoded compound library or DNA-only library is used in the evaluation, then both quantitative PCR and standard PCR amplification followed by DNA-sequencing analysis could be applied to determine the behavior of the controls. The control-evaluation and enrichment-optimization phases are dependent on determining the relative levels of the control compounds, and this process of optimization can be expedited through the application of quantitative PCR (Fig. 13.15). PCR primer sets can be designed that recognize common elements in all library members (i.e., including controls), and additional primer sets can be designed that are unique to specific control compound sequences. Primer sets that are common to all encoding DNAs allow for the total DNA concentration to be empirically determined, and primers unique to one control compound allow for the determination of the concentration of that specific compound. The rapid determination of enrichment values for the control compound can be determined by dividing its post screen frequency by its pre screen frequency.

DNA-encoded positive controls having confirmed affinity for the screening target (i.e., as determined by independent biophysical or biochemical methods) are predicted to be enriched through the affinity-screening process relative to nonbinding library members or DNA-encoded negative controls. Enrichment values of the positive control compounds under a preliminary set of screening conditions should be determined and represent a starting point for optimization prior to the screening of compound libraries. Conditions can be modulated to improve the recovery of the encoded positive control and to reduce the recovery of the nontarget-binding library, resulting in a net increase in the enrichment values and consequently a better signal-to-noise ratio. Blocking nucleic acids, soluble blocking proteins, salt concentrations, and detergent choice and concentrations are all components of the screen that can be altered and explored to improve the recovery and enrichment of target-binding encoded compounds.

A DNA-tagged control compound represents an individual library member of defined affinity present in the library at a defined frequency and serves both as an embedded affinity marker (i.e., internal control) to reference library compounds and as a process control to ensure that the multistep technique was effective. Including multiple control compounds that span a range of affinities for the target allows for a more precise prediction of the affinity of enriched library compounds. The incorporation of both positive and negative control molecules provides for the framing of the signal and noise for each library screen against a particular target under a defined set of conditions. The framing of the signal and noise is invaluable during further optimizations to reduce undesirable library interactions with the target and to increase the recovery of compounds specifically interacting with the target.

13.3 THE SCREEN

Affinity-based screening is elegant in its simplicity and powerful in the breadth of method modifications available to modulate compound recovery during the process of interrogating compound libraries. Practically, the ability to effectively segregate compounds of a DNA-encoded library as a function of their affinity for the target is facilitated by the ability to partition the target and bulk library to separate compartments. The immobilization of the target onto a solid phase allows for a phase separation to occur between library and target, thereby enabling a partitioning of compounds within the library based on their affinity for the target. The up-front development of an optimized system including a modified target, the identification of conditions supporting a functional target, an appropriately modified solid-phase capture matrix, encoded control compounds, and the optimization of conditions that maximize the enrichment of target-binding compounds ensures a process that is well suited to identify library compounds having affinity for the target.

13.3.1 The Screen: Mock Considerations

In an ideal affinity-based screening system, the affinity of a library compound for the target is the only factor affecting the enrichment of that library compounds. In reality, there are many structural and process-necessary components that are not the desired protein target that could have affinity for library members, resulting in the generation of "false-positive" hits. There are also linking and coding regions of the library members that are not the diversity elements (i.e., compounds) but could contribute to affinity toward target and nontarget components during the screening process.

When designing an affinity-based screening strategy, it is critical to recognize that there are many anticipated and unanticipated nontarget components that could result in the enrichment of library members. Although precautions are taken to reduce false positives through suppression of anticipated components, a general strategy is to routinely run a parallel "no-target" screen (i.e., a "mock" screen) to identify compounds enriched independent of target. The "mock" screen contains all components of the target screen with the exception of the target. The identification of library compounds that have affinity for components in the "mock" allows for their dismissal as "false positives" in the target screen through their identification and virtual removal from the postscreen compound-enrichment dataset. Designing a screening process that affects the suppression and identification of "false positives" allows for the unequivocal identification of library compounds having affinity specific to the target of interest.

The structural and process-necessary components of a chromatography-based affinity screen can be functionally diverse and separated into three major classes: the first are the plastics of labware including tubes, plates, and transfer equipment; the second are the solid-phase matrices such as agarose or plastic; and the third are target-capturing "association mediator" proteins and functionalities that are immobilized on the solid-phase matrix such as streptavidin or Ni^{2+}–NTA. All of these components are off-target but could have substantial affinity, specific or nonspecific, for members of a DNA-encoded library leading to their enrichment. During a screen, the recovery of library members that have affinity for nontarget components effectively increases the number of "false-positive" compounds, reduces the enrichment of "real" hits, and in extreme cases prevents the identification of target-binding compounds. Practically, these off-target interactions need to be suppressed while maintaining the desirable interactions between the library members and the target. The general application of nonionic detergents, salts, "blocking nucleic acids," and "blocking proteins" in the buffered solutions throughout the screens is effective at reducing library-compound affinity for labware and solid-phase matrices.

More directed approaches are necessary to limit the recovery of library compounds against capturing "association mediator" proteins and functionalities. Ni^{2+}–NTA presents a metal ion that could bind to certain library members (e.g., imidazole-containing functionalities). A simple method for suppressing these undesirable interactions is to include moderate levels (e.g., 20 mM) of free imidazole in the screening buffers. Streptavidin is an example of an "association mediator" protein that binds to and captures with high-affinity biotin-tagged target proteins. Streptavidin binds to a small molecule, and these small-molecule binding sites are what make association mediators of this sort vulnerable to generating false positives out of the libraries. Streptavidin is a tetrameric protein that has four equivalent biotin-binding sites and when included as a component of a screen is in its own right a protein target. To reduce the potential for identifying library members that have affinity for streptavidin, certain measures can be taken to render the most likely site (i.e., the biotin-binding sites) "blocked" from being available to bind to library compounds. The addition of saturating amounts of free biotin subsequent to the capture of the biotinylated target protein by the solid-phase streptavidin will block any remaining biotin-binding sites and minimize the enrichment of library compounds that bind to these off-target sites.

If the occurrence of "false-positive" hits is significant in a particular library and this behavior compromises the ability to identify "real" target-specific hits, there are effective methods that physically reduce or deplete these compounds from the library prior to the target screen. Physical reduction of compounds that have affinity for non-target components of the screening process is most effective when carried out on the library before the target-screening event. The library of compounds can be depleted of off-target compounds by performing a "mock" screen and recovering the library of compounds that "flows through" after the "mock" binding event. The off-target compounds in the library are retained on the components of the system that they have affinity for, and the depleted "flow through" represents the "precleared" library to be moved into the target screen. The "preclearing" process effectively reduces the off-target compounds with each "mock" exposure and in extreme cases must be repeated a number of times to deplete their levels below detection.

13.3.2 The Screen: DNA-Only Screens

A DNA-encoded library is composed of a combinatorial set of molecules, each having a compound component and a DNA component. By virtue of each unique compound being coded for by a unique DNA sequence, the library is not only made up of many different compounds but also many different DNA sequences. The design of the encoding DNAs should consider nucleotide composition and the length of random stretches toward minimizing the potential for DNA-driven (i.e., aptamer) target affinity. The encoding of libraries with double-stranded DNA minimizes the potential for three-dimensional DNA folding and dramatically reduces the frequency of DNA-directed affinity for most protein targets. If there is a concern that the target-based enrichment of specific library members during an affinity-based screen is not compound dependent and is DNA dependent, a DNA-only version of the library should be screened in parallel. The DNA-only library represents the diversity of the DNA in the encoded compound library without the corresponding chemical compound. If a screen against a target produces the enrichment of specific sequences from the DNA-only library and also in the encoded compound library, then the affinity for the target by those specific DNA sequences can be attributed to the DNA component of the encoded construct. These DNA sequences are being enriched as a result of affinity that the encoding DNA has for the target and should not be considered in the analysis to identify target-binding compounds.

13.3.3 The Screen: Library Exposure to Target

The primary objective of screening naïve libraries is to cast a "wide net" and recover all compounds that have appreciable affinity for the target. The goal of this "hit-identification phase" is common between traditional high-throughput and affinity-based library screening, but there is a critical parameter (e.g., compound concentration) that differentiates the two methods. For an HTS approach in which a single-compound assay occurs in a single well, a high compound concentration is routinely the component of the system that is driving the interaction with limiting target. Compound concentrations in biochemical or cell-based high-throughput assays can be as high as 10–30 μM. In an affinity-based screen, the compounds are presented to the target as mixtures where the concentration of each individual compound in the library can be very low (i.e., pM or less). The low concentration of each individual library compound dictates that the two primary determinants responsible for driving the interactions between the target and the compound during the screen are the compounds' affinity for the target and the target concentration. Although compound concentrations in complex libraries can be low and due to synthetic efficiency differences vary significantly from compound to compound, it is not a critical factor influencing recovery as long as the concentration of the compound does not fall below the detection limits imposed by PCR and DNA sequencing.

To ensure that all compounds within the library have an opportunity to be recovered under the screening conditions, it is important to ensure that the target is in stoichiometric excess over the compounds that bind to the target. Naïve libraries by design cover a breadth of chemical diversity and are expected to contain relatively few target-specific binding compounds. Even naïve libraries of high diversity and compound numbers contain relatively few hits, and except in extreme situations, there is a stoichiometric excess of target over target-binding compounds. In the extreme case where compound hits are in excess over the target, a competition situation will be set up where the higher-affinity compounds will have an advantage and be recovered preferentially over lower-affinity compounds. If there is a concern that a large number of hits might be in excess over the target, the screen should be repeated after adjusting the stoichiometry of the target to library.

In a properly optimized screen, the concentration of the target and the affinity of the individual library compounds for the target are the primary determinants of compound recovery. A solution-phase incubation of the library and the target allows for an accurate modulation of the target concentration to affect the affinity characteristics of the recovered target-binding compounds. An affinity-chromatography-based library screen is essentially a highly parallel binding assay. If the binding assay is based on the system reaching equilibrium, the occupancy of the library compounds on the target will be defined by both the target concentration and the affinity of the compound for the target. In this highly parallel binding assay, with the readout for each compound being its recovery (i.e., frequency relative to other library members) in the postscreen output, the modulation of the target concentration will affect the low-affinity threshold for compound recovery. In an ideal situation, if the target concentration is set to 10 μM, each library compound having a 10 μM K_D will be 50% bound to target under equilibrium conditions. Compounds of higher affinity will have a higher % bound to the target and will be recovered at a greater level, while compounds of lower affinity will have a lower % bound and will be recovered at a lower level. Setting the target concentration and exposing the library under equilibrium conditions will define the affinity characteristic for individual library members recovered. Higher target concentrations will result in the recovery of compounds having a wider range of affinities for the target (i.e., low and high affinity), while a reduction of the target concentration will reduce the recovery of the lower-affinity compounds and maintain the recovery of higher-affinity compounds.

Target	1× TBST/tRNA buffer
Recombinant human Bcl-x$_L$	50 mM Tris pH 7.6
Ser2-Arg212 with a C-terminal His$_6$ tag	150 mM NaCl
Molecular mass: 24 kDa	0.05% Tween-20
Stationary-phase matrix	1 mg/mL yeast tRNA
Ni^{2+} charged NTA-agarose	**Control-DNA conjugate**
beads loaded in a 5 μL bed	BIM BH3 peptide-DNA
volume pipet-tip column	**DNA-encoded library**
	1,000,000 unique compounds

1. **Bind:** In a volume of 50 μL combine Bcl-x$_L$ to 10 μM, DNA-encoded library to 20 μM, and control-DNA conjugate to 20 pM in 1× TBST/tRNA buffer
2. Incubate at room temperature for 1 h
3. **Capture:** Apply sample for 10 min to a column packed with 5 μL of Ni^{2+} charged NTA-agarose
4. Remove the "flow through" sample from the column
5. **Wash:** Wash the column five times with 100 μL 1× TBST/tRNA
6. **Elute:** Elute the column in 25 μL 72°C 1× TBST/tRNA for 10 min
7. Add Bcl-x$_L$ to a final concentration of 10 μM in a final volume of 50 μL in 1× TBST/tRNA
8. Repeat steps 2–6

Figure 13.16. A library screen is the application of an optimized process of target binding, solid-phase capture, solid-phase washing, and the elution of target-associated compounds.

The "binding step" is a point where the screen can be "tuned" to adjust the ultimate affinity-based recovery of library compounds.

An effective preparation phase prior to initiating a library screen provides the appropriate tools and conditions required to implement a successful screening campaign (Fig. 13.16). Each target-based screen begins with the introduction of well-characterized DNA-tagged control compounds into the library at a concentration that is equivalent to the concentration of a single library member. The library is then combined with the target under the predetermined buffer conditions (i.e., conditions supportive of target function and the minimization of nonspecific interactions between the library and the components of the system). The mixture is incubated for a time consistent with the system reaching equilibrium (e.g., between 30 and 60 min) to ensure that the correlation of affinity and compound recovery is affinity based (i.e., dependent on the K_D of the compound for the target). The purpose of this "binding step" is to allow equilibrium to be reached between the target and all library members to effect the recovery of each compound based on its affinity-based occupancy on the target.

With the target and library at equilibrium, the target now needs to be bound to a solid-phase matrix to enable the phase separation of solution-phase nonbinding compounds and target-associated binding compounds. The target contains a molecular handle enabling "capture" by a solid-phase matrix that is modified with the appropriate "association mediator." The incubation of the library–target mixture with the solid phase will affect the phase transfer of the target, and associated library compounds, from the solution phase to the solid phase. To ensure the recovery of all target molecules and associated library compounds, the amount of matrix should be such that the capacity for the target exceeds the target amount used in the "binding step."

A convenient format for holding small volumes of solid-phase matrix is in a pipet-tip-derived microcolumn (PhyNexus, Inc.). The column holds the matrix and allows for the screening solution to be easily and consistently passed over the solid phase. The incubation to effect efficient association of the tagged target (e.g., biotinylated or poly-histidine) to the capture matrix (e.g., streptavidin–agarose or Ni^{2+}–NTA agarose) can be as rapid as a few minutes but is typically 5–10 min to ensure complete capture.

The "binding step" and the "capture step" can also be reversed with the association of the target and the solid-phase matrix occurring prior to the incubation of library and target. In this configuration, the incubation of the library and target would occur with the target already associated with the solid phase. The precapturing of the target onto the solid-phase matrix prior to exposure to the library is a valuable approach if there is a fraction of the target that does not contain a capture tag. In a screen that captured the target subsequent to the binding step, the target molecules that do not have a capture tag, but are associated with library compounds, would be lost along with their bound library compounds. The precapture of the target effectively ensures that all target molecules used in the screen are associated with the solid phase. An additional advantage of having the target prebound to a solid phase is that the target, having reduced degrees of freedom as compared to a target in solution, is at a very high "effective local concentration" on the solid surface. The high local concentration of the target on the solid phase that can be obtained by presaturating the matrix with the target drives the interaction between the lower-affinity library members and the target. With the target on the solid phase, the library is incubated with (i.e., exposed to) the target for a time consistent with the system reaching equilibrium (e.g., between 30 and 60 min). As in the solution-phase equilibrium binding step, the solid-phase binding step functions to ensure that the correlation of affinity and compound recovery is affinity based (i.e., dependent on the K_D of the compound for the target).

Whether or not the library-to-target binding event occurs in solution or on the solid phase, both methods converge at a point when equilibrium is reached between target and library and the target is associated with the solid phase. With equilibrium reached between library and target, a partitioning can now occur between the solid-phase target-associated library compounds and the balance of the solution-phase compounds that do not bind to the target. The solution containing the library is separated from the solid-phase matrix and represents what is referred to in chromatography and affinity screening as the "flow-through" material. The "flow-through" material contains compounds that do not associate with the target under the conditions of the screen and are not of primary interest.

13.3.4 The Screen: Removing Weakly Associated Compounds

After the flow through is removed from the matrix, there is a void volume associated with the solid phase, and within this volume are compounds that do not have affinity for the target. To remove the compounds in the void volume, the solid-phase matrix is "washed" with the buffered solution used in the binding step of the screen. The goal of the "wash step" in this screen is to remove compounds that do not meet the target-affinity requirements of the screen while retaining the compounds on the solid-phase target that do meet the target-affinity requirements.

Monitoring the bulk levels of DNA-encoded library and the individual encoded controls during the process of washing the solid-phase target allows for sufficient washing to occur to remove weakly associated or nonspecific compounds while still retaining the target-associated specific compounds. DNA encoding of library compounds enables the use of sensitive tools to monitor the levels and to ultimately decode the compound composition of a mixture throughout the screening process. Specifically, quantitative PCR can be used to monitor the overall level of library DNA (e.g., if the primers complement library-constant regions), or it can more specifically be applied to determine the level of a specific encoding DNA (e.g., if the primers complement a compound-specific sequence). To keep with the "wide net" approach and a desire to recover all hits that meet minimum affinity criteria, washing of the solid-phase target should be kept to a minimum. The washing should continue and be monitored using quantitative PCR until the DNA-library concentrations coming off of the solid-phase target decrease rapidly to a stable and minimal level. The phase separation has enabled the target-associated compounds to be retained on the solid phase with the balance of the library compounds in the "flow through" or in subsequent volumes of "wash" buffer.

At this stage, the compounds in the library are at a binding equilibrium with the target, a phase separation has been implemented, and the weakly associated compounds have been washed from the solid-phase target. The compounds associated with the target need to be recovered, and their DNA tag decoded to determine the structures of the target-associated library compounds.

13.3.5 The Screen: Recovery of Target-Associated Compounds

With the target-binding compounds effectively partitioned to the solid phase and separated from the compounds not binding to the target, the next critical process is to release the desirable target-associated compounds from the solid-phase target by "elution" into the solution phase. The library members that are released, or eluted, from the target after a single screening enrichment cycle will have an altered distribution relative to that of the starting library. The frequency of compounds having affinity for the target will be increased in the elution relative to their frequency in the starting library, and the frequency of nonbinding compounds will be unchanged or reduced in the elution. At this point (i.e., screening round 1), a small fraction of the elution sample can be prepared for decoding analysis to determine the frequency of each library member, and the balance of the sample will move into a subsequent affinity-based screening cycle (i.e., screening round 2) if necessary.

The desired outcome of the elution process is to preferentially release the target-associated compounds from the solid phase. There is the potential during the screen for library compounds to bind to nontarget components of the system and be retained throughout the washing. Plastic of the reaction vessel, solid-phase matrix, and association mediators such as streptavidin all have the potential to bind to library members, and although measures have been taken to suppress this process, undesirable "false-positive" compounds can remain and could be released during the elution step. The elution of compounds associated with the nontarget components of the screen increases the noise in the system and reduces the enrichment of target-associated library compounds. The process of recovering library members from the solid phase is achieved if either the target is released from the solid phase or if the encoded compounds are released from the solid-phase target. To reduce the release of compounds not associated with the target, it is desirable to effect the release of the target and associated compounds from the matrix or the compounds from the solid-phase target using relatively mild conditions.

Target release to recover associated compounds is appropriate if the protein target is bound to the solid phase through transient association mediators such as polyhistidine targets and a solid-phase Ni^{2+}–NTA agarose. A His_6-tagged target and associated bound compounds can be released from the matrix in washing buffer with the addition of 250 mM imidazole or 100 mM EDTA. The relatively gentle target-release methods will effectively release the target with associated compounds and leave behind compounds associated with nontarget components of the system. An additional processing step, to remove the elution agents and the released target from the encoded library compounds, must be applied to the elution sample so as not to interfere with subsequent PCR amplification, decoding analysis, or additional screening steps.

When targets are bound to the solid phase covalently or via a streptavidin–biotin interaction, the high-affinity interaction precludes mild release of the target from the solid phase. Target release is possible under harsh conditions, but this would strip all target-associated and nontarget-associated compounds from the components of the system. A simple but elegant method to recover target-associated compounds from covalently immobilized or biotin-tagged targets is to induce a reduction of the compounds' affinity for the target while at the same time not releasing the target from the solid-phase matrix. The reduction of a compound's affinity for the target is enabled through a disruption of the binding surfaces of the target by raising the temperature above the melting temperature of the target protein. The increase in temperature induces a subtle disruption of the protein's tertiary or quaternary structure altering the available binding surface. A large percent of protein targets of therapeutic interest are the native, folded, and functionally relevant form of the protein. As the affinity of target-associated library compounds for the target is the result of key interactions with a conformation-dependent site, thermal denaturation of the target distorts the binding sites, resulting in reduced affinity for the target-associated compounds, thereby increasing their concentration in the target-free solution phase. The thermal elution method is effective against a majority of biological targets and does not require subsequent sample processing steps such as purification or neutralization prior to either a second round of affinity-based screening or sample amplification and decoding. The advantages of the thermal elution make it an attractive method not only for covalently coupled and biotinylated targets but also for tagged targets using association mediators like His_6 and Ni^{2+}–NTA.

There are a number of additional methods effective at eluting target-associated compounds, each requiring a post-elution processing step prior to either an additional round of screening or PCR amplification in preparation for DNA sequencing (Table 13.2).

TABLE 13.2. Additional elution methods for the release of target-associated library members

Target tag/matrix modification	Elution method	Elution condition	Effect	Post-elution processing
All/all	Low pH	200 mM Gly-HCl, pH 2.2	Denature target	Neutralization
All/all	Organic/high pH	Triethylamine pH 10	Denature target	Neutralization
All/all	High salt	2.5 M $MgCl_2$	Disrupt ionic	Desalt
All/all	Protein denaturant	6 M guanidine HCl	Denature target	Desalt
His_6/ metal-NTA	Histidine competitor	250 mM imidazole	Release His_6- tagged target	Desalt
GST/ glutathione	Glutathione competitor	10 mM reduced glutathione	Release GST- tagged target	Dialysis/size exclusion

The objective of the elution step is to apply a method that ensures a high level of target-associated compound recovery, minimizes the recovery of nontarget-associated compounds, and minimizes post-elution processing. The absolute number of each compound in a diverse DNA-encoded library is low (i.e., at 1 fmol), so minimizing loss throughout the process is critical. The "elution" step should be high yielding and specific. Minimizing sample processing of the elution sample reduces the potential for losing rare but functionally important library compounds.

To ensure sufficient enrichment of even relatively low-affinity molecules (i.e., in the µM range), multiple rounds of enrichment are routinely required. The elution output from the first round represents the input library for the second round of affinity-based screening and should be compatible with the screening conditions. Thermal denaturation as an elution method is performed in the same buffer used in the initial binding incubation step between target and library, and so it is compatible with subsequent screening steps without additional processing. After two rounds of affinity screening, the enrichments of target-specific compounds will be the product of the enrichments observed in the first and second rounds. The enrichments of target-associated compounds and the enrichments of nontarget-associated compounds will diverge throughout the screening process and be compounded as additional screening rounds are completed. Rounds of screening can effectively be continued until the concentration of DNA-encoded library compounds in the output of the last round is limited by PCR-based detection and signal recovery.

Ten percent of the elution from the first round of screening is retained for sequence analysis and 90% is used as the input library for the second round of screening. The second round of screening is set up identically to the first round with the exception that the input library sample is not naïve but is the round 1 elution sample (i.e., which has been through one round of affinity-based enrichment).

With DNA-encoded libraries, it is not possible to amplify the library between rounds of screening, so the potential for compound-enrichment artifacts due to library-amplification effects (i.e., the result of biased amplification and resynthesis of a subpopulation of library members) as the result of multiple screening and amplification cycles is not an issue. The compounds identified in the final elution of an affinity-based screen of a DNA-encoded library are the actual compounds that were synthesized during the original library synthesis. The lack of an amplification step to recreate the library between rounds helps to limit frequency changes of library compounds not associated with the affinity of the compound for the target.

13.3.6 Decoding of Screening Output

If there are compounds in the starting library that meet the criteria of the screen and bind to the target, then these compounds are present in the product of that screen (i.e., the elution). The physical process of partitioning compounds that have affinity for the target from those that do not is complete. The elution from a round of screening contains some reduced fraction of the compounds present in the naïve starting library, and the frequencies of these compounds are altered. Since by design there is a one-to-one correlation between a library compound (i.e., the product of the chemical synthesis encoded by the DNA) and its encoding DNA, amplification and subsequent reading of the DNA sequences present in the elution allows for the determination of the relative frequencies of each compound in the elution sample. The information captured (i.e., the synthetic history for each library compound) in the encoded DNAs of the elution needs to be read and translated to realize the output from the affinity screen.

DNA sequencing will ultimately be applied to read out the sequences encoding the compound structures in the elution, but the concentration of the encoding DNAs is too low to be sequenced directly and must be amplified. PCR is the preferred method of DNA amplification, and to efficiently amplify all library members equivalently, amplification primers need to be designed that are complementary to the constant regions of the DNA flanking the variable encoding regions. PCR amplification that is directed by such library-constant primers ideally amplifies all DNA equivalently, but to limit the potential for the PCR to alter the compound frequencies, it is prudent to limit the cycles of amplification by PCR to as few as possible to generate amounts sufficient to sequence.

In addition to serving as a method to amplify the output of a screen, the PCR must also generate product that is compatible with the sequencing methods being applied. The highly parallel "next-generation" DNA-sequencing methods (e.g., Illumina Inc.) are well suited to generate the millions of individual sequences to support encoded library analysis. The generation of DNA-sequencing-compatible products by PCR involves the incorporation of additional adapter sequence extensions that are unique to each sequencing method (Fig. 13.17).

With the high number of DNA sequences generated by the next-generation methods, it may not be necessary to use an entire sequencing run to analyze a single screening assay. Multiple independent affinity screens can be combined and analyzed simultaneously in a single "multiplexed" sequencing run. To enable a "multiplexed" sequencing run, multiple primer sets need to be created that are competent to amplify library, enable the DNA sequencing, and encode a unique sequence "identifier." The "identifier" would be several bases long, be unique, and would not interfere with the primary functions of

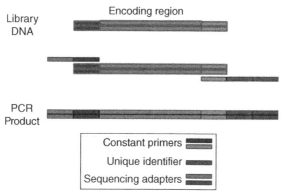

Figure 13.17. PCR-based amplification of library DNA with an appropriately designed primer set enables the creation of a product compatible with commercially available DNA-sequencing techniques. For color detail, please see color plate section.

the amplification primers. These unique PCR primers allow for the independent amplification of multiple unique elution samples and their subsequent mixing prior to DNA-sequencing analysis to more efficiently exploit the highly parallel next-generation sequencing technologies by combining samples.

Affinity-based screening methods, being fundamentally chromatographic methods, will alter the distribution of compounds in the library after the screen depending upon their individual affinity for the target. As DNA-encoded libraries are combinatorial, the chemical reactivity of different building block sets will vary and ultimately influence the final concentration of each final product compound. The potential for frequency variability from library member to library member in an encoded library makes it critical to empirically determine the values. A frequency profile for each library defines the abundance of each library member in the library before a screen (i.e., the prescreen), and when compared to the frequency profile after the screen (i.e., the postscreen), enrichment values can be calculated for each library member (Fig. 13.18).

The evolution of encoded library technologies has been paralleled by an increase in the capacity and reduction in the price of DNA-sequencing technologies. The current "next-generation" sequencing methods available from various vendors (e.g., Illumina Inc., PerkinElmer Inc., F. Hoffmann-La Roche Ltd.) allow for the generation of hundreds of millions of individual sequences. The vast number of sequences that can be generated to analyze the output from a screen allows for high confidence in the enrichment values for each library member.

If the diversity of the library is too large to confidently determine a frequency profile for each of the compounds within the library before the screen, it may be necessary to approximate the distribution of compound frequencies. With the approximation of initial frequencies, absolute enrichments cannot be determined, but sequencing of the postscreen output would allow for hit identification through the highlighting of compounds that deviate in their frequencies from the approximation.

Pre screen compound frequency

- Compound 1
- 1 million total library sequences obtained
- 100 copies of compound 1 observed in the initial library
- Frequency is 1 in 10,000 (0.0001)

Post screen compound frequency

- Compound 1
- 2 million total library sequences obtained
- 10,000 copies of compound 1 observed in the screened library
- Frequency is 1 in 200 (0.005)

Calculation of enrichment

Post screen frequency/pre-screen frequency = 0.005/0.0001 = 50-fold

Figure 13.18. DNA sequencing allows for the screen-dependent enrichment value (i.e., a function of the change in relative compound frequency within the library) to be determined simultaneously for all compounds.

13.4 TUNING HIT RECOVERY

Although an affinity-based screen is simple at its core, modifications can be applied to the basic process to generate additional information on the particular properties of the library compounds. Criteria can be applied to the screen that influence the distribution of compounds identified, such as competition with target-binding proteins, peptides, or small-molecule ligands, and modulation of the kinetics of binding. A basic affinity-based screen that casts a "wide net" and is correspondingly tuned to identify compounds that meet a minimum threshold affinity for the target is agnostic to the site of compound binding and identifies all compounds that satisfy the criteria. Adjusting the components or conditions of a screen and determining the effect of these changes on the target-specific enrichment of each library member have the potential to more extensively characterize each library compound. If a screen in which compounds were previously enriched is repeated in the presence of saturating levels of a known ligand and the enrichments of specific compounds are reduced, the conclusion is that the compounds that have been prevented from binding to the target are "competitive" with the known ligand. The compounds that are prevented from binding could be either directly competitive or indirectly competitive, and the compounds that are not prevented from binding are interacting with a remote site on the target not altered upon ligand binding. The change in the behavior of each library compound upon a conditional change to the screening process defines, at a finer resolution, how the compound is binding to the target and provides valuable additional information when evaluating each library compound. Competitive screens are particularly useful when a library screen yields many hits as additional information is generated for each hit compound that might help in the postscreening chemical synthesis prioritization.

An affinity-based screen is essentially a highly parallel binding assay performed at a single target concentration. In a traditional radioligand binding assay [22], the relationship between the ligand–target frequency and target concentration is determined and fit to an equation to calculate the affinity constant. It is possible to apply these principles to affinity-based library screens to differentiate between compounds of different affinities for the target. As the screen is replicated at reduced target concentrations, the recovery of lower-affinity compounds should be affected more dramatically and reduced with respect to the recovery of higher-affinity compounds. With the initial screen of naïve libraries casting a "wide net" to recover the majority of compounds having specific affinity for the target (i.e., the initial screen is carried out at high target concentrations), prioritization of high-affinity hits for subsequent chemical synthesis and characterization may be appropriate. The enrichments of higher-affinity compounds will be maintained at reduced target concentration, while the enrichments of low-affinity compounds will diminish. It is important to recognize that the library screen is applying the screening conditions to all library members in the same sample at the same time. The simultaneous evaluation of how the conditions of the screen affect all library members allows for a relative ranking to be applied to all library compounds in response to the changes in the conditions.

Altering the target concentration across replicate screens enables the differentiation of hit compounds based on the affinity of the compound for the target. The conditions of replicate screens can also be modulated to affect the recovery of compounds based on individual components of the binding kinetics for the target. Library compounds having a slower off-rate (i.e., the kinetic binding constant k_d) are of particular interest because long residence times on the target can be an advantage for many applications (e.g., an extended therapeutic effect or extended target coverage for an imaging application). A series of screens can be replicated while holding all conditions the same except the number of washes of the solid-phase target (i.e., the cumulative time of the washing). With more washes, the compounds that have longer off-rates will be preferentially retained over compounds having shorter off-rates, resulting in the long off-rate compounds being highlighted as a result of greater enrichment in the screen. In an additional application of the concept, a series of screens can be replicated while holding all conditions the same except the incubation time between the target and the library. The reduction of the incubation time should favor the recovery of library compounds that have a faster on-rate (i.e., the kinetic binding constant k_a), and their enrichment in a screen will be less affected than compounds having a slower on-rate.

The modulation of the conditions in subsequent screens can partition a set of compounds that satisfy the requirements of an initial "wide net" screen into subsets based on imposed parameters. Reducing the target concentration in a screen will affect the recovery of compounds as a function of the target-specific on-rate for the compound (i.e., the on-rate of the compound for the target has the units of $M^{-1}s^{-1}$, which is a function of target concentration and target-compound incubation time). Altering the time of library incubation with target or library–target washing will affect the recovery of compounds based on the compounds' target-specific on-rate and off-rate, respectively (i.e., the on-rate has the units of $M^{-1}s^{-1}$ and is a function of incubation time, while off-rate has the units of s^{-1} and is a function of wash time). The screening of libraries allows for the evaluation of large sets of compounds simultaneously and to determine how conditional changes to the screen alter the relative recovery of each compound. The additional information generated in subsequent screens can allow for more detailed characterization of individual library compounds and a more informed decision as to which compounds to prioritize for synthesis and characterization.

13.5 ENRICHMENT DATA AND HIT IDENTIFICATION

The goal of affinity-based screening is to interrogate encoded library mixtures to identify, if they are present, the rare set of molecules in that library that bind specifically to the particular target of interest. There is as with any screening platform a potential for the misidentification of compounds as positive (i.e., false positive), and a rigorous process should be put in place to minimize this outcome. Designating a compound that is present in the output from an affinity screen, a "positive hit" is dependent on several criteria being met (Table 13.3). The first criterion that must be met is that a compound must be enriched significantly relative to its initial frequency in the starting library. The receipt of sequencing data from a screen initiates a process that allows for the determination of compound counts for all library members present in the elution sample, and these counts are transformed into enrichment values (i.e., the change in frequency from the initial library sample to the postscreen elution sample) for all of the compounds. The enrichment value is inversely correlated with the affinity of the compound for the target, and there is a minimum positive enrichment value that is statistically significant relative to the majority of other compounds in the sample. The second criterion is that the enriched compound must be a member of a set of enriched compounds and within that set there must be structural relationships coincident with enrichment (i.e., SAR). Because of the combinatorial nature of the library, the synthesis creates compounds of varying degrees

TABLE 13.3. Compounds 2, 4, and 7–10 satisfy the 4 hit-defining criteria (e.g., target enrichment, SAR, mock enrichment, and target promiscuity) and would be designated as validated hit compounds

Compound #	Target enrichment	SAR	Mock enrichment	Target promiscuity	Validated hit compound
1	X		X	X	
2	X	X			X
3	X			X	
4	X	X			X
5	X			X	
6	X		X		
7	X	X			X
8	X	X			X
9	X	X			X
10	X	X			X

of relatedness, and it is highly unlikely that only a single structurally unique compound will be functional (i.e., have affinity for the target). The third criterion is that the compounds of interest must not be enriched to a significant level in the "mock" screen. Enrichment in the "mock" screen as well as in the target screen implies that the enriched compounds have affinity for a component of the system other than the target and these compounds are not of interest. The fourth criterion is that the enriched compounds must not be promiscuous in their binding and as such must not be enriched in an independent screen against an unrelated target (i.e., they must be target specific).

Additionally, the enrichment data generated from screens carried out in the presence of target ligands can be supportive of target-specific binding and can also provide mechanistic insight. The reduction in the enrichment of compounds in the presence of the target ligand (i.e., competition) is consistent with the compound binding competitively to the target, but the lack of competition is not inconsistent with target-specific binding at a site not affected by the ligand.

13.6 CLOSING REMARKS

A necessary property of any compound that exerts a direct effect on a specific target's function is affinity for that target. The compound must have affinity for the target and make critical contact interactions to affect with functional implications the target's activity. The affinity-based screening of highly diverse combinatorial mixtures is an extremely efficient and effective method for interrogating chemical diversity in search of functional compounds. The fundamental concept is elegant in its simplicity and draws practical methods from the purification techniques of basic biochemistry and the nucleic acid manipulation techniques of molecular biology. The ability to alter the conditions (e.g., target concentration, incubation times, wash duration, ligand competitions) of the screen and simultaneously monitor the behavior of all library members relative to each other is extremely valuable to enable the partitioning of compounds into various functional and nonfunctional groups. The advances in the capacity to sequence DNA of next-generation sequencing technologies can be exploited to accurately characterize a library's distribution of compounds. The ability to identify each member of a library and to determine its corresponding frequency before and after an affinity-based screen allows for enrichment values to be generated and compounds having affinity for the target to be identified. The thoughtful application of affinity-screening techniques to the screening of encoded libraries enables the rapid exploration of new and underexploited chemical diversity toward the identification of valuable compounds.

REFERENCES

1. Smith, G. P. (1985). Filamentous fusion phage: novel expression vectors that display cloned antigens on the virion surface. *Science 228*, 1315–1317.
2. Roberts, R. W., Szostak, J. W. (1997). RNA-peptide fusions for the in vitro selection of peptides and proteins. *Proc. Natl. Acad. Sci. USA 94*, 12297–12302.

3. Brenner, S., Lerner, R. A. (1992). Encoded combinatorial chemistry. *Proc. Natl. Acad. Sci. USA 89*, 5381–5383.

4. McGregor, L. M., Gorin, D. J., Dumelin, C. E., Liu, D. R. (2010). Interaction-dependent PCR: identification of ligand-target pairs from libraries of ligands and libraries of targets in a single solution-phase experiment. *J. Am. Chem. Soc. 132*, 15522–15524.

5. Kiss, M. M., Babineau, E. G., Bonatsakis, M., Buhr, D. L., Maksymiuk, G. M., Wang, D., Alderman, D., Gelperin, D. M., Weiner, M. P. (2011). Phage ESCape: an emulsion-based approach for the selection of recombinant phage display antibodies. *J. Immunol. Methods 367*, 17–26.

6. Starkenstein, E. V. (1910). Uber Fermentwirkung und deren Beeinflussung durch Neutralsalze. *Biochem. Z. 24*, 210–218

7. Roque, A. A, Lowe, C. R.(2008). Affinity chromatography: methods and protocols, vol. *421*: In: *Methods in Molecular Biology*, 2nd edition. M. Zachariou. Humana Press, Totowa.

8. Kroemer, G. (1997). The proto-oncogene Bcl-2 and its role in regulating apoptosis. *Nat. Med. 3*, 614–620.

9. Sattler, M., Liang, H., Nettesheim, D., Meadows, R. P., Harlan, J. E., Eberstadt, M., Yoon, H. S., Shuker, S. B., Chang, B. S., Minn, A. J., Thompson, C. B., Fesik, S. W. (1997). Structure of Bcl-x$_L$-Bak peptide complex: recognition between regulators of apoptosis. *Science 275*, 983–986.

10. Doorbar, J., Winter, G. (1994). Isolation of a peptide antagonist to the thrombin receptor using phage display. *J. Mol. Biol. 244*, 361–369.

11. Goodson, R. J., Doyle, M. V., Kaufman, S. E., Rosenberg, S. (1994). High-affinity urokinase receptor antagonists identified with bacteriophage peptide display. *Proc. Natl. Acad. Sci. USA 91*, 7129–7133.

12. Doyle, M. V., Doyle, L. V., Fong, S., Goodson, R. J., Panganiban, L., Drummond, R., Winter, J, Rosenberg, S. (1996). The utilization of platelets and whole cells for the selection of peptides ligands from phage display libraries. In R. Cortese (Ed.), *Combinatorial Libraries: Synthesis, Screening and Application Potential*. Walter de Gruyter, Berlin. pp. 158–174.

13. McCormick, D. B. (1965). Specific purification of avidin by column chromatography on biotin-cellulose. *Anal. Biochem. 13*, 194–198.

14. Hochuli, E., Bannwarth, W., Döbeli, H., Gentz, R., Stüber, D. (1988). Genetic approach to facilitate purification of recombinant proteins with a novel metal chelate adsorbent. *Nat. Biotechnol. 6*, 1321–1325.

15. Smith, M. C., Furman, T. C., Ingolia, T. D., Pidgeon, J. (1988). Chelating peptide-immobilized metal ion affinity chromatography: a new concept in affinity chromatography for recombinant proteins. *J. Biol. Chem. 263*, 7211–7215.

16. Smith, D. B., Johnson, K. S. (1988). Single-step purification of polypeptides expressed in Escherichia coli as fusions with glutathione S-transferase. *Gene 67*, 31–40.

17. di Guan, C., Li, P., Riggs, P. D., Inouye, H. (1988). Vectors that facilitate the expression and purification of foreign peptides in Escherichia coli by fusion to maltose-binding protein. *Gene 67*, 21–30.

18. Balbas, P. (2001). Understanding the art of producing protein and non-protein molecules in Escherichia coli. *Mol. Biotechnol. 19*, 251–267.

19. Bywater R. (1978). Elution of Immunoglobulins from Protein A-Sepharose Cl-4B columns. *Chromatogr. Synth. Biol. Polym. 2*, 337–340.

20. Porcher, C., Malinge, M. C., Picat, C., Grandchamp, B. (1992). A simplified method for determination of specific DNA or RNA copy number using quantitative PCR and an automatic DNA sequencer. *BioTechniques 13*, 106–114.

21. Walensky, L. D., Pitter, K., Morash, J., Oh, K. J., Barbuto, S., Fisher, J., Smith, E., Verdine, G. L., Korsmeyer, S. J. (2006). A stapled BID BH3 helix directly binds and activates BAX. *Mol. Cell. 24*, 199–210.

22. Cantor, C. R., Schimmel, P. R. (1980). Fluorescence polarization. In Biophysical chemistry Part II: techniques for the study of biological structure and function. W. H. Freeman & Company, San Francisco, pp. 454–465.

14

REPORTED APPLICATIONS OF DNA-ENCODED LIBRARY CHEMISTRY

Johannes Ottl

Centre for Proteomic Chemistry, Novartis Institutes for Biomedical Research, Novartis Pharma AG, Basel, Switzerland

This chapter will focus on published applications of DNA-Encoded Libraries (DELs) that utilize low-molecular-weight compound libraries for target-based affinity screening. Typically, such screening efforts are applied to identify hit, lead, or tool compounds for protein-based targets. The DELs themselves are usually assembled via chemical synthesis steps. Peptide DELs as applied, for example, in phage display technologies will not be covered here in any detail. They normally do not fulfill the criteria of low-molecular-weight libraries and are classically not established via chemical synthesis. However, it should nevertheless not be forgotten that phage display libraries are the historic basis for the successful application of DEL screening [1, 2].

There are several excellent recent reviews summarizing published applications of DEL chemistry and screening, and these also serve as the basis for this chapter [3–10]. The concept of applying DNA-encoded chemical libraries was first published by Lerner and Brenner in 1992 with their pioneering paper on the construction of bead-based peptidic libraries. Their proposal was to synthesize peptide-oligonucleotide conjugates on solid bead supports using a split-and-mix procedure to generate a library of different peptides, each encoded by the oligonucleotide to which it is conjugated

A Handbook for DNA-Encoded Chemistry: Theory and Applications for Exploring Chemical Space and Drug Discovery, First Edition. Edited by Robert A. Goodnow, Jr.

Figure 14.1. Scheme of the concept for DNA-encoded chemical libraries [11]. Peptides are synthesized via combinatorial split–mix strategy in conjunction with the corresponding DNA-coding oligonucleotides being amplifiable identification codes.

(Fig. 14.1). The DNA sequences are decoded by sequencing to identify the individual composition of the peptide [11].

In 1995, Kinoshita and Nishigaki introduced the concept of replacing the chemical oligonucleotide encoding with enzymatic steps that are also used in molecular biology approaches [12]. Such approaches have become the standard in today's DEL encoding strategies. DNA-encoded chemistry is a combination of fields of study that have traditionally been considered quite distinct – combinatorial chemical synthesis and molecular biology including deep sequencing. DEL chemistry and its applications can be grouped in many different ways. This chapter adopts the nomenclature used in the review of Mannocci et al. in 2011 [4]. The DEL application theme will be split into applications of single-pharmacophore libraries and dual-pharmacophore libraries. The single-pharmacophore libraries themselves will be subgrouped according to their synthesis strategy into classical split–mix synthesis and DNA-Templated Synthesis (DTS).

Table 14.1 in the following text summarizes low-molecular-weight DEL chemistry screening efforts with data published in peer-reviewed journals. The individual cases will be discussed in detail later, but it is impressive to note how in a few years DEL applications have moved from individual proof-of-concept studies with a few compounds to large-scale libraries containing a billion compounds or more. The strategies of library assembly and screening range from fragment-based screening or hit identification to affinity maturation.

14.1 SINGLE-PHARMACOPHORE LIBRARIES

Most DELs are assembled and applied as single-pharmacophore libraries. In such a DEL, individual chemical building blocks are successively assembled to generate an encoded chemical entity linked to its encoding DNA tag.

The DNA tag is assembled stepwise at each corresponding synthetic step by addition of the respective coding regions. Figure 14.2 shows a 3-cycle DEL where the small molecule is made up of building blocks A, B, and C. Before or after each synthetic step, the corresponding DNA-coding regions are added to a growing DNA strand that is physically linked to the chemical entity and allows decoding of the associated individual building blocks.

TABLE 14.1. Published hits resulting from selections using DELs

Reference	Library strategy	No. of DEL members	Representative hit example	Best hit affinity	Targets screened
2004 Halpin et al. [13]	Single-pharmacophore split–mix	1.00E+06		7.1 nM (positive control)	Monoclonal antibody 3-E7
2004 Gartner et al. [14]	Single-pharmacophore DTS	6.40E+01		Proof-of-concept study	Carbonic anhydrase

(Continued)

TABLE 14.1. (Continued)

Reference	Library strategy	No. of DEL members	Representative hit example	Best hit affinity	Targets screened
2004 Melkko et al. [15]	Dual-pharmacophore fragment assembly	1.38E+02		12 nM	Carbonic anhydrase, streptavidin, and HSA
2006 Dumelin et al. [16]	Dual-pharmacophore fragment assembly	4.77E+02		1.9 nM	Streptavidin
2007 Melkko et al. [17]	Dual-pharmacophore fragment assembly	6.20E+02		98 nM	Trypsin
2007 Wrenn et al. [18]	Single-pharmacophore split–mix	1.00E+08		16 µM	CrkSH3

Year / Reference	Method		Target affinity	Target
2008 Scheuermann et al. [19]	Dual-pharmacophore fragment assembly	5.50E+02	10 µM	MMP-3
2008 Dumelin et al. [20]	Dual-pharmacophore fragment assembly	6.19E+02	3.2 µM	HSA
2008 Mannocci et al. [21]	Single-pharmacophore split–mix	4.00E+03	11 µM	MMP-3, streptavidin, IgG
2009 Buller et al. [22]	Single-pharmacophore split–mix	4.00E+03	15 µM	TNF
2009 Clark et al. [23]	Single-pharmacophore split–mix	8.00E+08	7 nM	Aurora A and p38 MAP kinases

(Continued)

TABLE 14.1. (Continued)

Reference	Library strategy	No. of DEL members	Representative hit example	Best hit affinity	Targets screened
2009 Hansen et al. [24]	Single-pharmacophore DTS	1.00E+02		Proof-of-concept study	Anti-ENK mAb
2010 Melkko et al. [25]	Single-pharmacophore split–mix	4.00E+03		930 nM	Bcl-xL
2010 Mannocci et al. [26]	Single-pharmacophore split–mix	8.00E+03		3 nM	Trypsin

2010 Kleiner et al. [27]	Single-pharmacophore DTS	1.38E+04		680 nM	36 targets including Src kinase and VEGFR2
2011 Daguer et al. [28]	Dual-pharmacophore fragment assembly	6.25E+04		161 nM	Carbonic anhydrase
2011 Buller et al. [29]	Single-pharmacophore split–mix	1.00E+06		260 nM	CA IX

(Continued)

TABLE 14.1. (Continued)

Reference	Library strategy	No. of DEL members	Representative hit example	Best hit affinity	Targets screened
2012 Deng et al. [30]	Single-pharmacophore split–mix	4.00E+09		30 nM	ADAMTS-5
2012 Leimbacher et al. [31]	Single-pharmacophore split–mix	3.00E+04		2.5 μM	IL-2 and CA IX
2013 Podolin et al. [32]	Single-pharmacophore split–mix	Not mentioned		27 pM	Human epoxide hydrolase
2013 Disch et al. [33]	Single-pharmacophore split–mix	1.20E+06		1.3 nM	SIRT1, 2, 3

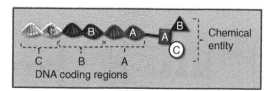

Figure 14.2. Assembly of a single-pharmacophore DEL.

The linkage between the chemical entity and the DNA tag must allow enough flexibility for the chemical entity to interact properly with the target protein and its binding pockets. The linker itself may link to only one or to both strands of the DNA.

14.2 SPLIT–MIX SYNTHESIS

Single-pharmacophore libraries present single chemical entities for interaction with the protein targets of interest. The total number of library members is the product of the number of building blocks used in each step. In the trimer DEL example (Fig. 14.2), a library using 100 building blocks for step A, 200 for step B, and 500 for step C would contain $100 \times 200 \times 500 = 10{,}000{,}000$ (10×10^6) members. Classically, most combinatorial efforts utilize split–mix synthesis. This approach allows the assembly of large libraries with few synthetic steps.

Figure 14.3 illustrates how such a split–mix library is assembled. Again, the trimer library example earlier with a total of 10,000,000 members are used. In step A, the first 100 chemical building blocks A1 to A100 are linked to their corresponding 100 DNA tags. After this step, all 100 individual vessels containing the 100 different Building Block A (bb A)–DNA conjugates will be mixed together. Purification of the library to remove unreacted material can happen either after the individual synthesis steps (e.g., using DNA precipitation) or as a mixture (e.g., using RP-HPLC). For the second synthetic step B, the mixture is again split into 200 vessels, and the second set of 200 chemical building blocks (B1–B200) are reacted with the preexisting chemical entities A. Before or after the chemical reaction step, the 200 DNA tags are added enzymatically to the already existing DNA strand from step A (see also Fig. 14.2). The library is again mixed together, and step C is conducted with its 500 building blocks and DNA tags (C1–C500) in a similar fashion as step B. Finally, the library consisting of 100 building blocks from step A, 200 from step B, and 500 from step C sums up to 10,000,000 individual DEL members.

It is important to keep in mind that the split–mix concept sounds quite easy to practice in theory. However, just as for classical combinatorial library synthesis, it is important to perform QC of all building blocks in on-DNA test reactions in order to assure the quality of the library. First one must optimize the reaction conditions and subsequently exclude building blocks from the DEL synthesis scheme that react

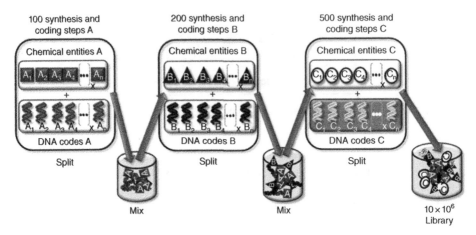

Figure 14.3. Stepwise assembly of a DEL via split–mix strategy.

poorly or lead to unintended products, including reaction with the DNA. Libraries assembled via the split–mix strategy are the most commonly used single-pharmacophore DEL chemistry applications. From the 15 published single pharmacophore DEL applications cited in Table 14.1, 12 utilized the split–mix approach to for library synthesis.

14.3 APPLICATIONS OF SINGLE-PHARMACOPHORE SPLIT–MIX LIBRARIES

One of the earliest applications of DELs was published by Halpin et al. in 2004 [13]. A library of 1,000,000 primarily nonnatural peptides was established and tested to identify a high-affinity ligand for the monoclonal antibody 3-E7. Three cycles of selection and amplification led reproducibly to the identification of the positive control Leu-enkephalin with binds to the antibody with an affinity of 7.1 nM.

Wrenn et al. published in 2007 the use of 100,000,000 8-mer peptoid DEL compounds for binding selection to the N-terminal SH3 domain of the proto-oncogene Crk, a member of an adapter protein family that binds to several tyrosine-phosphorylated proteins and is involved in several signaling pathways [18]. The huge size of this DEL originates from the relatively high number of synthetic steps involved, which also resulted in chemical entities of comparatively high molecular weight. After six affinity enrichments, the binder ligand population was down to a small number of novel SH3 domain ligands. Remarkably, the hits bind with affinities similar to those of peptidic SH3 ligands isolated from phage libraries of comparable complexity. The best peptoid showed affinities in the double-digit micromolar range.

In 2008, Buller et al. constructed a DEL with 4000 members via the Diels–Alder cycloaddition with a two-step split–mix strategy (20 dienes plus 200 maleimides) [34]. The focus of the Neri research group was the construction of relatively small libraries with a clear strategy to characterize the library intermediates by HPLC and LC-ESI-MS. In 2009, the authors investigated this library by panning it against streptavidin, the digestive system, Human Serum Albumin (HSA), B-Cell Lymphoma-Extra Large (Bcl-xL) (an anti-apoptotic Bcl-2 family member), and Tumor Necrosis Factor (TNF) (which is implicated in tumor regression). For TNF, a whole hit series with affinities between 10 and 50 µM was identified from this low-molecular-weight DEL [22, 34].

Another two-step, dimeric 4000-member DEL constructed and investigated by the Neri group was used to study the potential of the high-throughput DNA 454 sequencing technology for the decoding of DEL results [21].

For this study, they performed panning against streptavidin, Immunoglobulin G Antibody (IgG), and Matrix Metalloproteinase-3 (MMP-3, an enzyme involved in the breakdown of extracellular matrix). Binders were identified for each of the target proteins; for MMP-3, the most potent binder confirmed after resynthesis with an affinity of 11 µM.

In 2009, Clark et al. published data for an 800,000,000-member library assembled via the split–mix approach. This was the first publication of the production and screening of numerically large low-molecular-weight DELs [23]. Initially, the authors designed and screened a 7,000,000-member triazine library (DEL-A, Fig. 14.4), and encouraged by the success, they enlarged it with a similar library concept to an 800,000,000-member four-cycle library (DEL-B, Fig. 14.4).

This very large library was tested to identify binders for Aurora A (serine/threonine–protein kinase 6 is involved in mitosis and meiosis) and P38 Mitogen-Activated Protein Kinase (-38 MAP, which is involved in the cellular responses to cytokines and stress). The efforts yielded single-digit nanomolar ligands for Aurora A and binders for p38 MAP kinase with affinities as low as 270 nM. Selection in the presence of the non-DNA-bound ATP-competitive inhibitor VX-680, a known Aurora A inhibitor, revealed the mode of action for the hits and pointed to binding of the identified hits to the ATP site. Hit families that bound in the absence of VX-680 were no longer apparent when VX-680 was added during the affinity screen. Experiments that perform selections in presence and absence of molecules with a known mode of action highlight the power of DEL screening. The opportunity to run selections in parallel under various conditions without major effort enables one to identify compounds that bind also to deconvolute their mode of action or selectivity towards counter targets.

Figure 14.5 illustrates the p38 MAPK selection results for an identified hit family containing benzimidazole-5-carboxylate at Cycle 2 of the four-step synthesis of DEL-B. This hit class corresponds to a series of compounds where there is structural conservation at three positions, comprising a bicyclic tyrosine derivative at synthesis Cycle 1, a benzimidazole at Cycle 2, and tetrahydroquinoline at Cycle 4.

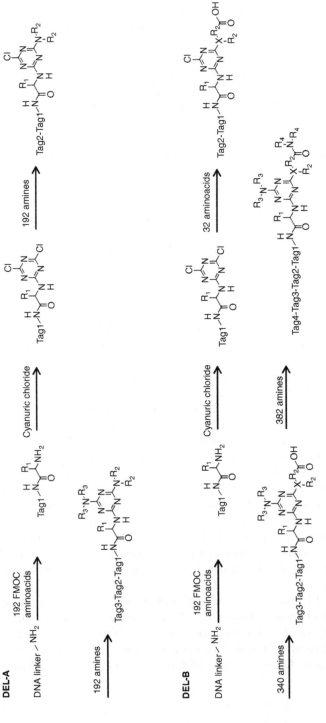

Figure 14.4. Synthetic schemes for DEL-A (top) and DEL-B (below). Alex Satz is acknowledged for this figure, which also appears in Figure 5.4.

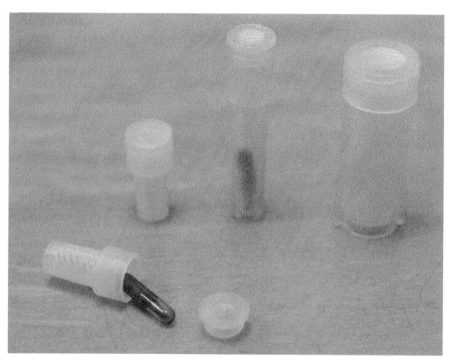

Figure 2.7. From left to right: a MicroKan® with a microfrequency tag and a cap, a closed MicroKan®, a MiniKan® containing a microfrequency tag, and a MacroKans®.

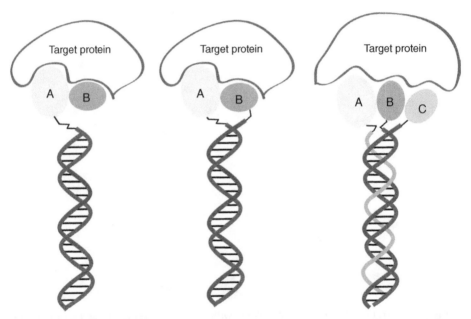

Figure 3.6. Alternative architectures of ESAC libraries are shown. The figure was kindly provided by Dr. Jörg Scheuermann.

Figure 6.1. Ethanol precipitation of DNA. By adding two volumetric equivalents of ethanol followed by 5 M NaCl to a solution of DNA, the DNA will precipitate from solution and can later be separated as a pellet by centrifugation.

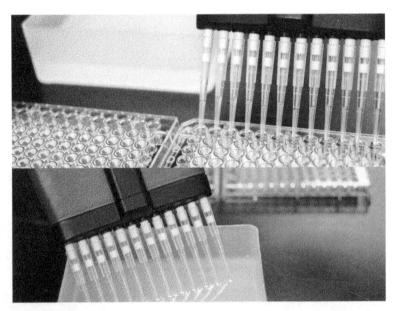

Figure 6.2. Split-and-pool. By splitting the growing library into a 96-well plate at the beginning of each ligation and synthesis cycle, a unique DNA tag is paired with a unique building block and 96 new encoded molecules can be generated. When the products of each cycle are collectively pooled into a single reservoir and split once more, the number of components in each well is multiplied by $96n$ (n = number of plates used per cycle).

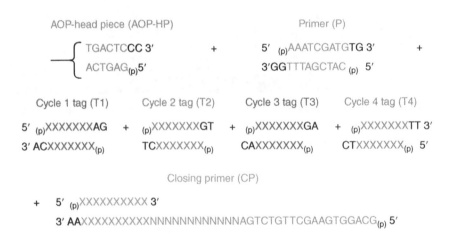

Figure 6.4. Example of a DNA tagging strategy [7]. Each segment will enzymatically ligate to the growing DEL structure via 2-base-pair "sticky ends" at the 3′-ends. The 5′-ends contain a free phosphate group (p). The X regions in the encoding tags (T1, T2, T3, and T4) represent undefined oligonucleotides that will encode the building blocks used in each synthesis cycle. The closing primer (CP) contains a long single-stranded segment, which includes a degenerate region **(N)**, that will later be filled in by Klenow reaction at the completion of the DEL.

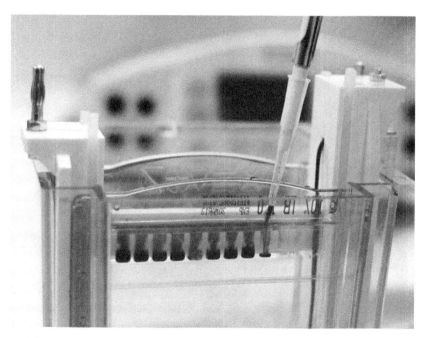

Figure 6.5. Loading a polyacrylamide gel. Performing gel electrophoresis on samples taken from representative wells after each ligation cycle allows you to follow the progression of the lengthening DEL and also look for problematic ligations.

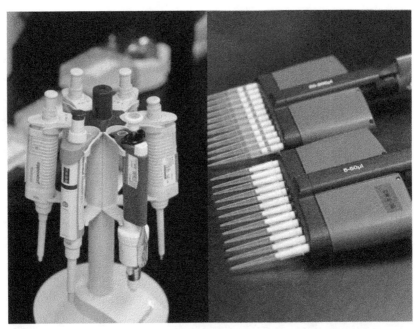

Figure 6.9. Essential tools for DEL synthesis. A collection of well-maintained single and multichannel pipettes will be used throughout the DEL production. Single pipettes will be needed for preparing building block solutions and reagents, while the multichannel pipettes will drive the split-and-pool plate syntheses.

TABLE 6.10. P ligation

AOP-HP   (1 component)

Step	P ligation				
1	**Add**	**Volume**	**Mols**	**Equiv.**	**Total volume**
1.1	AOP-HP (50 µM in H₂O)	50 mL	50 µmol	1	50 mL
1.2	10× ligation buffer*a*	20 mL	–	–	70 mL
1.3	T4 DNA ligase	2 mL	–	–	72 mL
1.4	H₂O	64 mL	–	–	136 mL
1.5	DNA P (1.0 mM in H₂O)	64 mL	64 µmol	1.28	200 mL
1.6	Stand overnight at room temperature				
1.7	Analyze by gel electrophoresis				
Step	**Ethanol crash**				
2	**Add**	**Volume**			**Total volume**
2.1	5 M NaCl(aq)	20 mL			220 mL
2.2	Cold EtOH	440 mL			660 mL
2.3	−78°C for ≥30 min or −20°C for 1 h				
2.4	Centrifuge and discard solvent				
2.5	Lyophilize DNA pellet				
2.6	Dissolve in approx. 50 mL of H₂O to make 1.0 mM solution				
2.7	Confirm conc. via OD measurement				

Table 6.10 outlines a step-by-step approach to attaching the P to the AOP-HP described in Table 6.4.
*a*500 mM Tris pH 7.5, 500 mM NaCl, 100 mM MgCl₂, 100 mM DTT, 20 mM ATP.

TABLE 6.11. Cycle 1

(4) 96 ligations
(6) 96 FMOC-AAs
(10) 10% piperidine

(1 plate = 96 components)

Step	Split				
3	Split 6 mL of AOP-HP P (1.0 mM) into 96 wells				

Step	Cycle 1— tag (T1) ligation				
4	**Add**	**Volume**	**Mols**	**Equiv.**	**Total volume (per well)**
4.1	AOP-HP P (1.0 mM in H_2O)	62.5 μL	62.5 nmol	1	62.5 μL
4.2	10× ligation buffer	25 μL	–	–	87.5 μL
4.3	T4 DNA ligase	2.5 μL	–	–	90 μL
4.4	H_2O	35 μL	–	–	125 μL
4.5	DNA tags (T1) (1.0 mM in H_2O)	125 μL	125 nmol	2	250 μL
4.6	Stand overnight at room temperature				
4.7	Analyze by gel electrophoresis				

Step	Ethanol crash			
5	**Add**	**Volume**		**Total volume (per well)**
5.1	5 M NaCl(aq)	25 μL		275 μL
5.2	Cold EtOH	550 μL		825 μL
5.3	−78°C for ≥30 min or −20°C for 1 h			
5.4	Centrifuge and discard solvent			

Step	Cycle 1—building block addition				
6	**Add**	**Volume**	**Mols**	**Equiv.**	**Total volume (per well)**
6.1	Sodium borate buffer (150 mM, pH 9.4)	62.5 μL	62.5 nmol	1	62.5 μL
6.2	Fmoc-AAs (200 mM in DMF)	12.5 μL	2.5 μmol	40	75 μL
6.3	DMT-MM (200 mM in H_2O)	12.5 μL	2.5 μmol	40	87.5 μL
6.4	Agitate for 2 h at 4°C				
6.5	Fmoc-AAs (200 mM in DMF)	12.5 μL	2.5 μmol	40	90 μL

(*Continued*)

TABLE 6.11. (cont'd)

6	Add	Volume	Mols	Equiv.	Total volume (per well)
6.6	DMT-MM (200 mM in H_2O)	12.5 µL	2.5 µmol	40	92.5 µL
6.7	Agitate overnight at 4°C				
6.8	Analyze by LC/MS				
Step	**Pool**				
7	Pool all 96 wells into a single vessel				
Step	**Ethanol crash**				

8	Add	Volume	Total volume (per well)
8.1	5 M NaCl(aq)	0.89 mL	9.77 mL
8.2	Cold EtOH	19.5 mL	29.3 mL
8.3	−78°C for ≥30 min or −20°C for 1 h		
8.4	Centrifuge and discard solvent		
8.5	Dissolve in H_2O to make 1.0 mM solution		
Step	**Purify**		
9	Purify by reverse-phase HPLC		
9.1	Lyophilize pooled fractions		
Step	**Remove Fmoc**		

10	Add	Volume	Total volume
10.1	10% piperidine in H_2O	36 mL	36 mL
10.2	Agitate for 1 h at room temperature		
10.3	Analyze by LC/MS		
Step	**Ethanol crash**		

11	Add	Volume	Total volume
11.1	5 M NaCl(aq)	3.6 mL	39.6 mL
11.2	Cold EtOH	80 mL	119.6 mL
11.3	−78°C for ≥30 min or −20°C for 1 h		
11.4	Centrifuge and discard solvent		
11.5	Lyophilize DNA pellet		
11.6	Dissolve in H_2O to make 1.0 mM solution		
11.7	Confirm conc. via OD measurement		

Table 6.11 outlines a step-by-step approach to performing the first cycle of a four-cycle DEL using a single 96-well plate.

TABLE 6.12. Cycle 2

(13) 96 ligations
(15) Cyanuric chloride/96 amines

(1 plate = 9,216 components)

Step	Split				
12	Split 5.76 mL of cycle 1 product (1.0 mM) into 96 wells				
Step	**Cycle 2—tag (T2) ligation**				
13	**Add**	**Volume**	**Mols**	**Equiv.**	**Total volume (per well)**
13.1	Cycle 1 product (1.0 mM in H₂O)	60 µL	60 nmol	1	60 µL
13.2	10 × ligation buffer	24 µL	–	–	84 µL
13.3	T4 DNA ligase	2.4 µL	–	–	86.4 µL
13.4	H₂O	33.6 µL	–	–	120 µL
13.5	DNA tags (T2) (1.0 mM in H₂O)	110 µL	110 nmol	1.8	230 µL
13.6	Stand overnight at room temperature				
13.7	Analyze by gel electrophoresis				
Step	**Ethanol crash**				
14	**Add**	**Volume**		**Total volume (per well)**	
14.1	5 M NaCl(aq)	23 µL		253 µL	
14.2	Cold EtOH	506 µL		759 µL	
14.3	−78°C for ≥30 min or −20°C for 1 h				
14.4	Centrifuge and discard solvent				
Step	**Cycle 2—building block addition**				
15	**Add**	**Volume**	**Mols**	**Equiv.**	**Total volume (per well)**
15.1	Sodium borate buffer (150 mM, pH 9.4)	60 µL	60 nmol	1	60 µL
15.2	Cyanuric chloride (200 mM in ACN)	3 µL	0.6 µmol	10	63 µL
15.3	Agitate for 1 h				
15.4	Cool plate to 4°C				

(*Continued*)

TABLE 6.12. (cont'd)

15	Add	Volume	Mols	Equiv.	Total volume (per well)
15.5	Amines (200 mM in DMA or ACN:H$_2$O (1:1))	15 µL	3 µmol	50	78 µL
15.6	Agitate for 40 h at 4°C				
15.7	Analyze by LC/MS				
Step	Pool				
16	Pool all 96 wells into a single vessel				
Step	Ethanol crash				
17	Add	Volume			Total volume
17.1	5 M NaCl(aq)	0.75 mL			8.24 mL
17.2	Cold EtOH	16.5 mL			24.7 mL
17.3	−78°C for ≥30 min or −20°C for 1 h				
17.4	Centrifuge and discard solvent				
17.5	Dissolve in approx. 5.76 mL H$_2$O to make 1.0 mM stock				
17.6	Confirm conc. via OD measurement				

Table 6.12 outlines a step-by-step approach to performing the second cycle of a four-cycle DEL using a single 96-well plate.

TABLE 6.13. Cycle 3

(19) 96 ligations
(21) 96 diamines (inc.60 Nvoc-diamines)
(24) 365 nm

(1 plate = 884,736 components)

Step	Split				
18	Split 5.76 mL of Cycle 2 product (1.0 mM) into 96 wells				
Step	Cycle 3—tag (T3) ligation				
19	Add	Volume	Mols	Equiv.	Total volume (per well)
19.1	Cycle 2 product (1.0 mM in H$_2$O)	60 µL	60 nmol	1	60 µL
19.2	10 × ligation buffer	24 µL	–	–	84 µL

(*Continued*)

TABLE 6.13. (cont'd)

19	Add	Volume	Mols	Equiv.	Total volume (per well)
19.3	T4 DNA ligase	2.4 µL	–	–	86.4 µL
19.4	H$_2$O	33.6 µL	–	–	120 µL
19.5	DNA tags (T3) (1.0 mM in H$_2$O)	180 µL	180 nmol	3	300 µL
19.6	Stand overnight at room temperature				
19.7	Analyze by gel electrophoresis				
Step	**Ethanol crash**				

20	Add	Volume		Total volume (per well)
20.1	5 M NaCl(aq)	30 µL		330 µL
20.2	Cold EtOH	660 µL		990 µL
20.3	−78°C for ≥30 min or −20°C for 1 h			
20.4	Centrifuge and discard solvent			
Step	**Cycle 3—building block addition**			

21	Add	Volume	Mols	Equiv.	Total volume (per well)
21.1	Sodium borate buffer (150 mM, pH 9.4)	60 µL	60 nmol	1	60 µL
21.2	Nvoc-diamines (200 mM in DMA or ACN:H$_2$O (1:1))	15 µL	3 µmol	50	75 µL
21.3	Agitate for 8 h at 80°C				
21.4	Analyze by LC/MS				
Step	**Pool**				
22	Pool all 96 wells into a single vessel				
Step	**Ethanol crash**				

23	Add	Volume	Total volume
23.1	5 M NaCl(aq)	0.72 mL	7.92 mL
23.2	Cold EtOH	15.8 mL	23.7 mL
23.3	−78°C for ≥30 min or −20°C for 1 h		
23.4	Centrifuge and discard solvent		
Step	**Nvoc deprotection**		
24	Dissolve in 3.6 mL AcOH aq. buffer (100 mM, pH 4.5)		
24.1	Transfer to flat-bottomed crystallizing dish		
24.2	Cool to 4°C and irradiate at 365 nm for 16 h		
24.3	Transfer to a centrifuge tube		

(*Continued*)

TABLE 6.13. (cont'd)

Step	Ethanol crash		
25	Add	Volume	Total volume
25.1	5 M NaCl(aq)	0.36 mL	3.96 mL
25.2	Cold EtOH	7.9 mL	11.86 mL
25.3	−78°C for ≥30 min or −20°C for 1 h		
25.4	Centrifuge and discard solvent		
25.5	Dissolve in H_2O		
Step	**Purify**		
26	Purify by reverse-phase HPLC		
26.1	Lyophilize pooled fractions		
26.2	Dissolve in H_2O to make 1.0 mM stock		
26.3	Confirm conc. via OD measurement		

Table 6.13 outlines a step-by-step approach to performing the third cycle of a four-cycle DEL using a single 96-well plate.

TABLE 6.14. Cycle 4

(28) 96 ligations
(30) 96 sulfonyl chlorides

(1 DEL = 84,934,656 components)

Step	Split				
27	Split 2.5 mL of Cycle 3 product (1.0 mM) into 96 wells				
Step	**Cycle 4—tag (T4) ligation**				
28	Add	Volume	Mols	Equiv.	Total volume (per well)
28.1	Cycle 3 product (1.0 mM in H_2O)	26 μL	26 nmol	1	26 μL
28.2	10 × ligation buffer	10.4 μL	–	–	36.4 μL

(Continued)

TABLE 6.14. (cont'd)

28	Add	Volume	Mols	Equiv.	Total volume (per well)
28.3	T4 DNA ligase	1.04 µL	–	–	37.4 µL
28.4	H_2O	14.6 µL	–	–	52.0 µL
28.5	DNA tag (T4) (1.0 mM in H_2O)	60 µL	60 nmol	2.3	112 µL
28.6	Stand overnight at room temperature				
28.7	Analyze by gel electrophoresis				

Step	Ethanol crash		
29	Add	Volume	Total volume (per well)
29.1	5 M NaCl(aq)	11.2 µL	123 µL
29.2	Cold EtOH	246 µL	369 µL
29.3	−78°C for ≥30 min or −20°C for 1 h		
29.4	Centrifuge and discard solvent		

Step	Cycle 4—building block addition				
30	Add	Volume	Mols	Equiv.	Total volume (per well)
30.1	Sodium borate buffer (150 mM, pH 9.4)	26 µL	26 nmol	1	26 µL
30.2	Sulfonyl chlorides (500 mM in ACN)	2.1 µL	1.05 µmol	40	28.1 µL
30.3	Agitate for 16 h at room temperature				
30.4	Analyze by LC/MS				

Step	Pool
31	Pool all 96 wells into a single vessel

Step	Ethanol crash		
32	Add	Volume	Total volume
32.1	5 M NaCl(aq)	270 µL	2.97 mL
32.2	Cold EtOH	5.93 mL	8.90 mL
32.3	−78°C for ≥30 min or −20°C for 1 h		
32.4	Centrifuge and discard solvent. Dissolve DNA pellet in H_2O		

Step	Purify
33	Purify by reverse-phase HPLC
33.1	Dissolve in H_2O to make 1.0 mM solution
33.2	Confirm conc. via OD measurement

Table 6.14 outlines a step-by-step approach to performing the last cycle of a four-cycle DEL using a single 96-well plate.

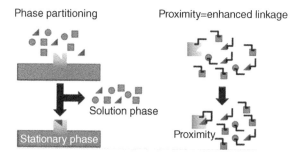

Figure 13.1. Identification of target-associated (▪) compounds from libraries of encoded compounds (▪▪▪) can be achieved by a physical phase separation or signal generation linked to a target-proximity-dependent event.

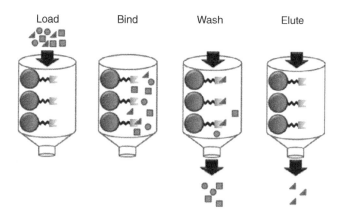

Figure 13.2. Affinity-based library screening applying a chromatographic phase-separation process is defined by the adsorption of target-binding library compounds (▪▪▪) to the solid-phase (●) target (▪) in the bind step, separation of weakly target-associated compounds in the wash step, and recovery of target-associated compounds in the elute step.

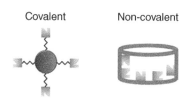

Figure 13.8. Direct linkage of the target (▪) to the solid-phase matrix can be through the covalent coupling to an activated resin bead (blue sphere) or through noncovalent interactions with the plastic surface blue well of a polystyrene plate.

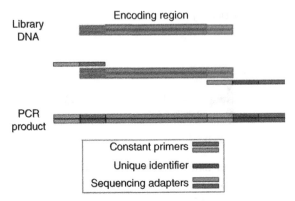

Figure 13.17. PCR-based amplification of library DNA with an appropriately designed primer set enables the creation of a product compatible with commercially available DNA-sequencing techniques.

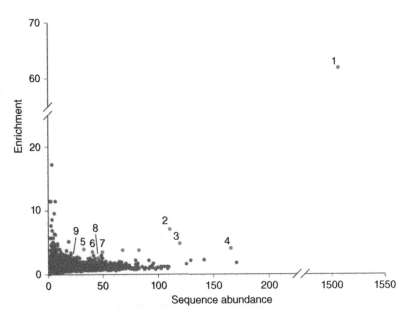

Figure 16.2. Analysis of high-throughput sequencing results. Plot of enrichment factor versus sequence abundance for library members after selection for binding to Src kinase [25].

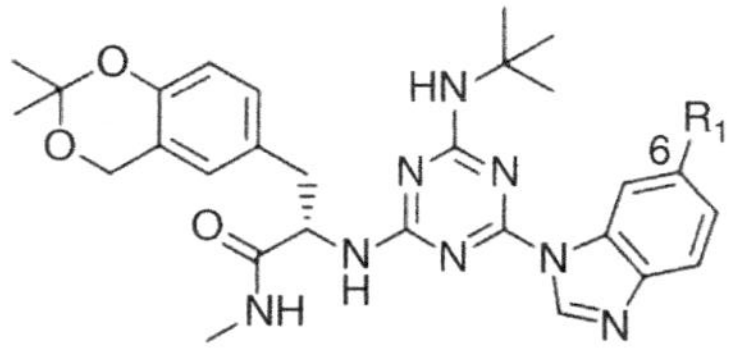

1

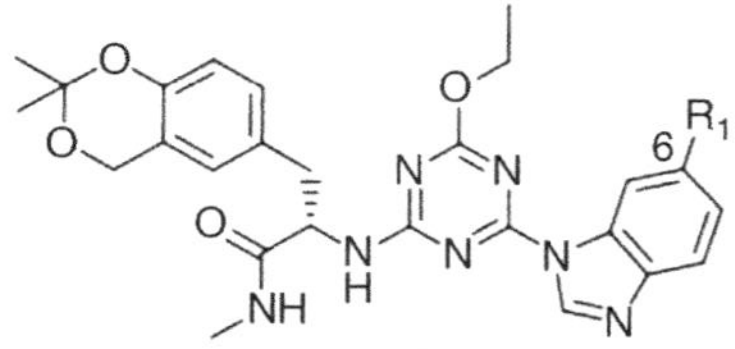

3

$EC_{50} = \leq 0.008\ \mu M$

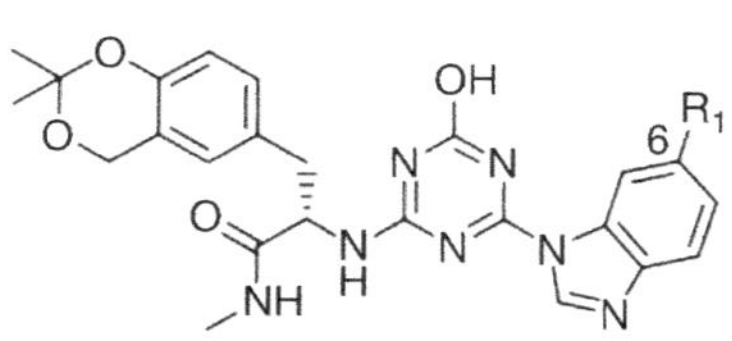

5

$EC_{50} = < 0.007\ \mu M$

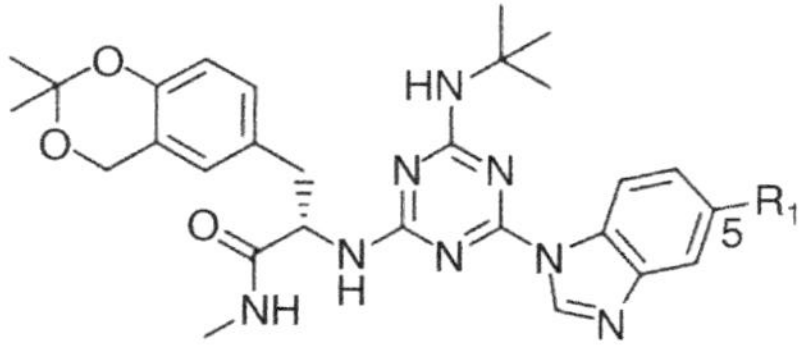

2

$EC_{50} = 0.25\ \mu M$

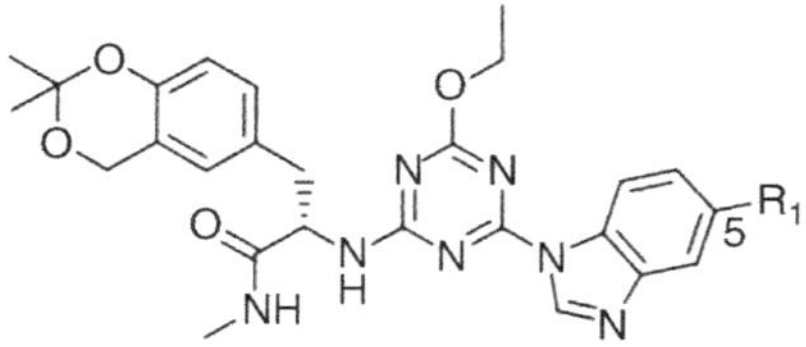

4

$EC_{50} = 2.7\ \mu M$

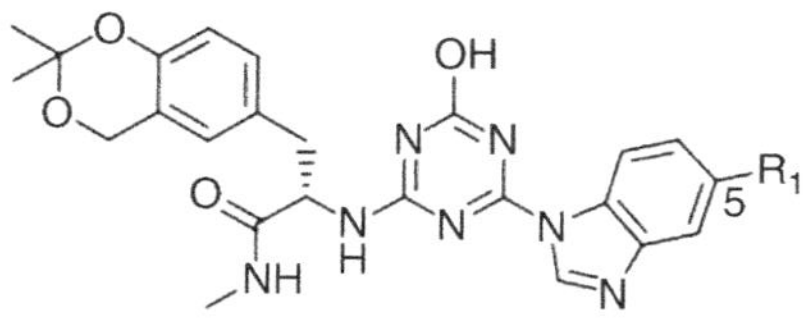

6

$EC_{50} = 5.2\ \mu M$

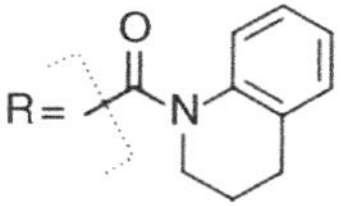

Figure 14.5. p38 MAPK hits and their activity in the functional assay as published by Clark et al. 2009 [23].

After resynthesis, a mixture of representative hit compounds (**1** and **2**) showed a potency of 250 nM in the functional p38 MAPK assay. Synthesis of the single isomers (**3**), (**4**), (**5**), and (**6**) indicated that the 6-substituted benzimidazoles are approximately 1000-fold more potent than the 5-substituted isomers. It turned out that these hits are potential side products that occurred during library synthesis.

Also interesting is the fact that the authors obtained 67,959 sequences in this study, but of these, 67,757 are not highly enriched. A total of only 355 sequences occurred at least twice and are enriched. These sequences encode only 153 unique structures.

Interestingly, a comparison between the 7-million-member DEL-A and the 800 million DEL-B indicates that DEL-A delivered hits for little beyond a few kinases, whereas the much larger DEL-B showed very good performance also for several other enzymes like hydrolases [8].

In 2010, Melkko et al. used the 4000-member dimer library published in 2008 by Mannocci et al. to identify binders to Bcl-x$_L$ (Fig. 14.6) [21, 25]. The identification of true, low-molecular-weight binders for a protein–protein interaction target like Bcl-xL by screening is challenging in general. Despite this, their relatively small DEL delivered several competitive binders with affinities in the high nanomolar range, a quite impressive achievement. Hits identified showed the intended mode of action and were competitive with the helical BH3 peptide, a domain derived from Bcl-2 Homologous Antagonist/Killer (BAK, a pro-apoptotic Bcl-2 family member which is dysregulated in some cancers).

The work of Mannocci et al. in 2010 establishing an 8000-member DEL yielded strong affinity maturation for the serine protease trypsin starting from a very weak benzamidine moiety [26]. This relatively small library led to the isolation of a trypsin inhibitor with an IC50 value of 3 nM. This is a >10,000-fold potency improvement compared to the parent benzamidine starting compound. The selectivity of the best hit over the related proteases, thrombin, Urokinase-Type Plasminogen Activator (uPA), and factor X, was >1000-fold.

In 2011, the Neri group expanded on their earlier Diels–Alder chemistry and constructed a trimer library with 1,000,000 members (three-step synthesis: 25 amino hexadienes/200 carbonyl chlorides, sulfonyl chlorides, isothiocyanates, carboxylic acid anhydrides, or carboxylic acids/200 maleimides) [29]. Affinity selection for carbonic anhydrase IX (CA IX), an enzyme upregulated in cancer, yielded a series of hits with nanomolar affinities for the target. The most promising binder identified was shown to localize in a selective manner to neoplastic lesions in mice, which had been implanted with LS174T and SK-RC-52 tumor xenografts.

The most recent single-pharmacophore library published by the Neri group dates from 2012. Leimbacher et al. show data from the testing of a 30,000-member dimer library against CA IX and the proinflammatory cytokine interleukin-2 (IL-2) [31]. As a first building block, they applied 100 amino acid-like derivatives synthesized beforehand by Suzuki, Sonogashira, Mitsunobu, and azide–alkyne

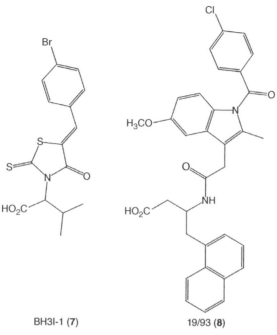

BH3I-1 **(7)**   19/93 **(8)**

Figure 14.6. Structure of the commercially available BH3I-1 **(7)** and the Bcl-xL DEL hit 19/93 **(8)** [25].

cycloaddition reactions in the absence of DNA. The authors' strategy was to introduce more complex chemistry before the actual library synthesis and to utilize rather standard and robust chemistry in their second DEL synthesis step. As the second step, the library was capped with 300 carboxylic acids. The resynthesized IL-2 binder hits (Fig. 14.7) showed K_d values in the low single-digit micromolar range. Tested compounds were shown to selectively inhibit T-cell proliferation. The off-DNA compounds did not exhibit any cytotoxicity in primary fibroblast cell cultures.

The hits derived from affinity screening against CA IX showed inhibitions in a functional assay with IC50 of 74 nM.

A recent example for the industrial use of DEL was published recently in 2012 by Deng et al. [30]. A 4,000,000,000-member library based on the triazine library concept reported earlier by the same group in Clark et al. 2009 [23] was established and tested. The library was assembled from 192 FMOC-amino acids acylated on the DNA headpiece, followed by applying the triazine scaffold. This scaffold was assembled via the addition of cyanuric chloride to the FMOC-deprotected amines from Cycle 1. After that, 479 different amines were reacted. The next synthesis cycle incorporated 96 diamines, 60 of them containing a photolabile NVOC protecting

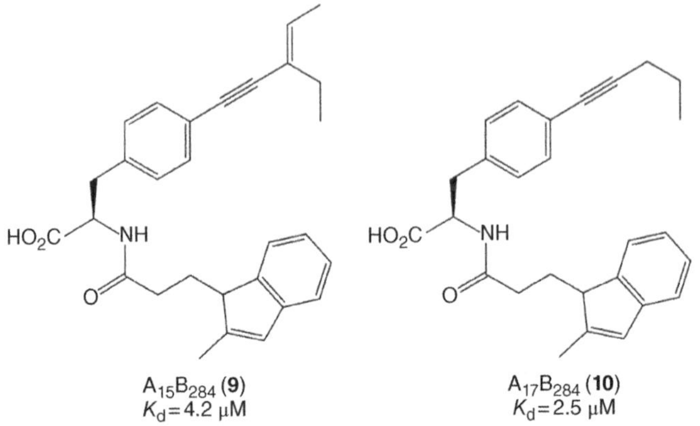

Figure 14.7. Human IL-2 hit compound structures $A_{15}B_{284}$ (**9**) and $A_{17}B_{284}$ (**10**) and their respective potencies in a fluorescence polarization binding assay data for IL-2 [31].

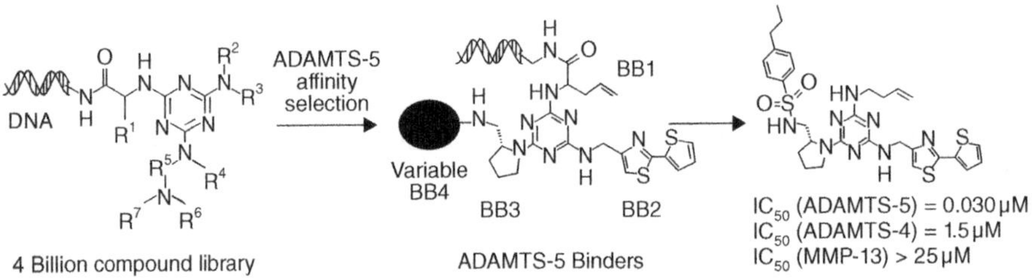

Figure 14.8. Strategy of library assembly by Deng et al. [30]. Reprinted with permission from Deng et al. [30]. Copyright 2012 American Chemical Society.

group. The latter amines were further acylated with 173 carboxylic acids, subjected to reductive amination with 94 aldehydes, sulfonylated with 107 sulfonyl chlorides, and reacted with 85 isocyanates.

Screening this library for ADAMTS-5 (aggrecanase-2, a member in the family of ADAMTS zinc metalloproteases cleaving the N-terminal interglobular domain of aggrecan) delivered multiple potent binders with affinities down to 30 nM (Fig. 14.8) paired with a clear structure–activity relationship of the hit series. The best hits showed strong selectivity for the ADAMTS-5 target compared to related metalloprotease enzymes like ADAMTS-1, ADAMTS-4, TACE, and MMP-13. Furthermore, they exhibited suppression of cytokine-mediated cartilage degradation in human osteoarthritis cartilage explants.

In 2013, two more papers from GSK highlighted the successful application of DEL screening in the pharmaceutical industry. Podolin et al. identified in GSK2256294A a highly potent inhibitor of human epoxide hydrolase with functional activity of 50 pM in a biochemical assay. *In vivo* validation of the compound showed significant reduction of pulmonary leucocytes by oral dosage in mice. GSK2256294A is described as a drug candidate for the treatment of cigarette smoke-induced diseases like Chronic Obstructive Pulmonary Disease (COPD) [32]. The second 2013 GSK publication by Disch et al. describes the discovery of inhibitors of sirtuins (histone deacetylases). The most potent compounds found from a 1,200,000-member DEL display single digit nanomolar activity for SIRT1, SIRT2 and SIRT3 [33].

14.4 DNA-TEMPLATED SYNTHESIS (DTS)

DTS is an interesting and elegant alternative approach to the assembly of DELs. In addition to the already described non-DTS DELs, for which the DNA tag serves the identification of the respective chemical entity, in DTS, the DNA tag drives the chemical reaction in addition to encoding it. This concept stems back to the principles that nature utilizes during the translation of genetic information into proteins via mRNA, tRNA, using the ribosomal machinery. Synthetic chemical libraries established via DTS will here be termed DNA-Templated Libraries (DTLs).

DTS utilizes the effect that complementary DNA strands bring two reactants linked to the corresponding paired DNA strands into close proximity. This pseudointramolecularity enables the formation of chemical bonds between the reactants at good rates even if they are only present at very low nominal concentrations. Gartner and Liu introduced this principle in 2001 by proposing amplifiable and evolvable libraries of nonnatural small molecules and developing methods to translate DNA into synthetic structures [35].

Figure 14.9 shows the principle of DTS. Two building blocks are each connected to a different DNA strand containing the corresponding coding sequence and a hybridization region. Building block A is connected to a donor strand, and building block B is connected to an acceptor strand. The linkage between building block B and the donor strand is cleavable and contains a handle to remove it, for example, a biotin tag. After the chemical reaction between the two chemical building blocks A and B, the donor DNA strand is cleaved from building block B followed by removal via its biotin moiety and avidin capture. The remaining acceptor ssDNA linked to the chemical entity A–B is now available for the next reaction step.

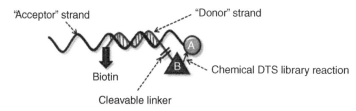

Figure 14.9. Concept of DTS as introduced by Gartner et al. [35].

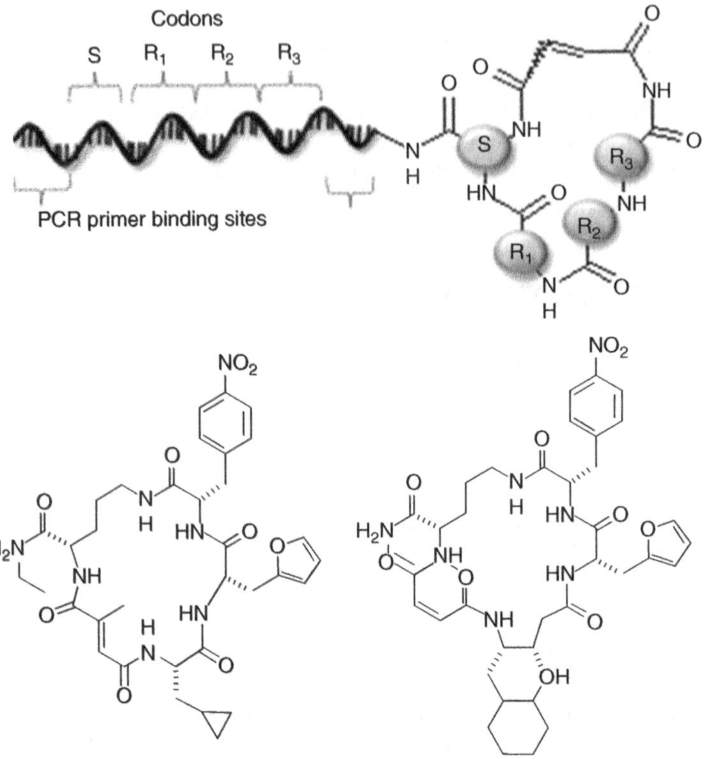

Figure 14.10. Top: DTS macrocycle library and encoding scheme reported in Kleiner et al. [27]. Bottom: two nanomolar binders of Src kinase identified during the study (**11** and **12**).

The DTS strategy allows iterative one-pot library synthesis and at least conceptually the amplification of stronger-binding library members by PCR and subsequent automatic resynthesis of the enriched binder species in the same pot. This concept is analogous to phage and ribosome displays, however, with the difference that it is not restricted to ribosomal synthesis of peptides, which is limited peptide-bonding chemistry. With DTS, a very different chemical space, the low-molecular-weight drug-discovery space, becomes accessible.

A pioneering paper for the application of DTL from Gartner, Liu, and coworkers dates back to 2004 [14]. In their study, a pilot library of 65 DNA-linked peptide–fumaride macrocycles containing a positive sulfonamide control was synthesized. This involved three DNA-templated amine acylations and one Wittig macrocyclization reaction. The DTL was subjected to one round of carbonic anhydrase affinity binding and amplification, resulting in the enrichment of a DNA sequence encoding the known sulfonamide ligand.

In 2008, the Liu group used the DTS principle to synthesize a larger 13,824-member macrocyclic library by DTL synthesis (Fig. 14.10, top) [36]. Two years later, in 2010,

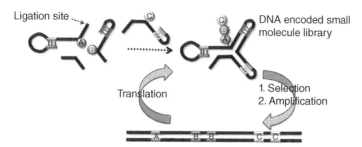

Figure 14.11. Vipergen's rolling translation principle published by Hansen et al. 2009 [24].

this library was published with data from affinity screening for 36 different target proteins, including 9 kinases, 3 phosphatases, 3 GTPases, 11 PDZ domains, 5 SH2 domains, and 3 other proteins (Bcl-xL, BIR3 (a caspase-9 inhibiting protein domain stopping apoptotic cell death), and the nuclear receptor PPARd (peroxisome proliferator-activated receptor transition factor)) [27]. This publication provides the largest applied assay data set for DTLs across many different targets and indeed in the whole field of published DELs. The library yielded macrocyclic ATP-competitive ligands with affinities down to 680 nM for Src kinase (proto-oncogene sarcoma tyrosine-protein kinase). Noncyclized analogues of the most potent ligands showed a drop of affinity to >100 μM. Two of the identified Src ligands showed strong selectivity for Src kinase compared to other closely related kinases (Fig. 14.10, bottom). One of the macrocyclic Vascular Endothelial Growth Factor Receptor 2 (VEGFR2) ligands of this library was found to activate, and not inhibit, VEGFR2 kinase.

Hansen et al. at Vipergen (Denmark) have developed DTS into a highly sophisticated approach they name rolling translation and Yoctoreactor [24]. Formation of three-way DNA hairpin junctions enables reactants to come into close proximity for reaction (Fig. 14.11). Reactant B is attached to a donor DNA oligonucleotide that binds and is ligated to the acceptor strand containing the reactant A. Because of their proximity, reagents A and B react and then the donor strand is cleaved and removed. The next synthesis step is initiated by introducing the donor strand containing building block C for the next reaction cycle. The system can be used for iterative cycles of affinity selection with enriched binder sequences that are reintroduced into the synthesis (translation).

In their proof-of-concept study with a 100-member peptide library, they showed >150,000-fold enrichment of Leu-enkephalin, a known binder to anti-ENK mAb.

14.5 DUAL-PHARMACOPHORE LIBRARIES

Dual-pharmacophore libraries offer a different strategy to assemble and expose DELs to their targets. Here, two chemical entities are exposed by the two hybridizing DNA strands at the same time. Dual-pharmacophore DNA-encoded chemical libraries are aimed at identifying pairs of molecules that bind cooperatively to target proteins of

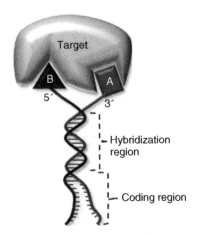

Figure 14.12. Principle of dual-pharmacophore ESAC libraries [15].

interest. Encoded Self-Assembling Chemical (ESAC) libraries as published by Melkko et al. [15] are assembled by the pairwise display of low-molecular-weight building blocks/fragments at the ends of self-assembling oligonucleotide regions having adjacent coding DNA sequence regions for the identification of the corresponding bound chemical fragments A and B as depicted in Figure 14.12. In this approach, one part of the chemical entity comprising the DEL is tethered to the 3' end of DNA tags and the other to the 5' end. Both parts have complementary DNA hybridization domains and individual, fragment-specific DNA-coding regions.

In addition to proof-of-concept studies with model selections, Melkko et al. [15] investigated the affinity maturation of binders for HSA and bovine carbonic anhydrase II identified via their ESAC strategy. Chemical linkage of a phenyl sulfonamide hit fragment with a second identified pharmacophore fragment improved potency by several orders of magnitudes to a 12 nM affinity as shown by ITC. The Neri research group has continued their work on ESAC dual-pharmacophore libraries and also used the concept to identify one high- and two low-affinity binding molecules to streptavidin in 2006 [16].

As mentioned earlier, ESAC libraries can also be used for affinity maturation of known binders, which serve as chemical starting points. The approach is based on the preferential enrichment of ESAC library compounds binding to the target protein during affinity screening. PCR amplification and sequencing of the recovered DNA tags yield the corresponding enriched hit fragment pairs. In 2007, Melkko et al. expanded their ESAC work by exploring such affinity maturation of compounds binding to the protease trypsin [17]. They started their approach with the weak trypsin binder benzamidine (IC50 ca. 100 μM) and finally yielded a series of inhibitors with IC50 as low as 98 nM. Their best identified inhibitor shows a strong selectivity among closely related serine proteases with a 40- and 6500-fold weaker potency towards thrombin and factor Xa, respectively, in comparison to the activity for trypsin.

Later, ESAC studies in 2008 led to the identification of novel inhibitors of strome-lysin-1 (MMP-3) [19]. After optimization of the chemical linkage between the identified two hit fragments, several novel inhibitors with potency down to 10 μM were found. It is important to note that this approach involved intensive investigation of the linkage moiety between the two fragments. Depending on the linker used, the affinities of the final molecules varied between 10 μM and >1 mM.

The same concept was used to identify binders to serum albumin, the most abundant human plasma protein, from a 619-member ESAC dual-pharmacophore library in 2008 by Dumelin et al. [16]. The identified and further optimized molecules showed affinities of 3.2 μM to HSA. Furthermore, studies have been carried out on *in vivo* circulatory half-life of imaging agents of pharmaceutical interest (fluorescein and Gd-DTPA) chemically modified with the most potent HSA binder identified via ESAC. *In vivo* experiments showed that such modified complexes had more than 100-fold improved *in vivo* circulatory half-life when injected into mice.

In 2011, Daguer et al. [28] published a method that utilizes the hybridization of a DNA template strand with two adjacent Peptide Nucleic Acid (PNA)–small molecule conjugates. Their work differs fundamentally from all previously described DELs. The authors utilize PNA to encode their chemotypes. PNAs emerged in the 1990s from the DNA-silencing field and are nonnatural analogues of DNA that have replaced the desoxyribose backbone by a peptidic backbone (Fig. 14.13, bottom). Both PNA and DNA share the common principle that complementing nucleotide pairs leads to the formation of double strands. The authors pair the PNA encoding of the chemical entity

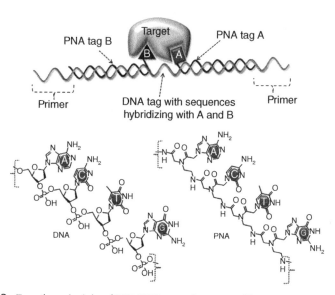

Figure 14.13. Top: the principle of PNA/DNA encoding as used by Daguer et al. [28]. Bottom: comparison of the chemical structure of DNA (left) and PNA (right).

with hybridizing DNA to form a double-stranded PNA/DNA conjugate. As PNA itself cannot be amplified with molecular biology methods as PCR, this is achieved via the DNA strand hybridized to the PNA.

PNA tagging offers some prominent advantages compared to DNA tags. They are chemically more stable and—as in the original concept of Brenner and Lerner [11]—are introduced chemically during library synthesis. As a result, they can be utilized in fully protected form, allowing a whole suite of chemical reactions to be applied during library synthesis, which typically is impossible in standard DEL synthesis as many reactions result in damage to the coding DNA [10, 37]. On the other hand, this approach requires the addition of multiple chemical synthesis steps to be conducted and controlled for the PNA coding itself. Additionally, PNAs are known to be quite insoluble compared to DNA, and the authors used solubilizing groups at the peptide backbone to diminish this unfavorable property of PNA.

Figure 14.13 (top) illustrates how Daguer et al. [28] utilized PNA/DNA conjugates to encode their fragment assembly library. As in ESAC, single-fragment building blocks are attached to their tag, in this case the PNA. 125 fragments A and 500 fragments B were coupled to the N- and C-terminus of coding PNA, respectively. This library yields $125 \times 500 = 62{,}500$ pairing combinations constituting the PNA-tagged combinatorial library. The pairing and templating DNA consists of the corresponding 62,500 hybridizing combinations of the PNA tags A and B. All possible combinations of the 125 A and 500 B tags are noncovalently linked to each other via their pairing DNA template, which contains both tag regions in a single strand in addition to PCR primer sites for amplification.

In their assay application, the 62,500 combinations of small-molecule fragment conjugates were affinity screened against carbonic anhydrase immobilized on magnetic beads. Affinity selection was followed by washing and release of the binders at 95°C. This heat denaturation yielded the enriched isolated DNA population from the binder. This was subjected to PCR utilizing primers that were 5′-biotinylated for the templating strand and modified at the 3′ end with the fluorescent dye Cy3 for the complementary strand. The PCR products could be captured on streptavidin resin via the biotinylation and converted to single-strand DNA to isolate the Cy3-labeled strand.

In order to initiate the next round of selection, the PNA-encoded library was added to allow the binder fragments to hybridize onto the corresponding DNA templates. After washing of nonbound PNA-library molecules, the PNA/DNA library was washed off the resin with a large excess of biotin, and several iterative rounds of target selection and binder isolation were performed. For each selection round, the respective Cy3-labeled isolated DNA strands could be used to visualize the selection results using DNA microarrays.

The iterative selection/amplification cycles provided convergence of the library to an arylsulfonamide on one PNA strand and several triazole units on the other PNA fragment. Chemical resynthesis of the corresponding hit compounds with a covalent linkage of the identified fragments yielded several high-affinity binders to carbonic anhydrase as verified by ITC measurements ($K_d = 161$ nM) [28].

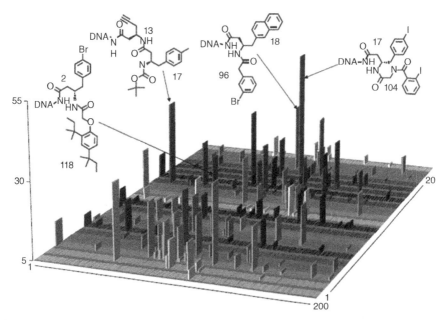

Figure 14.14. The plot represents the frequency/sequence counts of library members after selection on human MMP-3 as revealed by high-throughput 454 sequencing [21]. Chemical structures of representative enriched hits are shown. Reprinted with permission from Mannocci et al. [21]. Copyright 2008 National Academy of Sciences, USA.

A key aspect of the successful application of DEL is the technological boost that DNA sequencing has experienced over the past 10 years. Hit deconvolution in the early days of DNA-encoded library synthesis methods involved classical sequencing techniques having quite limited throughput. The biggest impact seen on DEL applications has come in the last few years with the availability of high-throughput sequencing technologies.

As mentioned earlier, the Neri group published in 2008 for the first time the use of high-throughput sequencing (454 technology) to decode DEL results [21]. Figure 14.14 shows the plot of sequences and their frequencies from a selection of their 4,000-member library for MMP-3 sequenced via the 454 technology, which yielded up to 40,000 sequences per sample. The plot shows that a few highlighted hit structures are enriched up to 55-fold compared to the background of the library. The authors compared the sequence patterns derived from the parent library before any selection, after exposure to unmodified resin, and after exposure to resin with attached target protein.

In 2010, the same group published a comparison of the 454 technology with the Illumina sequencing technology [38]. The Illumina sequencing technology utilizes the attachment of single-stranded DNA fragments to a solid surface, bridge amplification of the single-molecule DNA templates and subsequent sequencing by synthesis using reversible terminators. With this setup, the authors achieved more than 10×10^6 DNA sequences per chip flow lane (eight flow lanes per chip) and elaborated on the statistical analysis of such deep sequencing experiments and the power they offer for DEL.

The field is steadily changing, and as of today, the Illumina technology allows the reading of hundreds of millions of sequences per chip experiment. Such high data output is a prerequisite for statistically reliable sampling of very huge DELs. In 2012, the Harbury group published a modular approach to allow a robust construction and deconvolution of very large DELs [39, 40]. Parallelization of DEL chemistry with the use of mesofluidics and 384-well microtiter plates enables a reliable translation approach for huge DEL decks with potentially 10^{10}–10^{15} low-molecular-weight molecules. Confidently detecting ligands from such vast libraries bears many statistical challenges, and approaches like those reported by Weisinger et al. [39, 40] will be helpful to ensure high-quality data output from such libraries, particularly regarding industrialized drug discovery with DEL.

14.6 DISCUSSION

The DNA linkage of test molecules during the screen is often cited as a drawback of the DEL approach. However, the applications mentioned in this chapter clearly prove that the DNA tether by no means prevents the successful application of DEL in drug discovery. Sufficient library diversity allows enough flexibility of chemical moieties to bind productively to target proteins and their functional interaction pockets. Obviously, there are unsuitable targets with deeply buried pockets where the DNA tether will be a disadvantage.

Another popular criticism of DELs is the lack of enough chemical reactions to design and synthesize libraries appropriate for pharmaceutical discovery of low-molecular-weight drug discovery. The Liu group published an assembly of reactions that are compatible with DEL synthesis. Examples include amidation, sulfonamide formation, and reductive amination, as well as Pd-catalyzed cross coupling, S_NAr, dipolar cycloadditions, click chemistry, and other reaction schemes [41]. These examples are certainly not exhaustive and serve only to highlight those typically used to assemble DELs and used in published applications (Table 14.1). The strategy of PNA/DNA conjugates as used by the Winssinger group opens the door to an extensive chemistry portfolio as the PNA tag is assembled on solid support in its fully protected way; largely preventing chemical modification and harming of the coding nucleotides [28, 37]. Obviously, further developments in the DEL synthesis field will allow access to even greater chemistry space.

On the screening side, affinity screening is the most prominently applied DEL strategy to identify compounds binding to protein targets. A fundamental difference between DEL affinity screening and classical screening relates to the ratio of test compound to target protein. In functional assay readouts, for example, as applied in High-Throughput Screening (HTS), typically high concentrations of test compound compared to the target protein are applied in order to maximally saturate the binding of compounds to the target. This often brings unwanted properties of low-molecular-weight compounds into prominent play, and side effects like compound aggregation/insolubility, nonspecific target deactivation, or interference with the assay readout are possible occurring consequences [42]. In contrast to functional screens, DEL assays apply the target protein in excess to the small amounts of the individual test compounds present in the screen. The DNA tag permits the identification of even the lowest numbers of molecules after isolation from affinity screening. Testing compound at very low concentration prevents issues like compound insolubility and enables the identification of large numbers of different binders in the same screen in parallel. The compounds do not cross-compete for functional binding as the target protein is applied in great excess.

The ease of affinity screens also allows a setup of experiments that includes all kinds of controls: this can be the parallel screening of counter or selectivity targets, the investigation of mode of action with competition by ligands of known sites, screening of mutant proteins, or different activation/modification stages of the target. Affinity screening as applied in DEL approaches has several distinct advantages over more classical screening efforts. It must however be kept in mind that it is typically not readily applicable to screening readouts such as functional activity or phenotypic screening. This implies that hits identified via affinity screening have to be followed up in the respective project flowchart assays to ensure the desired functional mode of action of the compounds is verified.

The various cited applications of DNA-encoded chemical libraries have shown the breadth of DEL application that enables to address different challenges of drug discovery. Quite different philosophies and strategies have led to success. Single-pharmacophore DELs assembled via split–mix strategy can be used to assemble large libraries basically in few steps. This offers the opportunity to screen vast chemical libraries and efficiently sample a large chemical space. Dual-pharmacophore libraries on the other hand offer the possibility to explore new chemical space with self-assembling pairing fragments without the requirement of sophisticated library chemistry. However, the step to optimally link the identified fragments to final lead molecules needs to be investigated and bears some challenges.

DTS offers the unique advantage that the DNA tag not only encodes the individual chemical entities but drives the chemistry itself and conceptually offers the possibility to amplify and resynthesize in a single pot enriched library members. In the past, the ability to do this was limited to non-low-molecular library approaches like phage display. Judging which DEL size and chemistry to choose remains individual and cannot be generalized easily. Both small- and large-scale DELs have led to success as has been shown. On the one hand, rather small libraries and the use of only a few reaction steps enable very good control on QC of the library and the molecular properties of the compounds. They also offer the opportunity that full characterization of the parent library as well as oversampling of identified screening hits can be achieved with today's sequencing techniques. As many publications from the Neri group have shown, such comparatively small libraries can deliver new chemical matter for a variety of challenging targets and applications.

The size of low-molecular-weight DELs has increased in the last few years to unprecedented numbers, as publications like those from the GSK/Praecis efforts published by Morgan and Clark demonstrate. Libraries of 800 million or even 4 billion members have been investigated and tested in affinity screening for various targets. The chemical space and structure–activity relationships that can be tackled via such libraries are enormous.

Besides the impressive published results, such libraries have been tested in the past years for more than 150 targets at GSK/Praecis, and the first clinical candidate originating from DEL efforts has been identified (B. Morgan, DEL symposium, Zürich 2012).

Undoubtedly, DNA-encoded chemical libraries have evolved in the past 10 years from an exotic, initially academically driven phenomenon to an important tool of today's pharmaceutical drug-discovery efforts.

REFERENCES

1. Paul, S. M., Mytelka, D. S., Dunwiddie, C. T., Persinger, C. C., Munos, B. H., Lindborg, S. R., Schacht, A. L. (2010) How to improve R&D productivity: the pharmaceutical industry's grand challenge. *Nat. Rev. Drug Discov.*, *9*, 203–214.

2. McCafferty, J., Griffiths, A. D., Winter, G., Chiswell, D.J. (1990). Phage antibodies: filamentous phage displaying antibody variable domains. *Nature*, *348*, 552–554.

3. Kleiner, R. E., Dumelin, C. E., Liu, D. R. (2011). Small-molecule discovery from DNA-encoded chemical libraries. *Chem. Commun. (Camb.)*, *47*(48), 12747–12753.

4. Mannocci, L., Leimbacher, M., Wichert, M., Scheuermann, J., Neri, D. (2011). 20 years of DNA-encoded chemical libraries. *Chem. Commun.*, *47*, 12747–12753.

5. Kemp, M. M., Weiwer, M., Koehler, A. N. (2012). Unbiased binding assays for discovering small-molecule probes and drugs. *Bioorg. Med. Chem.*, *20*, 1979–1989.

6. Li, Y., Zhu, Z., Li, X. (2011). Drug discovery by DNA-encoded libraries, in *RSC Drug Discovery Series*, *5* (*New Frontiers in Chemical Biology*) (ed. Bunnage, M. E.) Royal Society of Chemistry, Cambridge, pp. 258–302.

7. Buller, F., Mannocci, L., Scheuermann, J., Neri, D. (2010). Drug discovery with DNA-encoded chemical libraries. *Bioconjug. Chem.*, *21*, 1571–1580.

8. Clark, M. A. (2010). Selecting chemicals: the emerging utility of DNA-encoded libraries. *Curr. Opin. Chem. Biol.*, *14*, 396–403.

9. Diezmann, F., Seitz O. (2011). DNA-guided display of proteins and protein ligands for the interrogation of biology. *Chem. Soc. Rev.*, *40*, 5789–5801.

10. Sadhu, K. K., Röthlingshöfer, M., Winssinger, N. (2013). DNA as a platform to program assemblies with emerging functions in chemical biology. *Isr. J. Chem.*, *53*, 75–86.

11. Brenner S., Lerner R. A. (1992). Encoded combinatorial chemistry. *Proc. Natl. Acad. Sci. U.S.A.*, *89*, 5381–5383.

12. Kinoshita, Y., Nishigaki, K. (1995). Enzymatic synthesis of code regions for encoded combinatorial chemistry (ECC). *Nucleic Acids Symp. Ser.*, *34*, 201–202.

13. Halpin, D. R., Harbury, P. B. (2004). DNA display. II. Genetic manipulation of combinatorial chemistry libraries for small-molecule evolution. *PLoS Biol.*, *2*, 1022–1030.

14. Gartner, Z. J., Tse, B. N., Grubina, R., Doyon, J. B., Snyder, T.M., Liu, D. R. (2004). DNA-templated organic synthesis and selection of a library of macrocycles. *Science, 305,* 1601–1605.

15. Melkko, S., Scheuermann, J., Dumelin, C. E., Neri, D. (2004.) Encoded self-assembling chemical libraries. *Nat. Biotechnol., 22,* 568–574.

16. Dumelin, C. E., Scheuermann, J., Melkko, S., Neri, D. (2006). Selection of streptavidin binders from a DNA-encoded chemical library. *Bioconjug. Chem., 17,* 366–370.

17. Melkko, S., Zhang, Y., Dumelin, C. E., Scheuermann, J., Neri, D. (2007). Isolation of high-affinity trypsin inhibitors from a DNA-encoded chemical library. *Angew. Chem. Int. Ed. Engl., 46,* 4671–4674.

18. Wrenn, S. J., Weisinger, R. M., Halpin, D. R., Harbury, P. B. (2007). Synthetic ligands discovered by in vitro selection. *JACS, 129,* 13137–13143.

19. Scheuermann, J., Dumelin, C. E., Melkko, S., Zhang, Y., Mannocci, L., Jaggi, M., Sobek, J., Neri, D. (2008). DNA-encoded chemical libraries for the discovery of MMP-3 inhibitors. *Bioconjug. Chem., 19* (3), 778–785.

20. Dumelin, C. E., Truessel, S., Buller, F., Trachsel, E., Bootz, F., Zhang, Y., Mannocci, L., Beck, S. C., Drumea-Mirancea, M., Seeliger, M. W., Baltes, C., Mueggler, T., Kranz, F., Rudin, M., Melkko, S., Scheuermann, J., Neri, D. (2008). A portable albumin binder from a DNA-encoded chemical library. *Angew. Chem. Int. Ed. Engl., 47,* 3196–3201.

21. Mannocci, L., Zhang, Y., Scheuermann, J., Leimbacher, M., De Bellis, G., Rizzi, E., Dumelin, C., Melkko, S., Neri, D. (2008). High-throughput sequencing allows the identification of binding molecules isolated from DNA-encoded chemical libraries. *PNAS, 105,* 17670–17675.

22. Buller, F., Zhang, Y., Scheuermann, J., Schäfer, J., Buhlmann, P., Neri, D. (2009). Discovery of TNF inhibitors from a DNA-encoded chemical library based on Diels-Alder cycloaddition. *Chem. Biol., 16,* 1075–1086.

23. Clark, M. A., Acharya, R. A., Arico-Muendel, C. C., Belyanskaya, S. L., Benjamin, D. R., Carlson, N. R., Centrella, P. A., Chiu, C. H., Creaser, S. P., Cuozzo, J. W., Davie, C. P., Ding, Y., Franklin, G. J., Franzen, K. D., Gefter, M. L., Hale, S. P., Hansen, N. J. V., Israel, D. I., Jiang, J., Kavarana, M. J., Kelley, M. S., Kollmann, C. S., Li, F., Lind, K., Mataruse, S., Medeiros, P. F., Messer, J. A., Myers, P., O'Keefe, H., Oliff, M. C., Rise, C. E., Satz, A. L., Skinner, S. R., Svendsen, J. L., Tang, L., van Vloten, K., Wagner, R. W., Yao, G., Zhao, B., Morgan, B. A. (2009). Design, synthesis and selection of DNA-encoded small-molecule libraries. *Nat. Chem. Biol., 5,* 647–654.

24. Hansen, M. H., Blakskjaer, P., Petersen, L. K., Hansen, T. H., Hoejfeldt, J. W., Gothelf, K. V., Hansen, N. J. V. (2009). A yoctoliter-scale DNA reactor for small-molecule evolution. *JACS, 131,* 1322–1327.

25. Melkko, S., Mannocci, L., Dumelin, C. E., Villa, A., Sommavilla, R., Zhang, Y., Gruetter, M. G., Keller, N., Jermutus, L., Jackson, R. H., Scheuermann, J., Neri, D. (2010). Isolation of a small-molecule inhibitor of the antiapoptotic protein Bcl-xL from a DNA-encoded chemical library. *ChemMedChem., 5,* 584–590.

26. Mannocci, L., Melkko, S., Buller, F., Molnar, I., Gapian Bianke, J. -P., Dumelin, C. E., Scheuermann, J., Neri, D. (2010). Isolation of potent and specific trypsin inhibitors from a DNA-encoded chemical library. *Bioconjug. Chem., 21,* 1836–1841.

27. Kleiner, R. E., Dumelin, C. E., Tiu, G. C., Sakurai, K., Liu, D. R. (2010). In vitro selection of a DNA-templated small-molecule library reveals a class of macrocyclic kinase inhibitors. *JACS, 132,* 11779–11791.

28. Daguer, J. P., Ciobanu, M., Alvarez, S., Barluenga, S., Winssinger, N. (2011). DNA-templated combinatorial assembly of small molecule fragments amenable to selection/amplification cycles. *Chem. Sci.*, *2*, 625–632.

29. Buller, F., Steiner, M., Frey, K., Mircsof, D., Scheuermann, J., Kalisch, M., Buhlmann, P., Supuran, C. T., Neri, D. (2011). Selection of carbonic anhydrase IX inhibitors from one million DNA-encoded compounds. *ACS Chem. Biol.*, *6*, 336–344.

30. Deng, H., O'Keefe, H., Davie, C. P., Lind, K. E., Acharya, R. A., Franklin, G. J., Larkin, J., Matico, R., Neeb, M., Thompson, M. M., Lohr, T., Gross, J. W., Centrella, P. A., O'Donovan, G. K., (Sargent) Bedard, K. L., van Vloten, K., Mataruse, S., Skinner, S. R., Belyanskaya, S. L., Carpenter, T. Y., Shearer, T. W., Clark, M. A., Cuozzo, J. W., Arico-Muendel, C. C., Morgan, B. A. (2012). Discovery of highly potent and selective small molecule ADAMTS-5 inhibitors that inhibit human cartilage degradation via encoded library technology (ELT). *J. Med. Chem.*, *55*, 7061–7079.

31. Leimbacher, M., Zhang, Y., Mannocci, L., Stravs, M., Geppert, T., Scheuermann, J., Schneider, G., Neri, D. (2012). Discovery of small-molecule interleukin-2 inhibitors from a DNA-encoded chemical library. *Chem. Eur. J.*, *18*, 7729–7737.

32. Podolin, P. L., Bolognesea, B. J., Foleya, J. F., Long III, E., Peck, B., Umbrecht, S., Zhang, X., Zhub, P., Schwartz, B., Xie, W., Quinn, C., Qic, H., Sweitzer, S., Chenc, S., Galope, M., Ding, Y., Belyanskaya, S. L., Israel, D. I., Morgan, B. A., Behm, D. J., Marino Jr., J. P., Kurali, E., Barnette, M. S., Mayera, R. J., Booth-Genthe, C. L., Callahan, J. F. (2013). In vitro and in vivo characterization of a novel soluble epoxide hydrolase inhibitor, *Prostaglandins Other Lipid Mediat.*, *104–105*, 25–31.

33. Disch, J. S., Evindar, G., Chiu, C. H., Blum, C. A., Dai, H., Jin, L., Schuman, E., Lind, K. E., Belyanskaya, S. L., Deng, J., Coppo, F., Aquilani, L., Graybill, T. L., Cuozzo, J. W., Lavu, S., Mao, C., Vlasuk, G. P., Perni, R. B. (2013). Discovery of thieno[3,2-d]pyrimidine-6-carboxamides as potent inhibitors of SIRT1, SIRT2, and SIRT3, *J. Med. Chem.*, *56*, 3666–3679.

34. Buller, F., Mannocci, L., Zhang, Y., Dumelin, C. E., Scheuermann, J., Neri, D. (2008). Design and synthesis of a novel DNA-encoded chemical library using Diels-Alder cycloadditions. *Bioorg. Med. Chem. Let.*, *18*, 5926–5931.

35. Gartner, Z. J., Liu, D. R. (2001). The generality of DNA-templated synthesis as a basis for evolving non-natural small molecules. *JACS*, *123*, 6961–6963.

36. Tse, B. N., Snyder, T. M., Shen, Y., Liu, D. (2008). Translation of DNA into a library of 13 000 synthetic small-molecule macrocycles suitable for in vitro selection. *JACS*, *130*, 15611–15626.

37. Chouikhi, D., Ciobanu, M., Zambaldo, C., Duplan, V., Barluenga, S., Winssinger, N. (2012) Expanding the scope of PNA-encoded synthesis (PES): Mtt-protected PNA fully orthogonal to Fmoc chemistry and a broad array of robust diversity-generating reactions. *Chem.: A Eur. J.*, *18*, 12698–12704.

38. Buller, F., Steiner, M., Scheuermann, J., Mannocci, L., Nissen, I., Kohler, M., Beisel, C., Neri, D. (2010). High-throughput sequencing for the identification of binding molecules from DNA-encoded chemical libraries. *Bioorgan. Med. Chem. Let.*, *20* (14), 4188–4192.

39. Weisinger, R. M., Marinelli, R. J., Wrenn, S. J., Harbury, P. B. (2012). Mesofluidic devices for DNA-programmed combinatorial chemistry. *PLoS One*, *7*, e32299.

40. Weisinger, R. M., Wrenn, S. J., Harbury, P. B. (2012). Highly parallel translation of DNA sequences into small molecules. *PLoS One, 7*, e28056.

41. Gartner, Z. J., Kanan, M. W., Liu, D. R. (2002). Expanding the reaction scope of DNA-templated synthesis. *Angew. Chem. Int. Ed. Engl., 41*, 1796–1800.

42. Coan, K. E. D., Ottl, J., Klumpp, M. (2011). Non-stoichiometric inhibition in biochemical high-throughput screening. *Expert Opin. Drug Discov. 6*, 405–417.

15

DUAL-PHARMACOPHORE DNA-ENCODED CHEMICAL LIBRARIES

Jörg Scheuermann and Dario Neri

Department of Chemistry and Applied Biosciences, ETH Zurich, Zürich, Switzerland

The isolation of specific ligands to target proteins of interest is a key step for the development of pharmaceutical agents or biochemical probes. Technologies that facilitate ligand discovery may substantially contribute to a better understanding of biological processes and to drug discovery.

Conventional technologies for the discovery of small-molecule ligands against a target protein of choice often feature the use of collections of chemical compounds ("compound libraries"), which are individually screened (i.e., one molecule at a time), based on enzymatic activity against the target protein or by displacement of labeled ligands. Such a High-Throughput Screening (HTS) approach for 10,000–1,000,000 compounds is complex and expensive in terms of library synthesis, management, logistics, and screening.

DNA-encoded chemical libraries consist of a set of organic molecules, which are individually coupled to a DNA fragment. This linkage of each library member to a tag allows the facile identification of binding molecules after affinity capture procedures on an immobilized target protein of choice. DNA-encoded chemical libraries allow the

A Handbook for DNA-Encoded Chemistry: Theory and Applications for Exploring Chemical Space and Drug Discovery, First Edition. Edited by Robert A. Goodnow, Jr.
© 2014 John Wiley & Sons, Inc. Published 2014 by John Wiley & Sons, Inc.

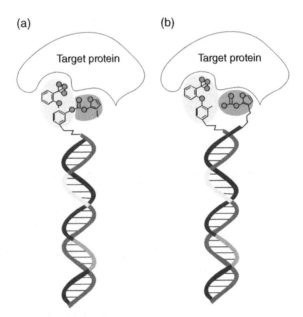

Figure 15.1. Schematic representation of DNA-encoded libraries displaying chemical compounds directly attached to oligonucleotides. (a) Schematic structure of compounds, generated by a stepwise combinatorial assembly and encoding process, resulting in a single-pharmacophore encoded chemical library. A single DNA fragment is covalently linked to a single organic molecule. (b) Schematic structure of compounds, yielding a dual-pharmacophore encoded chemical library. The two complementary strands of a DNA fragment each display a chemical molecule. These moieties can potentially synergize for target binding by the simultaneous engagement of adjacent nonoverlapping binding sites.

synthesis and screening of collections of chemical compounds of unprecedented size and diversity.

One type of DNA-encoded chemical library consists in the covalent attachment of individual compounds to unique DNA fragments. If a single organic molecule is displayed on each DNA fragment, the corresponding library can be termed a "single-pharmacophore library" (Fig. 15.1a). Alternatively, "dual-pharmacophore libraries" can be considered, which feature the display of pairs of chemical structures at the two extremities of complementary DNA strands (Fig. 15.1b). Our group has extensively worked in both fields [1–13], but in this chapter, we review the main concepts and applications related to Dual-pharmacophore DNA-encoded chemical libraries.

The attractive feature of dual-pharmacophore chemical libraries relates to the possibility that two chemical moieties, attached to the extremities of complementary DNA strands, synergize for target binding, by the simultaneous engagement of adjacent nonoverlapping binding sites. This "chelate effect" has been extensively studied in biological systems [14–17]. We have characterized the magnitude of the chelate effect in DNA-encoded dual-pharmacophore structures, using short DNA fragments in heteroduplex format to scaffold pairs of binding molecules with defined spatial arrangements,

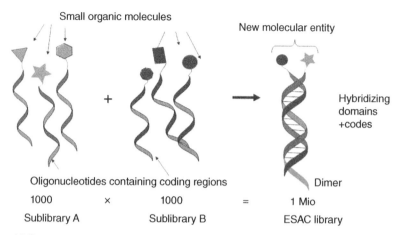

Figure 15.2. Principle of ESAC libraries. Small sublibraries A and B of individually purified oligonucleotide–compound conjugates can simply be combinatorially hybridized by mixing to form a heteroduplex dual-pharmacophore library (i.e., an ESAC library) of size A × B.

revealing apparent K_d improvement of ≥ 1000-fold for the bidentate binding mode compared to monodentate binding [12].

For this purpose, iminobiotin was coupled to the 5′- and 3′-extremities of DNA heteroduplexes by means of various bifunctional linkers, thus allowing bivalent binding to streptavidin. For comparison, iminobiotin was coupled only to the 5′-end (monovalent display). This analysis confirmed the strong superiority of dual display due to the chelate effect and demonstrated the suitability of the DNA heteroduplex as a scaffold for the identification of synergistic pairs of binding moieties, capable of a high-affinity interaction with protein targets by virtue of the chelate effect.

Dual-pharmacophore chemical libraries can be constructed by the combinatorial self-assembly of two complementary sublibraries, in which each DNA strand displays a small molecule. The first sublibrary contains chemical entities attached at the 5′-extremity of distinct single-stranded DNA fragments, whereas the second sublibrary displays chemical moieties at the 3′-extremity of unique DNA fragments, containing a hybridization domain for the first sublibrary (Fig. 15.2). Both sublibraries can be synthesized by the individual covalent coupling of chemical entities to modified DNA fragments (e.g., the coupling of a reactive moiety to amino-modified DNA fragment). The sublibraries can be HPLC purified to homogeneity after conjugation with a small molecule, ensuring an extremely high library quality. Additionally, large combinatorial libraries can be assembled by hybridization of two relatively small sublibraries. For example, the combinatorial self-assembly of two sublibraries containing 1,000 members each leads to the formation of an encoded library containing 1,000,000 members (Fig. 15.2).

The display of two neighboring chemical entities may enable the simultaneous binding of two compounds to adjacent nonoverlapping binding sites on a target protein. Subsequently, these individual binding entities have to be linked and assembled into a single organic molecule, in full analogy to fragment-based drug discovery approaches.

Dual-pharmacophore chemical libraries were originally constructed by the self-assembly of DNA oligonucleotides to heteroduplexes, each oligonucleotide providing a code associated with each binding moiety and displaying a small organic molecule (the "pharmacophore") [13]. Similar to antibody phage libraries, large Encoded Self-Assembling Chemical (ESAC) libraries can be "panned" (i.e., enriched for binders) on an immobilized target antigen. After the capture of the desired binding specificities on the target of interest, the binding code can be decoded by several experimental techniques (e.g., by PCR, followed by hybridization on oligonucleotide microarrays, or nowadays preferably by high-throughput DNA sequencing). For affinity maturation of known lead compounds, one of the codes can be omitted. After decoding, the DNA moiety of selected ESAC compounds is replaced by a suitable chemical linker that covalently connects the two (or more) pharmacophores, the characteristics of the linker (length, flexibility, geometry, chemical nature, and solubility) influencing the binding affinity and the chemical properties of the resulting binder.

Our group initially isolated several ligands of target proteins such as streptavidin (K_d 2 nM) [18], human serum albumin (K_d 3 μM) [9], and calmodulin (K_d 6 μM) (S. Melkko, C.E. Dumelin, J. Scheuermann and D. Neri, unpublished data) with small, single-pharmacophore libraries (620 library members). Moreover, the affinity maturation of known lead compounds with ESAC libraries led to affinity improvements of ligands to carbonic anhydrase (IC50 improved from 1 μM to 25 nM) [13], human serum albumin (K_d improved from 146 to 4 μM) [13], and trypsin (IC50 improved from 90 μM to 98 nM) [11]. The technology was later exploited for the isolation of portable albumin binders [9], the isolation of general IgG binders [8], and the isolation of MMP inhibitors [10].

Innovative methodologies have recently been implemented for the simultaneous synthesis and encoding of very large (>1,000,000) dual-pharmacophore chemical libraries, which are compatible with the use of high-throughput DNA sequencing, for the fragment-based *de novo* isolation of binders (Fig. 15.3).

In principle, one could consider alternative strategies for the construction and use of dual-pharmacophore DNA-encoded chemical libraries. Recently, in addition to the use of stable DNA heteroduplexes as scaffolds for displaying two pharmacophores, the use of two DNA- or PNA-encoded compounds has been proposed, which self-assemble on complementary DNA template strands, either immobilized on a microarray for drug discovery screening [19–21] or in solution, to scaffold two pharmacophores in an optimal orientation for high-affinity binding [22, 23].

The compounds used for the construction of dual-pharmacophore chemical libraries can be very small and may resemble building blocks used for Fragment-Based Drug Discovery (FBDD) projects [24–27]. The DNA scaffold, in addition to encoding, provides structural flexibility, which allows chemical moieties to reach adjacent binding sites on the target protein of interest. Compared to conventional fragment-based screening methodologies (for which the identification of initial hits can be problematic, due to the low affinity of small ligands), dual-pharmacophore encoded chemical library technology allows the identification of pairs of molecules, which may simultaneously engage in a binding interaction with the target proteins of interest at adjacent binding sites.

Once mutually complementary binding fragments have been identified using dual-pharmacophore chemical libraries, these chemical moieties need to be linked in

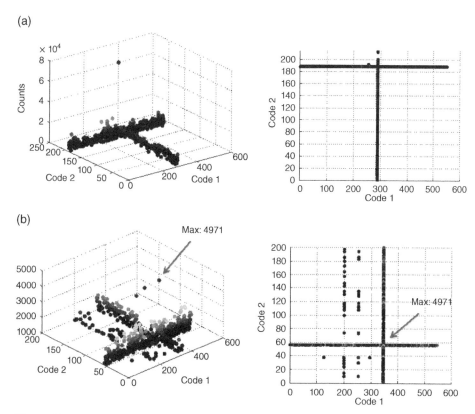

Figure 15.3. Decoding of an ESAC library by high-throughput DNA sequencing after affinity-based selections. Left panels: 3D plot of sequence counts versus code 1/code 2 combinations. Right panels: the enriched codes are depicted as lines in a 2D projection of the 3D plot. (a) and (b) show selections against two undisclosed target proteins, leading to individual target-specific "fingerprints." Most notably, the cross points of code 1 and code 2 lines show much greater enrichment (745-fold and 55-fold enrichment over background for (a) and (b), respectively), due to the chelate effect.

order to enjoy the benefit of the chelate effect in the absence of DNA and display their pharmaceutical potential. The choice of optimal linker (chemistry, length, rigidity, geometry) is still performed by the screening of different candidates, but it would be conceivable to use DNA-encoding technologies for the selection and optimization of linker structures. Whenever structural biology information is available, this may also be used to guide linker strategies and facilitate hit-to-lead conversion.

In summary, while the research field of dual-pharmacophore chemical libraries is still in rapid development, a few points have been firmly established. Libraries containing millions of compounds can be produced at acceptable costs, starting from a few thousand compounds and oligonucleotides. Selection and decoding procedures are in place, which provide information about preferential binders and their simultaneous engagement of adjacent epitopes on the target protein. Dual-pharmacophore DNA-encoded chemical libraries represent a nice complement to conventional fragment-based lead discovery strategies, and the next few years will reveal the real potential of this technology for the generation of pharmaceutically useful products.

REFERENCES

1. Leimbacher, M., Zhang, Y., Mannocci, L., Stravs, M., Geppert, T., Scheuermann, J., Schneider, G., Neri, D. (2012) Discovery of small-molecule interleukin-2 inhibitors from a DNA-encoded chemical library. *Chemistry 18*, 7729–7737.

2. Buller, F., Steiner, M., Frey, K., Mircsof, D., Scheuermann, J., Kalisch, M., Buhlmann, P., Supuran, C. T., Neri, D. (2011) Selection of carbonic anhydrase IX inhibitors from one million DNA-encoded compounds. *ACS Chem. Biol. 6*, 336–344.

3. Melkko, S., Mannocci, L., Dumelin, C. E., Villa, A., Sommavilla, R., Zhang, Y., Grutter, M. G., Keller, N., Jermutus, L., Jackson, R. H., Scheuermann, J., Neri, D. (2010) Isolation of a small-molecule inhibitor of the antiapoptotic protein Bcl-xL from a DNA-encoded chemical library. *ChemMedChem 5*, 584–590.

4. Mannocci, L., Melkko, S., Buller, F., Molnar, I., Bianke, J. P. G., Dumelin, C. E., Scheuermann, J., Neri, D. (2010) Isolation of potent and specific trypsin inhibitors from a DNA-encoded chemical library. *Bioconjug. Chem. 21*, 1836–1841.

5. Buller, F., Steiner, M., Scheuermann, J., Mannocci, L., Nissen, I., Kohler, M., Beisel, C., Neri, D. (2010) High-throughput sequencing for the identification of binding molecules from DNA-encoded chemical libraries. *Bioorg. Med. Chem. Lett. 20*, 4188–4192.

6. Buller, F., Zhang, Y., Scheuermann, J., Schafer, J., Buhlmann, P., Neri, D. (2009) Discovery of TNF inhibitors from a DNA-encoded chemical library based on Diels-Alder cycloaddition. *Chem. Biol. 16*, 1075–1086.

7. Scheuermann, J., Dumelin, C. E., Melkko, S., Zhang, Y., Mannocci, L., Jaggi, M., Sobek, J., Neri, D. (2008) DNA-encoded chemical libraries for the discovery of MMP-3 inhibitors. *Bioconjug. Chem. 19*, 778–785.

8. Mannocci, L., Zhang, Y., Scheuermann, J., Leimbacher, M., De Bellis, G., Rizzi, E., Dumelin, C., Melkko, S., Neri, D. (2008) High-throughput sequencing allows the identification of binding molecules isolated from DNA-encoded chemical libraries. *Proc. Natl. Acad. Sci. U. S. A. 105*, 17670–17675.

9. Dumelin, C. E., Trussel, S., Buller, F., Trachsel, E., Bootz, F., Zhang, Y., Mannocci, L., Beck, S. C., Drumea-Mirancea, M., Seeliger, M. W., Baltes, C., Muggler, T., Kranz, F., Rudin, M., Melkko, S., Scheuermann, J., Neri, D. (2008) A portable albumin binder from a DNA-encoded chemical library. *Angew. Chem. Int. Ed. Engl. 47*, 3196–3201.

10. Buller, F., Mannocci, L., Zhang, Y., Dumelin, C. E., Scheuermann, J., Neri, D. (2008) Design and synthesis of a novel DNA-encoded chemical library using Diels-Alder cycloadditions. *Bioorg. Med. Chem. Lett. 18*, 5926–5931.

11. Melkko, S., Zhang, Y., Dumelin, C. E., Scheuermann, J., Neri, D. (2007) Isolation of high-affinity trypsin inhibitors from a DNA-encoded chemical library. *Angew. Chem. Int. Ed. Engl. 46*, 4671–4674.

12. Melkko, S., Dumelin, C. E., Scheuermann, J., Neri, D. (2006) On the magnitude of the chelate effect for the recognition of proteins by pharmacophores scaffolded by self-assembling oligonucleotides. *Chem. Biol. 13*, 225–231.

13. Melkko, S., Scheuermann, J., Dumelin, C. E., Neri, D. (2004) Encoded self-assembling chemical libraries. *Nat. Biotechnol. 22*, 568–574.

14. Mack, E. T., Snyder, P. W., Perez-Castillejos, R., Bilgicer, B., Moustakas, D. T., Butte, M. J., Whitesides, G. M. (2012) Dependence of avidity on linker length for a bivalent ligand-bivalent receptor model system. *J. Am. Chem. Soc. 134*, 333–345.

15. Krishnamurthy, V. M., Semetey, V., Bracher, P. J., Shen, N., Whitesides, G. M. (2007) Dependence of effective molarity on linker length for an intramolecular protein-ligand system. *J. Am. Chem. Soc. 129*, 1312–1320.

16. Fersht, A. (2000) *Structure and mechanism in protein science: a guide to enzyme catalysis and protein folding*. W.H. Freeman & Company, New York.

17. Page, M. I., Jencks, W. P. (1971) Entropic contributions to rate accelerations in enzymic and intramolecular reactions and the chelate effect. *Proc. Natl. Acad. Sci. U. S. A. 68*, 1678–1683.

18. Dumelin, C. E., Scheuermann, J., Melkko, S., Neri, D. (2006) Selection of streptavidin binders from a DNA-encoded chemical library. *Bioconjug. Chem. 17*, 366–370.

19. Winssinger, N. (2012) DNA display of PNA-tagged ligands: a versatile strategy to screen libraries and control geometry of multidentate ligands. *Artif. DNA PNA XNA 3*, 105–108.

20. Huang, K. T., Gorska, K., Alvarez, S., Barluenga, S., Winssinger, N. (2011) Combinatorial self-assembly of glycan fragments into microarrays. *ChembioChem 12*, 56–60.

21. Daguer, J. P., Ciobanu, M., Barluenga, S., Winssinger, N. (2012) Discovery of an entropically-driven small molecule streptavidin binder from nucleic acid-encoded libraries. *Org. Biomol. Chem. 10*, 1502–1505.

22. Eberhard, H., Diezmann, F., Seitz, O. (2011) DNA as a molecular ruler: interrogation of a tandem SH2 domain with self-assembled, bivalent DNA-peptide complexes. *Angew. Chem. Int. Ed. Engl. 50*, 4146–4150.

23. Scheibe, C., Wedepohl, S., Riese, S. B., Dernedde, J., Seitz, O. (2013) Carbohydrate-PNA and aptamer-PNA conjugates for the spatial screening of lectins and lectin assemblies. *ChembioChem 14*, 236–250.

24. Zartler, E. R., Shapiro, M. J. (2005) Fragonomics: fragment-based drug discovery. *Curr. Opin. Chem. Biol. 9*, 366–370.

25. Murray, C. W., Rees, D. C. (2009) The rise of fragment-based drug discovery. *Nat. Chem. 1*, 187–192.

26. Erlanson, D. A. (2012) Introduction to fragment-based drug discovery. *Top. Curr. Chem. 317*, 1–32.

27. Hartenfeller, M., Schneider, G. (2011) De novo drug design. *Methods Mol. Biol. 672*, 299–323.

16

HIT IDENTIFICATION AND HIT FOLLOW-UP

Yixin Zhang

B CUBE, Center for Molecule Bioengineering,
Technische Universität Dresden, Dresden, Germany

16.1 INTRODUCTION: DNA-ENCODED CHEMICAL LIBRARIES. LARGE DIGITIZED DATASETS IN THE ERA OF "OMICS"

The biomedical and biological sciences are experiencing an era of "omics" or systems biology, which represents a collection of interdisciplinary research, aiming to analyze and to model complex parameters in large networks. Thanks to many advances in biochemical analyses, including high-throughput DNA sequencing technology and mass spectrometry, the capability to collect data is increasing exponentially. Furthermore, scientists are also increasing the dimensions of analysis from all perspectives. For example, genomics is beginning to characterize large human populations with different physiological and pathological conditions, while research in epigenetics and proteomics has started to tackle the temporal dimension of biology on a molecular level. Chemogenomics is analogous to classical genomics, aiming to develop high-throughput phenotypic screening of chemical libraries *in vitro* and *in vivo*. As in other "omics" approaches, advances in chemogenomics have led to high-throughput assays and large datasets. The increase of information has inevitably created new challenges for data analysis. The emerging field of DNA-encoded chemical libraries has increased chemogenomics information to an unprecedented level. In previous chapters, DNA-encoded

A Handbook for DNA-Encoded Chemistry: Theory and Applications for Exploring Chemical Space and Drug Discovery, First Edition. Edited by Robert A. Goodnow, Jr.
© 2014 John Wiley & Sons, Inc. Published 2014 by John Wiley & Sons, Inc.

chemical libraries have been discussed thoroughly from library design and construction to selection. This chapter will discuss current methods and challenges for analyzing large datasets from selection experiments for hit identification and hit follow-up.

As compared with other "omics" approaches, DNA-encoded chemical libraries represent collections of molecules that are purely artificially designed and synthesized. Therefore, although library size can reach billions of different compounds (as compared with millions of compounds in the largest commercial chemical libraries), the structural elements, as well as the DNA barcodes that encode them, are limited to fewer than a few thousand. The relatively focused structural diversity represents a disadvantage for DNA-encoded chemical libraries, compared to some other libraries. For example, natural products possess very high structural diversity, while other combinatorial synthetic methods, for example, solid-phase organic synthesis, can optimize reactions through methods not compatible with DNA [1–4]. However, structural focus has also its advantages. (i) A large database of billions of structures can be easily generated *in silico*. Such a database can be readily expanded and adapted to different computational methods, such as virtual screening. (ii) Although high-throughput sequencing reads of selection experiments could lead to billions of sequences, they can be easily analyzed and digitized, because the sequences are completely artificially designed, and the barcodes assigned to individual chemical building blocks are fewer than a few thousand unique, short sequences. (iii) Sequence reads of artificially designed codes can be made error-free. By designing codes with a minimal number of differences among any pairs of codes and adding a simple filtering function in the decoding program, errors resulting from sequencing can be completely excluded (as discussed in Section 16.5). (iv) The enrichment factor of each molecule is represented by the counts of an individual sequence, potentially indicating binding affinity to the target protein. It is a digitized readout by its origin, compared to other methods, which deal with analogue signals.

The simple structures of both the DNA codes and chemical compounds could result in a universal platform, incorporating scientists from different fields, including medicinal chemistry, biochemistry, computational biology, bioinformatics, and assay development. The synergy of these fields may one day lead to an integrated system to explore chemical space to discover specific and potent inhibitory compounds against various protein targets of medicinal interest. As antibody display technology has already proven that an efficient selection method based on a large library can lead to potent binding molecules against eventually any target [5–8], we have many reasons to believe that developments in DNA-encoded chemical libraries may lead to a powerful technology to discover small-molecule binders based on a larger collection of building blocks [9, 10].

Systems biology has led to an explosion of information, but in many fields it remains a challenge to comprehend the large amount of data. Such difficulties can be caused by network complexity and the chosen analytical methods. Both problems are encountered in hit identification from DNA-encoded chemical libraries. Network complexity can be exemplified by the similarity of proteins of the same family and the non-specific binding of DNA tag to proteins. Moreover, many biochemical parameters, including protein concentrations, library concentrations, the protein immobilization method, the resin of choice, and bias in PCR amplification, can all contribute to selection results.

Although the digitized counts of each library member using deep sequencing technology to decode large DNA-encoded chemical libraries has become the dominant method in this field, much knowledge was learnt based on experiments with relatively small library sizes and decoding methods before deep sequencing became available. These experiences with relatively simple libraries focusing on limited information have provided invaluable insight about hit identification and hit follow-up.

16.2 A COMPARISON OF PROTEIN/PEPTIDE DISPLAY AND DNA-ENCODED CHEMICAL LIBRARY IN HIT IDENTIFICATION

The concept of the DNA-encoded chemical library was inspired by protein/peptide display technology [11]. DNA sequencing has become the method of choice to identify the barcodes of the chemical compounds. In an ideal case, a few compounds will be exclusively enriched and only the barcode sequences corresponding to these structures would be obtained. However, the expected binding affinity and specificity of small organic hit compounds to a given target protein are, in general, much less than those for an antibody. Sequencing results from antibody selection experiments often do not reveal a perfect structure–activity relationship, because different antibodies can recognize different surface areas of a target protein. This approach has allowed researchers to obtain highly specific binding antibodies with different epitope recognition sites on an antigen. However, the binding of small organic compounds to a protein is quite different from that of macromolecular antibodies. For one thing, small organic compounds may be more promiscuous and lead to unspecific binding. The targetable surface area of a protein for desired specific binding and inhibition would be more limited for a small ligand than for an antibody. Therefore, a reasonable structure–activity relationship revealed from a selection experiment is important, not only for understanding and verifying the binding specificity but also for improving these binding features through medicinal chemistry. Unfortunately, except for some biased chemical libraries, using classical sequencing methods, tens to hundreds of DNA sequences would not lead to a clear structure–activity relationship. We might be able to identify a couple of needles in the haystack; however, without interrogating the entire pile, we will likely miss most of the other needles.

16.3 LESSONS FROM THE FIRST SELECTION AND DECODING EXPERIMENTS

The following was the situation before deep sequencing became readily available. The first proof-of-principle experiment with a DNA-encoded chemical library, with direct DNA–peptide conjugation, was carried out by Nielsen, Brenner, and Janda in 1993 [12]. Shortly afterwards, Needels et al. synthesized a DNA-encoded peptide library [13] of about one million compounds, where the oligonucleotides and peptides were covalently connected to the resin. The first two libraries were both synthesized on beads. Therefore, it was the beads that carried both chemical and genetic information. Although the size of the beads represents the ultimate limit for generating large chemical libraries and performing multiple selections, the generation of a library of one million different compounds with unique DNA barcodes was a great achievement at that time. Furthermore, the capability to decode the selection experiments lagged far behind the chemistry. Needels et al. performed the first proof-of-principle selection experiment using a Monoclonal Antibody (mAb) D32.39 and its preferred recognition sequence RQFKVVT of dynorphin B. The split-and-pool synthesis generated a library containing the RQFKVVT peptide. The beads that bound to the fluorescence-labeled antibody were sorted by FACS, while the presence of RQFKVVT peptide in solution abolished the fluorescent signal of all beads. Twelve beads with high fluorescence intensity were sequenced, but the DNA sequence corresponding to RQFKVVT peptide was not found among the 12 sequenced beads. The 12 DNA sequences revealed through sequencing that the barcodes on the beads were all different from each other, thus the results represented the simple digital readout of either one or zero for each chemical compound. Such a simple digital readout cannot distinguish a good binder from a moderate binder. Furthermore, the low sampling rate is very likely to miss some good binders, for example, the RQFKVVT peptide. Although some negatively selected beads as controls resulted in peptides with no substantial binding to the mAb, the 12 identified peptides showed a wide range of binding affinity from 0.29 nM to nonbinding.

As a model system, the binding pair of a mAb and its preferred peptide sequence should be the ideal model system, because the binding between an antibody and its antigen is more specific and more potent than most protein–ligand interactions. Although the identified sequences with high affinity to mAb D32.39 showed a clear sequence similarity to RQFKVVT, the missing of the original sequence as well as the presence of false positives indicate the limitation of this method. Furthermore, to distinguish good binders from moderate binders, oversampling could provide a solution. However, when using a DNA-encoded library on beads, the number of redundant beads in one selection experiment would need to be increased to investigate whether we can identify the same compound on many different beads from a selection. Large numbers of sequence reads represented a technological limit in the early 1990s, while increasing the number of beads in a selection experiment is practically impossible, if we are aiming at larger library size as well as multiple and parallel selection against many different proteins.

In 2004, Halpin, Harbury, and coworkers published three papers on the construction of a large DNA-encoded chemical library in solution [14–16]. Together with several works published by other groups at around that time, changing the DNA-encoded chemical libraries from a pool of covalently modified beads to a collection of conjugated molecules in solution revolutionized not only the chemical synthesis but also the biochemical methodology for hit identification. For a chemical library on beads, the phenotypic and genetic units are the beads with size of tens to hundreds of μm, whereas the direct conjugation between DNA and chemical compound in solution reduced the size trillions of times to the dimension of a molecule. Obviously, with this new approach, researchers can generate libraries of large size, which is not possible for a bead-based chemical library. The advances in the chemical synthesis of DNA–peptide conjugates resulted in a library of 100 million distinct peptoids. In 2007, Wrenn et al. published a selection experiment of this large library against the SH3 domain of Crk [17]. DNA sequencing technology had advanced and the researchers obtained 840 full-sequence reads from 960 clones, containing 215 unique sequences, 112 of which appeared more than once. Although the increased information from the decoding reads provided more insight into the binding of peptides to protein, 4 of the 10 selected compounds exhibited no binding to the target protein. Six of the 10 selected compounds showed binding affinity in the range of 10–100 μM. The low sampling rate appeared to be a drawback for the DNA sequencing-based readout, which can be partially overcome by DNA array technology.

16.4 DECODING BEFORE DEEP SEQUENCING: FROM DIGITAL TO ANALOGUE

Before deep sequencing became the major tool for decoding selection experiments of DNA-encoded chemical libraries, the use of DNA arrays was the most promising method to analyze the binding features of a relatively large collection of DNA-conjugated small organic compounds. The first *de novo* selection experiments of DNA-encoded chemical libraries against three different protein antigens were carried out by Dario Neri and coworkers [18], and the resulting amplicons were decoded using DNA arrays. Code design was challenging because cross talk among sequences increases exponentially with increased library size. For example, for a library of 20 compounds, the number of potential pairings is 190, while for a library of 200 compounds, the number increases to 19,900. The number of mismatches between two DNA barcodes may possess a wide distribution, resulting in a large difference in their annealing temperatures. It is practically impossible to narrow this distribution, considering that even for a small library size of about 1000 compounds, the number of potential pairs would be in the range of half a million. Because of these problems, decoding using DNA arrays has shown differences among the codes in the control experiment, when the library was subjected to the analysis before selection. Nevertheless, before the era of deep sequencing, DNA arrays were the most promising method available. Although it is difficult to compare the absolute fluorescence intensity among different barcode sequences, the difference of a given code sequence before and after selection is eventually significant enough to identify hit compounds. Most importantly, many decoding experiments using DNA arrays showed a clear structure–activity relationship, which provides the essential information for hit identification. Nevertheless, because of the aforementioned disadvantages and the limited size and high cost of DNA arrays, they can be used only for relatively small chemical libraries. David Liu and coworkers also successfully used this technology to decode selection experiments of DNA-encoded chemical reactions [19]. By mixing 13 different DNA-encoded chemical structures with another 13 different DNA-encoded chemical structures, they were able to discover new chemical reactions under five different conditions among the 169 potential reaction pairs.

Array technology has led to a number of interesting selection experiments against various protein targets, in the formats of either a single-pharmacophore chemical library [20] or a double-pharmacophore affinity maturation chemical library [21, 22]. From the early selection experiments decoded by the array method to the current selection experiments decoded by deep sequencing, a lasting question in the researchers' minds is whether the DNA double strands contribute to the binding, thus causing background noise. Fortunately, if the target proteins do not possess a DNA-binding domain, strong promiscuous binding of DNA to target proteins has not been a problem reported to date. Moreover, the high similarity of these artificially designed DNA sequences has also prevented the potential selective binding of target protein to a particular DNA strand. Nevertheless, the influence of DNA must always be kept in mind, and it can be shown in a single example of a selection experiment against one of the most abundant proteins, Human Serum Albumin (HSA) [23].

Analysis of a selection against HSA using DNA arrays revealed a number of compounds with similar structures, containing a haloaromatic ring and a short aliphatic chain. Because of the nature of the array technology mentioned previously, it was difficult to rank the compounds. In order to select compounds from these potential hits to determine the binding affinity to HSA, ^{32}P-labeled DNA conjugates of the compounds were subjected to an HSA affinity column. Although the enrichment factors from the array analyses are different from those in the affinity capture experiments, all enriched compounds observed on the DNA array showed a remarkable increase in retention on the HSA-modified matrix, as compared to control compounds. However, when the four best binders identified by affinity chromatography were tested for binding to albumin in the absence of DNA, no binding of any of the compounds was found. It seemed that all of the structures were false positive. However, considering that albumin binds to many fatty acids with moderate to weak affinity, the researchers wondered whether the C6 linker and the phosphate group on the 5′ end of the oligonucleotide might represent an analogue of fatty acid and contribute to the enrichments of the compounds in the selection experiment as well as in the affinity chromatography (Fig. 16.1). Interestingly, when the selected compounds were conjugated to the ε-carbon of lysine, moderate to potent binding of all these compounds was observed in fluorescence polarization experiment as well as in label-free Isothermal Titration Calorimetry (ITC) measurements. One of these albumin binders led to a number of interesting conjugates with either diagnostic reagents or proteins with remarkably extended half-lives in blood. Furthermore, we can learn two important lessons for the selection experiment with DNA-encoded chemical libraries in general (*vide infra*).

16.4.1 A Weak Binding Moiety Can Contribute to Bidentate Interaction

A very weak interaction between protein and ligand (e.g., with a dissociation constant in the high μM to low mM range) is rarely considered as specific binding and is generally ignored in most cases. However, when two moieties are specific to two different areas of the target protein, the bidentate effect can lead to a remarkable increase of binding affinity, as in the case of the C6 linker at the 5′ end of the DNA and the chemical moiety conjugated to the linker. It is reasonable to think that many proteins have weak affinity for either DNA or DNA modifiers used for constructing the DNA-encoded chemical library. In general, these background binding events may be neglected when interrogating selection results. However, such effects should be always kept in mind

1- Conjugate structure in DNA-encoded library

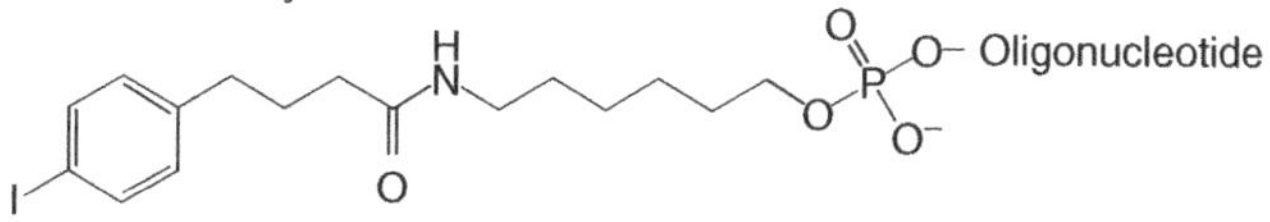

2- Building block for library synthesis

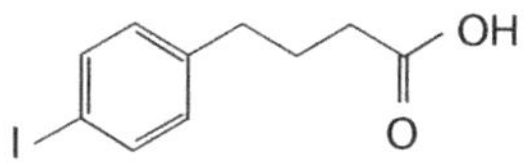

3- Redesigned albumin binder

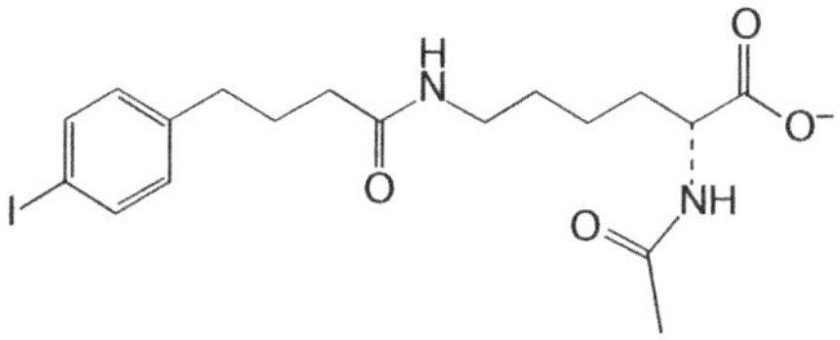

Figure 16.1. Albumin binding of selected DNA-conjugate (**1**) is dependent on the linker as well as the phosphate group on the oligonucleotide [23]. The building block 2 has no detectable affinity to albumin. The redesigned compound 3 possesses a negative charge as well as a linker to mimic the structure of 1, resulting in a K_d value of 3.2 µM to human serum albumin.

when inconsistencies are found between DNA-on and DNA-off experiments. Obviously, as a fragment-based drug discovery approach, such an effect can be further explored through constructing a dual-pharmacophore affinity maturation library, as discussed by Scheuermann and Neri in the previous chapter.

On the other hand, when an interaction occurs between the matrix used for protein immobilization and a number of library members, this interaction will very likely cause a waste of sequence reads, though not necessarily lead to false positives. For sample, polyphenolic compounds interact weakly with many polysaccharide matrices. In spite of the low affinity, given that the amount of matrix is much higher than the protein immobilized on it, the polyphenolic compounds may be highly enriched in both protein-loaded matrix and protein-free control matrix. Thus, many sequence reads from a selection experiment will be discarded because they are also present in the control selection due to their binding to matrices.

16.4.2 Affinity Chromatography Provides a Ranking of Hit Compounds Superior to the Array Readout

As a method to verify selection results, affinity chromatography represents a cleaner approach and leads to a ranking of hits more reliable than fluorescence detection on an array chip. As discussed previously, because cross talk among different codes is an intrinsic noise of the array-based method, affinity chromatography involves only one compound in each measurement and thus avoids the complication of the heterogeneity of the DNA-complementary binding associated with the array method [23]. Nevertheless, affinity chromatography can be used only for verifying the binding of a few selected compounds but cannot provide thermodynamic and kinetic information such as k_{on}, k_{off}, and K_d values regarding the binding.

The dramatic increase in library size and the use of deep sequencing in decoding in recent years have inevitably led to more complex readouts from selection experiments. Therefore, one of the major challenges in the near future will be to develop a relatively high-throughput method to verify and to rank a few hundred selected compounds and to provide more insightful information regarding binding kinetics and thermodynamics.

16.5 DEEP SEQUENCING IN DECODING: FROM ANALOGUE TO DIGITAL

In 2008, Neri and coworkers [24] reported the first use of deep sequencing in decoding a 4000-compound library. After that, other groups also used this approach in library decoding [25, 26]. The method has since become the dominant technique for decoding selection experiments, though both analogue and digital readouts have their advantages and disadvantages. In this section, a number of examples will be given to illustrate how researchers treated selection data resulting from deep sequencing to identify hit compounds.

16.5.1 The First Decoding Using Deep Sequencing

Mannocci et al. synthesized a DNA-encoded chemical library with 4000 compounds [24]. The 20 × 200 combinatorial library was generated through the split-and-pool method and encoding was carried out through Klenow polymerization. Although many chemical reactions have been proven compatible with oligonucleotides in aqueous solution, peptide bond formation mediated with EDC/sulfo-NHS remains one of the most reliable reactions for the incorporation of building blocks with relatively high structural diversity. Selection experiments were performed against the protein targets streptavidin, polyclonal human antibody IgG, and Matrix Metalloproteinase 3 (MMP3).

By using the high-throughput sequencing technology 454, the binding of a large DNA-encoded chemical library to various protein targets could be presented for the first time as full digital plots containing every compound. Importantly, 100% of the 4000 possible library sequences were detected at least once. In all selection experiments, correlations of the chemical structures and sequence counts were observed. The authors identified an IgG binder, which could be used to generate an IgG affinity column, as well as low μM MMP3 binders. However, the chemical structures and sequence counts could not be translated to a clear structure–activity relationship. For example, the 4000 compound library was doped with a potent streptavidin binder desthiobiotin. Although the binding of desthiobiotin to streptavidin is not as strong as that of biotin, the low nM K_d was expected to lead to high enrichment over most compounds. The 2D mapping of streptavidin binders showed a selective enrichment of compounds with an alkyl thioester-containing building block. In fact, all selected hits were shown to bind to streptavidin with K_d values in the range of 350 nM to 11 µM. Interestingly, in spite of its high binding affinity, the sequence corresponding to desthiobiotin was significantly enriched, but remarkably lower than some hit compounds with low µM affinity. Therefore, it was not possible to exclude the possibility that there are some potent binders that were treated as false negatives. In the following years, high-throughput sequencing has become the routine technology for DNA-encoded chemical libraries. Researchers from many different laboratories can now synthesize libraries containing more than a billion compounds in a relatively short time. They can perform parallelized selection experiments against dozens of protein targets and under different selection conditions. Full-scale chemical genetics profiling, from selection and sequencing to decoding and data analysis, can be carried out in a very short time. However, like the simple observation of the underrepresented desthiobiotin in the first selection experiment against streptavidin, hit identification might become a more serious a challenge, as DNA-encoded chemical libraries have been rapidly increasing in terms of both size and structural diversity, as compared to the first 4000 compound library.

16.5.2 Selection with a DNA-Encoded Macrocycle Library

Due to their drug-likeness, macrocycles are of particular interest in the field of drug discovery. Given that intramolecular cyclization is often the most critical step in synthesizing macrocyclic compounds, it is more challenging to generate a macrocycle library than to generate its linear counterpart. David Liu and coworkers have pioneered the field of DNA-templated chemistry [27], which provides an elegant solution for the generation of otherwise difficult-to-synthesize macrocycle libraries on DNA. DNA-templated chemical reactions occur at concentrations much lower than those required for most conventional organic reactions in solution, by bring two reactants to proximity through hybridizing the complementary strands. Therefore, intermolecular reactions and intramolecular reactions could be considered as equivalent. Using a macrocycle library containing 13,824 compounds, Kleiner et al. performed selections against 36 different protein targets [25], including enzymes (kinases, phosphatases, and GTPases) and protein domains (PDZ domains and SH3 domains). A barcode was added to each selection during PCR amplification and high-throughput sequencing produced 12,400,000 sequence reads.

The authors applied a novel approach to analyze the selection data. Among the 12.4 million sequence reads, they obtained about 200,000 sequence reads for each selection and, very importantly, 1.7 million sequence reads of eight library samples before selection. Two hundred thousand sequence reads per selection for a library of 13,824 compounds produce sufficient oversampling, and results for the selected compounds are statistically significant. However, the unequal abundance of the codes revealed in the preselection samples made it difficult to choose hits according to either the sequence abundance or the enrichment factor. The unequal abundance of the background may be caused by many different factors, including the yields of the chemical reactions, liquid handling of samples, biochemical reactions such as hybridization and PCR amplification, and high-throughput sequencing. Therefore, a plot of enrichment factor versus sequence abundance was generated (Fig. 16.2). Enrichment factors of low-abundance sequences

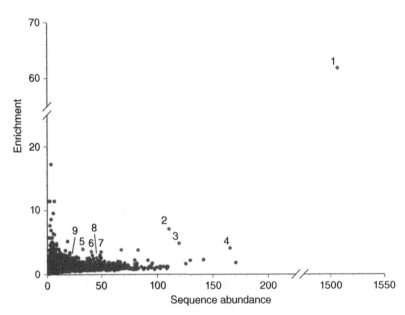

Figure 16.2. Analysis of high-throughput sequencing results. Plot of enrichment factor versus sequence abundance for library members after selection for binding to Src kinase [25]. For color detail, please see color plate section.

vary widely because of statistical undersampling. Thus, the enrichment factor was used as the major criteria for hit identification, if the abundance of the sequence is above a certain level. For example, in the selection against Src kinase, the sequence that has the second highest count (~170) possesses only twofold enrichment and so was not considered as a hit. In contrast, some sequences with very high enrichment factors (>10) were also excluded because of their extremely low abundances in the preselection sample. Although it remains very difficult to score each compound from a selection experiment, the multicriteria hit identification strategy used by Liu and coworkers and increasing the oversampling rate represent two general approaches for improving the data analysis. Nevertheless, such approaches meet limitations when the library size becomes very large.

16.5.3 Large DNA-Encoded Chemical Libraries

Nowadays, DNA-encoded chemical libraries can reach sizes comparable to protein/peptide display libraries. Although the DNA-encoded chemical library is inspired by its biological analogue, there are a number of differences between these two classes of large libraries. Hit identification and structure–activity relationship are not very important for protein/peptide display. In contrast, information regarding how individual building blocks contribute to protein binding is at the heart of medicinal chemistry and drug discovery. A selected antibody from protein/peptide display may be improved using a process called affinity maturation, through creating a second library based on the selected antibody. To create a new chemical library based on the selected compound may be difficult in an on-DNA-chemistry format because of the limited chemical tools in library design and synthesis. Therefore, it is currently more straightforward to achieve the improvement through medicinal chemistry. Furthermore, people have obtained enormous knowledge about drug-likeness and pharmacokinetics in the past century. Medicinal chemists can combine knowledge about protein crystal structure as well as structure–activity relationships revealed from selection experiments to design new and more potent compounds.

One of the most successful large DNA-encoded chemical libraries was reported by Clark and coworkers from GlaxoSmithKline [28]. They constructed a three-building block library of about seven million compounds and a four-building block library of about 800 million compounds. As a proof-of-principle experiment, they doped the seven-million-compound library with a 20 nM inhibitor of Aurora A kinase and performed selections against the same protein. The positive control occurred 971 times in 66,201 sequence reads. If the sequencing had been carried out using conventional method with 100 sequence reads, there would have been about a 20% probability that the positive control could be found in the selection experiment. Using the deep sequencing approach, a 100,000-fold enrichment of the positive control was achieved. Nevertheless, for large DNA-encoded chemical libraries, for example, the aforementioned libraries containing 10^8–10^9 compounds, it is not yet clear whether oversampling would be practically feasible and necessary, whereas for relatively small library sizes (10^4–10^6 compounds), oversampling can be easily achieved and various statistical methods can be used to assist in hit identification (as discussed in Section 16.8).

Although the design of barcodes for large libraries is not the most challenging part of DNA-encoded chemical library technology, the encoding and decoding of a large pool of artificially designed DNA sequences could provide great knowledge in encoding and decoding a large amount of information in general. To design n different 10-base codes, we can set a number m as the minimal difference among any pairs of the n codes. Obviously, a higher m value will lead to a lower maximal number of possible codes. For example, in the extreme cases, when $m = 1$, the number of codes that can be generated is 4^{10}, whereas only four codes can be generated when $m = 10$. A low m value leads to high information storage density, while a high m value results in high fidelity for encoding and decoding. The number m in code design determines whether a sequence read will be treated as an error if it does not match perfectly with a designated code. For example, for a single-base mismatch in a sequence read, it can be tolerated and corrected if $m = 4$, while it must be discarded if $m = 2$. For selection experiments, if oversampling is carried out in sequencing and decoding, a relatively lower m number can be used in code design. In contrast, if the number of sequence reads is lower than the library size, a high m value should be applied in code design. DNA code design has applications beyond the DNA-encoded chemical library itself. For instance, a number of recent inventions use DNA and deep sequencing for information storage. The knowledge gained from designing billions of DNA barcodes for DNA-encoded chemical libraries may provide much valuable experience for code design in general. It would allow us to store information efficiently, and the information may be written and read with high fidelity [29, 30].

16.6 INFLUENCE OF SELECTION CONDITIONS ON HIT IDENTIFICATION

The equilibrium between protein/ligand complex and free ligand is determined by the dissociation constant and the concentrations of the two interacting binding partners. However, how to apply the principles of physical chemistry and thermodynamics to the selection and hit identification of a DNA-encoded chemical library remains unexplored. On the one hand, there are myriads of different potential binding compounds in solution, which can bind to the target protein with specific or unspecific interactions; on the other hand, immobilization of proteins on a surface causes uncertainty on aspects of both protein concentration and protein native structure. Nevertheless, experiments with selections and model selections in the past few years have provided some guidance on how to approach this complex system, especially regarding protein immobilization. For example, using biotinylated or His-tagged protein may be superior to immobilization with cyanogen bromide-activated sepharose. Although the final washing steps on a solid support in a selection experiment cannot be avoided, using biotinylated or His-tagged proteins and capturing by resin may cause less disturbance of protein native structures. Moreover, the selection procedure carried out partially in solution also reduces the influence of the solid surface.

The loading density of protein on solid support has been shown to have a dramatic effect on selection results. Melkko et al. performed a model selection experiment using benzamidine-modified double-stranded DNA and trypsin-coated sepharose beads [21]. Benzamidine is a weak inhibitor of the protease trypsin with an IC_{50} value of 35 μM. The DNA–benzamidine conjugate was radiolabeled and affinity capture experiments were carried out with four different loading densities of protein. Interestingly, at low loading density, the trypsin resin was not able to affinity capture the weak binder, whereas an increase of the loading density enhanced the retention of the modified oligonucleotide on the solid support remarkably, compared to the unmodified DNA. In contrast, it is very easy to capture strong binding molecules such as biotinylated DNA with low-density streptavidin resin. Therefore, it has been proposed that high protein loading on resin should be used for selection experiments in which only weakly binding molecules are expected in the library. There are many proteins that have been considered to be difficult to target with small organic molecules, for example, globular proteins without a catalytic pocket. On the other hand, if researchers expect to discover potent binders from a library, for example, a biased library with privileged structural motifs as library building blocks against the protein target, a solid support with low loading density of target protein should be used.

A low protein-loading resin may be used to distinguish strong binders from moderate binders, while a high protein-loading resin may be applied to select a moderate binder from weak binders. Leimbacher et al. synthesized a DNA-encoded chemical library of 30,000 compounds [31], including 200 compounds containing a phenyl sulfonamide building block, a strong inhibitor of carbonic anhydrase. Interestingly, at high protein-loading density (40 μg), the 200 phenyl sulfonamide derivatives were barely distinguished from many other library members. Decreasing the protein-loading density (8 μg) in the selection experiment resulted in remarkably enhanced enrichment factors for the phenyl sulfonamide compounds, compared to other library members. A further decrease of the protein-loading density (2 µg) as well as an increase in the washing stringency led to not only a clear selection of sulfonamide derivatives but also a significant difference in the enrichment factors among the different putative binders. With the same library, Leimbacher et al. also performed a selection experiment against IL-2. IL-2 is a global cytokine and has been considered to be a difficult target for discovering small-molecule binders. Therefore, a very high loading density was applied for the selection experiments. Using these experimental conditions, the authors were able to identify IL-2 binders with low µM dissociation constants.

16.7 HIT FOLLOW-UP: FROM ON-DNA TO OFF-DNA

From an organic chemistry point of view, there are some apparent drawbacks to synthesize a combinatorial chemical library on DNA. On the one hand, the structural diversity is largely limited because of the limited choices of reactions compatible with DNA in aqueous solution. On the other hand, analytical tools to characterize the stereochemistry and configuration of the resulting products are also lacking, and separation of a reaction mixture is a challenge. For hit validation, a DNA-encoded compound (on-DNA) discovered in a selection is synthesized and tested as a small-molecule organic compound (off-DNA). In some cases, it is not straightforward to convert an on-DNA compound to an off-DNA compound.

Buller et al. synthesized two DNA-encoded chemical libraries based on the Diels–Alder reaction [32, 33]. Diels–Alder reactions can lead to formation of a variety of structural isomers and stereoisomers (enantiomers and diastereomers). Because of the various dienes and maleimides used as building blocks in the library synthesis, the resulting stereochemistry could not be easily predicted and analyzed on-DNA. To bypass the structural complexity, the researchers replaced the core structure resulting from the Diels–Alder reaction with a phthalimide moiety. This replacement resulted in reduction of structural complexity, while maintaining the rigidity of the core structure and the steric distance between the pharmacophoric elements, and led to products that are more water soluble than the original compounds.

The final step of the aforementioned DNA-encoded macrocyclic library produced by Liu and coworkers was carried out by a Wittig macrocyclization reaction [25]. After synthesizing the selected compounds in solution, the authors were able to show that inhibition of the target protein Src kinase is dependent on the building blocks as well as the *cis/trans* configuration of the double bond resulting from the Wittig reaction (Fig. 16.3). The most potent Src kinase inhibitor *trans*-A10-B1-C5-D6 discovered in this study possesses an IC_{50} value of 0.68 µM, 10 times lower than its *cis* counterpart. Moreover, for another pair of isomers, *trans*- and *cis*-A11-B1-C5-D7, the *trans* compound was the second most potent compound discovered in the study, and its IC_{50} value (0.96 µM) was 12 times lower than the *cis* conformer.

DNA-templated chemical reactions have allowed researchers to explore organic reactions using genetic methods. In this context, on-DNA chemical reactions are not only tools for exploring chemical space to discover potent compounds for biomedical application. New reaction discovery and understanding of reaction mechanisms

cis A11-B1-C5-D7 IC_{50} = 0.96 µM 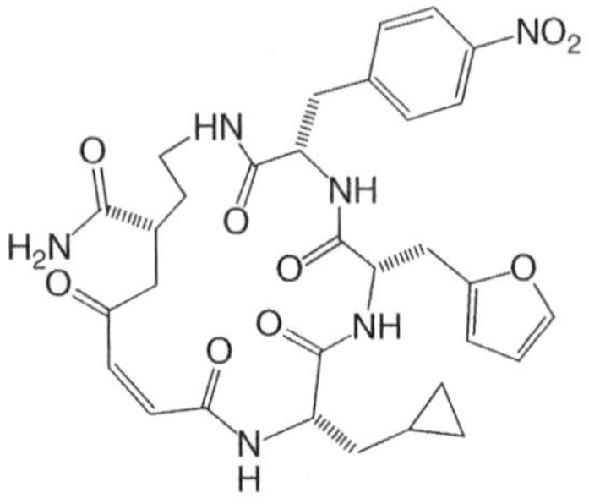   *cis* A10-B1-C5-D6 IC_{50} = 7.4 µM

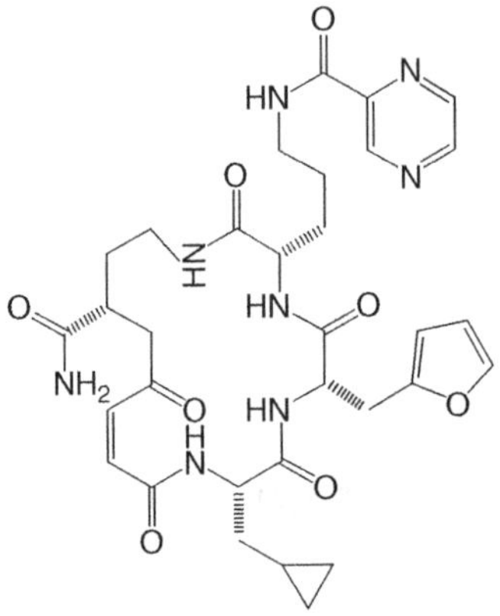

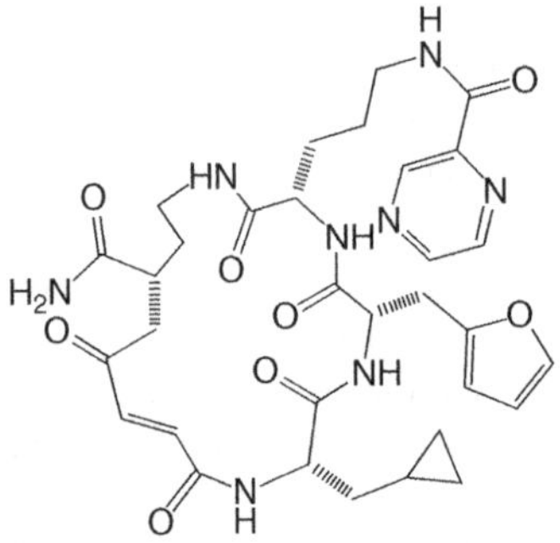

trans A11-B1-C5-D7 IC_{50} = 12 µM 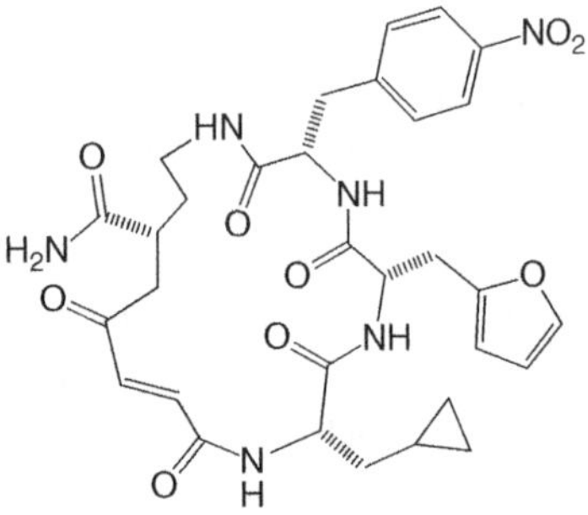   *trans* A10-B1-C5-D6 IC_{50} = 0.68 µM

Figure 16.3. Inhibition of Src kinase activity by compounds discovered from a 13,824-member small-molecule macrocycle library and hit identification through off-DNA organic synthesis [25].

represent central problems in organic chemistry. Using DNA-templated synthesis, Liu and coworkers were able to test 13 × 13 pairs of reactants under five different conditions in parallel [19]. Through an on-DNA reaction discovery to an off-DNA reaction study, they demonstrated the Pd(II)-catalyzed intermolecular oxidative coupling of alkynamides and alkenes to provide α,β-unsaturated ketones with high stereo- and regioselectivity [34].

16.8 CHALLENGES IN HIT IDENTIFICATION

Decoding translates the relative abundance of a pool of DNA-encoded compounds after selection to different counts of individual DNA barcodes generated by deep sequencing. The translation from analogue to digital information can be affected by many factors, though a high count of an individual code indicates a strong binding affinity of the encoded chemical structure to the target protein.

First of all, the relative concentration of DNA-encoded compounds before selection is governed by the yield of the multistep organic reactions in the synthetic sequence, which can differ from compound to compound dramatically. The reaction yield cannot be evaluated by sequencing the pool of library before selection, because the sequencing process is independent of the small organic compound on the DNA strand. Therefore, to optimize reaction conditions on DNA to improve the yield of each reaction is probably the only general solution. Leimbacher et al. carried out an exhaustive coupling protocol [31] based on Harbury's method on ion-exchange resin [16]. Furthermore, they avoided performing organic chemical reactions other than peptide bond formation on DNA by preparing the purified first building blocks using classical organic chemistry. After the split-and-pool procedure of DNA-conjugates with the first building blocks, the coupling of the second building block using Harbury method in the library construction cannot be chemically characterized. Therefore, they optimized the reaction conditions by repeating the coupling on solid support six times. Many structures were shown to be conjugated with high yields through this procedure and have been fully characterized as model reactions. Nevertheless, such exhaustive procedures cannot be applied to construct large libraries with more than two building blocks.

Second, the relative abundance of compounds captured by immobilized protein on matrix may be correlated to binding affinity only in a certain concentration window, which is dependent on the protein-loading density, library concentration, selection conditions, and binding constants. Although we know that we need to use low protein-loading density in order to select strong binders and high protein-loading density to capture weak binders, this information would not be available before we perform the subsequent cumbersome hit validation experiments. Furthermore, it is also difficult to determine the amount of protein with native conformation on a solid support.

Third, PCR amplification may have bias, because different codes could have different GC contents. Furthermore, high-throughput sequencing technologies have been developed emphasizing the capability to analyze read large numbers of DNA sequences, instead of treating all the sequences with equal efficacy. Buller et al. performed a statistical analysis of the decoding results of a 4000 compound library against five different protein targets [33]. Given that the sequencing experiment can be considered as a series of independent events, a negative binomial distribution was suggested. The same model was been successfully used to analyze a library of 500,000 compounds and a library of one million compounds [32].

16.9 OUTLOOK

A drug screening and discovery campaign may start with high-throughput methods, such as a DNA-encoded chemical library, and aim to identify one compound for clinical application. As shown in Figure 16.4, the size of the funnel increases quickly, exploring both chemical space and biochemical information. One major challenge in drug discovery is to identify hits along the path going down the funnel, moving high-throughput

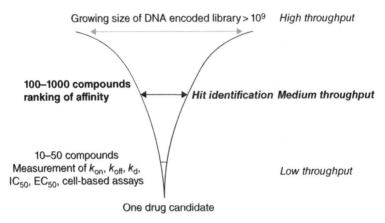

Figure 16.4. Medium-throughput screening in hit identification.

screening to low-throughput characterization of a number of selected hit compounds. To synthesize pure candidate compounds and to set up and perform various biochemical assays are knowledge-intensive and labor-intensive tasks.

For hit identification using a DNA-encoded chemical library, the selection of tens of compounds from the selection experiment with libraries of millions to billions of compounds will often involve many arbitrary criteria. Changing the selection conditions, such as washing with buffer or changing the protein-loading density, may provide more information, though it will also lead to further complications in decision-making. Therefore, the development of medium-throughput methodology, which can serve as the linker between high-throughput screening and low-throughput characterization, will provide invaluable information for selecting hit compounds. A noncovalent interaction between protein and ligand is the basis of the selection experiments, though fast and reliable measurement of the dissociation constants (K_d) as well as kinetic constants (k_{on} and k_{off}) in solution remain a challenge. ITC [35] represents the gold standard for characterizing binding energy, though the assay is time-consuming and thus very difficult to carry out for a medium-throughput screening. To develop medium-throughput methods, which can characterize hundreds to thousands of identified hit compounds from a large library, will lead to a refined structure–activity relationship and help the researcher to choose candidates for further biochemical studies and *in vivo* investigations. As discussed previously, the affinity capture of individual radiolabeled candidate conjugates could provide a ranking superior to the DNA array, though this approach cannot provide information regarding binding constants.

One of the dominant technologies for determining binding constants between biomolecules is chip-based biosensors, mainly using instrumentation based on Surface Plasmon Resonance (SPR) but also including other sensing technologies such as interferometry and Quartz Crystal Microbalance (QCM). These methods have been widely used to perform screening of protein–protein (e.g., antibody–antigen) interactions. Many of the biosensor instruments have been constructed to be able to perform hundreds to thousands of assays. It will be very attractive to apply such technologies in hit identification, serving as a medium-throughput platform between high-throughput selections using DNA-encoded chemical libraries and low-throughput characterizations of selected hit compounds.

16.9.1 Direct Analysis of Binding in the Sequencing of DNA Tags

Second-generation sequencing machines, such as the Illumina Genome Analyzer, build millions of distinct clusters on a flow cell. Each cluster, consisting of a few hundred DNA molecules, is sequenced *in situ* using a fluorescence-based detection method. In principle, such instrumentation can also be applied to perform a direct binding assay, where each cluster is used to immobilize one of the encoded compounds, while the second fluorescence-labeled compound (e.g., a protein) is in the mobile phase.

Deep sequencing instruments have been used recently to determine the DNA affinity landscape of transcription factors [36]. In spite of the analogy to an addressable chemical library on a chip, it is difficult to apply this approach in conventional screening campaigns. On the one hand, it would involve difficult and cumbersome chemistry to immobilize a large chemical library; on the other hand, addressability of each cluster in a flow cell system presents a daunting challenge. However, just as modern sequencing technology has found an unexpected application in drug screening through DNA-encoded chemical libraries, the direct affinity measurement on a second-generation sequencer can be implemented to improve the hit identification process. Given that each member of a DNA-encoded chemical library has a DNA tag, they can be specifically immobilized on a cluster through hybridization. The addressability problem does not exist either, because the identity of the individual compound is not revealed through its physical location, but through sequencing its DNA strand.

A binding assay on a chip may provide not only a comparison of different compounds for binding to the target protein but also an evaluation of the dissociation constants. For instance, the DNA affinity landscape study characterized about half a billion binding events of 8–12-mer oligonucleotides to a basic leucine zipper transcription factor Gcn4p. Approximately 100,000 reads revealed binding constants between 10 nM and 1 μM, which would also be the most interesting affinity range expected for a selection experiment using a DNA-encoded chemical library. This approach will enable researchers to correlate two completely different sets of information: the digital counts of each DNA barcode and the analogue reads regarding the affinity. Although both approaches could involve a number of biases associated with high-throughput methodology, a systematic analysis and comparison of them, in combination with the statistical method mentioned previously, will allow us to focus on a few interesting hits for further biochemical characterization.

16.9.2 A Simple Database for DNA Encoding

The simplicity of a database is essential for broad application across many different research fields. One of the most successful examples is the PDB file of protein structures. In addition to their importance in structural biology, the simple data format of PDB file is one of its essential features, which makes protein structure databases universal resources used by all disciplines of biological sciences. DNA-encoded chemical libraries, due to their artificial design, possesses simplicity that does not occur with many other technologies. Information can be, on the one hand, easily digitized. The chemical structures of billions of compounds and their binding profiles against many different proteins can be presented as simple datasets. On the other hand, such information can be easily circulated among and used by people from many different fields from medicinal chemistry to bioinformatics.

If we can generate a database that can assemble all information, including the chemical structures of DNA-encoded libraries and selection results against various proteins from researchers in different labs, with a simple and universal data structure containing the digitized information of each experiment, the resulting platform will lead to collaborations of many different and distant disciplines. The discovery or design of specific binding molecules represents a great challenge not only for the pharmaceutical sciences but also for chemistry and biochemistry in general. A synergy of knowledge-based approaches such as structural biology and bioinformatics and technology-based approaches such as high-throughput screening and deep sequencing could assist in the hit identification process from DNA-encoded chemical libraries in particular and deepen our understanding of molecular recognition in general.

REFERENCES

1. Schreiber, S. L. (2000) Target-oriented and diversity-oriented organic synthesis in drug discovery. *Science* 287(5460):1964–1969.

2. Kuruvilla, F. G., Shamji A. F., Sternson S. M., Hergenrother P. J., & Schreiber S. L. (2002) Dissecting glucose signalling with diversity-oriented synthesis and small-molecule microarrays. *Nature* 416(6881):653–657.

3. Wetzel, S., Bon R. S., Kumar K., & Waldmann H. (2011) Biology-oriented synthesis. *Angew. Chem. Int. Ed. Engl.* 50(46):10800–10826.

4. Wu, X., Schultz P. G. (2009) Synthesis at the interface of chemistry and biology. *J. Am. Chem. Soc.* 131(35):12497–12515.

5. Smith, G. P. (1985) Filamentous fusion phage: novel expression vectors that display cloned antigens on the virion surface. *Science* 228(4705):1315–1317.

6. Nissim, A., Hoogenboom H. R., Tomlinson I. M., Flynn G., Midgley C., Lane D., & Winter G. (1994) Antibody fragments from a 'single pot' phage display library as immunochemical reagents. *EMBO J.* 13(3):692–698.

7. Neri, D., Bicknell R. (2005) Tumour vascular targeting. *Nat. Rev. Cancer* 5(6):436–446.

8. Hoogenboom, H. R. (2002) Overview of antibody phage-display technology and its applications. *Methods Mol. Biol.* 178:1–37.

9. Mannocci, L., Leimbacher M., Wichert M., Scheuermann J., & Neri D. (2011) 20 years of DNA-encoded chemical libraries. *Chem. Commun. (Camb)* 47(48):12747–12753.

10. Kleiner, R. E., Dumelin C. E., & Liu D. R. (2011) Small-molecule discovery from DNA-encoded chemical libraries. *Chem. Soc. Rev.* 40(12):5707–5717.

11. Brenner, S. Lerner R. A. (1992) Encoded combinatorial chemistry. *Proc. Natl. Acad. Sci. U. S. A.* 89(12):5381–5383.

12. Nielsen, J., Brenner S., & Janda K. D. (1993) Synthetic methods for the implementation of encoded combinatorial chemistry. *J. Am. Chem. Soc.* 115(21):9812–9813.

13. Needels, M. C., Jones D. G., Tate E. H., Heinkel G. L., Kochersperger L. M., Dower W. J., Barrett R. W., & Gallop M. A. (1993) Generation and screening of an oligonucleotide-encoded synthetic peptide library. *Proc. Natl. Acad. Sci. U. S. A.* 90(22):10700–10704.

14. Halpin, D. R., Harbury P. B. (2004) DNA display I. Sequence-encoded routing of DNA populations. *PLoS Biol.* 2(7):1015–1021.

15. Halpin, D. R., Harbury P. B. (2004) DNA display II Genetic manipulation of combinatorial chemistry libraries for small-molecule evolution. *PLoS Biol.* 2(7):1022–1030.

16. Halpin, D. R., Lee J. A., Wrenn S. J., & Harbury P. B. (2004) DNA display III Solid-phase organic synthesis on unprotected DNA. *PLoS Biol.* 2(7):1031–1038.

17. Wrenn, S. J., Weisinger R. M., Halpin D. R., & Harbury P. B. (2007) Synthetic ligands discovered by in vitro selection. *J. Am. Chem. Soc.* 129(43):13137–13143.

18. Melkko, S., Scheuermann J., Dumelin C. E., & Neri D. (2004) Encoded self-assembling chemical libraries. *Nat. Biotechnol.* 22(5):568–574.

19. Kanan, M. W., Rozenman M. M., Sakurai K., Snyder T. M., & Liu D. R. (2004) Reaction discovery enabled by DNA-templated synthesis and in vitro selection. *Nature* 431(7008): 545–549.

20. Dumelin, C. E., Scheuermann J., Melkko S., & Neri D. (2006) Selection of streptavidin binders from a DNA-encoded chemical library. *Bioconjug. Chem.* 17(2):366–370.

21. Melkko, S., Zhang Y., Dumelin C. E., Scheuermann J., & Neri D. (2007) Isolation of high-affinity trypsin inhibitors from a DNA-encoded chemical library. *Angew. Chem. Int. Ed. Engl.* 46(25):4671–4674.

22. Scheuermann, J., Dumelin C. E., Melkko S., Zhang Y., Mannocci L., Jaggi M., Sobek J., & Neri D. (2008) DNA-encoded chemical libraries for the discovery of MMP-3 inhibitors. *Bioconjug. Chem.* 19(3):778–785.

23. Dumelin, C. E., Trussel S., Buller F., Trachsel E., Bootz F., Zhang Y., Mannocci L., Beck S. C., Drumea-Mirancea M., Seeliger M. W., Baltes C., Muggler T., Kranz F., Rudin M., Melkko S., Scheuermann J., & Neri D. (2008) A portable albumin binder from a DNA-encoded chemical library. *Angew. Chem. Int. Ed. Engl.* 47(17):3196–3201.

24. Mannocci, L., Zhang Y., Scheuermann J., Leimbacher M., De Bellis G., Rizzi E., Dumelin C., Melkko S., & Neri D. (2008) High-throughput sequencing allows the identification of binding molecules isolated from DNA-encoded chemical libraries. *Proc. Natl. Acad. Sci. U. S. A.* 105(46):17670–17675.

25. Kleiner, R. E., Dumelin C. E., Tiu G. C., Sakurai K., & Liu D. R. (2010) In vitro selection of a DNA-templated small-molecule library reveals a class of macrocyclic kinase inhibitors. *J. Am. Chem. Soc.* 132(33):11779–11791.

26. Hansen, M. H., Blakskjaer P., Petersen L. K., Hansen T. H., Hojfeldt J. W., Gothelf K. V., & Hansen N. J. (2009) A yoctoliter-scale DNA reactor for small-molecule evolution. *J. Am. Chem. Soc.* 131(3):1322–1327.

27. Gartner, Z. J., Tse B. N., Grubina R., Doyon J. B., Snyder T. M., & Liu D. R. (2004) DNA-templated organic synthesis and selection of a library of macrocycles. *Science* 305(5690): 1601–1605.

28. Clark, M. A., Acharya R. A., Arico-Muendel C. C., Belyanskaya S. L., Benjamin D. R., Carlson N. R., Centrella P. A., Chiu C. H., Creaser S. P., Cuozzo J. W., Davie C. P., Ding Y., Franklin G. J., Franzen K. D., Gefter M. L., Hale S. P., Hansen N. J., Israel D. I., Jiang J.,

Kavarana M. J., Kelley M. S., Kollmann C. S., Li F., Lind K., Mataruse S., Medeiros P. F., Messer J. A., Myers P., O'Keefe H., Oliff M. C., Rise C. E., Satz A. L., Skinner S. R., Svendsen J. L., Tang L., van Vloten K., Wagner R. W., Yao G., Zhao B., & Morgan B. A. (2009) Design, synthesis and selection of DNA-encoded small-molecule libraries. *Nat. Chem. Biol.* 5(9):647–654.

29. Church, G. M., Gao Y., & Kosuri S. (2012) Next-generation digital information storage in DNA. *Science* 337(6102):1628.

30. Goldman, N., Bertone P., Chen S., Dessimoz C., LeProust E. M., Sipos B., & Birney E. (2013) Towards practical, high-capacity, low-maintenance information storage in synthesized DNA. *Nature* 494(7435):77–80.

31. Leimbacher, M., Zhang Y., Mannocci L., Stravs M., Geppert T., Scheuermann J., Schneider G., & Neri D. (2012) Discovery of small-molecule interleukin-2 inhibitors from a DNA-encoded chemical library. *Chemistry* 18(25):7729–7737.

32. Buller, F., Steiner M., Frey K., Mircsof D., Scheuermann J., Kalisch M., Buhlmann P., Supuran C. T., & Neri D. (2011) Selection of carbonic anhydrase IX inhibitors from one million DNA-encoded compounds. *ACS Chem. Biol.* 6(4):336–344.

33. Buller, F., Zhang Y., Scheuermann J., Schafer J., Buhlmann P., & Neri D. (2009) Discovery of TNF inhibitors from a DNA-encoded chemical library based on diels-alder cycloaddition. *Chem. Biol.* 16(10):1075–1086.

34. Momiyama, N., Kanan M. W., & Liu D. R. (2007) Synthesis of acyclic alpha,beta-unsaturated ketones via Pd(II)-catalyzed intermolecular reaction of alkynamides and alkenes. *J. Am. Chem. Soc.* 129(8):2230–2231.

35. Ladbury, J. E., Klebe G., & Freire E. (2010) Adding calorimetric data to decision making in lead discovery: a hot tip. *Nat. Rev. Drug Discov.* 9(1):23–27.

36. Nutiu, R., Friedman R. C., Luo S., Khrebtukova I., Silva D., Li R., Zhang L., Schroth G. P., & Burge C. B. (2011) Direct measurement of DNA affinity landscapes on a high-throughput sequencing instrument. *Nat. Biotechnol.* 29(7):659–664.

17

USING DNA TO PROGRAM CHEMICAL SYNTHESIS, DISCOVER NEW REACTIONS, AND DETECT LIGAND BINDING

Lynn M. McGregor and David R. Liu

Department of Chemistry and Chemical Biology and Howard Hughes Medical Institute, Harvard University, Cambridge, MA, USA

17.1 INTRODUCTION

Advances in the life sciences have led to a rapid increase in the number of protein targets of interest to the biomedical community. Chemical probes or lead compounds that modulate the activity of these targets illuminate biology and provide a foundation for the development of new therapeutics. Selections on DNA-encoded chemical libraries offer a high-throughput, low-cost alternative to traditional target-oriented screening. In contrast to a screen, a selection simultaneously evaluates all library members so that the time, effort, and expense required to assay a compound library are independent of the size of the library. Furthermore, selections require only basic laboratory equipment (plastic tubes, pipettes, magnets, PCR) rather than the more sophisticated liquid handling and imaging infrastructure required for traditional target-oriented screens.

In a DNA-encoded chemical selection, the library of compounds displayed on DNA is incubated with a target protein that is immobilized either before or after incubation with the compound library. Unbound compounds are washed away, and bound compounds are eluted from the target protein, commonly using heat. This selection procedure may be iterated to increase the degree of enrichment of target-binding molecules [1].

A Handbook for DNA-Encoded Chemistry: Theory and Applications for Exploring Chemical Space and Drug Discovery, First Edition. Edited by Robert A. Goodnow, Jr.
© 2014 John Wiley & Sons, Inc. Published 2014 by John Wiley & Sons, Inc.

After the final elution, the DNA sequences corresponding to active library members are amplified by PCR and analyzed, typically by massively parallel high-throughput sequencing.

DNA-encoded chemical libraries can be constructed either using DNA-programmed reactivity, in which the chemical fate of a molecule is directly determined by its DNA sequence, or in a DNA-recorded fashion in which DNA serves as a tag to record a compound's reaction history. DNA-programmed library construction methods enable the possibility of library "retranslation," a process by which mixtures of unknown DNA sequences encoding molecules surviving selection are transformed into their corresponding chemical library members *en masse*. Retranslation makes possible multiple complete rounds of DNA-programmed reactivity, selection, and PCR amplification. In principle, retranslation also makes possible diversification steps in which DNA sequences surviving a selection are mutated before retranslation to enable true evolution of small molecules with desirable properties. DNA-programmed reaction methods include DNA-Templated Synthesis (DTS) [2–6], DNA routing [7–10], strand displacement-mediated reactivity [11], and mRNA display of ribosomally translated peptides [12–18]. Recently, several methods for one-pot, multistep DNA-programmed syntheses have been developed [19–22], some of which are amenable to the construction of DNA-encoded chemical libraries.

In contrast, DNA-recorded chemistry methods do not require the manipulation of chemical reactivity or access to reagents in a DNA sequence-specific manner and are therefore more amenable to the construction of very large libraries (e.g., greater than 10^6 members). DNA-recorded methods include ligation-mediated construction [23] and split–react–tag–pool [24–30]. DNA hybridization has also been used to generate Encoded Self-Assembled Chemical (ESAC) libraries [31], in which the DNA tags encode noncovalently associated groups of pharmacophores. DNA-recorded synthesis strategies generally do not allow DNA sequences to be translated *en masse* into corresponding synthetic compounds, precluding the evaluation of these libraries using iterated cycles of *in vitro* selection, PCR amplification, and translation [31–35].

17.2 DNA-RECORDED SYNTHESIS

The first examples of DNA-encoded combinatorial libraries were generated through alternating rounds of peptide or DNA synthesis followed by pooling and splitting [36–38]. The requirement for each synthetic step to be compatible with both peptide and nucleic acid synthesis significantly limited the scope of chemical reactions used to build the small molecules. To remove the requirement for compatibility with DNA synthesis, DNA tags are now incorporated by ligation [24, 25], primer extension [26, 28, 39], or a combination of the two [27]. This strategy allows the use of any reactions that do not severely damage DNA. As a result, the split–react–tag strategy has been used to make the largest small-molecule libraries reported to date.

From a 30,000-member split–pool DNA-recorded synthesis library, Leimbacher and coworkers discovered two compounds with 3–4 μM activity against IL-2 and an

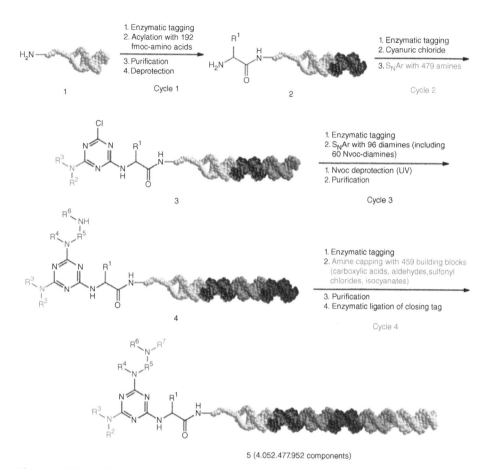

Figure 17.1. Split–pool–ligate method of assembling DNA-recorded chemical libraries. Reprinted with permission from Deng et al. [25]. Copyright 2012 American Chemical Society.

IC_{50} of 30–70 μM in a T-cell proliferation assay [26]. More recently, Buller and coworkers reported a one-million-member split–pool library constructed by alternating chemical reactions and elaboration of the DNA tag, which was used to discover inhibitors of CA IX with an IC_{50} of 240–920 nM [27]. Clark and coworkers have reported an 800-million-member library that was panned against p38 MAP kinase to reveal inhibitors with EC_{50} values between 7 nM and 5.2 μM [24]. Subsequently, Deng and coworkers reported the construction of a four-billion-member DNA-encoded library built around a triazine core (Fig. 17.1) [25]. This library was selected for binding to the ADAMTS-5 protease to reveal potent inhibitors ($K_i = 30$–50 nM) with good selectivity (37–1000×) over related zinc metalloproteases including ADAMTS-4, ADAMTS-1, TACE, and Matrix Metalloproteinase 13 (MMP-13) [25]. For structures, see Figure 17.18.

17.3 NONCOVALENT DNA-ASSEMBLED LIBRARIES

Libraries displaying noncovalent combinations of pharmacophores have been generated by hybridization of encoding strands to each other (Fig. 17.2) [31] or to a template (Fig. 17.3) [33]. These self-assembled libraries are typically smaller than split–pool libraries because each pharmacophore library must be individually conjugated to its encoding DNA and split–pool libraries often have more than two diversification steps. Dual-pharmacophore libraries are particularly amenable to affinity maturation, in which small fragments are evaluated for their ability to improve the binding of a known ligand. These libraries have been applied to the discovery and subsequent affinity maturation of MMP-3 inhibitors [32], yielding an inhibitor with $IC_{50} = 9.9\,\mu M$ for MMP-3 and little activity against the related proteases CPA and uPA(Fig. 17.18).

Peptide Nucleic Acids (PNAs) have also been used to encode self-assembled libraries [33]. Daguer and coworkers generated multiple libraries of PNA-displayed small molecules that combinatorially hybridize to a DNA template (Fig. 17.3) [34]. Assembly of multiple pharmacophores on a single DNA template allows the amplification of sequences corresponding to active library members and the hybridization-based retranslation of the library to support additional rounds of selection. This method was applied to the assembly of a PNA-encoded glycan library used in the affinity maturation of a known carbohydrate ligand of gp120 [35]. Svensen and coworkers performed selections on CCR6-overexpressing cells using a library of 10,000 PNA-encoded peptides. The selection was evaluated either in a solution binding assay with PCR amplification of DNA encoding active library members [29] or by hybridizing the PNA to a custom

Figure 17.2. ESAC can be used to prepare libraries displaying two or three pharmacophores or can be used for affinity maturation experiments in which multiple (left) or only one of the pharmacophores is encoded (right).

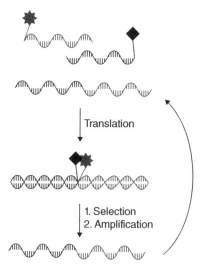

Figure 17.3. PNA-linked library members can be hybridized to a single DNA template for the DNA-encoded display of multiple pharmacophores.

DNA microarray and visualizing the locations of DAPI-stained cells [30]. Hybridization between PNA and DNA templates has also been used to template reactions including peptide transfer [40] and PNA polymerization by reductive amination [41].

17.4 EFFECTIVE MOLARITY: A BASIS FOR DNA-PROGRAMMED REACTIVITY

The ability to predict the strength of interactions between self-assembled DNA strands based on their primary sequences has enabled the field of DNA nanotechnology to flourish [42–44] and is also a foundation of DNA-programmed chemistry. The hybridization of two DNA strands increases the effective molarity of groups covalently or noncovalently associated with those strands. Intramolecular [45] and multivalent [46, 47] protein–ligand binding have been used to connect empirical observations of effective molarity with theoretical predictions of effective concentration. Effective molarity, defined as $M_{eff} = K_d^{inter}/K_d^{intra}$, is an empirical term that relates kinetics or equilibria of an intermolecular system with that of an analogous intramolecular system (Fig. 17.4). The related term effective concentration is typically used to describe the likelihood that two ends of a polymer would occupy space within a small distance of each other [45]. Effective concentration is proportional to the inverse cube of a parameter that increases with increasing length or stiffness of the linker between the two ends of a molecule [45].

In ring-closing macrocyclizations, the effective molarity and effective concentration are expected to have equivalent values. With the goals of testing whether intramolecular protein–ligand binding agrees with this observation and of determining which of several models of intramolecular binding were the most accurate (Fig. 17.5), Krishnamurthy

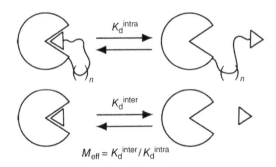

Figure 17.4. Effective concentration is defined as the ratio of the dissociation constants for intermolecular versus intramolecular binding. Reprinted with permission from Krishnamurthy et al. [45]. Copyright 2007 American Chemical Society.

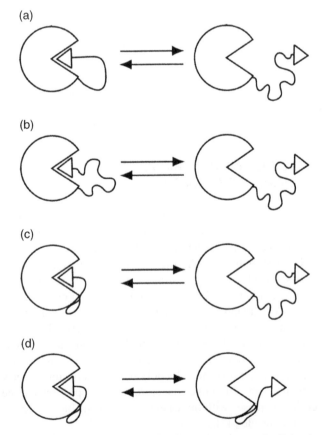

Figure 17.5. In an intramolecular protein–ligand binding event, the linker between protein and ligand may (a, d) or may not (b, c) be highly ordered and may (c, d) or may not (a, b) make stabilizing contacts with the protein. Reprinted with permission from Krishnamurthy et al. [45]. Copyright 2007 American Chemical Society.

and coworkers systematically varied the length of an ethylene glycol linker between human Carbonic Anhydrase II (CAII) and a sulfonamide ligand [45]. Isothermal titration calorimetry performed on a CAII–ethylene glycol conjugate lacking any ligand verified that there was no interaction between the linker and the protein, suggesting that the linker's contribution to effective molarity is entropic, rather than enthalpic, eliminating models C and D (Fig. 17.5) [45]. With an optimal linker length, the effective molarity was measured as 26 mM. As the linker length was further extended, the effective molarity decreased only tenfold to 2 mM. When an equation for effective concentration from theoretical polymer chemistry was fitted to the empirical effective molarity data, the model fitted quite well; however, two empirical parameters had values that significantly differed from their predicted values, leading the conclusion that predictions of effective concentration using theory alone are likely to be inaccurate. Significantly, the data of Krishnamurthy and coworkers suggest that a flexible and slightly too long linker is more likely to be successful than one that is too stiff or too short [45].

Gargano and coworkers studied the binding of multivalent ligands to a multivalent receptor and proposed a simple formula for estimating the Binding Enhancement (BE) conferred by multivalency: $F[sK_a(10^{-2})]^{(n-1)}$, where n is the number of binding sites on the receptor or ligand, $s = (30/\text{distance in Angstroms})$, and F is a system-specific statistical factor [47]. This estimate is applicable if (i) the binding sites are equivalent, (ii) binding is not cooperative, (iii) a flexible linker of optimal length is used, (iv) no linker–receptor interactions exist, and (v) BE is due to intramolecular binding. To support this formula, they compare the inhibition of Shiga toxin, a hexameric AB_5 toxin with five identical B subunits, using monovalent ($K_a = 1 \times 10^3 \, \text{M}^{-1}$) and multivalent trisaccharide ligands. Based on their model, the BE for binding a pentameric receptor, when $F = 1$, would be expected to be $BE = F[s(10)]^4 = 10^4$ per ligand. Indeed, Shiga toxin inhibition by monovalent ($K_i = 5 \, \text{mM}$) and multivalent ($K_i = 800–900 \, \text{nM}$) ligands revealed a BE of $>5 \times 10^3$, which is similar to the value predicted by their model [47].

While these studies focused on either ethylene glycol linkers [45] or alkyl/acrylamide linkers [47], many DNA-encoded syntheses and subsequent in vitro selections use single- or double-stranded nucleic acids as linkers. Therefore, nucleic acid stiffness could have a dramatic effect on the effective molarity of groups linked to hybridized DNA strands. The persistence length (the shortest length of a polymer in which the two ends can still meet) has been measured for both single-stranded DNA (ssDNA) and double-stranded DNA (dsDNA) under various salt conditions both in solution and on solid support. FRET-based studies of the persistence length of an oligo-(dT) polymer revealed that the persistence length of oligo-(dT) is 3 nm in 25 mM NaCl or 1.5 nm in 2 M NaCl [48]. These measurements are consistent with those obtained by Atomic Force Microscopy (AFM) [49, 50], transient electric birefringence [51], and equilibrium DNA hairpin melting profiles [52]. Based on the bond lengths from several crystal structures, Murphy and coworkers estimate the average interphosphate distance as 6.3 Å, giving a ssDNA persistence length of approximately 5 nt in 25 mM NaCl [48]. In contrast, the persistence length of dsDNA, as measured by AFM, DNA ligation [53], or winding of DNA around a F1 ATPase spool [54], has been measured as about 50 nm or 80 nt, depending on sequence, salt, or temperature conditions. A recent FRET assay has observed looping of dsDNA strands as short as 67 nt [55].

The approximately tenfold higher persistence length (higher stiffness) of dsDNA provides a clear explanation for the observation that even short segments of dsDNA between two DNA-linked reagents hybridized to the same DNA template can prevent a reaction between them [2, 56].

Empirical measurements of intramolecular kinetics and equilibria imposed by hybridization of 15–20 nt DNA strands suggest that hybridization results in effective molarities in the high micromolar or low millimolar range for templated reactions [2, 57] and for annealing of other DNA strands [58].

17.5 DNA-PROGRAMMED SYNTHESIS

Each of the DNA-programmed synthesis methods represents a different strategy to translate a nucleic acid sequence into a functional molecule while maintaining a link between genotype (nucleic acid sequence) and phenotype (DNA-programmed synthesis product). Ribosome display and mRNA display link ribosomally translated peptides to their encoding mRNA, either noncovalently, by omitting the stop codon, cooling the complexes, and adding Mg^{2+} [59] to increase the stability of the mRNA–ribosome–peptide complex, or covalently, by modifying the 3′ end of an mRNA transcript with the antibiotic puromycin, which forms a covalent bond with the nascent peptide chain (Fig. 17.6) [16]. The size of a library generated by *in vitro* translation is limited either by the number of ribosomes or by the number of DNA templates, and libraries with up to 10^{13} members have been reported [12, 16]. For comparison, limitations on transformation efficiencies typically restrict peptide libraries encoded in phage, bacteria, or yeast to $\sim 10^9$ members [12].

Nonproteinogenic amino acids can impart unique properties in peptides. Chemical acylation of tRNAs [13, 14] and the use of promiscuous Aminoacyl tRNA Synthetases

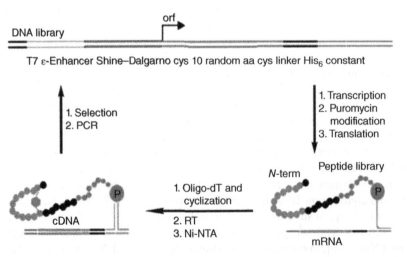

Figure 17.6. Scheme for one round of translation of an mRNA-displayed library of cyclic peptides. Reprinted with permission from Schlippe et al. [16]. Copyright 2012 American Chemical Society.

(AARSs) [15] and flexizymes [60] have significantly improved the feasibility of incorporating nonproteinogenic amino acids into *in vitro*-translated libraries. Flexizymes are DNAzymes capable of charging any tRNA with a broad array of nonproteinogenic amino acids activated as 3,5-dinitrobenzyl esters [61]. Combining flexizyme-charged tRNAs with fully recombinant *in vitro* translation systems (the Protein Synthesis Using Recombinant Elements (PURE) system) [62], in which specific natural tRNAs or amino acids can be omitted, eliminates competition between incorporation of natural and non-proteinogenic amino acids, enabling increased yield and fidelity of peptide libraries containing many nonproteinogenic amino acids [16, 63].

These strategies were applied to the *in vitro* translation of mRNA-displayed peptoids and peptoid–peptide hybrids with functionalized or beta-branched side chains [64] and of naturally occurring bioactive *N*-methyl peptides [17]. Furthermore, translation and selection of a cyclic peptide library of 10^{12} unique members, containing about 25% *N*-methylated amino acids, led to the identification of an inhibitor of the E6AP E3 ubiquitin ligase with $K_d = 0.60$ nM [65]. Schlippe and coworkers reported the discovery of two potent inhibitors ($K_i = 23$ nM or 35 nM) of thrombin containing 40–50% nonpro-teinogenic amino acids [16]. Compared to DNA-encoded chemical libraries, mRNA-displayed libraries offer advantages of (i) a high-yielding translation reaction, enabling the translation of molecules containing many building blocks; (ii) ease of library retrans-lation, enabling up to ten rounds of selection; and (iii) translation of libraries of up to >10^{13} members. On the other hand, the chemical diversity of libraries that require ribosomal biosynthesis is inherently limited to building blocks that are capable of being efficiently charged onto tRNA and that are accepted by the ribosomal machinery.

To stabilize the encoding nucleic acid of an mRNA-displayed library, reverse tran-scription is generally performed prior to selection, leaving a double-stranded cDNA/mRNA hybrid [12, 16]. In a method called "cDNA display," puromycin is added to the mRNA template by ligation of a puromycin-linked DNA hairpin rather than through hybridization. Reverse transcription on this template results in a cDNA strand that is covalently linked to the puromycin, and therefore the encoded peptide, enabling removal of the RNA by RNase H digestion [66]. Alternatively, several groups have used oil-in-water emulsions to perform *in vitro* transcription and translation of peptide libraries fused to proteins that tightly (biotin/streptavidin) [67] or irreversibly (5′-GG-5-fluorodeoxycytidine/C-3′-HaeIII DNA methyltransferase) [68] bind to modified encod-ing DNA (Fig. 17.7). After *in vitro* transcription, translation, and protein–DNA linkage, the oil-in-water emulsion is broken and *in vitro* selection is performed [69].

DNA routing is a DNA-programmed method of split–pool encoded library construction in which hybridization of a DNA template to one of several anticodon col-umns automates the "splitting" step (Fig. 17.8). The split templates are then transferred to DEAE columns, which nonspecifically adsorb DNA, allowing excess chemical reactants to be easily added to and removed from the reaction column [70]. A key strength of the DNA routing platform is that chemical reactions can be performed under conditions in which DNA is insoluble or incapable of hybridizing. The DNA is then eluted from the solid support and subjected to a new set of anticodon/splitting resins to generate a library of 10^6 peptides [7]. DNA routing was next applied to the translation of a 100-million-member peptoid library, which was selected for affinity for the Crk SH3 domain. Surviving sequences were subjected to four additional rounds of translation and selection, after which six peptoids with affinity for Crk SH3 were discovered ($K_d = 16–97\,\mu M$) (Fig. 17.18) [8]. The ease of retranslation of DNA-routed libraries makes them particularly amenable to selections on covalent inhibitors, in which elution of the small-molecule–template conjugate is not possible and retranslation is required for multiple rounds of selection.

Using custom mesofluidic devices, Weisinger and coworkers expanded DNA routing to a 384-well format in which each cycle consists of hybridization to an "anticodon array" of resins functionalized with complementary oligonucleotides, followed by Southern blotting to an anion-exchange array on which spatially separated chemical reactions are performed [10]. Routed synthesis of one tripeptoid followed by LC/MS analysis revealed that 59% of the DNA was modified with the desired peptoid, 22% was modified with side products, and 9% was not recovered. Unrouted synthesis of the same

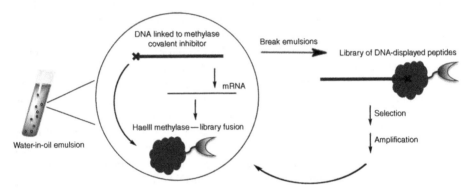

Figure 17.7. In a DNA-displayed library, *in vitro* compartmentalization is used to accomplish transcription, translation, and covalent conjugation to DNA the encoding each library member.

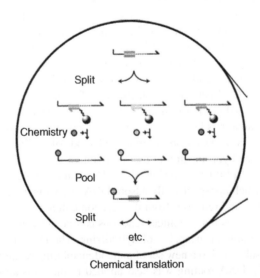

Chemical translation

Figure 17.8. Translation of DNA sequences into synthetic molecules through DNA routing. Reprinted with permission from Wrenn et al. [8]. Copyright 2007 American Chemical Society.

tripeptoid occurred with 66% yield. Mesofluidic DNA routing was applied to a proof-of-concept library synthesis in which 13,000-fold enrichment was demonstrated for a library member containing biotin [9]. As with other DNA-routed assemblies, the mesofluidic translation device can also be used for retranslation of enriched library members, enabling several rounds of selection and effectively unbounding the possible enrichment of active library members. The nature of the mesofluidic system used in DNA routing, however, limits the overall amount of material that can be produced to approximately 50 pmol or approximately 10^{13} molecules, limiting the practical library sizes to approximately 10^{10} if each library member will be present several thousand times [10].

17.6 DNA-TEMPLATED SYNTHESIS

DTS (Fig. 17.9), demonstrated as generally applicable by Gartner and coworkers in 2001, harnesses DNA hybridization to dramatically accelerate the rate of reaction (~200-fold) between alpha-iodoacetamides, alpha-bromoacetamides, maleimides, and vinyl sulfones with thiols or amines conjugated to complementary oligonucleotides [2]. Subsequently, Gartner and coworkers reported several DNA-templated reactions,

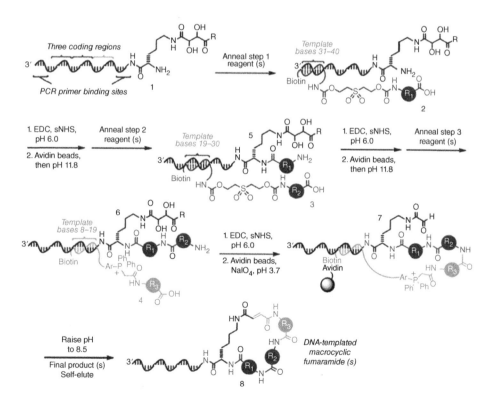

Figure 17.9. Translation of libraries of DNA into corresponding libraries of synthetic small molecules using DTS. Reprinted with permission from Tse et al. [6]. Copyright 2008 American Chemical Society.

including amine acylation, reductive amination, nitro-aldol, nitro-Michael, Wittig olefination, and 1,3-dipolar cycloaddition [71]. The presence of water-soluble Pd precatalysts enables DNA-templated Heck couplings between aryl iodides and olefin-containing templates including maleimide, acrylamide, vinyl sulfone, and cinnamide [71]. Additional DNA-compatible reactions include nucleophilic substitution of thiols for bromides [72], aldehyde thiazolidination [73], palladium-catalyzed enyne coupling [3, 73], and formation of boronic esters [74]. In the presence of pyrrolidine analogs or acyclic diamine catalysts, a DNA-templated aldol condensation was used to produce a hemicyanine dye that could in principle be used for biodetection [75]. A DNA-linked reagent was used to catalytically promote aldol reactions with unlinked ketones [76, 77]. DNA-templated triphenylphosphine-mediated azide reductions have been used to unmask amines, thiols, and carboxylic acids [78] and to release functional small molecules such as estradiol or doxorubicin [79]. Similarly, DNA-linked imidazole catalytically cleaved a DNA-linked p-nitrophenyl ester, releasing p-nitrophenol, as a model of prodrug release by DTS [80].

DNA-templated reactions also support the formation of metallosalen conjugates [81] such as Ni–salen complexes capable of cleaving DNA at dG residues when treated with an oxidant followed by piperidine [82]. Hybridization between a DNA template, a PNA-linked catalyst, and a PNA-linked substrate resulted in the catalytic cleavage of a pyridinyl ester by a Cu(II) complex [83]. Recently, a DNA-templated click reaction was applied to the fluorescent detection of Cu(II) ions at concentrations greater than 300 nM [84].

Although DTS generally requires conditions that support DNA hybridization, Rozenman and coworkers adapted DTS to organic solvents by first hybridizing the template and reagent in aqueous buffer before transferring the prehybridized mixture to a solution containing ≥95% organic solvent [85]. The yields of DNA-templated Wittig olefination were improved by using 95% acetonitrile as the reaction solvent [85]. Recently reported DNA-encoded chemical libraries have been constructed using amine acylation [6, 8, 9], Diels–Alder reactions [26], sulfonylation [25], and triazine substitution [24, 25].

Mechanistic studies of DTS revealed that the apparent increase in reaction rate was roughly equivalent whether the two reactants were separated by 1 or 30 nt or whether the linker between the reactants was comprised of deoxynucleotides, abasic deoxynucleotides, ethylene glycol, short alkane chains, or longer alkane chains. dsDNA between the reactants substantially decreased reaction yield, likely due to the increased rigidity of dsDNA compared to the other linkers [53]. This surprising distance independence of DTS reaction rates suggested that DNA hybridization was the rate-limiting step of the reaction. This hypothesis was tested by decreasing the concentration of DNA-linked reagents, which was expected to slow the rate of DNA hybridization, but was not expected to affect the rate of the chemical reaction, in which the effective molarity is constant after DNA hybridization has occurred. As the concentration of the DNA-linked reagent was decreased, the rate of reaction decreased dramatically, demonstrating that DNA hybridization is the rate-limiting step in the DNA-templated reactions tested [2].

This observation suggested that mutually competitive reactions could be performed simultaneously in the same vessel without formation of appreciable side products. Calderone and coworkers demonstrated the ability of DTS to perform amine acylation, conjugate addition, reductive amination, and Wittig olefination simultaneously in one solution, despite the need for distinct reagents and the fact that the second-order rate

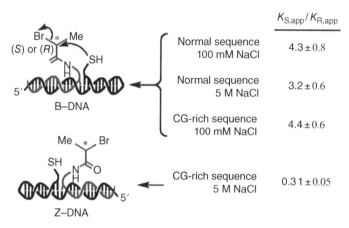

$K_{S,app} / K_{R,app}$

Normal sequence 100 mM NaCl	4.3 ± 0.8
Normal sequence 5 M NaCl	3.2 ± 0.6
CG-rich sequence 100 mM NaCl	4.4 ± 0.6
CG-rich sequence 5 M NaCl	0.31 ± 0.05

Figure 17.10. The handedness of the DNA helix can impart stereoselectivity on DTS reactions. Reprinted with permission from Li and Liu [72]. Copyright 2003 American Chemical Society.

constants of the reactions differed by orders of magnitude [3]. When performed in dilute solutions, the high effective molarity of reactants linked to hybridized oligonucleotides governs reactivity even in the presence of competing reactants.

The ability of the chirality of the DNA double helix to impart stereoselectivity onto DNA-templated reactions was studied by comparing the rate of DNA-templated nucleophilic substitution of a DNA-linked (S)- or (R)-2-bromopropionamide with a DNA-linked thiol [72]. Whether the reagents were conjugated to the DNA strands in the middle or at the end of the helix, the relative reaction rate between the thiol and 2-bromopropionamide was three- to fivefold faster for the (S)-isomer [72]. Replacing the ssDNA linker with achiral linkers or chiral abasic phosphoribose linkers reduced stereoselectivity, demonstrating that base stacking interactions induce stereoselectivity of DTS reactions [72]. Sequences with (5-Me-C)G-rich regions are capable of adopting right-handed (B-form) helices in low salt but adopt left-handed helices (Z-form) in the presence of 5M NaCl (Fig. 17.10) [72]. When these reactions were templated by (5-Me-C)G-rich sequences, the thiol preferentially reacted with the (S)-bromide in low salt (100 mM NaCl) but favored the (R)-bromide in the presence of 5 M NaCl. A change in the handedness of the DNA helix resulted in inversion of stereoselectivity, consistent with the model that the conformation of the template and reagent can determine stereoselectivity in a DNA-templated reaction (Fig. 17.10).

Some reagents of interest, including formates and isocyanates, would be difficult to couple to DNA. DTS-mediated functional group transformation could be used to unmask functional groups that could subsequently react with reagents not linked to DNA while maintaining a record of such reactions. Sakurai and coworkers developed a DTS-mediated azide-to-amine reduction using DNA-linked triphenylphosphine to perform a Staudinger reaction [78]. This DTS Staudinger reaction was also used to reveal caged carboxylic acids and thiols. The reduction products were subsequently reacted with reagents that would have been difficult to tether to DNA including dansyl chloride, ethyl chloroformate, 4-methoxy-phenyl isocyanate, and 6-morpholino-pyridin-3-yl isothiocyanate [78].

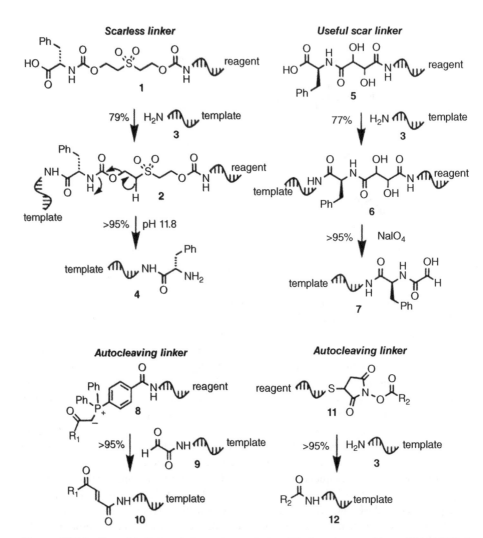

Figure 17.11. Cleavable linkers between reagents and their codons enable multistep DTS. A variety of linkers with different properties were developed. Reprinted with permission from Gartner et al. [4]. Copyright 2002 American Chemical Society.

Multistep DTS required the development of linkers capable of cleaving the DNA-linked reagent from the DNA prior to the next round of DTS [4], including a "scarless linker" comprised of a carbamoylethyl sulfone, a "useful scar linker" comprised of a sodium periodate-cleavable diol linker that reveals an aldehyde functional group, and two "self-cleaving linkers" in which the encoding DNA was linked through the triphenylphosphine group for a Wittig reaction or the succinimide of an activated N-hydroxy-succinimide ester. Multistep DTS reactions also required the removal of unreacted templates and reagents after each round of synthesis (Fig. 17.11). The reagent-linked anticodons were biotinylated to enable capture of templates that reacted with reagents linked through "scarless" and "useful scar" linkers or the removal of all anticodons with "self-cleaving" linkers.

(a)   (b)

End-of-helix architecture   Omega architecture   T architecture

Figure 17.12. DTS can be carried out with various template architectures. (a) The omega architecture improves reactivity of codons at the distal end of the template. Reprinted with permission from Snyder et al. [95]. Copyright 2008 American Chemical Society. (b) The T architecture enables two reactions to occur in succession without any intervening steps. Reprinted with permission from Li et al. [87]. Copyright 2004 American Chemical Society.

As a proof of concept, two separate three-step DTS translations were reported, using either three rounds of amide bond formation with the "scarless linker" or using amide bond formation, Wittig olefination, and conjugate addition with the "useful scar," "self-cleaving," and "scarless" linkers, respectively [4]. These linkers were subsequently applied to a model diversifying combinatorial DTS translation in which multiple incompatible reactions were successfully performed simultaneously on different templates, due to the ability of DNA hybridization to govern reactivity of DNA-linked reactants [73].

Gartner and coworkers also developed a T architecture in which two oligonucleotides with a 5'-linked reagent or a 3'-linked reagent simultaneously hybridize to the same template, placing the two reactants in close proximity and enabling a two-reaction cascade without intervening purification steps (Fig. 17.12) [86]. Li and coworkers applied the T architecture to the DTS of an N-acyloxazolidine. The T architecture enabled immediate N-acylation of the oxazolidine, without intervening purification steps that would have led to oxazolidine decomposition [87].

Snyder and coworkers applied the observation that dsDNA linkers prevent DTS reactions to perform a sequential, three-step DTS reaction without any intervening annealing or purification steps aside from a progressive increase in temperature [56]. Three DNA-linked reactants were annealed to a template such that all but the first reagent is separated from the template-linked reactant by dsDNA. As the temperature is increased, the first reagent dissociates, exposing a region of ssDNA between the template and the second reagent and enabling the second reagent to react with the template. This strategy was used to synthesize a DNA-programmed triolefin using three Wittig olefination reactions and a tripeptide linked by NHS-activated esters, which was produced in substantially higher yield than a previous DTS tripeptide (21% vs. 3%) [56].

Reducing the number of manipulations between coupling steps has dramatically improved the yield of multistep DTS [6, 56], enabling DTS involving five or more successive templated reactions [11]. These methods typically require the researcher to add new reagents at each step, but do not require intermediate purification steps. A DNA nanowalker capable of moving sequentially between multiple stations on a DNA track through successive cycles of walker translocation, DNAzyme-catalyzed cleavage of RNA bases, and dissociation of the cleaved fragment was recently developed [88] and applied to a three-step DTS (Fig. 17.13a) [19]. In this DNA-based ribosome mimetic,

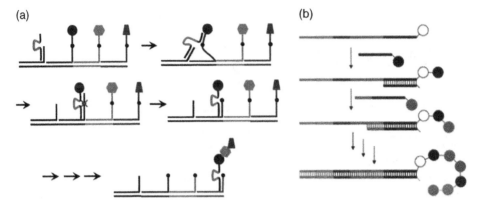

Figure 17.13. (a) A DNA nanowalker capable of moving along a DNA track was adapted to perform a one-pot multistep DTS reaction. (b) Strand displacement can be used to program a one-pot multistep DTS reaction with up to six consecutive steps.

the template track is analogous to mRNA, and the reactant-linked substrate strands are analogous to charged tRNAs. The DNA walker is prehybridized to a truncated substrate initiator, and in a separate tube, each of the "tRNA" substrates is annealed to the track. Upon mixing of the two solutions, the walker–initiator complex hybridizes to the first position on the track, initiating a sequence of walker translocation, DNA-templated amine acylation, DNAzyme-catalyzed cleavage, and dislocation of the 5′ fragment of the expended substrate. This cycle repeats spontaneously until the walker rests at the final station, after having completed three successive DTS reactions without any intervention by the researcher with higher overall efficiency than other multistep DTS translation systems [19].

McKee and coworkers developed a strategy for a one-pot multistep DNA-templated reaction in which two DNA-linked reagents are brought into close proximity upon annealing of a long complementary region while a variable region at the end of each DNA-linked reagent remains single stranded [20]. After the reaction occurs, a "remover strand" capable of annealing to the entire reagent strand is added to sequester the spent reagent by "toehold displacement" (Fig. 17.14a) [20]. The next DNA-linked reagent is then added and hybridizes to the now single-stranded reagent linked to the first two building blocks. Milnes and coworkers applied this method to perform nine successive DNA-programmed Wittig olefinations resulting in an olefin polymer of defined length with 23–30% overall yield [21]. McKee and coworkers also developed a strategy using a DNA adapter to bring together two DNA-linked reactants in a T configuration, promoting a reaction between them (Fig. 17.14b) [22]. After the reaction has occurred, a "remover strand" complementary to the DNA adapter releases the DNA-linked growing chain to hybridize with a new adapter and new reaction partner. The growing molecule either remains attached to the same DNA strand or is passed from one strand to the next, termed the "alternate-strand method" (Fig. 17.14b). In the "alternate-strand method," the growing chain is transferred to the incoming DNA-linked monomer so that each reaction will have equivalent geometry, regardless of how many monomer units have

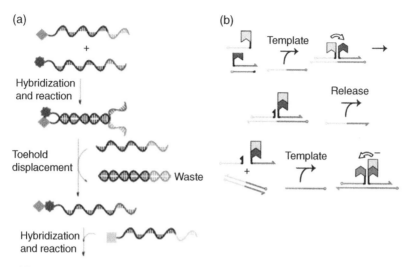

Figure 17.14. (a) In a one-pot multistep DNA-programmed reaction, a "remover strand" displaces expended reagents. (b) Another strategy for a one-pot multistep DNA-programmed reaction involves passing the growing macromolecule between alternating strands. Reprinted with permission from McKee et al. [22]. Copyright 2011 American Chemical Society.

been added, and the overall yield of the reaction should depend only on the yield of each individual reaction [22]. Collectively, these methods allow the simple, high-yielding synthesis of DNA-programmed macromolecules, but they cannot be directly applied to the synthesis of an encoded library of molecules because the DNA strand associated with the final product does not encode that product.

To address this potential limitation, He and coworkers designed a multistep one-pot DTS in which coding information remains attached to the final product (Fig. 17.13b) [11]. DNA-linked reagents anneal to a long template that encodes the growing molecule. Addition of the next reagent, linked to a DNA strand encoding the current and previous reactions, results in strand displacement removal of the expended DNA-linked reagent from the previous step and a reaction between the new reagent and the growing molecule. When the reagents and anticodons are linked by self-cleaving groups, no cleavage or purification steps are necessary. Other than adding each reagent successively, no researcher intervention is required. Because each anticodon extends to the 5′ end of the template, the reactive building blocks remain in close proximity with the growing molecule, regardless of the number of synthetic steps. Strand displacement translation was used to perform six reactions in a single solution with an overall yield of 35% (average yield per reaction 85%), representing the longest and highest yielding DNA-templated multistep small-molecule synthesis to date [11]. One limitation in translating a library by strand displacement is that each reagent must be linked to DNA strands encoding the current reaction and all combinations of prior DNA-encoded reactions. Advances in chip-based parallel oligonucleotide synthesis have made such materials more readily accessible [89–91].

17.7 TEMPLATE DESIGN FOR DNA-ENCODED LIBRARIES

Template sequences for DNA-encoded chemical libraries must support template construction, library synthesis, PCR amplification, and decoding. PCR amplification requires primer binding sites $\geq \sim 10\,nt$ [92]. Selections decoded by microarray require all coding regions to have equivalent melting temperature (T_m) values [93]. Decoding by massively parallel high-throughput sequencing, in contrast, can readily discriminate single-nucleotide differences, although DNA is sufficiently information dense that DNA-encoded libraries typically use codons that are least two mismatches away from any other codon [6, 28]. For DNA-recorded libraries, template construction occurs during library synthesis and is relatively straightforward, but the templates for DNA-programmed libraries must be constructed prior to translation. To expedite this process, templates often include constant regions used to assemble the full-length template by pooled ligation or PCR reactions [6, 8, 73, 94]. Ribosomal translation of peptide libraries typically requires DNA libraries containing a T7 RNA Polymerase (RNAP) promoter, as well as ribosome binding sites, and adapter sequences used to attach puromycin in the case of mRNA display [12]. For libraries translated by DTS, thoughtful template design is essential for high-fidelity and high-yielding library synthesis.

Faithful translation of a DTS library requires that each reaction must proceed with good yield, regardless of the codon's position along the DNA template. The "omega" (Ω) template architecture contains a short constant region near the 5′ end of the template that can increase the yield of distance-dependent DTS reactions (Fig. 17.12). Reagent anticodons hybridize with the 5′ constant region after hybridizing to their corresponding codon, bringing the DNA-linked reactants into closer proximity than without constant-region hybridization [86]. The Ω architecture substantially improved the yields of distally located Wittig olefinations, 1,3-dipolar cycloadditions, and reductive aminations.

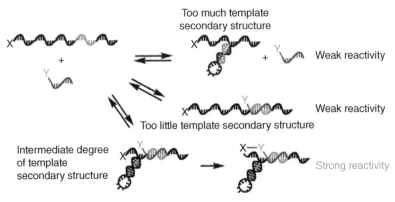

Figure 17.15. DNA templates with an intermediate degree of secondary structure result in the strongest reactivity between two DNA-linked reagents. Reprinted with permission from Snyder et al. [95]. Copyright 2008 American Chemical Society.

The secondary structure of the DNA template can also affect the yield of DTS reactions, especially for codons at the distal end of the template. Snyder and coworkers found that templates with intermediate secondary structures ($\sim -7\,\text{kcal/mol} > \Delta G > \sim -3\,\text{kcal/mol}$) produced higher-yielding DTS amine acylations or reductive aminations than either highly structured or completely unstructured templates (Fig. 17.15) [95]. It was reasoned that templates with intermediate amounts of secondary structure are unlikely to fold in ways that obscure reagent binding sites but may adopt conformations with internal base pairing between the reagent and the end of the helix, which help pay for the entropic cost of forming the Ω architecture.

17.8 TRANSLATION AND SELECTION OF A DNA-TEMPLATED LIBRARY OF PEPTIDE MACROCYCLES

To increase the yields of multistep DTS, Tse and coworkers reduced the number of purification steps required between reactions (Fig. 17.16). After the first two DTS reactions, unreacted library members are capped with acetic anhydride, rather than removed by affinity capture and washing. Second, by performing the first reactions at distal codons, expended reagent oligonucleotides can remain annealed to the template without impeding subsequent DNA-templated reactions. The double-stranded regions also reduce competition for annealing subsequent reagents by preventing distal codons from annealing to internal codons. Integrating these advances with an Ω architecture template library of intermediate secondary structure enabled the synthesis of a library of 13,824

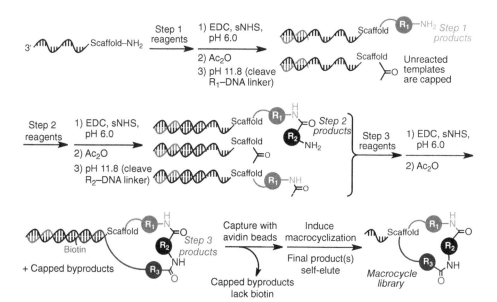

Figure 17.16. DTS-based translation of a library of 13,824 peptide macrocycles. Reprinted with permission from Tse et al. [6]. Copyright 2008 American Chemical Society.

peptide macrocycles [6]. This class of molecules was chosen because their macrocyclic nature may partially offset the entropic cost of protein binding and may impart improved pharmacokinetic properties [96].

To ensure high-fidelity translation of the library, each codon was designed with at least three mismatches to all other codons in the library. Each codon–anticodon pair was vetted with computational hybridization and pilot reactions to ensure at least 40% yield for matched codons and at most 8% yield with mismatched codons [6]. Demonstration of the fidelity and building block tolerance of the library translation was accomplished by exhaustive studies of sublibraries culminating with the translation of a 1728-member sublibrary in which the scaffold building block was held constant and all possible combinations of DNA-templated reactions were performed. LC/MS analysis of this sublibrary revealed that 95% of the expected unique masses were observed with an overall yield of 1.0% [6]. Synthesis of the library of 13,824 macrocycles was performed three times with an average yield of 1.6%, in total producing 375 pmol of material, sufficient for hundreds of *in vitro* selection experiments [6]. To date, over 36 *in vitro* selections have been performed with this library, resulting in the discovery of compounds that selectively inhibit several therapeutically relevant kinases and proteases. A highly selective Src inhibitor with $IC_{50} = 680$ nM directly emerged from this library (Fig. 17.18) [97]. Second-generation macrocycles based on positives emerging from selections have potencies as high as $IC_{50} \leq 4$ nM [98]. Mechanistic studies revealed that these macrocycles are both ATP competitive and peptide substrate competitive and that they lock Src into an inactive conformation that is incapable of binding peptide substrates [98].

17.9 *IN VITRO* SELECTIONS FOR PROTEIN BINDING USING DNA-ENCODED CHEMICAL LIBRARIES

Because DNA can be readily replicated *en masse* and sequenced, DNA-encoded chemical libraries have the distinct advantage of being amenable to assay by selections rather than screens. In a screen, each molecule must be individually assayed for activity. In a selection, each member of the library is assayed simultaneously, significantly reducing cost and effort in terms of quantities of compound, target protein, and time required to evaluate binding of the library [99]. Selections on DNA-encoded small molecules are typically performed by immobilizing a target protein to a solid support, such as a Sepharose or magnetic bead, using affinity tags such as His_6 or Glutathione S-Transferase (Fig. 17.17) [1]. The DNA-encoded library is incubated with the immobilized target protein to allow binding to occur, and then several washes are performed to remove inactive library members. The library recovered from the washing steps can be retained for use in subsequent selections. Leimbacher and coworkers reported that selections performed with less target protein and using washes with lower volumes and higher detergent concentrations showed better discrimination between hits and inactive compounds. Conversely, washes using large volumes decreased the ability of a selection to discriminate between active and inactive library members [26]. A single round of selection is completed by the elution of active library members, either by heating the solution or by eluting the target protein from the beads. The eluted material can then be

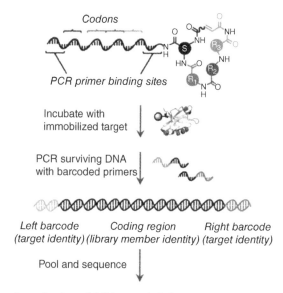

Figure 17.17. *In vitro* selection of DNA-encoded chemical libraries. Reprinted with permission from Kleiner et al. [97]. Copyright 2012 American Chemical Society.

introduced into additional round(s) of selection or subjected to PCR amplification of sequences encoding active library members.

The resulting amplified sequences are analyzed either by microarray or massively parallel high-throughput DNA sequencing. For libraries with ≤10^6 members, enrichment factors (occurrences after selection divided by occurrences in the starting library) can be calculated for each member of the library, and compounds with enrichment factors higher than an empirically determined threshold are considered "hits." For libraries with an equal abundance of all compounds, the variation in sequence counts observed before selection can be fit to a negative binomial distribution. Deviations from this distribution after selection can be analyzed, and a one-sided test for significance can be performed to identify likely "hit" compounds [100]. Because modern high-throughput sequencing routinely returns 10^7–10^8 sequence reads, it becomes impractical for libraries larger than approximately 10^6 members to obtain enough sequences to observe every member of the library with sufficient coverage to calculate enrichment factors. In these cases, the probability of observing any particular sequence becomes small enough that the distribution of sequence counts is likely to follow the Poisson distribution [101], which is a special case of the binomial distribution for low-probability events. In a Poisson process, the mean number of observed counts is close the variance of observed counts. The probability of observing an arbitrary number of counts, k, is a function of k and the mean number of counts observed, λ. The probability of observing a particular number of counts, k, is given by $P(k)=\lambda k e - \lambda/k!$. In these cases, molecules that contain a significant number of occurrences (determined by fitting the data to a Poisson distribution) are considered "hits." *In vitro* selections on DNA-encoded chemical libraries have resulted in the discovery of numerous compounds with inhibitory activity towards proteins involved in disease (Fig. 17.18).

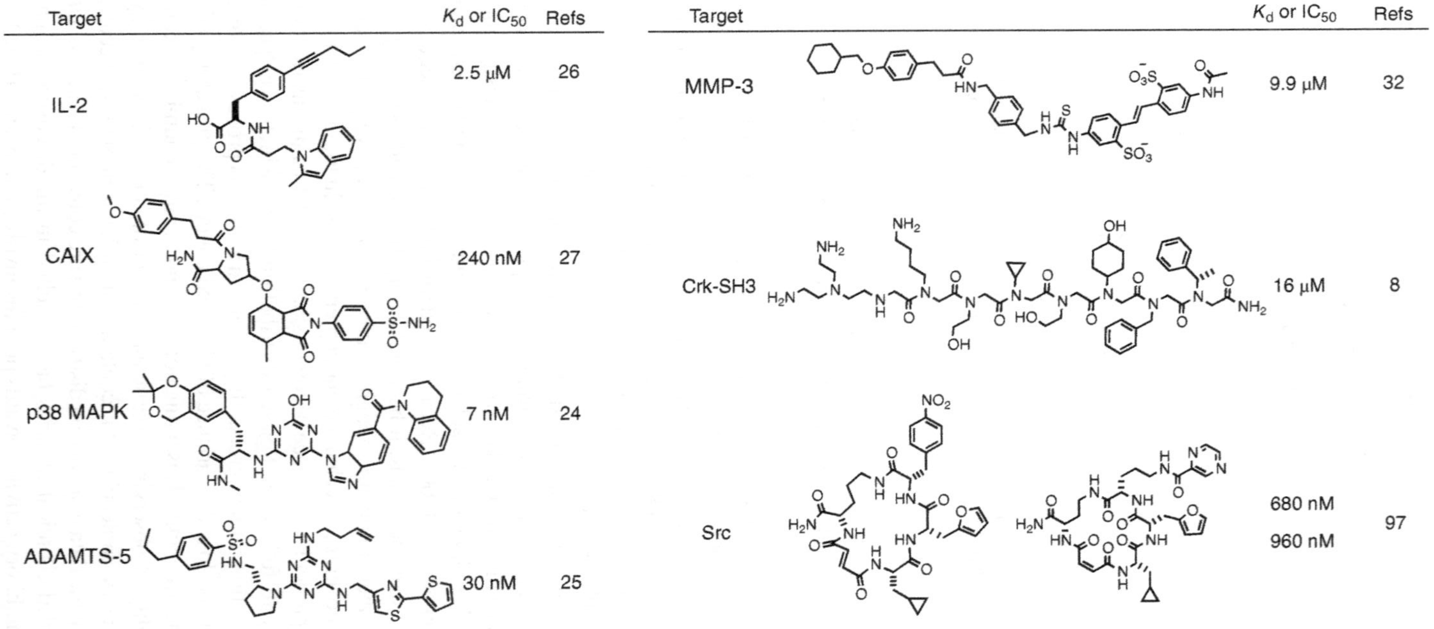

Figure 17.18. Structures of bioactive compounds discovered by *in vitro* selection on DNA-encoded libraries.

17.10 REACTION DISCOVERY USING DTS

The discovery of new reactions is typically approached by studying a particular substrate or transformation, but a more unbiased search could yield unexpected transformations. To this end, screens for new reactions have been performed by LC/MS [102], resulting in the discovery of addition reactions of 1,3-dicarbonyls and a tandem Friedel–Crafts addition, and by GC/MS, resulting in the discovery of an α-amino C–H arylation reaction [103]. Quinton and coworkers recently developed a sandwich immunoassay-based reaction discovery system in which reactants are linked to small-molecule tags capable of binding a surface-immobilized antibody or an enzyme-linked detection antibody [104]. Each candidate reaction must be screened separately because both small-molecule tags are shared, but the readout of heterocoupling products may be significantly streamlined in comparison to screens analyzed by LC/MS [102] or GC/MS [103]. A screen of 3360 different reactions lead to the discovery of copper-mediated coupling of phenols and thioureas and of alkynes and N-hydroxy-thioureas. In addition to enabling the discovery of bioactive synthetic molecules, DNA-encoded libraries can also be used to discover new chemical reactions by performing selections for reactivity, rather than for interaction with targets of interest.

Kanan and coworkers developed an unbiased, DNA-encoded *in vitro* selection for simultaneously evaluating many substrate combinations for bond formation (Fig. 17.19a) [57]. When a library of n substrates encoded on strand A is combined with a library of m substrates encoded on strand B at nanomolar concentration, members of pool B and pool A hybridize, increasing the effective molarity of templated reagents to the millimolar range. Nontemplated pairs have nanomolar molarities and should not react at a significant rate. DNA templating therefore enables multiple reactions to be screened discretely in one solution and allows the suppression of homocoupling products. After the combined pool A and pool B libraries hybridize, they are incubated under various reaction conditions before a reduction step that cleaves the b substrate and biotin affinity tag from strand B. If covalent bond formation occurs between a and b, strand A will remain linked to biotin. If bond formation does not occur, the biotin affinity tag will be lost. After enrichment of strand A sequences encoding bond-forming pairs by avidin pulldown, the identity of reactive substrates is analyzed using a two-color microarray. Control experiments in which the library was incubated with Cu(I) or with (1-(3-dimethylamino) propyl)-3-ethylcarbodiimide hydrochloride (EDC-HCl) revealed the known reactions between a terminal alkyne and an azide or between a carboxylic acid and an amine.

A combined pool A and pool B library encoding 168 substrate combinations was exposed to various catalysts. Incubating the combined library with Na_2PdCl_4 for 1 h at 37°C results in heterocouplings between aryl iodide and acrylamide, between an olefin and a boronic ester, and between a terminal alkyne and a terminal alkene. The Pd(II)-mediated alkyne–alkene macrocyclization was performed in a non-DNA-templated format in aqueous or organic solvent and could be performed using catalytic (5 mol%) Na_2PdCl_4 with 1 equiv. $CuCl_2$ as an oxidant. This represents the first reported example of enone formation from a simple alkyne–alkene precursor and demonstrates that reactions discovered in a DNA-encoded format can also proceed in a nontemplated format in aqueous or organic solvent [57].

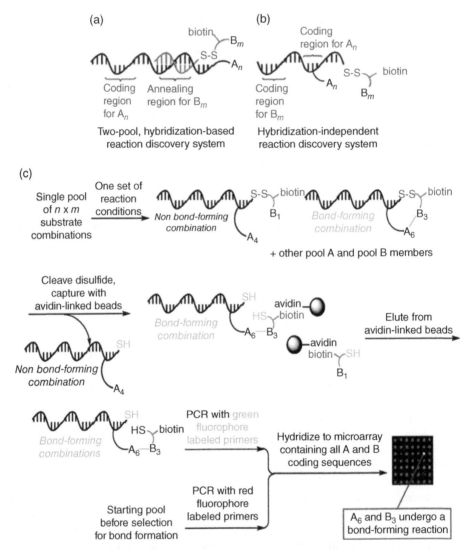

Figure 17.19. (a) A hybridization-based DNA-templated *in vitro* selection for reaction discovery. (b) A hybridization-independent DNA-encoded *in vitro* selection for reaction discovery. (c) Both DNA-templated and DNA-encoded selections for reactivity can be analyzed using a microarray. Reprinted with permission from Rozenman et al. [105]. Copyright 2007 American Chemical Society.

Expanding upon this method, Rozenman and coworkers developed a hybridization-independent DNA-encoded selection for reaction discovery in organic solvent (Fig. 17.19b, c) [105]. In the hybridization-independent method, a single DNA strand encodes and is linked to both reactants. To generate this library, primer extension is used to add strand B to a strand A/reactant *a* conjugate. Next, an oligo modified with reactant

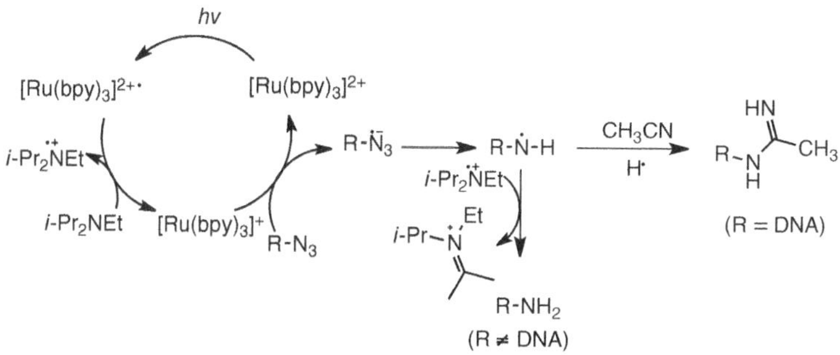

hv

$[Ru(bpy)_3]^{2+\cdot}$ $[Ru(bpy)_3]^{2+}$

$i\text{-Pr}_2\overset{+\cdot}{N}Et$

$i\text{-Pr}_2NEt$ $[Ru(bpy)_3]^{+}$

$R\text{-}N_3$

$R\text{-}\ddot{N}_3 \longrightarrow R\text{-}\dot{N}\text{-}H$

$i\text{-Pr}_2\overset{+\cdot}{N}Et$

$\xrightarrow{\underset{H^{\cdot}}{CH_3CN}}$

$R\text{-}N\overset{HN}{\underset{H}{\diagup}}CH_3$

$(R = DNA)$

$i\text{-Pr}\text{-}\overset{Et}{\underset{\cdot}{N}}$

$R\text{-}NH_2$
$(R \neq DNA)$

Figure 17.20. Proposed mechanism for a Ru(II)-catalyzed azide reduction discovered using DNA-encoded reaction discovery.

b is ligated to strand AB. A disulfide links reactant b and a biotin affinity tag to strand AB. If a covalent bond forms, the biotin handle is retained after disulfide reduction and the encoding DNA is enriched using an avidin pulldown before PCR amplification and analysis using a two-color microarray. This method was used to investigate the reactivity of Au(III) salts, revealing a novel reaction between an indole and a styrene [105].

Hybridization-independent DNA-encoded reaction discovery was next used to screen approximately 45 reaction conditions on 225 substrate pairs, or a total of approximately 10,000 reactions [106]. The screens led to the discovery of a mild, Ru(II)-catalyzed azide reduction induced by visible light (Fig. 17.20). The reaction is remarkably chemoselective and is compatible with alcohols, phenols, acids, alkynes, alkenes, aldehydes, alkyl halides, alkyl mesylates, and disulfides. The azide reduction also showed remarkable biocompatibility, selectively reducing DNA-linked arylazides and alkylazides while sparing disulfides, including those of a protein enzyme requiring disulfide bonds for activity [106].

17.11 PCR-BASED DETECTION OF COVALENT AND NONCOVALENT BOND-FORMING EVENTS

In contrast to the DNA-encoded reaction discovery methods described earlier, Reactivity-Dependent PCR (RDPCR) uses effective molarity to assay bond formation rather than to direct chemical reactivity. RDPCR relies on the observation that an intramolecularly hybridized duplex has a higher T_m than an intermolecularly hybridized duplex (Fig. 17.21a) [107]. Each member of pool A is conjugated to a strand containing a primer binding site, code A, and a region complementary to pool B. When the T_m of this complementary region is lower than the assay temperature, hybridization will occur only if covalent bond formation has occurred between the pool A and pool B library members. Covalent bond formation leads to the formation of a self-priming hairpin that can be rapidly amplified during PCR, enriching sequences corresponding to covalent bond-forming pairs (Fig. 17.21b). Unreacted templates can only hybridize intermolecularly and will be inefficiently amplified during PCR. The "selection" step of RDPCR is

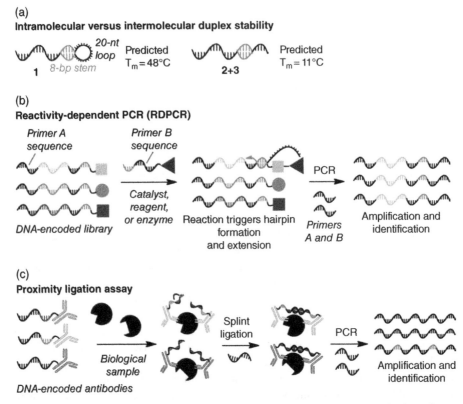

Figure 17.21. (a) Due to their higher effective molarity, intramolecular DNA duplexes have a higher T_m than intermolecular duplexes. (b) In RDPCR, covalent bond formation increases the effective molarity of two ssDNAs, promoting formation of a self-priming hairpin that can be rapidly amplified during PCR. (c) In PLA, formation of a ternary complex between two DNA-linked antibodies and their target protein promotes splint ligation of the DNA tags. The product of the splint ligation can be rapidly PCR amplified and subsequently decoded. (a) and (b) are reprinted with permission from Gorin et al. [107]. Copyright 2009 American Chemical Society.

therefore performed by PCR, rather than by the capture, washing, and elution steps of an avidin pulldown.

RDPCR was validated by demonstrating that oligos linked to an amine and a carboxylic acid or an aryl iodide and an alkene were rapidly amplified by PCR after incubation in the presence of required activators or catalysts, but not in their absence [107]. RDPCR can also be configured to detect bond cleavage. Combining a DNA-linked peptide and a DNA-linked carboxylate in the presence of the coupling reagent DMT-MM led to rapid PCR amplification of the cognate DNA only if the peptide was first treated with the protease subtilisin (generating a peptide with a free amino terminus), but not if either the protease or DMT-MM were omitted [107]. In principle, RDPCR could be adapted to assay any bond cleavage event that reveals a functional group capable of participating in a DNA-compatible reaction. Since the sequences encoding

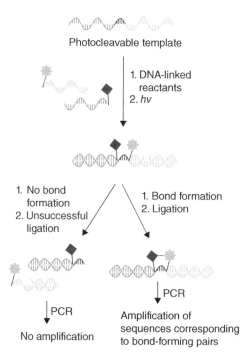

Figure 17.22. DNA-linked reactants hybridize to a template containing a photocleavable group. If covalent bond formation occurs between the reactants, the template can be relegated after photocleavage and serves as a PCR template encoding the reactive pair.

reactive substrates are linked during the selection step, there is no need to generate a template for each substrate pair, greatly reducing the effort required to build a substrate library compared to the affinity-based reaction discovery methods.

Li and coworkers recently reported a DNA-templated reaction discovery system that uses DNA both to bring together substrate library members and to report on bond formation [108]. Previous DNA-based reaction discovery methods use DNA hybridization either to increase effective molarity of two substrates or to report the increased effective molarity of the encoding strands after covalent bond formation, but not both. In this system, a DNA template modified near one end with a photocleavable base hybridizes to two DNA-linked substrates, bringing them into close proximity. After allowing the substrates to react, the photolabile base is cleaved and the template will dissociate into two segments unless a covalent bond has been formed. If covalent bond formation has joined the two substrate sequences, this linked oligo will serve as a splint for a ligation reaction that will restore the cleaved template sequence and regenerates a PCR template containing two primer binding sites (Fig. 17.22). In this way, DNA can be used both to bring reactants together and serve as a nucleic acid signal for productive bond-forming events. DNA has previously been used as a molecular ruler, and this system was used as a "reaction ruler" to probe the spatial requirements of a three-component reductive amination [108].

While RDPCR uses DNA hybridization to assay covalent bond-forming events, the formation of noncovalent protein complexes can also be converted to a rapidly amplified nucleic acid signal. Using the Proximity Ligation Assay (PLA), formation of a ternary complex between two DNA aptamers or two DNA-linked antibodies and their target antigen promotes splint ligation of probe sequences (Fig. 17.21c) [109]. The ligated product contains two primer binding sites for rapid PCR amplification [109]. In an alternate protocol, referred to as a Proximity Extension Assay (PEA), a DNA polymerase performs a primer extension reaction to link together two DNA probes that hybridize after formation of a ternary complex including two probes and a target protein [110].

In principle, aptamers could be generated for any protein target of interest. However, due to the wide availability of antibodies for most proteins relevant to disease and the relative ease of generating protein–DNA conjugates, the proximity ligation and extension assays were adapted to use DNA-linked antibody probes. Probe oligos must be conjugated to either two different antibodies for the same target or to two separate aliquots of a polyclonal antibody [111, 112]. Antibody-mediated PLA was used for multiplexed quantification of cancer biomarkers from human serum [112].

PLA has also been applied to a variety of other detection problems. The P-LISA assay enables immunohistochemical detection of protein complexes comprised of two or three proteins. P-LISA proximity probes serve as splints for the ligation of a circular-izable probe that is circularized only if all of the proximity probes are located within 30 nm of each other [113]. One of the proximity probes also primes the circular product for Rolling Circle Amplification (RCA) by Phi29 polymerase. In 1 h, RCA is capable of amplifying the circularized product 1000-fold enabling detection with fluorophore-linked pairing oligonucleotides. P-LISA was shown to detect complexes between c-Myc/Max or between c-Myc/Max/RNAPII in fixed cells and was used to assess the ability of several small molecules to perturb the c-Myc/Max interaction [113]. In contrast to ge-netic methods for studying protein–protein interactions *in vivo*, PLA can be performed on unmodified proteins from clinical samples and has a very high signal-to-noise ratio due to signal amplification by RCA. When PLA-RCA is performed on transmembrane receptors expressed on the surface of live cells, fluorescence from the probes hybridized to the RCA product can be used to perform FACS sorting on cells with receptor complexes [114]. PLA has also been developed for detection of three antibodies bound to one or more proteins [115] and for the detection of four antibody–receptor binding events on the surface of prostasomes associated with prostate cancer [116].

Recently, proximity ligation was adapted to read out all pairwise protein–protein interactions within a larger set of proteins as well as the abundance of each protein in the set. Multiplexed readout of PLA was accomplished by hybridization to a microarray, in which the microarray probe serves as a primer for RCA. The abundance of the RCA product is imaged by incubation with a complementary fluorophore labeled oligo. Proof of principle was demonstrated by comparing interactions in lysates transfected with p50, RelB, and IkBa or with p50, RelA, and IkBa [117].

PLA was also used to study molecules capable of disrupting the interaction between VEGF-A and either of its receptors, VEGFR-1 or VEGFR-2 [118]. Proximity probes for the VEGF-A homodimer were incubated with a mixture of VEGF-A, VEGFR-1 or

VEGFR-2, and molecules thought to interrupt the VEGF/VEGFR interaction, including a VEGF aptamer, a VEGF neutralizing monoclonal antibody, a monoclonal antibody for VEGFR-2, and GFA-116, a small-molecule ligand of VEGFR-2 [118]. The DNA aptamer and the neutralizing antibody both decreased the PLA signal for VEGF-A in a dose-dependent manner by competing with binding of the proximity probes. The VEGFR-2 antibody and ligand increased the PLA signal by disrupting the VEGF/VEGFR-2 complex and increasing the concentration of free VEGF capable of binding the proximity probes [118].

DNA-encoded chemical libraries have dramatically increased the throughput of evaluating protein–ligand interactions by enabling selections, in which all members of a library are simultaneously assayed for activity. Compared to screens, selections use substantially smaller quantities of compound, target protein, and assay time [25, 99]. Regardless, each protein of interest must be assayed separately. Selections performed on immobilized proteins require several steps (e.g., binding, washing, elution) that introduce artifacts associated with immobilization. The high local concentration of immobilized protein allows disproportionate enrichment of compounds with high k_{off} values because dissociated compounds may easily rebind the solid support. A solution-phase selection capable of simultaneously evaluating all ligand–protein pairs from combined libraries would overcome each of these limitations. Interaction-Dependent PCR (IDPCR) is a solution-phase selection method capable of evaluating interactions between DNA-linked small molecules and DNA-linked targets [119]. If binding occurs between a small molecule and a target, the DNA tags will be brought into close proximity, inducing annealing of a short complementary region (Fig. 17.23).

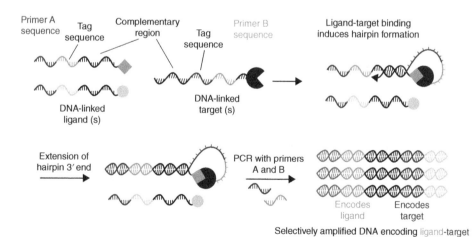

Figure 17.23. Formation of a noncovalent bond between a DNA-linked small molecule and DNA-linked protein results in hybridization of their ssDNA tags. The hybridized region is extended by a DNA polymerase, producing a PCR template that encodes both the ligand and its target. Reprinted with permission from McGregor et al. [119]. Copyright 2010 American Chemical Society.

Addition of a DNA polymerase results in extension of the self-priming hairpin to produce a longer DNA product that contains codes identifying both the ligand and target and contains two primer binding sites, enabling its rapid amplification by PCR. In a model selection on an equimolar 261-member DNA-ligand library and an equimolar 259-member DNA-target library containing five known protein–ligand interactions, the most highly enriched sequences corresponded to the known interactions, despite the fact that their affinities spanned five orders of magnitude [119].

17.12 OTHER SOLUTION-PHASE SELECTIONS FOR PROTEIN–LIGAND BINDING

Binding between a protein and a small molecule linked to the 3′ end of DNA protects the oligonucleotide from digestion by Exonuclease I (ExoI). When combined with various DNA detection methods, this observation has enabled the development of several methods of assaying protein–small-molecule binding [120]. One method of DNA detection is association with Single-Walled Nanotubes (SWNTs) (Fig. 17.24a). Charge–charge repulsion between the DNA–SWNTs and a densely carboxy-modified gold electrode prevents adsorption of the DNA–SWNTs onto the gold electrode, whereas SWNTs that are not complexed with DNA can associate with the electrode. Differential pulse voltammetry measurements show strong peaks when SWNTs adsorb onto the carboxy-modified gold electrode but only very weak peaks if SWNT adsorption is prevented by association with DNA. The concentration of the protein receptor affects the degree of DNA protection from hydrolysis, giving a quasilinear relationship for the interaction of folate and the folate receptor between 10 pM and 1 nM [120].

The nucleic acid signal associated with DNA protected from ExoI digestion by protein–small-molecule binding can also be amplified by Nicking Endonuclease-Assisted Amplification (NEA). When the DNA surviving the selection is used to nick DNA modifying a gold electrode, allowing an electroactive species ($[Fe(CN)_6]^{3-/4-}$) to interact with the electrode, this DNA signal can be measured as a change in impedance or current of the electrode in a differential pulse voltammetry assay [121]. Although the proof-of-concept report did not discuss use in a library format, in principle, a set of gold electrodes modified with a library of acceptor oligonucleotides could be used to analyze binding to a DNA-encoded chemical library, with the noted disadvantage that each electrode would be used only once because NEA is a destructive amplification process. Binding between a protein target and a DNA-linked small molecule also protects the DNA from digestion by the dsDNA exonuclease, Exonuclease III (ExoIII) (Fig. 17.24b) [122]. DNA surviving digestion by ExoIII can be detected using double-strand-specific DNA dyes.

Compared to selections on immobilized proteins, exonuclease-mediated selections do not require washing or elution steps. Unlike solid-phase selections, in which the library members, including a large fraction of active binding molecules, flow through the selection and can be recycled in subsequent selections, the libraries used in exonuclease-mediated selections would be single use because the DNA sequences corresponding to nonbinding active and inactive compounds would be destroyed.

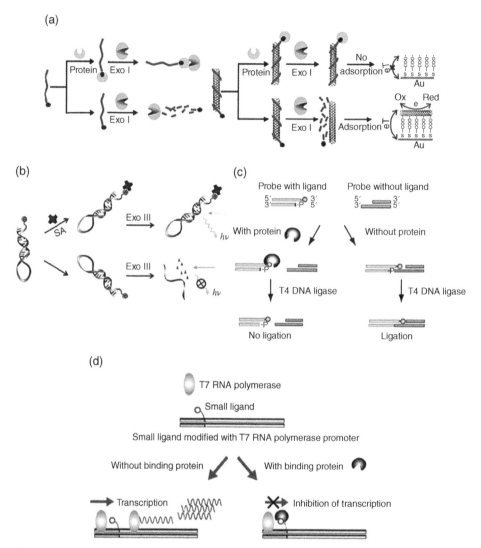

Figure 17.24. (a) Protein binding to a DNA-linked ligand protects the DNA from digestion with Exol, preserving it for detection through association to SWNTs. Reprinted with permission from Wu et al. [120]. Copyright 2009 American Chemical Society. (b) Protein binding to a DNA-linked ligand protects a DNA hairpin from digestion with ExoIII, enabling subsequent detection with a dye specific for dsDNA. Reprinted with permission from Wu et al. [122]. Copyright 2011 American Chemical Society. (c) Binding of a protein to a DNA-linked ligand prevents DNA ligase from performing a ligation. Reprinted with permission from Sugita et al. [123]. Copyright 2010 Springer. (d) Binding of a protein to a DNA-linked ligand located in the binding site of T7 RNAP prevents RNA transcription at that location. Reprinted from Publication title Mie et al. [124]. Copyright 2010, with permission from Elsevier.

Protein–ligand interactions can also be used to prevent access of ligases and polymerases. Sugita and coworkers developed an assay in which binding between a protein and a small molecule linked to one strand of an asymmetric DNA duplex prevents ligation of the duplex (probe A) to another duplex (probe B) with a complementary sticky end (Fig. 17.24c) [123]. If binding does not occur, the two duplexes are ligated together, linking two primer binding sites into one strand and generating a product that can be rapidly amplified by PCR. By conjugating probe A strands of different length to biotin or digoxigenin, binding between the two ligands and their respective antibodies was detected in a single solution and analyzed by gel electrophoresis [123]. Alternately, conjugation of a small molecule to the DNA backbone in the T7 RNAP initiation site inhibits transcription only when a protein binds the small molecule (Fig. 17.24d) [124]. Incubation of DNA modified with biotin or methotrexate with increasing quantities of antibodies for biotin or methotrexate or human dihydrofolate reductase decreased the amount of RNA transcribed by T7 RNAP. Competition with unlabeled biotin or methotrexate disinhibited T7 RNAP transcription in a concentration-dependent manner, demonstrating the specificity of the method [124]. By conjugating the small molecules to unique DNA sequences, this method could be used to evaluate the binding of a library of molecules to one protein target, however, with the significant disadvantage that sequences corresponding to inactive, rather than active, compounds are enriched by the selection.

17.13 CONCLUSIONS

The predictability of Watson–Crick base pairing enables researchers to design changes in effective molarity that can either direct or detect chemical reactions between DNA-linked reagents. The hybridization of complementary DNA strands increases the effective molarity of pendant reactants such that they can experience millimolar or higher effective concentrations, even when their absolute concentrations in solution are many orders of magnitude lower. This phenomenon is the basis of the DTS of DNA-encoded chemical libraries and is also used to govern combinations of substrates in DNA-templated reaction discovery, resulting in the discovery of new transformations. The ease of designing DNA sequences with defined properties also enabled the development of PCR-based selection methods for detecting covalent and noncovalent bond-forming events. In general, these methods rely on formation of a complex to increase the effective molarity of two or more oligonucleotides so that they can be enzymatically linked, resulting in a longer, rapidly amplifiable DNA sequence that encodes the identity of reacting or interacting partners.

Whether a DNA-encoded chemical library is assembled by a DNA-programmed or DNA-recorded strategy, the ability of each member of a DNA-encoded chemical library to be simultaneously assayed for activity using *in vitro* selection dramatically increases the throughput with which the library members can be evaluated for activity (e.g., binding or reactivity). Over the past few years, these selections have resulted in the discovery of new classes of synthetic small-molecule ligands against a variety of protein targets including several associated with human disease. Given the ever-increasing number of proteins and nucleic acids implicated in human disease, DNA-encoded libraries and their rapid *in vitro* selection are likely to play an increasingly important role in the discovery of small molecules with the potential to probe therapeutically relevant biological pathways or to serve as leads for the development of new medicines.

ACKNOWLEDGMENT

The authors gratefully acknowledge support by the Howard Hughes Medical Institute and the NIH/NIGMS (R01GM065865).

REFERENCES

1. Doyon, J.B., Snyder, T.M., Liu, D.R. (2003). Highly sensitive in vitro selections for DNA-linked synthetic small molecules with protein binding affinity and specificity. *J. Am. Chem. Soc.* 125, 12372–12373.

2. Gartner, Z.J., Liu, D.R. (2001). The generality of DNA-templated synthesis as a basis for evolving non-natural small molecules. *J. Am. Chem. Soc.* 123, 6961–6963.

3. Calderone, C.T., Puckett, J.W., Gartner, Z.J., Liu, D.R. (2002). Directing otherwise incompatible reactions in a single solution by using DNA-templated organic synthesis. *Angew. Chem. Int. Ed.* 41, 4104–4108.

4. Gartner, Z.J., Kanan, M.W., Liu, D.R. (2002). Multistep small-molecule synthesis programmed by DNA templates. *J. Am. Chem. Soc.* 124, 10304–10306.

5. Gartner, Z.J., Tse, B.N., Grubina, R., Doyon, J.B., Snyder, T.M., Liu, D.R. (2004). DNA-templated organic synthesis and selection of a library of macrocycles. *Science* 305, 1601–1605.

6. Tse, B.N., Snyder, T.M., Shen, Y., Liu, D.R. (2008). Translation of DNA into a library of 13000 synthetic small-molecule macrocycles suitable for in vitro selection. *J. Am. Chem. Soc.* 130, 15611–15626.

7. Halpin, D.R., Harbury, P.B. (2004). DNA display II. Genetic manipulation of combinatorial chemistry libraries for small-molecule evolution. *PLoS Biol.* 2, 1022–1030.

8. Wrenn, S.J., Weisinger, R.M., Halpin, D.R., Harbury, P.B. (2007). Synthetic ligands discovered by in vitro selection. *J. Am. Chem. Soc.* 129, 13137–13143.

9. Weisinger, R.M., Wrenn, S.J., Harbury, P.B. (2012). Highly parallel translation of DNA sequences into small molecules. *PLoS One* 7, e28056.

10. Weisinger, R.M., Marinelli, R.J., Wrenn, S.J., Harbury, P.B. (2012). Mesofluidic devices for DNA-programmed combinatorial chemistry. *PLoS One* 7, e32299.

11. He, Y., Liu, D.R. (2011). A sequential strand-displacement strategy enables efficient six-step DNA-templated synthesis. *J. Am. Chem. Soc.* 133, 9972–9975.

12. Wilson, D.S., Keefe, A.D., Szostak, J.W. (2001). The use of mRNA display to select high-affinity protein-binding peptides. *Proc. Natl. Acad. Sci. U. S. A.* 98, 3750–3755.

13. Li, S., Millward, S., Roberts, R. (2002). In vitro selection of mRNA display libraries containing an unnatural amino acid. *J. Am. Chem. Soc.* 124, 9972–9973.

14. Frankel, A., Millward, S.W., Roberts, R.W. (2003). Encodamers: unnatural peptide oligomers encoded in RNA. *Chem. Biol.* 10, 1043–1050.

15. Josephson, K., Hartman, M.C.T., Szostak, J.W. (2005). Ribosomal synthesis of unnatural peptides. *J. Am. Chem. Soc.* 127, 11727–11735

16. Schlippe, Y.V.G., Hartman, M.C.T., Josephson, K., Szostak, J.W. (2012). In vitro selection of highly modified cyclic peptides that act as tight binding inhibitors. *J. Am. Chem. Soc.* 134, 10469–10477.

17. Kawakami, T., Ohta, A., Ohuchi, M., Ashigai, H., Murakami, H., Suga, H. (2009). Diverse backbone-cyclized peptides via codon reprogramming. *Nat. Chem. Biol.* 5, 888–890.

18. Hipolito, C.J., Suga, H. (2012). Ribosomal production and in vitro selection of natural product-like peptidomimetics: the FIT and RaPID systems. *Curr. Opin. Chem. Biol.* 16, 196–203.

19. He, Y., Liu, D.R. (2010). Autonomous multistep organic synthesis in a single isothermal solution mediated by a DNA walker. *Nat. Nanotechnol.* 5, 778–782.

20. McKee, M.L., Milnes, P.J., Bath, J., Stulz, E., Turberfield, A.J., O'Reilly, R.K. (2010). Multistep DNA-templated reactions for the synthesis of functional sequence controlled oligomers. *Angew. Chem. Int. Ed.* 49, 7948–7951.

21. Milnes, P.J., McKee, M.L., Bath, J., Song, L., Stulz, E., Turberfield, A.J., O'Reilly, R.K. (2012). Sequence-specific synthesis of macromolecules using DNA-templated chemistry. *Chem. Commun.* 48, 5614–5616.

22. McKee, M.L., Milnes, P.J., Bath, J., Stulz, E., O'Reilly, R.K., Turberfield, A.J. (2012). Programmable one-pot multistep organic synthesis using DNA junctions. *J. Am. Chem. Soc.* 134, 1446–1449.

23. Hansen, M.H., Blakskjaer, P., Petersen, L.K., Hansen, T.H., Hojfeldt, J.W., Gothelf, K.V., Hansen, N.J.V. (2009). A yoctoliter-scale DNA reactor for small-molecule evolution. *J. Am. Chem. Soc.* 131, 1322–1327.

24. Clark, M.A., Acharya, R.A., Arico-Muendel, C.C., Belyanskaya, S.L., Benjamin, D.R., Carlson, N.R., Centrella, P.A., Chiu, C.H., Creaser, S.P., Cuozzo, J.W., et al. (2009). Design, synthesis and selection of DNA-encoded small-molecule libraries. *Nat. Chem. Biol.* 5, 647–654.

25. Deng, H., O'Keefe, H., Davie, C.P., Lind, K.E., Acharya, R.A., Franklin, G.J., Larkin, J., Matico, R., Neeb, M., Thompson, M.M., et al. (2012). Discovery of highly potent and selective small molecule ADAMTS-5 inhibitors that inhibit human cartilage degradation via encoded library technology (ELT). *J. Med. Chem.* 55, 7061–7079.

26. Leimbacher, M., Zhang, Y., Mannocci, L., Stravs, M., Geppert, T., Scheuermann, J., Schneider, G., Neri, D. (2012). Discovery of small-molecule interleukin-2 inhibitors from a DNA-encoded chemical library. *Chem. Eur. J.* 18, 7729–7737.

27. Buller, F., Steiner, M., Frey, K., Mircsof, D., Scheuermann, J., Kalisch, M., Bühlmann, P., Supuran, C.T., Neri, D. (2011). Selection of carbonic anhydrase IX inhibitors from one million DNA-encoded compounds. *ACS Chem. Biol.* 6, 336–344.

28. Mannocci, L., Zhang, Y., Scheuermann, J., Leimbacher, M., de Bellis, G., Rizzi, E., Dumelin, C.E., Melkko, S., Neri, D. (2008). High-throughput sequencing allows the identification of binding molecules isolated from DNA-encoded chemical libraries. *Proc. Natl. Acad. Sci. U. S. A.* 105, 17670–17675.

29. Svensen, N., Díaz-Mochón, J.J., Bradley, M. (2011). Decoding a PNA encoded peptide library by PCR: the discovery of new cell surface receptor ligands. *Chem. Biol.* 18, 1284–1289.

30. Svensen, N., Diaz-Mochon, J.J., Bradley, M. (2011). Encoded peptide libraries and the discovery of new cell binding ligands. *Chem. Commun.* 47, 7638–7640.

31. Melkko, S., Scheuermann, J., Dumelin, C.E., Neri, D. (2004). Encoded self-assembling chemical libraries. *Nat. Biotechnol.* 22, 568–574.

32. Scheuermann, J., Dumelin, C.E., Melkko, S., Zhang, Y., Mannocci, L., Jaggi, M., Sobek, J., Neri, D. (2008). DNA-encoded chemical libraries for the discovery of MMP-3 inhibitors. *Bioconjug. Chem.* 19, 778–785.

33. Gorska, K., Huang, K.-T., Chaloin, O., Winssinger, N. (2009). DNA-templated homo- and heterodimerization of peptide nucleic acid encoded oligosaccharides that mimick the carbohydrate epitope of HIV. *Angew. Chem. Int. Ed.* 48, 7695–7700.

34. Daguer, J.P., Ciobanu, M., Alvarez, S., Barluenga, S., Winssinger, N. (2011). DNA-templated combinatorial assembly of small molecule fragments amenable to selection/amplification cycles. *Chem. Sci.* 2, 625–632.

35. Ciobanu, M., Huang, K.-T., Daguer, J.P., Barluenga, S., Chaloin, O., Schaeffer, E., Mueller, C.G., Mitchell, D.A., Winssinger, N. (2011). Selection of a synthetic glycan oligomer from a library of DNA-templated fragments against DC-SIGN and inhibition of HIV gp120 binding to dendritic cells. *Chem. Commun.* 47, 9321–9323.

36. Brenner, S., Lerner, R.A. (1992). Encoded combinatorial chemistry. *Proc. Natl. Acad. Sci. U. S. A.* 89, 5381–5383.

37. Needels, M.C., Jones, D.G., Tate, E.H., Heinkel, G.L., Kochersperger, L.M., Dower, W.J., Barrett, R.W., Gallop, M.A. (1993). Generation and screening of an oligonucleotide-encoded synthetic peptide library. *Proc. Natl. Acad. Sci. U. S. A.* 90, 10700–10704.

38. Nielsen, J., Brenner, S., Janda, K.D. (1993). Synthetic methods for the implementation of encoded combinatorial chemistry. *J. Am. Chem. Soc.* 115, 9812–9813.

39. Buller, F., Mannocci, L., Zhang, Y., Dumelin, C.E., Scheuermann, J., Neri, D. (2008). Design and synthesis of a novel DNA-encoded chemical library using Diels-Alder cycloadditions. *Bioorg. Med. Chem. Lett.* 18, 5926–5931.

40. Erben, A., Grossmann, T.N., Seitz, O. (2011). DNA-instructed acyl transfer reactions for the synthesis of bioactive peptides. *Bioorg. Med. Chem. Lett.* 21, 4993–4997.

41. Brudno, Y., Birnbaum, M.E., Kleiner, R.E., Liu, D.R. (2010). An in vitro translation, selection and amplification system for peptide nucleic acids. *Nat. Chem. Biol.* 6, 148–155.

42. Zhang, D.Y., Seelig, G. (2011). Dynamic DNA nanotechnology using strand-displacement reactions. *Nat. Chem.* 3, 103–113.

43. Endo, M., Sugiyama, H. (2009). Chemical approaches to DNA nanotechnology. *ChemBioChem* 10, 2420–2443.

44. Seeman, N.C. (2007). An overview of structural DNA nanotechnology. *Mol. Biotechnol.* 37, 246–257.

45. Krishnamurthy, V.M., Semetey, V., Bracher, P.J., Shen, N., Whitesides, G.M. (2007). Dependence of effective molarity on linker length for an intramolecular protein–ligand system. *J. Am. Chem. Soc.* 129, 1312–1320.

46. Kramer, R.H., Karpen, J.W. (1998). Spanning binding sites on allosteric proteins with polymer-linked ligand dimers. *Nature* 395, 710–713.

47. Gargano, J.M., Ngo, T., Kim, J.Y., Acheson, D.W.K., Lees, W.J. (2001). Multivalent Inhibition of AB5 Toxins. *J. Am. Chem. Soc.* 123, 12909–12910.

48. Murphy, M.C., Rasnik, I., Cheng, W., Lohman, T.M., Ha, T. (2004). Probing single-stranded DNA conformational flexibility using fluorescence spectroscopy. *Biophys. J.* 86, 2530–2537.

49. Rechendorff, K., Witz, G., Adamcik, J., Dietler, G. (2009). Persistence length and scaling properties of single-stranded DNA adsorbed on modified graphite. *J. Chem. Phys.* 131, 095103.

50. Rivetti, C.C., Walker, C., Bustamante, C. (1998). Polymer chain statistics and conformational analysis of DNA molecules with bends or sections of different flexibility. *J. Mol. Biol.* 280, 41–59.

51. Mills, J.B., Vacano, E., Hagerman, P.J. (1999). Flexibility of single-stranded DNA: use of gapped duplex helices to determine the persistence lengths of poly(dT) and poly(dA). *J. Mol. Biol.* 285, 245–257.

52. Kuznetsov, S.V., Shen, Y.Q., Benight, A.S., Ansari, A. (2001). A semiflexible polymer model applied to loop formation in DNA hairpins. *Biophys. J.* 81, 2864–2875.

53. Geggier, S., Kotlyar, A., Vologodskii, A. (2011). Temperature dependence of DNA persistence length. *Nucleic Acids Res.* 39, 1419–1426.

54. You, H., Iino, R., Watanabe, R., Noji, H. (2012). Winding single-molecule double-stranded DNA on a nanometer-sized reel. *Nucleic Acids Res.* 40, e151.

55. Vafabakhsh, R., Ha, T. (2012). Extreme bendability of DNA less than 100 base pairs long revealed by single-molecule cyclization. *Science* 337, 1097–1101.

56. Snyder, T.M., Liu, D.R. (2005). Ordered multistep synthesis in a single solution directed by DNA templates. *Angew. Chem. Int. Ed.* 44, 7379–7382.

57. Kanan, M.W., Rozenman, M.M., Sakurai, K., Snyder, T.M., Liu, D.R. (2004). Reaction discovery enabled by DNA-templated synthesis and in vitro selection. *Nature* 431, 545–549.

58. Greschner, A.A., Toader, V., Sleiman, H.F. (2012). The role of organic linkers in direct DNA self-assembly and significantly stabilizing DNA duplexes. *J. Am. Chem. Soc.* 134, 14382–14389.

59. Lipovsek, D., Pluckthun, A. (2004). In-vitro protein evolution by ribosome display and mRNA display. *J. Immunol. Methods* 290, 51–67.

60. Murakami, H., Ohta, A., Ashigai, H., Suga, H. (2006). A highly flexible tRNA acylation method for non-natural polypeptide synthesis. *Nat. Methods* 3, 357–359.

61. Ohuchi, M., Murakami, H., Suga, H. (2007). The flexizyme system: a highly flexible tRNA aminoacylation tool for the translation apparatus. *Curr. Opin. Chem. Biol.* 11, 537–542.

62. Shimizu, Y., Inoue, A., Tomari, Y., Suzuki, T., Yokogawa, T., Nishikawa, K., Ueda, T. (2001). Cell-free translation reconstituted with purified components. *Nat. Biotechnol.* 19, 751–755.

63. Ohta, A., Yamagishi, Y., Suga, H. (2008). Synthesis of biopolymers using genetic code reprogramming. *Curr. Opin. Chem. Biol.* 12, 159–167.

64. Kawakami, T., Murakami, H., Suga, H. (2008). Ribosomal synthesis of polypeptoids and peptoid-peptide hybrids. *J. Am. Chem. Soc.* 130, 16861–16863.

65. Yamagishi, Y., Shoji, I., Miyagawa, S., Kawakami, T., Katoh, T., Goto, Y., Suga, H. (2011). Natural product-like macrocyclic N-methyl-peptide inhibitors against a ubiquitin ligase uncovered from a ribosome-expressed de novo library. *Chem. Biol.* 18, 1562–1570.

66. Yamaguchi, J., Naimuddin, M., Biyani, M., Sasaki, T., Machida, M., Kubo, T., Funatsu, T., Husimi, Y., Nemoto, N. (2009). cDNA display: a novel screening method for functional disulfide-rich peptides by solid-phase synthesis and stabilization of mRNA-protein fusions. *Nucleic Acids Res.* 37, e108.

67. Yonezawa, M., Doi, N., Kawahashi, Y., Higashinakagawa, T., Yanagawa, H. (2003). DNA display for in vitro selection of diverse peptide libraries. *Nucleic Acids Res.* 31, e118.

68. Bertschinger, J., Neri, D. (2004). Covalent DNA display as a novel tool for directed evolution of proteins in vitro. *Protein Eng. Des. Sel.* 17, 699–707.

69. Bertschinger, J., Grabulovski, D., Neri, D. (2007). Selection of single domain binding proteins by covalent DNA display. *Protein Eng. Des. Sel.* 20, 57–68.

70. Halpin, D.R., Lee, J.-A., Wrenn, S.J., Harbury, P.B. (2004). DNA display III. Solid-phase organic synthesis on unprotected DNA. *PLoS Biol.* 2, e175.

71. Gartner, Z.J., Kanan, M.W., Liu, D.R. (2002). Expanding the reaction scope of DNA-templated synthesis. *Angew. Chem. Int. Ed.* 41, 1796–1800.

72. Li, X., Liu, D.R. (2003). Stereoselectivity in DNA-templated organic synthesis and its origins. *J. Am. Chem. Soc.* 125, 10188–10189.

73. Calderone, C.T., Liu, D.R. (2005). Small-molecule diversification from iterated branching reaction pathways enabled by DNA-templated synthesis. *Angew. Chem. Int. Ed.* 44, 7383–7386.

74. Martin, A.R., Barvik, I., Luvino, D., Smietana, M., Vasseur, J.-J. (2011). Dynamic and programmable DNA-templated boronic ester formation. *Angew. Chem. Int. Ed.* 50, 4193–4196.

75. Huang, Y., Coull, J.M. (2008). Diamine catalyzed hemicyanine dye formation from nonfluorescent precursors through DNA programmed chemistry. *J. Am. Chem. Soc.* 130, 3238–3239.

76. Tang, Z., Marx, A. (2007). Prolinmodifizierte DNA als Katalysator der Aldolreaktion. *Angew. Chem.* 119, 7436–7439.

77. Grossman, T.N., Strohbach, A., Seitz, O. (2008). Achieving turnover in DNA-templated reactions. *ChemBioChem* 9, 2185–2192.

78. Sakurai, K., Snyder, T.M., Liu, D.R. (2005). DNA-templated functional group transformations enable sequence-programmed synthesis using small-molecule reagents. *J. Am. Chem. Soc.* 127, 1660–1661.

79. Gorska, K., Manicardi, A., Barluenga, S., Winssinger, N. (2011). DNA-templated release of functional molecules with an azide-reduction-triggered immolative linker. *Chem. Commun.* 47, 4364–4366.

80. Ma, Z., Taylor, J.-S. (2000). Nucleic acid-triggered catalytic drug release. *Proc. Natl. Acad. Sci. U. S. A.* 97, 11159–11163.

81. Czlapinski, J.L., Sheppard, T.L. (2001). Nucleic acid template-directed assembly of metallo-salen-DNA conjugates. *J. Am. Chem. Soc.* 123, 8618–8619.

82. Czlapinski, J.L., Sheppard, T.L. (2004). Site-specific oxidative cleavage of DNA by metallo-salen-DNA conjugates. *Chem. Commun.* 21, 2468–2469.

83. Brunner, J., Mokhir, A., Kraemer, R. (2003). DNA-templated metal catalysis. *J. Am. Chem. Soc.* 125, 12410–12411.

84. Shen, Q., Tang, S., Li, W., Nie, Z., Liu, Z., Huang, Y., Yao, S. (2012). A novel DNA-templated click chemistry strategy for fluorescent detection of copper(II) ions. *Chem. Commun.* 48, 281–283.

85. Rozenman, M.M., Liu, D.R. (2005). DNA-templated synthesis in organic solvents. *ChemBioChem* 7, 253–256.

86. Gartner, Z.J., Grubina, R., Calderone, C.T., Liu, D.R. (2003). Two enabling architectures for DNA-templated organic synthesis. *Angew. Chem. Int. Ed.* 42, 1370–1375.

87. Li, X., Gartner, Z.J., Tse, B.N., Liu, D.R. (2004). Translation of DNA into synthetic N-acyloxazolidines. *J. Am. Chem. Soc.* 126, 5090–5092.

88. Tian, Y., He, Y., Chen, Y., Yin, P., Mao, C. (2005). A DNAzyme that walks processively and autonomously along a one-dimensional track. *Angew. Chem. Int. Ed.* 44, 4355–4358.

89. Tian, J., Gong, H., Sheng, N., Zhou, X., Gulari, E., Gao, X., Church, G. (2004). Accurate multiplex gene synthesis from programmable DNA microchips. *Nature* 432, 1050–1054.

90. Lee, C.-C., Snyder, T.M., Quake, S.R. (2010). A microfluidic oligonucleotide synthesizer. *Nucleic Acids Res.* 38, 2514–2521.

91. Svensen, N., Diaz-Mochon, J.J., Bradley, M. (2011). Microarray generation of thousand-member oligonucleotide libraries. *PLoS One* 6, e24906.

92. SantaLucia, J. (2007). Physical principles and visual-OMP software for optimal PCR design. In *Methods in Molecular Biology*, Volume 402: PCR Primer Design, A. Yuryev, ed. (Totowa, NJ: Humana Press), pp. 3–33.

93. Dumelin, C.E., Scheuermann, J., Melkko, S., Neri, D. (2006). Selection of streptavidin binders from a DNA-encoded chemical library. *Bioconjug. Chem.* 17, 366–370.

94. Halpin, D.R., Harbury, P.B. (2004). DNA display I. Sequence-encoded routing of DNA populations. *PLoS Biol.* 2, e173.

95. Snyder, T.M., Tse, B.N., Liu, D.R. (2008). Effects of template sequence and secondary structure on DNA-templated reactivity. *J. Am. Chem. Soc.* 130, 1392–1401.

96. Driggers, E.M., Hale, S.P., Lee, J., Terrett, N.K. (2008). The exploration of macrocycles for drug discovery—an underexploited structural class. *Nat. Rev. Drug Discov.* 7, 608–624.

97. Kleiner, R.E., Dumelin, C.E., Tiu, G.C., Sakurai, K., Liu, D.R. (2010). In vitro selection of a DNA-templated small-molecule library reveals a class of macrocyclic kinase inhibitors. *J. Am. Chem. Soc.* 132, 11779–11791.

98. Georghiou, G., Kleiner, R.E., Pulkoski-Gross, M., Liu, D.R., Seeliger, M.A. (2012). Highly specific, bisubstrate-competitive Src inhibitors from DNA-templated macrocycles. *Nat. Chem. Biol.* 8, 366–374.

99. Zhu, Z., Cuozzo, J. (2009). Review article: high-throughput affinity-based technologies for small-molecule drug discovery. *J. Biomol. Screen.* 14, 1157–1164.

100. Buller, F., Steiner, M., Scheuermann, J., Mannocci, L., Nissen, I., Kohler, M., Beisel, C., Neri, D. (2010). High-throughput sequencing for the identification of binding molecules from DNA-encoded chemical libraries. *Bioorg. Med. Chem. Lett.* 20, 4188–4192.

101. DeGroot, M.H., Schervish, M.J. (2012). *Probability and Statistics*, 4th Edition, (New York: Addison-Wesley).

102. Beeler, A.B., Su, S., Singleton, C.A., Porco, J.A. (2007). Discovery of chemical reactions through multidimensional screening. *J. Am. Chem. Soc.* 129, 1413–1419.

103. McNally, A., Prier, C.K., MacMillan, D.W.C. (2011). Discovery of an a-amino C-H arylation reaction using the strategy of accelerated serendipity. *Science* 334, 1114–1117.

104. Quinton, J., Kolodych, S., Chaumonet, M., Bevilacqua, V., Nevers, M.-C., Volland, H., Gabillet, S., Thuery, P., Creminon, C., Taran, F. (2012). Reaction discovery by using a sandwich immunoassay. *Angew. Chem. Int. Ed.* 51, 6144–6148.

105. Rozenman, M.M., Kanan, M.W., Liu, D.R. (2007). Development and initial application of a hybridization-independent, DNA-encoded reaction discovery system compatible with organic solvents. *J. Am. Chem. Soc.* 129, 14933–14938.

106. Chen, Y., Kamlet, A.S., Steinman, J.B., Liu, D.R. (2011). A biomolecule-compatible visible-light-induced azide reduction from a DNA-encoded reaction-discovery system. *Nat. Chem.* 3, 146–153.

107. Gorin, D.J., Kamlet, A.S., Liu, D.R. (2009). Reactivity-dependent PCR: direct, solution-phase in vitro selection for bond formation. *J. Am. Chem. Soc.* 131, 9189–9191.

108. Li, Y., Zhang, M., Zhang, C., Li, X. (2012). Detection of bond formations by DNA-programmed chemical reactions and PCR amplification. *Chem. Commun.* 48, 9513–9515.

109. Fredriksson, S., Gullberg, M., Jarvius, J., Olsson, C., Pietras, K., Gustafsdottir, S.M., Ostman, A., Landegren, U. (2002). Protein detection using proximity-dependent DNA ligation assays. *Nat. Biotechnol.* 20, 473–477.

110. Lundberg, M., Eriksson, A., Tran, B., Assarsson, E., Fredriksson, S. (2011). Homogeneous antibody-based proximity extension assays provide sensitive and specific detection of low-abundant proteins in human blood. *Nucleic Acids Res.* 39, e102.

111. Gullberg, M., Gustafsdottir, S.M., Schallmeiner, E., Jarvius, J., Bjarnegård, M., Betsholtz, C., Landegren, U., Fredriksson, S. (2004). Cytokine detection by antibody-based proximity ligation. *Proc. Natl. Acad. Sci. U. S. A.* 101, 8420–8424.

112. Fredriksson, S., Dixon, W., Ji, H., Koong, A.C., Mindrinos, M., Davis, R.W. (2007). Multiplexed protein detection by proximity ligation for cancer biomarker validation. *Nat. Methods* 4, 327–329.

113. Soderberg, O., Gullberg, M., Jarvius, M., Ridderstråle, K., Leuchowius, K.-J., Jarvius, J., Wester, K., Hydbring, P., Bahram, F., Larsson, L.-G., et al. (2006). Direct observation of individual endogenous protein complexes in situ by proximity ligation. *Nat. Methods* 3, 995–1000.

114. Leuchowius, K.-J., Weibrecht, I., Landegren, U., Gedda, L., Soderberg, O. (2009). Flow cytometric in situ proximity ligation analyses of protein interactions and post-translational modification of the epidermal growth factor receptor family. *Cytometry* 75A, 833–839.

115. Schallmeiner, E., Matyi-Tóth, A., Ericsson, O., Spångberg, L., Eriksson, S., Stenman, U.-H., Pettersson, K., Landegren, U. (2006). Sensitive protein detection via triple-binder proximity ligation assays. *Nat. Methods* 4, 135–137.

116. Tavoosidana, G., Ronquist, G., Darmanis, S., Yan, J., Carlsson, L., Wu, D., Conze, T., Ek, P., Semjonow, A., Eltze, E., et al. (2011). Multiple recognition assay reveals prostasomes as promising plasma biomarkers for prostate cancer. *Proc. Natl. Acad. Sci. U. S. A.* 108, 8809–8814.

117. Hammond, M., Nong, R.Y., Ericsson, O., Pardali, K., Landegren, U. (2012). Profiling cellular protein complexes by proximity ligation with dual tag microarray readout. *PLoS One* 7, e40405.

118. Gustafsdottir, S.M., Wennstrom, S., Fredriksson, S., Schallmeiner, E., Hamilton, A.D., Sebti, S.M., Landegren, U. (2008). Use of proximity ligation to screen for inhibitors of interactions between vascular endothelial growth factor A and its receptors. *Clin. Chem.* 54, 1218–1225.

119. McGregor, L.M., Gorin, D.J., Dumelin, C.E., Liu, D.R. (2010). Interaction-dependent PCR: identification of ligand-target pairs from libraries of ligands and libraries of targets in a single solution-phase experiment. *J. Am. Chem. Soc.* 132, 15522–15524.

120. Wu, Z., Zhen, Z., Jiang, J.-H., Shen, G.-L., Yu, R.-Q. (2009). Terminal protection of small-molecule-linked DNA for sensitive electrochemical detection of protein binding via selective carbon nanotube assembly. *J. Am. Chem. Soc.* 131, 12325–12332.

121. Cao, Y., Zhu, S., Yu, J., Zhu, X., Yin, Y., Li, G. (2012). Protein detection based on small molecule-linked DNA. *Anal. Chem.* 84, 4314–4320.

122. Wu, Z., Wang, H., Guo, M., Tang, L.-J., Yu, R.-Q., Jiang, J.-H. (2011). Terminal protection of small molecule-linked DNA: a versatile biosensor platform for protein binding and gene typing assay. *Anal. Chem.* 83, 3104–3111.

123. Sugita, R., Mie, M., Funabashi, H., Kobatake, E. (2010). Evaluation of small ligand–protein interaction by ligation reaction with DNA-modified ligand. *Biotechnol. Lett.* 32, 97–102.

124. Mie, M., Sugita, R., Endoh, T., Kobatake, E. (2010). Evaluation of small ligand-protein interactions by using T7 RNA polymerase with DNA-modified ligand. *Anal. Biochem.* 405, 109–113.

18

THE CHANGING FEASIBILITY AND ECONOMICS OF CHEMICAL DIVERSITY EXPLORATION WITH DNA-ENCODED COMBINATORIAL APPROACHES

Robert A. Goodnow, Jr.

Chemistry Innovation Centre,
AstraZeneca Pharmaceuticals LP, Waltham, MA, USA
GoodChem Consulting, LLC
Gillette, NJ, USA

18.1 AN OUTLOOK FOR APPLICATIONS OF DNA-ENCODED LIBRARIES

In the preceding chapters, the authors have laid out a diverse set of abilities necessary for a drug discovery scientist to engage in the generation and use of DNA-encoded chemical libraries. It should be clear that several of the technical methods (e.g., PCR, combinatorial chemistry, oligonucleotide synthesis) existed prior to the introduction of DNA-Encoded Library (DEL) technology. DEL technology has been aided by advances in next-generation sequencing. The ready availability of next-generation sequencing is also changing the way in which many life science researchers conduct research. As costs decrease and convenience increases for running this technology, scientists are able to answer questions that were previously unapproachable. Routinely sequencing an

A Handbook for DNA-Encoded Chemistry: Theory and Applications for Exploring Chemical Space and Drug Discovery, First Edition. Edited by Robert A. Goodnow, Jr.
© 2014 John Wiley & Sons, Inc. Published 2014 by John Wiley & Sons, Inc.

individual's genome as part of a personalized health-care approach is now conceivable. Therefore, it should come as no surprise that the momentum of change in sequencing capabilities would have an impact on the way chemists utilize this approach in maximizing their work. The understanding and capabilities of the existing methods that are necessary for the practice of DEL technology (e.g., PCR, deep sequencing) have greatly accelerated the development of DEL technology. As with many clever innovations, there is a building on existing technologies, methods, and understanding. One can imagine that future applications will arise from the intersection of DEL technology with other technologies to drive innovation in various disciplines.

To date, the majority of applications for DNA-encoded chemical libraries have supported the identification of high-affinity ligands for exploring chemical/biological target space. However, there are a few examples of extending this technique in exploring new chemical reactions. As discussed in Chapter 17, the Liu laboratories report using DNA-directed methods in exploring reaction space; this hints at possible extensions to the technology. Great potential also exists for creative applications of DNA-encoded methods in fields other than drug discovery. To date, the popular application of DNA-encoded chemical libraries has been finding new ligands for biological targets. It is a quick way to explore multiple binding conditions. Relatively small quantities of proteins are needed, and recombinant protein expression readily supports the installation of affinity tags. The agnostic nature of affinity-mediated selection may aid in the identification of allosteric modulators. The convenience of simultaneously running multiple selection conditions is a significant advantage, and among others, these features will continue to drive interest in this approach in identifying new hits and leads for biological targets. The development of selection methodologies against whole cells and membrane preparations will extend the exploration of chemistry space with DELs to a new set of drug targets. Further, new ways of target immobilization to solid support, for example, via antibodies, will also facilitate such exploration. Conversely, fully solution-based selection methods will likely circumvent concerns about the effect of an affinity tag on target proteins.

The continuing need for new ligands with drug-like properties will drive the use of DELs as a tool in drug discovery. In analyzing the success rate of pharmaceutical companies in finding useful leads for discovery programs [1], about 60% were successful at this stage. Similar rates of success were estimated for other phases of the drug discovery process from target proposal to drug launch, making the overall process highly inefficient (~1–2% success). The average costs of launching a new molecular entity has recently been estimated for 100 drug discovery organizations [2]. This estimate showed that for companies that have launched more than three drugs, the median cost is US\$4.2 billion per drug! Of course, such analyses are dependent on factors that may be unique to each drug discovery operation. However, they suggest the reality for most drug discovery scientists: they are faced with few choices for high-quality targets and compounds for drug discovery campaigns. Technologies that offer an advantage, if only for some targets and/or molecules, will continue to be valued and maintained. As an established, albeit somewhat specialized, approach, the use of a DEL becomes a complementary option for hit and lead finding [3] (see Chapter 10).

In another consideration, the application of selection methodologies and sequencing of a DNA-tagged small molecule that has crossed a cellular membrane has potential for addressing issues related to siRNA delivery. In the siRNA therapeutic approach, the drug substance (siRNA) selectively targets a specific message of mRNA for ablation based on the complementary hybridization of the antisense strand to that message [4]. Such ablation results in a "knockdown" of that message and reduces the expressed levels of protein translated from the message in the cell. The siRNA molecules require stabilization against endo- and exonucleases and transport across cellular membrane. Although the stabilization of siRNA while maintaining knockdown potency is routinely achieved *in vitro*, the effective and safe transport across cellular membrane *in vivo* has been an enduring problem. If both stabilization and transport of siRNA to the appropriate cell and subcellular compartment can be achieved, the siRNA effect can be catalytically efficient and has significant potential for tackling the so-called undruggable targets. In one approach to this delivery challenge, scientists have attached small molecules (e.g., cholesterol [5]) to the 5′-terminus of the sense strand showing enhanced uptake of such derivatized siRNAs into hepatocytes. Still, other targeting small molecules are needed to deliver siRNA selectively and with greater efficiency to cells of interest (e.g., tumor cells). The often laborious synthetic process to derivatize siRNA with small molecules creates an interesting and appealing situation to apply DEL technology. DNA-encoded chemical libraries are already composed of small-molecule–oligonucleotide conjugates and may be selected for the delivery of the encoding oligonucleotide into various subcellular compartments. The potentially huge diversity in DELs provides models of how large molecules such as siRNA may tap into receptor-mediated cell surface binding as well as differential subcellular compartmental internalization.

Following sequencing of the human genome, there is a great opportunity to ascertain what biological targets exist. The challenge is to identify the right targets for therapeutic research. Putative targets of interest can be validated by a range of means. Rapid access to specific target-binding small molecules can greatly facilitate such efforts and may also support the advancement of projects beyond this stage by provision of tool compounds. Chemical biology is one response to this challenge [6]. In this approach, drug discovery scientists use chemical methods to study biological pathways and putative targets within those pathways. There are strategies by which chemical methods can be applied for such study, but these strategies still require the availability of high-affinity small molecules. DNA-encoded chemistry methods hold promise for conveniently delivering millions of probe molecules to begin identifying high-affinity small molecules for new target and pathway characterization.

18.2 FUTURE INNOVATION GUIDED BY POTENTIAL COST ADVANTAGE OF DNA-ENCODED CHEMISTRY LIBRARY TECHNOLOGY

An additional appeal in developing and applying DEL technology relates to cost of preparation and use. Although a scientist may initially hesitate to associate innovation with cost drivers, it is important to note that aspects of research cost have a profound and persistent influence on what research can be undertaken, on what scale, and by how many people. Research that is relatively inexpensive and convenient to start will be attempted and adapted by more scientists. Conversely, the types of research that are cost prohibitive will be practice by fewer scientists.

Cost analysis of any research activity is difficult, subjective, and results from a complex mixture of budget categories: materials, labor, infrastructure, overhead, etc. Costs may also vary as a function of geographic location and scale. For example, an organization that is structured to make and assay tens of thousands of single molecules per year will be structured differently than an organization focused on traditional medicinal chemistry optimization of identified hits. Consequently, costs associated with each organization differ. Similarly, an organization that makes multiple DELs can engineer processes to reduce costs. Several features of the DEL process are comparable to traditional drug discovery (e.g., cost of protein production and hit follow-up assays); other costs are incomparable (e.g., next-generation sequencing is not used in standard medicinal chemistry for hit finding). Despite necessary, qualifying statements, a sense of the order of magnitude of the cost of reagents for producing a DEL is useful to consider how this approach may present a significant cost advantage.

To begin the analysis, one must establish a likely basis of comparison: collections of single small molecules assembled for high-throughput screening (HTS) campaigns. Many large pharmaceutical organizations have large collections of small molecules (>1 million) accumulated through historical processes and strategies. Developing a cost estimate for such actual pharmaceutical collections presents complexity beyond the scope of this chapter. However, for the sake of the analysis, we begin with the assumption that if one were to construct a new million-compound library, small molecules would come from three major sources: (i) commercially available single compounds, (ii) compounds derived from combinatorial and/or parallel chemistry libraries, and (iii) compounds generated by bespoke or custom-designed and one-at-a-time synthesis. Each category of molecules has particular attributes. For each of these categories, one can estimate a range of costs in their creation (Table 18.1). Citations for estimates are not easily found; rather, the numbers are based on years of experience in outsourcing and compound library building. Once again, it is important to note that the costs mentioned here are indeed estimates and the actual costs are the result of certain variables not being taken into consideration in the current analysis.

Many small-molecule samples used in building compound libraries can be purchased from commercial sources. Molecule costs vary depending on quantity, complexity, purity, and exclusivity. Nowadays, purity levels of >90% according to UV_{220nm} are standard. For example, 10 mg of a commercially available molecule may cost in the range of US$50–250 per sample. The number of commercially available compounds has been estimated to be 21 million compounds. The ZINC database contains the structures of over 21 million purchasable compounds (http://zinc.docking.org/) [7]. There are two important considerations. First, many of these compounds may not be ready "off the shelf," but rather are only available on demand. Second, the availability of structures that are commercially available changes. (One source of commercially available molecules is eMolecules, from which there are some five million structures "on the shelf" [8].)

The cost of synthesizing samples of single small molecules of bespoke design in milligram to gram quantities should then be considered. Again, the costs depend on the format in which the molecules are produced, whether produced as libraries or as

TABLE 18.1. Cost estimates in creating a one-million-compound collection for use in HTS campaigns

Cost estimates	Costs category estimates per sample (US$)			Costs for 333,333 compounds (US$)			Total cost for one-million-compound library	Cost per compound
	Commercially available single (10 mg)	Custom synthesis (mg to g)	Purified compound from parallel or combinatorial methods (10 mg)	Commercially available (10 mg)	Custom synthesis (g)	Purified compound from parallel or combinatorial methods (10 mg)		Average of all categories
Lower	50	1000	100	16,666,650	333,333,000	33,333,300	383,332,950	383
Higher	250	5000	500	83,333,250	1,666,665,000	166,666,500	1,916,664,750	1917

one-at-a-time single compounds. An exclusive, custom-synthesized single compound will cost substantially more, often ranging between US$1000 and 5000 for a sample of several grams to tens of grams. Compounds produced by high-throughput chemistry methods cost somewhere between the two other categories, for example, US$100–500 per sample. Usually, the parallel processing efforts applied in making arrays of compounds offer an order-of-magnitude advantage in productivity and thus in cost. Again, such costs vary based on scale, complexity, contractual arrangements, and the geographic location where the compounds are synthesized.

The order-of-magnitude difference in these estimates is instructive. For the sake of argument, if one were to create a million-compound library of one-third commercially available compounds, one-third combinatorial libraries, and one-third exclusive compounds, the anticipated financial resources needed would be in the range of US$370 million to 1.9 billion. The average cost per compound would be US$380–1900. The actual quantity of a compound used in a high-throughput screen is small, and as a result, the actual cost of the screen based on consumed compound would be orders of magnitude less. However, to have such a collection on hand, expenditures in this range should be anticipated.

In contrast, the chemistry to produce a DEL results in many more compounds owing to the split-and-pool nature of the process. Each compound is produced on a much smaller scale, typically subpicomoles. The largest costs are associated with the preparation of DNA tags and purchase synthesis reagents. Additionally, the cost of DNA oligo production varies with scale and purity levels. Producing approximately 1 μmol of a desalted oligonucleotide without HPLC purification costs approximately US$1–2 per base (e.g., GenScript [9]). Production of larger quantities and/or provision for HPLC purification substantially increases cost. Using modified oligonucleotides also increases cost. The cost of chemical reagents varies with the reagent itself and the quantity. A minimum amount of reagent (e.g., bottle) will need to be purchased even though the actual quantity of consumed reagent is very small. For this exercise, the average cost of a reagent is estimated at US$100. For the purposes of this analysis, rough estimates are simulated to produce DNA-Encoded Library-B (DEL-B) as reported in the supplemental information by Clark et al. [10] (see Table 18.2). DEL-B is constructed of four chemistry and tagging cycles presumably composed of 1491 oligo tags in addition to the headpiece and any closing tags. Further, in two of the cycles of the chemistry, the same type of reagent is used (i.e., amine); therefore, it can be assumed the same reagent may have been used in two cycles but purchased once. For this analysis, 608 reagents would be purchased. In this library, several cycles of reagents were multiply tagged; thus, for each reagent, there were more than two associated tags. Simple multiplication of the combinations results in the following library: 802,160,640 compounds encoded by 19,251,855,360 unique tag combinations. The numbers are derived from the following combinations: 192Cycle 1 Fmoc-amino acids, 32Cycle 2 bifunctional acids, 340Cycle 3 amines, and 384Cycle 4 amines encoded by 384Cycle 1, 384Cycle 2, 340Cycle 3, and 384Cycle 4 tags. It is simulated that this library would cost approximately US$131,000 or 0.02¢ per encoded structure.

It cannot be overemphasized that these costs are simulations of actual and total expenses and will vary with economy of scale and according to the detail of the work.

TABLE 18.2. Simulation of reagent costs to produce DEL-B as reported by Clark et al. [10]

Item	Purity category	Count of oligos	Bases for duplex tag	Quantity used	Synthesis scale category	Cost per base (US$)	Cost (US$)	Chemistry reagent category	Count of reagents	Reagent costs (US$100 per bottle)	Total	Cost per compound (US$)
Headpiece	HPLC purified	1	14	43 μmol	10 μmol	315	18,963					
Cycle A	Desalted	384	40	220 nmol	1 μmol	2	30,720	Amino acid	192	19,200		
Cycle B	Desalted	384	18	116 nmol	1 μmol	2	13,824	Bifunctional acid	32	3,200		
Cycle C	Desalted	340	18	162 nmol	1 μmol	2	12,240	Amine	340			
Cycle D	Desalted	384	18	41 nmol	1 μmol	2	13,824	Amine	384	38,400		
Closing tag	Desalted	1	48	200 nmol	1 μmol	2	96					
Totals							89,667			60,800	150,467	
Combination totals		19,251,855,360							802,160,640			0.0002

423

Headpiece cost is particularly difficult to estimate as it is a custom-modified oligonucle-otide (see Fig. 4.3). Although the cost of this reagent is likely higher than that shown in Table 18.2, it would probably be of a similar order of magnitude. Despite these cautioning qualifications, as with the previous analysis, rough estimates can be instructive. In both exercises, costs such as labor associated with the distribution of compounds and reagents as well as capital infrastructure are not included in production and handling estimates.

In summary, the reagents needed to produce a DEL of 800 million compounds cost approximately US$150,000. This is cheaper by 2500–12,000-fold relative to the creation of a conventional one million single-compound library. The cost of reagents needed to create a DEL compound is cheaper by some 2–10 million-fold.

It is important to stress other qualifying considerations. Confirmation and follow-up of hits generated by DELs still require single-compound synthesis, but in a conventional one-million-compound library, samples may be immediately available at no additional cost. Given the DEL method of synthesis, the chemistry is likely to be readily amenable to high-throughput chemistry procedures, and therefore, follow-up synthesis of individual hits should be routine. The quantities mentioned for the conventional HTS library (tens of milligrams to grams) will support HTS campaigns for decades, given the current miniaturization technologies. By contrast, it is reported for DEL-B that the "Final yield was 4.6 μmol, sufficient material for 920 selection experiments at 5 nmol per selection" [10].

Despite the caution in using this costing analysis, the economic advantage in displaying a large number of compounds before a protein target of interest should be evident. It

TABLE 18.3. Comparison of DEL and HTS advantages and limitations

Aspect	DEL technology	HTS
Compound numbers	Hundreds of millions	One to two million
Compound sources	Combinatorial libraries	Natural products, historical compounds from medicinal chemistry research, commercial collections
Chemical diversity	Diversity around specific combinatorial chemistry chemotypes, limitation with current DNA-compatible methods	Multiple, diverse chemotypes possible from a broad diversity of chemistries
Assay formats	Largely binding affinity based	Multiple assay types
Assay readout	PCR sequences of high-affinity binders	Multiple and diverse assay systems
HTS campaign durations	1–2 days for selection	~1 month
Identification of hits	1–2 weeks for sequencing and data processing	Immediate
Follow-up assays	Requires off-DNA resynthesis of hit	Reassay of same hit compound from inventory

would not be efficient in terms of cost, nor possible in terms of operational feasibility, to perform 800 million single-compound well HTS campaigns. The costs of HTS versus selection are also different. Estimates exist for US$0.07–0.20 per well for HTS [11]; thus, a one-million-compound campaign would cost US$100,000–200,000. As described in Chapter 13, a selection of many millions of DNA-encoded compounds is conducted at one time with a single filter tip and microtiter plate. Other aspects of comparison are listed in Table 18.3.

In HTS, protein consumption is often 10–100 mg, depending on the assay type. A DEL selection requires only a few micrograms; however, the development work must be taken into account for selections and confirmatory assays. Thus, a prudent estimate would be 1–5 mg of protein. See Ref. [3] of Chapter 9.

18.3 SUMMARY COMMENT

Ultimately, the cost estimates and the potential for deriving value from innovation must be put into context: finding a high-quality, high-affinity hit with potential for transformation into a *drug* is the ultimate litmus test in value-for-cost consideration of an innovation. Pharmaceutical companies have shown their willingness to invest in a variety of methods to achieve success in the discovery of new drugs and therapies. All told, DEL technology has become a compelling, complementary approach in identifying molecules that create a positive impact in drug discovery and other chemistry-related sciences.

REFERENCES

1. Brown, D., Superti-Furga, G. (2003) Rediscovering the sweet spot in drug discovery. *Drug Discov. Today*, *8*, 1067–1077.
2. Herper, M. (2013) How much does pharmaceutical innovation cost? A look at 100 companies. *Forbes* August 11, 2013, http://www.forbes.com/sites/matthewherper/2013/08/11/the-cost-of-inventing-a-new-drug-98-companies-ranked/ (accessed on November 27, 2013).
3. Goodnow, R. A. (2006) Hit and lead identification: integrated technology-based approaches. *Drug Discov. Today*, *3*, 367–374.
4. Castanotto, D., Rossi, J. J. (2009) The promise and pitfalls of RNA-interference-based therapeutics. *Nature*, *457*, 426–433.
5. Wong, S. C., Klein, J. J., Hamilton, H. L., Chu, Q., Frey, C. L., Trubetskoy, V. S., Hegge, J., Wakefield, D., Rozema, D. B., Lewis, D. L. (2012) Co-injection of a targeted, reversibly masked endosomolytic polymer dramatically improves the efficacy of cholesterol-conjugated small interfering RNAs in vivo. *Biomaterials*, *33*, 8893–8905.
6. Schenone, M., Dančík, V., Wagner, B. K., Clemons, P. A. (2013) Target identification and mechanism of action in chemical biology and drug discovery. *Nat. Chem. Biol.*, *9*, 232–240.
7. Irwin, J. J., Shoichet, B. K. (2005) ZINC – a free database of commercially available compounds for virtual screening. *J. Chem. Inf. Comput. Sci.*, *45*, 177–182.
8. http://www.emolecules.com/ (accessed on November 27, 2013).

9. http://www.genscript.com/DNA_Oligo.html (accessed on November 27, 2013).

10. Clark, M. A., Acharya, R. A., Arico-Muendel, C. C., Belyanskaya, S. L., Benjamin, D. R., Carlson, N. R., Centrella, P. A., Chiu, C. H., Creaser, S. P., Cuozzo, J. W., Davie, C. P., Ding, Y., Franklin, G. J., Franzen, K. D., Gefter, M. L., Hale, S. P., Hansen, N. J. V., Israel, D. I., Jiang, J., Kavarana, M. J., Kelley, M. S., Kollmann, C. S., Li, F., Lind, K., Mataruse, S., Medeiros, P. F., Messer, J. A., Myers, P., O'Keefe, H., Oliff, M. C., Rise, C. E., Satz, A. L., Skinner, S. R., Svendsen, J. L., Tang, L., van Vloten, K., Wagner, R. W., Yao, G., Zhao, B., Morgan, B. A. (2009) Design, synthesis and selection of DNA-encoded small-molecule libraries. *Nat. Chem. Biol.*, *5*, 647–654.

11. Burbaum, J. J. (1998) Miniaturization technologies in HTS: how fast, how small, how soon? *Drug Discov. Today*, *3*, 313–322.

19

KEEPING THE PROMISE? AN OUTLOOK ON DNA CHEMICAL LIBRARY TECHNOLOGY

Samu Melkko and Johannes Ottl

Centre for Proteomic Chemistry, Novartis Institutes for Biomedical Research, Novartis Pharma AG, Basel, Switzerland

19.1 THE PROMISE OF DNA-ENCODED LIBRARIES

DNA-encoded chemical libraries are collections of small organic molecules covalently linked to DNA moieties that serve as identifier tags. Several different methods have been put into practice to synthesize such libraries, which are described in different chapters of this book. The isolation of binders to a given target protein from that mixture of compounds is very similar to phage display selections or—to take a nonscientific comparison—to fishing with a bait. The target protein—either immobilized on solid support or containing a tag that allows the efficient capture of the protein—is incubated with the DNA-Encoded Library (DEL), and the library members capable of binding to the protein are captured and separated from the rest. Minute amounts of isolated DNA codes can then be amplified, identified, and quantified by Polymerase Chain Reaction (PCR) combined with deep sequencing. The process sounds facile and fast.

The following promises were made when DEL technology was introduced [1–8]:

- Enormous, unprecedented sizes of libraries.
- Selections are superior to screening.

A Handbook for DNA-Encoded Chemistry: Theory and Applications for Exploring Chemical Space and Drug Discovery, First Edition. Edited by Robert A. Goodnow, Jr.
© 2014 John Wiley & Sons, Inc. Published 2014 by John Wiley & Sons, Inc.

- The technology is facile and fast.
- No assay is needed for the isolation of binders.
- It's chemical evolution.

In this chapter, we'd like to critically evaluate this list and offer some personal thoughts on what needs to be done in order to keep the promise. After that, we'd like to lay out a vision of a streamlined hit-finding campaign with DEL technology that contains a new set of promises for the technology.

19.2 LIBRARY SIZE

Split-and-pool methods using multiple rounds of synthesis allow, in principle, the synthesis of libraries of enormous size. Also, the number of sequences that can be sequenced with modern deep sequencing instrumentation has risen significantly in recent years, giving access to hundreds of millions of sequences in a single run at moderate costs [9]. These advances now allow the efficient decoding of selections with libraries of enormous size. There are several issues with such library sizes, however.

First, in split split-and-pool synthesis, it is difficult to control the quality of the synthesis after pooling and hence the quality and the real size of the library. Varying and insufficient reaction yields will lead to truncated products and significantly smaller library sizes than the theory suggests. It is impossible to determine the real number, because quality control of the synthesis of mixtures of compounds is difficult and the individual members of such a library are only present in tiny amounts not amenable to direct chemical QC. In the seminal publication by Clark et al. [10], in which the performance of large DNA-encoded split split-and-pool libraries was convincingly demonstrated, a highly illustrative example is shown: in a selection campaign with a four-cycle library aimed at identifying p38 MAP kinase inhibitors, hit validation efforts showed that the most potent inhibitor of p38 MAP was actually a compound lacking one of the four building blocks, suggesting that this truncated species was the actual species present and selected in the library.

Second, sheer numbers do not necessarily correlate to meaningful chemical diversity. The composition of libraries depends on the available building blocks and the available chemical reactions for library synthesis. It is easy to create large libraries using building blocks that have small variations from each other (such as aromatic ring systems with different patterns of halogen decorations on different positions), which will lead to collections of molecules in a narrow but densely populated chemical space. It is apparent that it would be desirable to create diverse libraries, but it is less clear what actually constitutes chemical diversity [11, 12]. Having a better idea of the concept of meaningful chemical diversity of compound libraries could potentially make DEL technology more powerful than it is today, especially as the synthesis of DELs is fast compared to the replacement of a conventional collection of compounds for high-throughput screening. Looking into the future, computational chemists will skillfully guide the choosing of building blocks and chemistries that result in truly diverse chemical libraries.

Third, adding more synthesis cycles in a split-and-pool library will make the compounds larger and less likely to have drug-like or lead-like properties [13]. Using fewer synthesis cycles leads to smaller compounds with more favorable properties but will also lead to smaller library sizes. Compared to the challenge of creating libraries containing a high degree of chemical diversity, the knowledge about favorable chemical space for lead-likeness is better understood.

In conclusion, while it is in principle possible to create and decode enormously large libraries with DEL technology and deep sequencing, there are good reasons to limit the size. The arguments against overly large library sizes (library quality, chemical space, compound properties) have already been made in the context of combinatorial chemistry as it was introduced in the mid-1990s in the pharmaceutical industry, and it is worthwhile learning from that experience [14]. The methods to identify hits from libraries, however, are very different with both methods. While hits in classical combinatorial libraries are found by traditional screening, DEL technology relies on affinity selections.

19.3 SELECTIONS ARE SUPERIOR TO SCREENING

In the context of this book chapter, the terminologies "selection" and "screening" mean quite different things. In screening, compounds of a chemical library are assayed individually or in mixtures of low complexity (e.g., 10) in wells of microtiter plates. The process usually requires the storage of compound libraries distributed into plates and sophisticated logistics and robotic systems for the transfer of library compounds into assay plates, dispensing of assay components into the wells of the assay plates, and running of the assay including assay readout detection and quantification. In contrast, the selection process in DELs is similar to phage display selections. The library compounds are present as a mixture in a single tube and compete with each other for binding to the target protein that is incubated with the library compounds all at once. The selection procedures can be carried out on a normal laboratory bench with standard laboratory equipment without the need for sophisticated and expensive instruments. Running several selections in parallel with different selection conditions is little extra effort, while running a screening campaign with several assay conditions in parallel multiplies work and expenses and is rarely done with large compound collections. Running selections in parallel with different conditions, such as concentration of target protein, protein mutants/family members or counter targets, different buffer conditions, and presence or absence of cofactors, competitors, or interaction partners of the target protein, can help direct the isolation of binders with diverse modes of interaction with the target protein and gives additional information compared to a selection with only one condition.

As in conventional screening, selections will produce false-positive hits, and there is the risk that not every binder in the library will be detected, that is, of false negatives. The rates of false positives and false negatives likely depend on many different factors such as the precise selection methods (which typically contains immobilization of the target protein, stringent washing steps, and possibly rebinding to the same target after elution), the behavior of the target protein in the selection environment, the library size, and the decoding method. The enormous advances in deep sequencing are certainly helpful, as they allow digging deeper into the sequence space representing the output of a selection experiment. The enrichment factor is a crucial parameter in this context. It describes the enrichment of a binding library member compared to nonbinders during the selection procedure. A high enrichment factor allows the detection of few binding entities from a large excess of nonbinding library members; therefore, large enrichment factors are highly desirable. Unfortunately, there is little quantitative data published about enrichment factors and the effect of optimization efforts of selection procedures on the enrichment factors. In order to further improve DEL technology, more quantitative work on the influence of improved selection procedures on enrichment factors is necessary. The properties and behavior of the target protein are certainly an important aspect, which deserves thorough investigation. A high enrichment factor allows working with even bigger library sizes and would give access to a larger fraction of the chemical universe. To add a further level of complexity, studies on enrichment factors in selections comparing binders with different binding affinities, as opposed to only binders versus nonbinders, would be important in order to get a better feel of the granularity of the selection process [15]. In conclusion, we think that there is a lack of published work on quantitative aspects of the selection process in DEL technology, such as enrichment factors, false-positive rates, false-negative rates, and the influence of selection procedures, experimental conditions, and ingredients on these parameters. A better understanding of these parameters will allow more rational choices for the best application of DEL technology and help in further improving the technology. In contrast, a wealth of publications and manuals exist for the critical quantitative analysis and practice of traditional high-throughput screening.

19.4 FACILE AND FAST

After selection, sequencing, and data analysis, the identified hits require validation, which in practice means synthesizing and testing the hits individually in assays. Compared to classical high-throughput screening campaigns, in which typically thousands of hits from the primary screening campaign can be followed up by cherry-picking selected hits from a compound storage, the need to resynthesize hits limits the amount of hits that can be followed up after a selection campaign with a DEL. In our vision of a future hit-finding campaign using DELs, we propose the synthesis of thousands of compounds for hit validation, which is unfortunately not the reality at the moment. A practical consideration for choosing building blocks for the construction of the original DEL is to make sure that sufficient amount of each building block is available for resynthesis for hit validation. While the selection itself is facile and fast, the subsequent steps of deep sequencing, code analysis, and hit validation are a serious effort and take time but can certainly be streamlined. Furthermore, if a DEL selection campaign is successful, the desired outcome will be validated hits, which typically need classical medicinal chemistry efforts for further optimization into drug candidates. Even with enormously large libraries, it seems farfetched to envisage that a selection expedition could directly lead into a drug candidate.

19.5 NO ASSAY NEEDED

While the primary selection campaign relies on the binding of library compounds to the target protein and does not require an assay, hit validation does require an assay, such as an enzyme activity assay (if the target is an enzyme), a displacement assay (in which the ability of a hit to interfere with the binding of a labeled probe to the target protein is assessed), or binding assays using other readouts (Surface Plasmon Resonance (SPR), calorimetry, NMR). A possibility to include a label into the synthesized compounds for hit validation could facilitate performing binding assays, for example, a fluorine atom for fluorine-observed NMR [16]. Furthermore, as the original selection campaign should result in hits that are able to interact with the target protein irrespective of the binding site or possible activity, it would be desirable to run assays for hit validation that can differentiate between different modes of interacting with the target protein.

19.6 CHEMICAL EVOLUTION

DEL technology certainly contains some evolutionary principles [17, 18]. In a selection process, molecules with different properties compete with each other for binding to the target protein, and the "fittest" library compounds can be isolated (it "survives") if the selection procedure works. The molecules even contain the information ("genotype") that is responsible for their activity ("phenotype"). However, the power of Darwinian evolution is the iterative process of adding variations on the progenitors of the survivors of the selection process over generations. This feature of evolution (variation) is hard to incorporate into DEL technology. Furthermore, the natural selection process in Darwinian evolution is multidimensional and does not focus on one single trait. In drug discovery, a multidimensional selection process would not only select for compounds able to interact with the target of choice but also select for other desirable properties such as absence of binding to off-targets, good pharmacokinetic behavior, ability to cross a cell membrane, metabolic stability, and minimized toxicity. It is a considerable challenge to emulate this multidimensional process in a simple selection experiment. Even when concentrating on a single trait, another difficulty would be that progenitors of a parent compound may contain compounds that have only small differences in their desired activity (like binding affinity to the target protein). The selection process should then be able to differentiate compounds with small differences in their binding affinities to the target protein. Selection procedures with such fine granularity, allowing the differentiation between good and the slightly better, will be difficult to set up.

19.7 VISION

Despite many open questions, DEL technology has proven to deliver useful hits against target proteins that are valuable for drug discovery, and it continues to hold enormous promise for the future. We offer a vision of an implementation of DEL technology

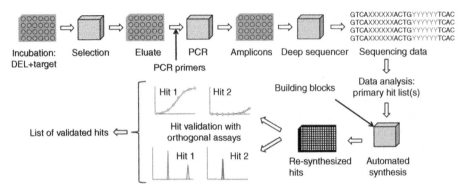

Figure 19.1. Vision of how a streamlined target-based hit-finding campaign using DNA-encoded chemical library technology could look in the future. In the first step, the target protein and the DEL are incubated in individual wells of a 96-well microtiter plate, using different conditions in each well. The difference between the conditions may be the use of different DELs in different wells (2,3,4-cycle split split-and-pool libraries, ESAC-based fragment libraries, macrocycle libraries, natural product-like libraries, special libraries aimed at a specific chemical space), different concentrations of the target protein, different buffer conditions, and the absence or presence of competitors, inhibitors, cofactors, etc. Then, automated multiparallel selections are performed, for example, using a magnetic bead separator or capillary electrophoresis. The eluates after the selection, containing the library members capable of binding to the target protein, are mixed with PCR reaction mixture and coded DNA primers (which can be formatted in microtiter plates, allowing efficient processing), and a PCR reaction is performed in a suitable microtiter plate. The resulting amplicons are then mixed and subjected to deep sequencing. After data analysis of the raw sequence data for each selection condition, primary hit lists are generated and automated synthesis of a selection of thousands of primary hits is triggered, for example, using solid-phase synthesis similar to DNA oligonucleotide or peptide synthesis robots. The resynthesized hits are then tested in two orthogonal assays, for example, a biochemical enzyme activity assay and a biophysical binding assay, which results in a list of validated hits. With streamlined processes, we envision that this process could be run within a few weeks.

that would make it an indispensable part of target-based hit-finding campaigns in the pharmaceutical industry. The vision is depicted in Figure 19.1, and it contains the streamlining of existing processes in combination with applications that hopefully will become available in the next years.

In our ideal scenario, the scientists running the hit-finding campaign have access to different high-quality DELs. These could be different split-and-pool libraries with diverse properties (e.g., libraries built up in 2, 3, or 4 cycles; natural product-like libraries; libraries populating different parts of the chemical space), fragment libraries based on ESAC technology, or macrocycle libraries. One of the most powerful features of DEL technology is the ability to perform many selections in parallel. In the presented scenario, we envision running, for example,

96 different selections with different conditions in parallel using automation such as a magnetic bead separator (in which the target protein is immobilized on magnetic beads) or capillary electrophoresis (which allows selections in solution and separates protein-bound from nonbound library members) for running multiparallel selections. The conditions in the 96 selections would be different and adjusted according to the specific scientific questions related to the target protein of interest. It may contain the use of the different libraries, different forms of the target protein (e.g., in the presence or absence of protein domains, if the target protein is a multidomain protein), and the presence or absence of competitors, cofactors, inhibitors, etc. It would also be possible to run selections against other proteins such as closely related analogs or counter targets in parallel. The ability to run selections with many different conditions in parallel is a unique feature of DEL technology, that cannot be matched with competing technologies at present. After automated selection, the following steps of amplicon generation by PCR and deep sequencing can be streamlined, for example, by formatting of coded PCR primers in 96-well plates. The primers contain a code for each position of the 96-well plate, thus encoding the different selection conditions. The cross-comparison of the sequencing data from different selections conditions provides a lot of information on the target and will be important for the generation of the primary hit lists. A major bottleneck for DEL technology as it is practiced today is the efficient resynthesis of compounds for hit validation. In our vision, the resynthesis of hits would be facilitated by automated solid-phase synthesis, in analogy to DNA oligonucleotide synthesis or peptide synthesis. Compared to DNA oligonucleotide synthesis or peptide synthesis, the resynthesis of hits from DEL selections comprises fewer synthesis cycles, but it requires access to hundreds or thousands of building blocks compared to only a handful. In our vision, the automated synthesis allows the preparation of thousands of compounds for validation with high quality and in a sufficient amount to run two orthogonal assays for hit validation. These orthogonal assays could be on the one hand a biochemical assay (e.g., an enzyme activity assay or a displacement assay) and on the other hand a biophysical binding assay (e.g., SPR or binding by NMR). The result of the campaign would be a list of validated hits and ideally a lot of information on the druggability of the target protein and SAR knowledge on the validated hit series. If the described process is streamlined, it could be run in a couple of weeks.

The most important requirement for a successful implementation of this vision is the availability of high-quality DELs, as discussed earlier in this chapter. Further development of the automation of selection procedures and the automation of compound resynthesis will also be essential to make this vision a reality. The best way to design different selection strategies will depend largely on the scientific problem to be solved and the creativity of scientists and the accumulated experience that will be made with DEL selections. The data analysis of the sequence raw data from the different selection conditions and its interpretation will certainly be challenging and requires scientific creativity and experience with the technology, as well. With all these caveats, we think that this is a powerful vision for the future of DEL technology, and once available, we think that any organization involved in target-based hit finding will want to have access to such a process.

REFERENCES

1. Brenner, S., Lerner, R. A. (1992) Encoded combinatorial chemistry. *Proc. Natl. Acad. Sci. U. S. A. 89*, 5381–5383.

2. Halpin, D. R., Harbury, P. B. (2004) DNA display I. Sequence-encoded routing of DNA populations. *PLoS Biol. 2*, E173.

3. Halpin, D. R., Harbury, P. B. (2004) DNA display II. Genetic manipulation of combinatorial chemistry libraries for small-molecule evolution. *PLoS Biol. 2*, E174.

4. Halpin, D. R., Lee, J. A., Wrenn, S. J., Harbury, P. B. (2004) DNA display III. Solid-phase organic synthesis on unprotected DNA. *PLoS Biol. 2*, 1031–1038.

5. Gartner, Z. J., Tse, B. N., Grubina, R., Doyon, J. B., Snyder, T. M., Liu, D. R. (2004) DNA-templated organic synthesis and selection of a library of macrocycles. *Science 305*, 1601–1605.

6. Melkko, S., Scheuermann, J., Dumelin, C. E., Neri D. (2004) Encoded self-assembling chemical libraries. *Nat. Biotechnol. 22*, 568–574.

7. Buller, F., Mannocci, L., Scheuermann, J., Neri, D. (2010) Drug discovery with DNA-encoded chemical libraries. *Bioconjug. Chem. 21*, 1571–1580.

8. Clark, M. A. (2010) Selecting chemicals: the emerging utility of DNA-encoded libraries. *Curr. Opin. Chem. Biol. 14*, 396–403.

9. Mardis, E. R. (2013) Next-generation sequencing platforms. *Annu. Rev. Anal. Chem. 6*, 287–303.

10. Clark, M. A., Acharya, R. A., Arico-Muendel, C. C., Belyanskaya, S. L., Benjamin, D. R., Carlson, N. R., Centrella, P. A., Chiu, C. H., Creaser, S. P., Cuozzo, J. W., Davie, C. P., Ding, Y., Franklin, G. J., Franzen, K. D., Gefter, M. L., Hale, S. P., Hansen, N. J., Israel, D. I., Jiang, J., Kavarana, M. J., Kelley, M. S., Kollmann, C. S., Li, F., Lind, K., Mataruse, S., Medeiros, P. F., Messer, J. A., Myers, P., O'Keefe, H., Oliff, M. C., Rise, C. E., Satz, A. L., Skinner, S. R., Svendsen, J. L., Tang, L., van Vloten, K., Wagner, R. W., Yao, G., Zhao, B., Morgan, B. A. (2009) Design, synthesis and selection of DNA-encoded small-molecule libraries. *Nat. Chem. Biol. 5*, 647–654.

11. Roth, H. J. (2005) There is no such thing as 'diversity'! *Curr. Opin. Chem. Biol. 9*, 293–295.

12. Gorse, A. D. (2006) Diversity in medicinal chemistry space. *Curr. Top. Med. Chem. 6*, 3–18.

13. Hann, M. M., Oprea T. I. (2004) Pursuing the leadlikeness concept in pharmaceutical research. *Curr. Opin. Chem. Biol. 8*, 255–263.

14. Fotouhi, N., Gillespie, P., Goodnow, R. Jr. (2008) Lead generation: reality check on commonly held views. *Expert Opin. Drug Discov. 3*, 733–744.

15. Melkko, S., Dumelin, C. E., Scheuermann, J., Neri, D. (2006) On the magnitude of the chelate effect for the recognition of proteins by pharmacophores scaffolded by self-assembling oligonucleotides. *Chem. Biol. 13*, 225–231.

16. Dalvit, C., Fagerness, P. E., Hadden, D. T., Sarver, R. W., Stockman, B. J. (2003) Fluorine-NMR experiments for high-throughput screening: theoretical aspects, practical considerations, and range of applicability. *J. Am. Chem. Soc. 125*, 7696–7703.

17. Rozenman, M. M., McNaughton, B. R., Liu, D. R. (2007) Solving chemical problems through the application of evolutionary principles. *Curr. Opin. Chem. Biol. 11*, 259–268.

18. Sadhu, K. K., Röthlingshöfer, M., Winssinger, N. (2013) DNA as a platform to program assemblies with emerging functions in chemical biology. *Isr. J. Chem. 53*, 75–86.

INDEX

A Handbook for DNA-Encoded Chemistry: Theory and Applications for Exploring Chemical Space and Drug Discovery, First Edition. Edited by Robert A. Goodnow, Jr.
© 2014 John Wiley & Sons, Inc. Published 2014 by John Wiley & Sons, Inc.

Printed and bound by CPI Group (UK) Ltd, Croydon, CR0 4YY

16/04/2025

14658530-0004